LARGE-SCALE ATMOSPHERE–OCEAN DYNAMICS, I

The Isaac Newton Institute of Mathematical Sciences of the University of Cambridge exists to stimulate research in all branches of the mathematical sciences, including pure mathematics, statistics, applied mathematics, theoretical physics, theoretical computer science, mathematical biology and economics. The research programmes it runs each year bring together leading mathematical scientists from all over the world to exchange ideas through seminars, teaching and informal interaction.

LARGE-SCALE ATMOSPHERE–OCEAN DYNAMICS

Volume I

Analytical Methods and Numerical Models

edited by

John Norbury

University of Oxford

and

Ian Roulstone

Met Office

PUBLISHED BY THE PRESS SYNDICATE OF THE UNIVERSITY OF CAMBRIDGE
The Pitt Building, Trumpington Street, Cambridge, United Kingdom

CAMBRIDGE UNIVERSITY PRESS
The Edinburgh Building, Cambridge CB2 2RU, UK
40 West 20th Street, New York, NY 10011-4211, USA
477 Williamstown Road, Port Melbourne, VIC 33207, Australia
Ruiz de Alarcón 13, 28014 Madrid, Spain
Dock House, The Waterfront, Cape Town 8001, South Africa

http://www.cambridge.org

Printed in the United Kingdom at the University Press, Cambridge

Typeset in 11pt Computer Modern

A catalogue record for this book is available from the British Library

ISBN 0 521 80681 X hardback

To the memory of Rupert Ford

Contents

Contributors

J.S. Allen, College of Oceanic and Atmospheric Sciences, Oregon State University, Corvallis, Oregon, USA
jallen@oce.orst.edu

A. Babin, Department of Mathematics, University of California, Irvine, CA, 92697, USA
ababine@math.uci.edu

S. Baigent, Centre for Nonlinear Dynamics and its Applications, University College London, Gower Street, London WC1E 6BT, UK
s.baigent@ucl.ac.uk

M.J.P. Cullen, Department of Mathematics, University of Reading, Whitenights, Reading RG6 6AX, UK
Mike.Cullen@ecmwf.int

R.J. Douglas, Department of Mathematics, University of Wales, Aberystwyth, Ceredigion SY23 3BZ, UK
rsd@aber.ac.uk

Darryl D. Holm, Theoretical Division and Center for Nonlinear Studies, Los Alamos National Laboratory, MS B284, Los Alamos, NM 87545, USA
dholm@lanl.gov

J.C.R. Hunt, Centre for Polar Observation and Modelling, University College London, 17, Gordon Street, London WC1H 0AH, UK
jcrh@mssl.ucl.ac.uk

A. Mahalov, Department of Mathematics, Arizona State University, Tempe, AZ 85287, USA
alex@taylor.la.asu.edu

P.A. Newberger, College of Oceanic and Atmospheric Sciences, Oregon State University, Corvallis, Oregon, USA
newberg@oce.orst.edu

B. Nicolaenko, Department of Mathematics, Arizona State University, Tempe, AZ 85287, USA
byn@stokes.la.asu.edu

J. Norbury, Lincoln College, University of Oxford, Oxford OX1 3DR, UK
john.norbury@lincoln.ox.ac.uk

I. Roulstone, Met Office, Joint Centre for Mesoscale Meteorology, Meteorology Building, University of Reading, Reading RG6 6BB, UK
ian.roulstone@metoffice.com

A. White, Met Office, Bracknell RG12 2SZ, UK
andy.white@metoffice.com

Preface

These two volumes provide an up-to-date account of the mathematics and numerical modelling that underpins weather forecasting, climate change simulations, dynamical meteorology and oceanography. The articles are a combination of teaching/review material and present results from contemporary research. The subject matter will be of interest to mathematicians and meteorologists, from graduate students to experts in the field. The articles have been written with the intention of providing accessible, interdisciplinary, accounts. The Introduction, which appears in both volumes, provides a guide to, and a perspective on, the subject matter and contents, and draws some tentative conclusions about the possible directions for future research.

The volumes are the result of the stimulus provided by the programme on *The Mathematics of Atmosphere and Ocean Dynamics* held at the Isaac Newton Institute for Mathematical Sciences in 1996, together with a follow-up meeting there in December 1997. The mathematical, scientific and computational challenge behind weather forecasting is why should we be able to forecast at all when the dynamical equations, the heat/moisture processes, and the billions of arithmetical calculations on 10–100 million unknowns involved in global forecasting each have associated instabilities and the potential for chaos? The overarching idea was to identify the stabilising principles and represent them effectively in mathematics that would lead to successful and efficient computation. Certain geometrical ideas are found to characterise the essential controlling physical principles, and the interplay of geometry and analysis makes for interesting new mathematics and helps to explain why computation of useful information becomes possible in the presence of chaos.

For obvious reasons, with over four years having elapsed since the conclusion of the original Programme, the subject matter has advanced as a result of work undertaken in the intervening period, both exploring ideas that were originally conceived in the Programme and developing new approaches. This has enabled new directions to be explored and this is reflected in the contributions. The Editors are indebted to all the contributors for both their perseverance and patience which has brought the project to fruition. While bringing the contributions together, we have received valuable help and encouragement from a number of people in addition to the support from the Met Office, Bracknell, and Lincoln College, Oxford. In particular, we would like to thank Terry Davies, Raymond Hide, Brian Hoskins and Emily Shuckburgh for reading various articles and providing useful comments. We would also like to thank Julian Hunt for his unstinting support for the programme from its inception, and John Toland for suggesting a programme on this subject matter in the first place! Much of the hard work in organizing the Programme was borne by the staff at the Newton Institute, and we would like to thank them, and the Director, Keith Moffatt, for valuable advice and assistance. Finally, David Tranah of the University Press showed us how to bring it all together.

Both the Programme and the volumes are forward looking, and history will decide on their success. However, it is with deep sadness that we record here the tragic passing of our much respected colleague Dr. Rupert Ford of Imperial College, who fell ill and died in March 2001 at the age of thirty-three. Although not making a written contribution to these volumes, Rupert was one of the most active and enthusiastic participants in the Programme itself, and a tremendous stimulus to us all. He stood astride the several disciplines that the organizers of the Programme sought to bring together. By the time of his death he had already published several outstanding contributions toward solving the problems with which the Programme was concerned. He will be sorely missed throughout a wide research community and we, the Editors and contributors, dedicate these volumes to his memory.

Ian Roulstone
Met Office

John Norbury
University of Oxford

Introduction and Scientific Background

J.C.R. Hunt, J. Norbury and I. Roulstone

Because of the importance and excitement of recent developments in research on large scale atmosphere-ocean dynamics, in 1996 an intense programme was held at the Isaac Newton Institute in Cambridge bringing together about 300 scientists from a wide range of specialisms. The articles in these two volumes consist of reviews, up to date research findings, and challenging statements about problems for future research. These are based on presentations made during the programme and more recent developments in the research, resulting from the vigorous and continuing interactions between many of the participants.

Numerical weather prediction and ocean modelling are successful applications of mathematical physics and numerical analysis. Their scientific methodology is essentially reductionist, because it involves reducing the calculations of a complex environmental process into constituent parts, each of which can be understood scientifically and modelled (Hunt 1999). This involves combining quantitative representation at every point in space and time of physical processes, governing phase changes, radiation and molecular diffusion, with the mathematical modelling of fluid mechanics on a wide range of scales from thousands of kilometres to centimetres. In order that the predictions cover all the aspects of practical importance, as well as increasing their accuracy year on year, regular improvements are needed in the models of key processes and mechanisms; some are well understood such as phase changes and low amplitude waves, but others such as radiation and turbulence can only be approximately parameterised or modelled, using the latest research as it develops. Once these large systems of mathematical equations and boundary conditions have been fixed in any particular model, they are then further approximated by some form of discretisation, so as to be suitable for computation. Additional mathematical algorithms are introduced for the iterative recalculation of the equations for the 'assimilation' of the observational data as it continually arrives. Numerical analysis, mathematical and physical compromises are all necessary in these stages of the development of an accurate and practical operational system.

Typically 10^{10}–10^{11} equations have to be calculated in the operations of national and international meteorological organisations when they produce their regular forecasts for the global weather. They utilize both the largest computers in the world and 100 million observations per day which, according to the World Meteorological Organisation, now cost more than \$1 billion per year. The question of how to optimally incorporate satellite observations of particular atmospheric features, together with the more traditional ground and ship based observations, is one of growing importance both scientifically and economically. One could say that this effort has 'paid-off' because the errors, which increase with the number of days ahead for the forecasts, have been steadily decreasing, so that a 3-day forecast today is by many measures as accurate as a 1-day forecast 20 years ago. Forecasts for up to 7 days are now regularly issued and found to have useful accuracy on continental scales. However, to maintain this downward trend in errors, continuing research is essential.

In the 1980s prediction of global ocean currents began to be developed based on similar types of mathematical and computational methods, and fluid mechanics, but the models had to allow for the quite different thermodynamics and mixing processes of the watermass. Also the boundary conditions of the oceans at the surface, coasts and ocean floor are obviously different from those of the atmosphere. Although soundings from ships and buoys are now being supplemented by satellite borne measurements at the ocean surface, regular observations for initialising ocean models are only available over limited regions of the world. Nevertheless useful forecasts for global ocean temperatures and currents are produced every few days. Furthermore now that these models are working, it is possible to develop global climate models by coupling the atmospheric and ocean models together, and then to take up the challenge of predicting aspects of variability on seasonal timescales and climate change over the continents, oceans and icecaps for periods of the order of 100 years and beyond. As the models improve, their spatial discrimination is becoming finer.

On long climatic timescales processes have to be modelled that, on the shorter timescale of weather or ocean forecasts, either can be neglected, such as chemical reactions whose effects on weather are only significant over a period of months, or can be considered to be fixed boundary conditions, such as ice-sheets which change relatively slowly. On the climate timescale these otherwise neglected effects, such as the chemistry of the ozone hole, grow and decay significantly and affect the whole globe. As J.–L. Lions (1995) has pointed out, the mathematical properties of the governing equations may be transformed so substantially by the introduction of certain effects, such as modelling the dynamics of ice sheets, that it is no longer possible to prove an existence theorem! Despite such mathematical doubts, climate change computations converge to the same equilibrium state even over quite a wide range of initial conditions. The results for the key parameters, such as global temperature, now agree with measurements of the global climate taken over the past 150 years within the natural fluctuations of the system. Governments have accepted the reliability of these models as a basis for their policies to mitigate the effects of increases over the next century of global temperature and sea level because of their likely effects on human life and economic activities.

By the end of the nineteenth century, the equations of motion, of thermodynamics, and of transport of moisture, that are the essential components of any model for forecasting the weather, had been worked out. However, it was also clear to those interested in such endeavours, for example Vilhelm Bjerknes (1914) and Lewis Fry Richardson (1922), that the problem of finding and computing solutions was extremely difficult. Although it was only 30-odd years between Richardson writing about a 'mere dream' of machines capable of performing such tasks and the advent of the first numerical forecasts (Charney, Fjørtoft and Von Neumann 1950), the intervening years witnessed the creation of ingenious methods for studying and analysing the atmospheric and oceanic flows that are still important in the context of weather and climate forecasting. Examples include fronts, ocean eddies and mid-latitude cyclones — such as the low pressure systems that cross the Atlantic and bring 'weather' to northern Europe.

The key idea behind these advances is to study the solutions of much simpler dynamical systems, whose solutions stay close for finite, but useful, time intervals, to those of the full fluid and thermodynamic equations. Indeed, much of modern dynamical meteorology is based on such studies, beginning with the pioneering work of Rossby (1936, 1940), Charney (1947, 1948) and Eady (1949). These approximate models usually correspond to some mathematical asymptotic state in which there is a dominant 'geostrophic' balance between the Coriolis, buoyancy and pressure-gradient forces on fluid particles so that the effects of acceleration of the particles (in the rotating frame of reference of the Earth) are relatively small. The asymptotic state arises from the rapid rotation and strong stratification of the Earth's atmosphere. Here, geostrophic balance (at its simplest) means horizontal flow around the pressure contours (Buys-Ballot's law), and this is coupled to the changes in the buoyancy force (hydrostatic balance between the vertical pressure gradient and gravity) in the vertical. Such approximations to Newton's second law are commonly referred to as *balanced models*. The Navier-Stokes equations for rotating, compressible, stratified fluid flow together with the equations of state and thermodynamics, commonly known in meteorology as the primitive equations, are the basis for numerical models used for atmospheric and oceanic predictions, and are therefore the starting point for the derivation of balanced models.

In the mid to late nineteenth century, classical hydrodynamics centred on the mathematical theorems of vortex motion, discovered by Helmholtz and Kelvin. The most notable of these governed the strength (or 'circulation'), the movement and the stability of vortices. Vortices persist even when their surroundings are quite disturbed or turbulent, as one observes by a simple experiment in one's bath. Vortices can move dangerously as tornadoes and swirling tropical storms, and last a long time over hundreds or thousands of rotation periods. Helmholtz' and Kelvin's theorem was formulated for a barotropic fluid in which the pressure is a function of the density alone, and therefore is too restrictive to represent air or sea water in motion because of the lack of thermodynamics.

Vilhelm Bjerknes in 1897 (Friedman 1989) first made the link between theoretical fluid mechanics and meteorology, by generalising the circulation theorem to include the usual atmospheric and oceanic situations where vorticity is generated or destroyed by the variation of buoyancy forces involving temperature changes in the vertical. The application of these results to *synoptic* meteorology in the ensuing years is, perhaps, the most important advance in the subject (Petterssen 1956). However Bjerknes and his son Jakob are more famous for their observational description in the 1920s of how cyclonic disturbances develop, with converging air flow leading to the formation of fronts and the triggering of rain bands along the fronts. Through their advocacy and organisation of rapid international exchange of meteorological measurement, their ideas featured in public weather forecasts in the 1930s (Friedman 1989). Qualitative elements of frontal analysis and the further dynamical analysis of regions of convergence and divergence by Sutcliffe (1947) and his contemporaries provided the conceptual basis of practical forecasting until the 1990s. Rossby (1936, 1940) and Ertel (1942) provided the next important conceptual development in meteorology and oceanography with the unifying concept of 'potential vorticity' (PV). PV is proportional to the vertical component of the vorticity of a fluid parcel per unit mass, and is approximately conserved when the effects of friction and external heating are slow compared to the other changes that are occurring in an air mass as it moves horizontally and vertically, e.g. over another air mass or mountains. This dynamical insight about changing meteorological conditions constrained by the conservation of a scalar quantity was connected to the earlier ideas of geostrophic balance through the pioneering work of Charney (1947) on quasi-geostrophic theory and by Kleinschmidt (1950a,b; 1951) on the dynamics of cyclones. However, exploitation of this new variable (PV) had to wait until the introduction of super-computers and the greater availability of upper-air data in the 1980s. The concept of PV has become a useful tool in practical forecasting because this one scalar field determines (via so-called 'inversion') the wind, pressure, temperature and density fields. This is a conceptual simplification because the changing weather (or even errors in weather patterns) can be described very economically (and errors corrected) using this one variable at different levels (Hoskins, McIntyre and Robertson 1985). The mathematical significance of potential vorticity conservation is not only that it is a 'governing' variable, but also that its properties reflect the underlying symmetries of the fluid-dynamical system which, in turn, determine conservation properties in both the infinite-dimensional, and numerical finite-dimensional, approximate 'models' of such systems.

In recent years a new appreciation has emerged of the central role, in controlling the behaviour of the equations and their solutions, of conservation laws of dynamical systems. This has been achieved by connecting them with the intrinsic geometric structure of the underlying equations of motion regarded as a hamiltonian dynamical system (i.e. one defined by its integral properties such as mass, energy, potential vorticity). Recent research in mechanics and dynamical systems using this powerful concept is often not familiar to those working in theoretical fluid dynamics, meteorology and oceanography. Modern hamiltonian mechanics provides a natural framework for understanding phenomena such as nonlinear stability, integral invariants and constrained dynamical systems (such as balanced models), and also for developing improved numerical schemes that have reduced errors because the schemes reflect the intrinsic geometrical properties of the analytical equations (Budd and Iserles 1999). The interplay of geometry and analysis will have many applications in geophysical mechanics; forecasting and climate modelling being prime examples here. The Newton Institute programme was designed to help advance this understanding.

The lectures in these volumes explain why simplifications to Newton's second law applied to the complex motions in the atmosphere and oceans are needed to understand and solve the equations. Since the early work of Runge (1895), Kutta (1901) and Richardson (1911), mathematical analysis has enabled the accuracy of such approximations to be assessed systematically on what are now large scale computations. However, whereas meteorologists have sought patterns in the weather for over 300 years, mathematicians have only recently begun to use geometrical thinking to understand the structure behind the governing equations and their approximate forms. Here constrained hamiltonian mechanics, transformation groups, and convex analysis are used to control the potentially chaotic dynamics in the numerical simulations, and to suggest optimal ways to exploit observational data. Many of the chapters [1] included in these volumes describe studies of the governing systems of equations, with all their complexities and approximations, although the main emphasis was on simpler systems whose integral properties and detailed solutions can be derived exactly. The approximations involved in deriving these idealised systems are controversial and have not always been mathematically consistent. Recent research, such as Cullen [I, 4], has centred on quantifying these approximations, by making full use of the latest results from the theory of stratified, rotating fluid dynamics. This book and its companion show how geometry and analysis quantify the concepts behind the fluid dynamics, and thus facilitate new solution strategies.

Any selection of contributions from an extensive subject such as weather and ocean forecasting necessarily reflects a particular viewpoint concerning both the historical significance of certain developments and their implications for future progress. The following brief commentary indicates the viewpoint taken and supplies a setting for the individual papers. However, the emphasis is always on large-scale atmosphere and ocean dynamical models that are useful in predicting changing weather patterns and climatic trends.

[1] They are designated hereafter by a number in square brackets [ ], with the volume number first, where needed. Other references are referred to by their date e.g. (1999).

Introduction to Volume 1 — Analytical Methods and Numerical Models

The article *A View of the Equations of Meteorological Dynamics and Various Approximations* by White [1], is a pedagogical introduction to the mathematics of meteorological fluid dynamics, which includes the derivation of the governing equations from those for the conservation of mass, momentum, thermodynamics etc., making further suitable approximations consistent with the asymptotic regimes to be modelled. White reviews the problem of deriving simplified balance equations which, as he explains, requires certain assumptions. This article has been written for mathematicians and physicists who desire a compact introduction to the subject rather than the more extensive treatments to be found in good contemporary textbooks on meteorology. Attention is also paid to various recent developments which have received little exposure outside the research literature yet. The approximated models studied include the hydrostatic primitive equations, the shallow water equations, the barotropic vorticity equation, several approximately-geostrophic models and some acoustically-filtered models which permit buoyancy modes. Conservation properties and frame invariance are given special emphasis. A straightforward problem of small-amplitude wave motion in a rotating, stratified, compressible atmosphere is addressed in detail, with particular attention paid to the occurrence or non-occurrence of acoustic, buoyancy and planetary modes in these models. The concluding section contains a short discussion of basic issues in numerical model construction.

The motion of a rotating, stratified fluid governed by the hydrostatic primitive equations is studied by Allen *et al.* [2]. The hydrostatic approximation, as discussed by White, reflects the high degree of stratification in the atmosphere and oceans. Approximate models are derived from the hydrostatic primitive equations for application to mesoscale oceanographic problems. The approximations are made within the framework of Hamilton's principle using the Euler–Poincaré theorem for ideal continua (see Holm *et al.* [II, 7]). In this framework, the resulting eulerian approximate equations satisfy Kelvin's theorem, conserve potential vorticity of fluid particles and conserve a volume-integrated energy. In addition, Allen *et al.* assess the accuracy of the model equations through numerical experiments involving a baroclinically unstable oceanic jet.

Roulstone and Norbury (1994) describe how one particular balanced model, the so-called semi-geostrophic (SG) equations, can be formulated in a manner similar to the Euler equations in two dimensions. Balanced evolution, which in this model entails the complete absence of fast inertia-gravity waves, is generated by a hamiltonian such that the solution is a sequence of minimum energy states, in a certain sense. Hoskins and Bretherton (1972) showed that the SG equations may be expressed in terms of lagrangian conservation laws. Thence a stable manifold within the dynamical system of the atmosphere is defined by using a convexity principle to minimize the energy. An extra advantage of this principle is that it applies to variables which have discontinuities. Furthermore, Hoskins and Bretherton (1972) showed that there exists a transformation of coordinates under which the motion of the fluid parcels is exactly geostrophic. For this reason such coordinates are sometimes referred to as geostrophic coordinates. Singularities of this differentiable map can be interpreted as fronts.

For a solution to the semi-geostrophic equations on a plane rotating with constant angular velocity — a so-called f-plane — the *Cullen–Norbury–Purser* principle (Cullen *et al.* 1991) states that at each fixed time, the fluid particles arrange themselves to minimise energy. Rewriting the equations in terms of the so-called geostrophic coordinates (Sewell [II,5]), this principle yields a constrained variational problem (where the constraint evolves with time): at each fixed time t, minimize the energy over all possible fluid configurations, given that values of the geostrophic transformation are known on particles. The minimizer, if it exists and is unique, gives the actual state of the fluid (in terms of the geostrophic transformation) at time t. Assuming the geostrophic energy is finite, it has been proved (Douglas 1998) that there is a unique minimizer, equal to the gradient of a convex function. In this way, solutions can be viewed as a sequence of minimum energy states. The set of possible states is described by a set of rearrangements; the unique minimizer is the *monotone rearrangement* (see Brenier 1991).

Douglas [5] presents some mathematical ideas on rearrangements of fluid volumes that have found application in meteorology, and that promote the lagrangian viewpoint. An intuitive idea of when two functions are rearrangements is as follows. Let f be a function, defined on a bounded region, such as temperature or moisture content. Imagine that the bounded region is a continuum of infinitesimal particles, and suppose that we exchange the particle positions with each particle retaining its value of f, that is, we conserve the temperature or moisture on fluid particles. This yields a new function g, which describes the temperature or moisture at the new locations, which is a 'rearrangement' of f. The concept of rearranging a function can be applied to both scalar and vector valued functions, and Douglas [5] develops the theory for both cases. Examples are given to illustrate the key ideas. Essentially, rearrangements allow us to conserve quantities on fluid masses as the masses are transported through the atmosphere using a lagrangian rather than eulerian viewpoint.

We can rewrite the energy minimization problem as a 'Monge mass transfer problem', for which there already exists a significant mathematical theory and numerical solution procedure. We then find that the monotone rearrangement is the optimal mapping. Thus, the geostrophic energy-minimising arrangement of fluid masses can be related to local stability conditions that require convexity of certain pressure (or geopotential) surfaces in the atmosphere. Failure to satisfy these conditions is usually associated with a breakdown in balance and rapid change of atmospheric conditions, including storms. An introduction to the theory of rearrangements, together with a discussion of their application to the semi-geostrophic equations, is given by Douglas [5]. An alternative interpretation of this theory based on probability ensembles, considering in what sense maximum likelihood states are equivalent to the Cullen–Norbury–Purser principle in semi-geostrophic theory, is given in Baigent and Norbury [6].

Following the seminal work of Vilhelm Bjerknes, the method of numerical weather prediction (NWP) was first worked out by L.F. Richardson in 1922 (see, for example, Nebeker 1995). He anticipated that sufficient measurement of data would become available and that computations would become sufficiently fast and comprehensive that the accuracy of weather forecasts should eventually equal those for the stellar and planetary positions recorded annually in the Nautical Almanac. This presumption was essentially questioned by Lorenz (1963), who showed that even much simpler mathematical representations of fluid flow (3 coupled non-linear, first-order, differential equations) are intrinsically prone to errors, so that however small their initial value the magnitudes of errors generally grow. His broad conclusions have had a major influence on the interpretation of weather forecasts ever since, the first being that there is much more sensitivity to errors in some states of a system (e.g. near saddle points in the phase plane) than in others. The second is that errors can grow *exponentially.* The latter conclusion has been bowdlerised in much popular comment as implying that since errors grow rapidly the weather is so chaotic that it cannot be forecast at all! Reasons why this might not be true for large scale weather evolution were advanced during the Isaac Newton Institute programme and have been the basis of significant follow-up work. First it is necessary to think carefully about what is meant by forecast error. A new approach to the evaluation of weather forecast error is to decompose the error into a combination of displacement error and difference in qualitative features. Douglas [5], and Cullen [4], demonstrate this idea and give a precise formulation using rearrangements of functions.

Directly or indirectly, the papers in this volume show why useful predictions can be made in the presence of chaos. Cullen [4] explains how the errors for more complex systems than those considered by Lorenz often grow more slowly, one of the reasons being that typical atmosphere and ocean weather events have a localised or vortex nature rather than a wave-like form (Hunt 1999). Other papers (see Arnol'd 1998) show that whether the systems are simple or complex, whatever their growth rate over the first few days, the errors are limited because the range of possible solutions for small initial errors tend to be confined within certain 'basins of attraction' in the phase planes of the system. This geometrical interpretation reflects recent mathematical research in which the results of geometrical analysis of differential systems leads to a clearer definition of their 'global' (in the mathematical sense) properties. Babin, Mahalov and Nicolaenko [3] give a detailed derivation of the errors involved in the balanced dynamics in the different asymptotic regimes of interest in atmospheric and oceanic dynamics. Babin *et al.* derive a new theorem for these error limits, and show how some of the standard approximations based on 'balance' and the neglect of the nonlinear time averaged effects of 'unbalanced' motion may be significant — reflecting perhaps the practical meteorologist's well known concern with waves on fronts, another example of further instability.

Weather forecasts are routinely computed for up to 10 days ahead, based on large quantities of wind, temperature and humidity data that are collected continuously, at random locations around the globe, and used to modify the computations. The data are of course insufficient to determine the exact state of the atmosphere. Since the data are very expensive to obtain there is a premium on their optimal exploitation. Therefore it is of the highest importance for numerical weather prediction to identify the dominant processes and flow features that determine how the large scale weather patterns develop. By ensuring that the continuous assimilation of data is consistent with these features the accuracy of the forecasts is greatly increased. Ocean modelling is beginning to develop similar data assimilation techniques. Cullen [4] explains how we can think of the atmosphere as evolving close to a dynamical system with high predictability which both explains the current success of operational predictions, and suggests that further useful progress can be made by exploiting this closeness more fully in the design of numerical prediction systems. Furthermore, using the notion of balance, and the associated transformation theory described by Sewell [II,5], Cullen suggests ways of using the incomplete observational data in more efficient ways, by exploiting the information implicit in the balance conditions to project the data onto the model grid in ways that respect the prevailing synoptic conditions. Babin *et al.* [3] provide a rigorous account of the asymptotic validity of these simpler systems. Cullen [4] argues that some recent results presented during the programme, from both atmosphere and ocean models, suggest that it is well worth making efforts to reduce the generation of spurious solutions arising from model and computational errors. Recent work supports the aim of building better numerical models that naturally support the desired simpler solutions.

Atmosphere and ocean models include approximate representations of sub grid scale processes and physical forcing; their best mathematical representation is not certain. Considerable progress is being made in showing how certain turbulence mixing processes that have been represented by diffusion-like terms can better be represented as effective advective transport terms. This could even affect conclusions about the large scale atmosphere and ocean circulation. Furthermore this change affects the form of the overall mathematical model, since these 'transports' have to be properly integrated with the rest of the dynamics. This issue too is discussed in the article by Cullen [4].

Introduction to Volume 2 — Geometric Methods and Models

Salmon (1983, 1985, 1988) pioneered the systematic derivation of balanced models within the framework of Hamilton's principle. The rationale is to make approximations to the lagrangian without disturbing the symmetry properties of the functional, thereby ensuring that the resulting model retains approximations to the conservation laws of the primitive equations. The derivation and understanding of balanced models from the hamiltonian point of view was one of the key themes of the Newton Institute programme. The chapter *Balanced models in geophysical fluid dynamics: hamiltonian formulation, constraints and formal stability* by Bokhove [1], gives a step by step account of the basics of hamiltonian mechanics and proceeds to demonstrate how hamiltonian formulations of balanced models can be constructed such that fast inertio-gravity waves can be eliminated by imposing certain constraints.

Most fluid systems, such as the three-dimensional compressible Euler equations, are too complicated to yield general analytical solutions, and approximation methods are needed to make progress in understanding aspects of particular flows. Bokhove reviews derivations of approximate or reduced geophysical fluid equations which result from combining perturbation methods with preservation of the variational or hamiltonian structure. Preservation of this structure ensures that analogues of conservation laws in the original 'parent' equations of motion are preserved. Although formal accuracy in terms of a small parameter may be achieved with conservative asymptotic perturbation methods, asymptotic solutions are expected to diverge on longer time scales. Nevertheless, perturbation methods combined with preservation of the variational or hamiltonian structure are hypothesised to be useful in a climatological sense because conservation laws associated with this structure remain to constrain the reduced fluid dynamics. Variational and hamiltonian formulations, perturbative approaches based on 'slaving', and several constrained variational or hamiltonian approximation approaches are introduced, beginning with finite-dimensional systems because they facilitate a more succinct exposition of the essentials. (The more technical mathematical aspects of infinite-dimensional hamiltonian systems are not considered, see e.g. Marsden and Ratiu 1994.) The powerful energy-Casimir method which can be used to derive stability criteria for steady states of (canonical) hamiltonian systems is introduced and the hamiltonian approximation approaches to various fluid models starting from the compressible Euler equations and finishing with the barotropic quasi-geostrophic and higher-order geostrophically balanced equations is presented. The presentation of fluid examples runs in parallel with the general finite-dimensional treatment which facilitates a clear understanding of the methods involved.

An illustration of the concept of balance within the framework of a finite-dimensional system is provided by Lynch [2]. The linear normal modes of the atmosphere fall into two categories, the low frequency Rossby waves and the high frequency gravity waves. The elastic pendulum is a simple mechanical system having low frequency and high frequency oscillations. Its motion is governed by four coupled nonlinear ordinary differential equations. Lynch studies the dynamics of this system, drawing analogies between its behaviour and that of the atmosphere. The linear normal mode structure of the system is analysed, the procedure of initialization is described and the existence and character of the slow manifold is discussed. This allows non-specialists to see, in a very simple example, what is performed routinely with the enormous systems of equations in modern numerical weather prediction and why.

Balmforth and Morrison [4] develop a hamiltonian description of shear flow, including the dynamics of the continuous spectrum. Euler's equation linearized about a shear flow equilibrium is solved by means of a novel invertible integral transform that is a generalization of the Hilbert transform. The integral transform provides a means for describing the dynamics of the continuous spectrum that is well-known to occur in this system. The results are interpreted in the context of hamiltonian systems theory, where it is shown that the integral transform defines a canonical transformation to action-angle variables.

Many balanced models do not support gravity waves, indeed the elimination of these waves from the solutions is usually the aim in defining an appropriate balance. Caillol and Zeitlin [3] point out that although internal gravity waves are not normally associated with 'weather' (see also Cullen [I,4]), they play an important role in energy transport in atmosphere and ocean dynamics. In [3], Caillol and Zeitlin study statistically steady states of an ensemble of interacting internal gravity waves and the corresponding energy spectra. They derive a kinetic equation for a system of weakly nonlinear plane-parallel internal gravity waves in the Boussinesq approximation and solve them to find stationary energy spectra for wave packets propagating in the direction close to vertical. The result is a Rayleigh–Jeans energy equipartition solution and a Kolmogorov-type solution of the form $\epsilon_k \sim k_1^{-(3/2)} k_3^{-(3/2)}$ corresponding to a constant energy flux through the wave spectrum.

The canonical vortex structures, their interaction and slow evolution, may be described, in the semi-geostrophic model, by solutions to the non-standard (Monge mass transfer) optimization problem described by Cullen [I,4] and Douglas [I,5]. It has been shown, by Chynoweth and Sewell (1989) for example, that singularities arise from the convexifications of multivalued Legendre dual functions, such as the swallowtail, with a typical singular surface being identified with a weather front. Sewell [5] reviews many aspects of transformation theory including Legendre duality and other types, and of lift transformations and canonical transformations. Applications are mentioned in several branches of mechanics. A straightforward style is adopted, so that the paper is accessible to a wide readership. Developments in the semi-geostrophic theory of meteorology in the last fifteen years have prompted this review, but it draws upon earlier work in, for example, plasticity theory, gas dynamics, shallow water theory, catastrophe theory, hamiltonian mass-point mechanics, and the theory of maximum and minimum principles. Singularities need to be described in transformation theory, and the swallowtail catastrophe is one such example. The intimate relation between lift transformations and hamiltonian structures is described. New exact solutions in a semi-geostrophic central orbit theory are given and properties of constitutive surfaces in gas dynamics and shallow water theory are described.

Purser [6] demonstrates that, using transformation theory, one can construct different versions of the semi-geostrophic equations for the purposes of modelling non-axisymmetric vortices on an f-plane and hemispheric (variable-f) dynamics. Both formulations retain a Legendre duality — a feature which is central to the construction of lagrangian finite-element methods. Note also that McIntyre and Roulstone [8] ask whether higher-order corrections to semi-geostrophic theory may be constructed while retaining some of the mathematical features that facilitate the integration of the equations both analytically and numerically.

For semi-geostrophic theories derived from the hamiltonian principles suggested by Salmon it is known (e.g. Purser and Cullen 1987) that a duality exists between the physical coordinates and geopotential, on the one hand, and isentropic geostrophic momentum coordinates and geostrophic Bernoulli function, on the other hand. The duality is characterized geometrically by a *contact structure* as described by Sewell [5]. This enables the idealized balanced dynamics to be represented by horizontal geostrophic motion in the dual coordinates, while the mapping back to physical space is determined uniquely by requiring each instantaneous state to be the one of minimum energy with respect to volume-conserving rearrangements within the physical domain.

Purser [6] shows that the generic contact structure permits the emergence of topological anomalies during the evolution of discontinuous flows. For both theoretical and computational reasons it is desirable to seek special forms of semi-geostrophic dynamics in which the structure of the contact geometry prohibits such anomalies. Purser proves that this desideratum is equivalent to the existence of a mapping of geographical position to a euclidean domain, combined with some position-dependent additive modification of the geopotential, which results in the semi-geostrophic theory being manifestly Legendre-transformable from this alternative representation to its associated dual variables.

Legendre transformable representations for standard Boussinesq f-plane semi-geostrophic theory and for the axisymmetric gradient-balance version used to study the Eliassen vortex are already known and exploited in finite element algorithms. Here, Purser re-examines two other potentially useful classes of semi-geostrophic theory: (i) the *non*-axisymmetric f-plane vortex; (ii) hemispheric (variable-f) semi-geostrophic dynamics. We find that the imposition of the natural dynamical and geometrical symmetry requirements together with the requirement of Legendre-transformability makes the choice of the f-plane vortex theory unique. Moreover, with modifications to accommodate sphericity, this special vortex theory supplies what appears to be the most symmetrical and consistent formulation of *variable-f* semi-geostrophic theory on the hemisphere. The Legendre-transformable representations of these theories appear superficially to violate the original symmetry of rotation about the vortex axis. But, remarkably, this symmetry is preserved provided the metric of the new representation is interpreted to be a pseudo-euclidean *Minkowski* metric. Rotation-invariance of the dynamical formulation in physical space is then perceived as a formal *Lorentz-invariance* in its Legendre-transformable representation.

Motivated by the remarkable mathematical structure of balanced models formulated in terms of a variational principle and their use in solving this class of problems, the last two articles consider more general and more accurate models of balanced atmospheric dynamics. The contributions by Holm, Marsden and Ratiu [7], and McIntyre and Roulstone [8], present recent developments in the theory of hamiltonian balanced models. Holm *et al.* [7] show how a number of models can be written in Euler–Poincaré form, and they propose a new modification of the Euler–Boussinesq equations which adaptively filters high wavenumbers and thereby enhances stability and regularity.

Recent theoretical work has developed the Hamilton's-principle analogue of Lie–Poisson hamiltonian systems defined on semidirect products. The main theoretical results presented in [7] are twofold: (i) Euler–Poincaré equations (the lagrangian analogue of Lie–Poisson hamiltonian equations) are derived for a parameter dependent lagrangian from a general variational principle of Lagrange–d'Alembert type in which variations are constrained; (ii) an abstract Kelvin–Noether theorem is derived for such systems. By imposing suitable constraints on the variations and by using invariance properties of the lagrangian, as one does for the Euler equations for the rigid body and ideal fluids, Holm *et al.* cast several standard eulerian models of geophysical fluid dynamics (GFD) at various levels of approximation into Euler–Poincaré form and discuss their corresponding Kelvin–Noether theorems and potential vorticity conservation laws. The various levels of GFD approximation are related by substituting a sequence of velocity decompositions and asymptotic expansions into Hamilton's principle for the Euler equations of a rotating stratified ideal incompressible fluid. They emphasize that the shared properties of this sequence of approximate ideal GFD models follow directly from their Euler–Poincaré formulations. New modifications of the Euler–Boussinesq equations and primitive equations are also proposed in which nonlinear dispersion adaptively filters high wavenumbers and thereby enhances stability and regularity without compromising either low wavenumber behaviour or geophysical balance.

The final article — epitomising the open-endedness of the Programme and the ongoing research it has stimulated — describes an unfinished journey, as well as presenting background tutorial material. Semigeostrophic theory and its contact structure and other formal properties are first of all reviewed in the simplest nontrivial context, f-plane shallow-water dynamics in $\mathbf{R}^2 = \{x,\, y\}$. A number of these properties are remarkably simple and elegant, and mathematically important. The authors ask which of those properties might generalize to more accurate hamiltonian models of balanced vortex motion. Many of the properties are intimately associated with the special canonical coordinates $(X,\, Y)$ discovered by Hoskins (1975). The jacobian $\partial(X,\, Y)/\partial(x,\, y)$ of these coordinates with respect to the physical space coordinates $(x,\, y)$ is equal to the absolute vorticity measured in units of the Coriolis parameter f; and Hoskins' transformation $(x,\, y) \mapsto (X,\, Y)$ is, in a natural sense, part of an explicitly invertible contact transformation (see also Sewell [5]). The invertibility is associated with a symmetric generating function. Unlike the flow in physical space $\{x, y\}$, the flow in the space $\{X, Y\}$ space is solenoidal, and its streamfunction $\Phi(X, Y, t)$ is obtainable by solving an elliptic Monge–Ampère equation expressing 'potential vorticity invertibility'. There are also certain Legendre duality and convexity properties, which make the model well-behaved, both mathematically and numerically, even when phenomena like frontal discontinuities occur (see also Cullen [I,4], Purser [6] and Sewell [5]).

No such canonical coordinates were known in simple analytical form for any other balanced model until the recent — and to fluid dynamicists very surprising — discovery by McIntyre and Roulstone (1996) of *complex-valued* canonical coordinates $(X,\, Y)$ in a certain class of hamiltonian balanced models, some of which are more accurate than semigeostrophic theory. The general way in which these models and their canonical coordinates are systematically derived by constraining an unbalanced 'parent dynamics' (hence 'splitting' the parent velocity field into two or more different fields) is discussed, following the method of Salmon (1988). The coordinates $(X,\, Y)$ are such that $\partial(X,\, Y)/\partial(x,\, y)$ is still real, and still equal to the absolute vorticity in units of f. The models include Salmon's L_1 dynamics and a new family of '$\sqrt{3}$ models' that are formally the most accurate possible of this class. The authors pursue the question thus raised: do these new models, or any subset or superset of them, share significant properties with semigeostrophic theory beyond the underlying hamiltonian dynamical structure and the special canonical coordinates $(X,\, Y)$ and their association with vorticity? The answer seems to be yes to the extent that the flow in (complex!) (X, Y) space is solenoidal — so that a complex streamfunction $\Phi(X, Y, t)$ must exist — and that elliptic Monge–Ampère equations expressing potential vorticity invertibility occur in all the new models, as well as in semigeostrophic theory. Otherwise, the answer is no. For instance the transformation $(x,\, y) \mapsto (X,\, Y)$ is no longer part of a contact transformation. However, the 'conjugate' transformation $(x,\, y) \mapsto (X,\, \bar{Y})$, where $\bar{Y}$ is the complex conjugate of Y, is, by contrast, part of an explicitly invertible contact transformation with a symmetric generating function and a new transformed potential $\hat{\Phi}(X, \bar{Y}, t)$. This fact, discovered by Roubtsov and Roulstone (2001), implies connections with hyper-Kähler geometry. The pair of transformations — taking (x, y) into (X, Y) and relating to vorticity, potential vorticity and elliptic Monge–Ampère equations, on the one hand, and taking (x, y) into $(X, \bar{Y})$ and relating to contact structure on the other — reveals that the structure underlying the whole picture is just that of a hyper-Kähler space or manifold, which in turn is part of a twistor space. The implications of this remain to be explored.

Conclusions

The Newton Institute programme successfully brought together ideas from geometry, analysis and dynamical systems theory, and showed their many benefits in numerical prediction used for weather forecasting, ocean and climate modelling. Various papers in these volumes advance the programme; and they suggest new problems and avenues for research in the theory of constrained dynamical systems, in particular for strongly stratified and rotating fluid flows where chaotic dynamics may be minimized.

From the material presented in these volumes we hope to gain new insights into the important issues surrounding various questions about the description of weather systems, on the large scale in both the atmosphere and the oceans, described by constrained variational principles. These issues include 'potential vorticity inversion' — the relationship between the potential vorticity and the balanced wind and temperature fields as described earlier in this Introduction — which usually involves solving a nonlinear elliptic problem (as in semi-geostrophic theory, for example). The convergence and practical stability of numerical schemes, and the relationship between stability of the flow and ellipticity of the operators, is far from being completely understood (for example, see comments in Ziemianski and Thorpe 2000 and also Knox 1997). For instance, Cullen [I,4] conjectures that 'elliptic PV inversion' constrains the enstrophy cascade, and hence controls the decay of fluid motions to turbulence. One direction in which work on these issues is proceeding is demanding more in terms of non-smooth analysis and ideas from rearrangement theory, as well as promoting a lagrangian view of fluid dynamics. Convex analysis plays a key role in many of the applications discussed here; in fact for the semi-geostrophic model, convexity, ellipticity and stability are directly related. From a purely mathematical perspective in terms of the lagrangian description of infinite-dimensional systems, and from a physical point-of-view relating to the stability of large-scale flows, convexity appears to limit chaotic dynamics.

There is increasing evidence to suggest that a major application of the hamiltonian dynamical aspects would be useful in numerical weather prediction. Numerical models based on Hamilton's equations pose a challenge to the numerical analyst working in partial differential equations and to the theorist who needs to find a hamiltonian formulation of the relevant constrained equations of motion and their associated conservation properties. In particular, balance conditions are an important constraint for the new generation of data assimilation schemes which seek to minimise cost functions based on the fit of observations to four-dimensional integral curves of the equations of motion (Courtier and Talagrand 1990). The new techniques proposed here may therefore have a major impact on our ability to provide accurate and appropriately balanced initial conditions for numerical weather prediction. Such study of meteorological problems may also promote further insight into the theory of dynamical systems.

We draw three main conclusions for practical computation from the papers presented here. First, new approaches are now available for reducing errors in numerical schemes by considering local integral properties; secondly, the standard assumptions of geophysical fluid dynamics describing how flows are in approximate geostrophic balance can be used to reduce significant errors in certain forecasting situations, especially by making better use of assimilated data in each application; and thirdly, the growth rate and maximum level of errors caused by data uncertainty, when analysed using realistic local dynamics and global dynamics respectively, differ quantitatively and conceptually from those inferred from Lorenz's much simpler chaotic systems.

We conclude by noting that the past few years have witnessed a number of exciting parallel developments in both the mathematical aspects and the phenomenology of stratified, rotating fluid dynamics, with the promise of practically important spinoffs including improved analyses and prediction of weather systems. Recent mathematical advances have brought a new geometric viewpoint to these problems, in particular a new appreciation of the central role of potential vorticity and its connection with the symplectic geometric structure of the underlying equations of motion regarded as a hamiltonian dynamical system.

Weather forecasting and climate modelling are excellent examples of how these mathematical advances have practical applications in solving problems where there is a strong interplay of geometry and physics.

References

Arnol'd, V. (1998) *Problèmes mathematiques de l'hydrodynamique et de la magnetohydronamique.* Procédes Journée Annuelle, S.M.F. Paris.

Bjerknes, V. (1914) Meteorology as an exact science. *Mon. Wea. Rev.*, **42**, 11–14.

Brenier, Y. (1991) Polar factorisation and monotone rearrangement of vector-valued functions. *Commun. Pure and Appl. Math.*, **44**, 375–417.

Budd, C.J. and Iserles, A. (1999) Geometric integration: numerical solution of differential equations on manifolds. *Phil. Trans. R. Soc. Lond. A*, **357**, 943–1133.

Charney, J.G. (1947) The dynamics of long waves in a baroclinic westerly current. *J. Meteorol.*, **4**, 135–162.

Charney, J.G. (1948) On the scale of atmospheric motions. *Geofys. Publ.*, **17**, No. 2, 17pp.

Charney, J.G., Fjørtoft, R. and von Neumann, J. (1950) Numerical integration of the barotropic vorticity equation. *Tellus*, **2**, 237–254.

Chynoweth, S. and Sewell, M.J. (1989) Dual variables in semi-geostrophic theory. *Proc. R. Soc. Lond.* A **424**, 155–186.

Courtier, P. and Talagrand, O. (1990) Variational assimilation of meteorological observations with the direct and adjoint shallow-water equations. *Tellus* **42A**, 531–549.

Cullen, M.J.P., Norbury, J. and Purser, R.J. (1991) Generalized Lagrangian solutions for atmospheric and oceanic flows. *SIAM J. Appl. Math.*, **51**, 20–31.

Douglas,R.J. (1998) Rearrangements of vector valued functions, with application to atmospheric and oceanic flows. *SIAM J. Math. Anal.*,**29**, 891–902.

Eady, E.T. (1949) Long waves and cyclone waves. *Tellus*, **1**, 33–52.

Ertel, H. (1942) Ein Neuer hydrodynamischer Wirbelsatz. *Met. Z.*, **59**, 271–281.

Friedman, R.M. (1989) *Appropriating the Weather: Vilhelm Bjerknes and the Construction of a Modern Meteorology. Cornell University Press.* 251pp.

Hoskins, B.J. (1975) The geostrophic momentum approximation and the semi-geostrophic equations. *J. Atmos. Sci.*, **32**, 233–242.

Hoskins, B.J. and Bretherton, F.P. (1972) Atmospheric frontogenesis models: mathematical formulation and solution. *J. Atmos. Sci.*, **29**, 11–37.

Hoskins, B.J., McIntyre, M.E. and Robertson, A.W. (1985) On the use and significance of isentropic potential vorticity maps. *Q.J.R. Meteorol. Soc.*, **111**, 877–946.

Hunt, J.C.R. (1999) Environmental forecasting and turbulence modelling. *Physica D*, **133**, 270–295.

Kleinschmidt, E. (1950a) Über Aufbau und Entstehung von Zyklonen. *Met. Runds.*, **3**, 1–6.

Kleinschmidt, E. (1950b) Über Aufbau und Entstehung von Zyklonen. *Met. Runds.*, **3**, 51–64.

Kleinschmidt, E. (1951) Über Aufbau und Entstehung von Zyklonen. *Met. Runds.*, **4**, 89–96.

Knox, J.A. (1997) Generalized nonlinear balance criteria and inertial stability. *J. Atmos. Sci.*, **54**, 967–985.

Kutta, W. (1901) Beitrag zur näherungsweisen Integration totaler Differentialgleichen. *Zeits. Math. u. Phys.*, **46**.

Lions, J.-L. (1995) 'Mathematics of climate' ICIAM lecture, Hamburg, in *Proceedings of the Third International Congress on Industrial and Applied Mathematics.* K. Kirchgässner, O. Mahrenholtz and R. Mennicken (eds.) 487pp. Akademie Verlag.

Lorenz, E.N. (1963) Deterministic nonperiodic flow. *J. Atmos. Sci.*, **20**, 130–141.

Marsden, J.E. and Ratiu, T. (1994) *Introduction to Mechanics and Symmetry.* Springer-Verlag. Second edition 1999.

McIntyre, M.E. and Roulstone, I. (1996) Hamiltonian balanced models: constraints, slow manifolds and velocity splitting. *Forecasting Research Scientific Paper*, **41**,

Met Office, UK Corrected version in revision for *J. Fluid Mech.*; the full text and corrections are available at the web site: `http://www.atm.damtp.cam.ac.uk/people/mem/`

Nebeker, F. (1995) *Calculating the Weather: Meteorology in the 20th Century.* Academic Press. 255pp.

Petterssen, S. (1956) *Weather Forecasting and Analysis*, 2nd ed., (2 volumes). McGraw Hill.

Purser, R.J. and Cullen, M.J.P. (1987) A duality principle in semi-geostrophic theory. *J. Atmos. Sci.* **44**, 3449–3468.

Richardson, L.F. (1911) The approximate arithmetical solution by finite differences of physical problems involving differential equations, with an application to the stresses in a masonry dam. *Phil. Trans. R. Soc. Lond. A* **210**, 307–357.

Richardson, L.F. (1922) *Weather Prediction by Numerical Process.* Cambridge University Press. 236pp.

Rossby, C.-G. (1936) Dynamics of steady ocean currents in the light of experimental fluid mechanics. *Papers in Physical Oceanography and Meteorology*, **5**, 1–43. Massachusetts Institute of Technology and Woods Hole Oceanographic Institution.

Rossby, C.-G. (1940) Planetary flow patterns in the atmosphere. *Q.J.R. Met. Soc.* **66**, Suppl., 68–87.

Roubtsov, V.N., Roulstone, I. (1997) Examples of quaternionic and Kähler structures in Hamiltonian models of nearly geostrophic flow. *J. Phys. A.* , **30**, L63–L68.

Roubtsov, V.N., Roulstone, I. (2001) Holomorphic structures in hydrodynamical models of nearly geostrophic flow. *Proc. R. Soc. Lond. A* **457**, 1519–1531.

Roulstone, I. and Norbury, J. (1994) A Hamiltonian structure with contact geometry for the semi-geostrophic equations. *J. Fluid Mech.*, **272**, 211–233.

Runge, C. (1895) Über die numerische Auflösung vor Differentialgleichungen. *Math. Ann.* **46**.

Salmon, R. (1983) Practical use of Hamilton's principle. *J. Fluid Mech.*, **132**, 431–444.

Salmon, R. (1985) New equations for nearly geostrophic flow. *J. Fluid Mech.*, **153**, 461–477.

Salmon, R. (1988) Semi-geostrophic theory as a Dirac bracket projection. *J. Fluid Mech.*, **196**, 345–358.

Sutcliffe, R.C. (1947) A contribution to the problem of development. *Q.J.R. Meteorol. Soc.*, **73**, 370–383.

Ziemianski, M.Z. and Thorpe, A.J. (2000) The dynamical consequences for tropopause folding of PV anomalies induced by surface frontal collapse. *Q.J.R. Meteorol. Soc.*, **126**, 2747–2764.

A View of the Equations of Meteorological Dynamics and Various Approximations

A.A. White

1 Introduction

One of the attractions of meteorology is its many-faceted character. It invites study by mathematicians and statisticians as well as by physicists of either practical or theoretical disposition. Amongst other fields, its concerns border or overlap those of oceanography, geophysics, environmental science, biological science, agriculture and human physiology, and impinge on those of economics, politics and psychology. (Climatology, for present purposes, is counted as part of meteorology.) Its breadth can lead to a perception that meteorology is a 'soft' science. This article focuses on part of the subject's 'hard' core: the equations governing atmospheric flow, and the approximate forms used by many numerical modellers and theorists.

A discussion (in section 3) of the basic equations of meteorological dynamics is preceded by a glance at a pre-Newtonian but fundamental subject: fluid kinematics (section 2). Some of the conservation laws which the basic equations express or imply are examined in section 4. Subsequent sections deal with approximate versions of the basic equations. Consistent approximation is one of the mathematical challenges of meteorology, and the sheer range of possible (and permissible?) approximations can be a bewildering feature. The hydrostatic approximation, the hydrostatic primitive equations (HPEs) and the shallow water equations (SWEs) are considered in section 5. The HPEs are the basis of many of the numerical models used worldwide in weather forecasting and for climate simulation, and the SWEs are widely studied as a testbed for further approximations and for numerical schemes.

We pause in section 6 to discuss various vertical coordinate systems, and various approximations of Coriolis effects and the Earth's sphericity beyond those associated with the HPEs. The geostrophic approximation is considered in a diagnostic (non-evolutionary) sense in section 7. Atmospheric wave motion is discussed in linear analytical terms in section 8 – we identify acoustic, gravity (buoyancy) and Rossby (planetary) waves and note the existence of special tropical modes.

Approximations of the HPEs which result in the removal of gravity waves as well as acoustic waves are considered in section 9; the shallow water equations are a convenient vehicle for most of this discussion. The quasi-geostrophic model, QG1, is singled out for particular attention in section 10. QG1 is one of the coarsest of those models that allow time-evolution of synoptic-scale weather systems (the 'Lows' and 'Highs' of the weather forecaster's chart), but it succeeds in representing most of the physical content of more quantitatively accurate models. Its importance in the conceptual development of meteorological dynamics can hardly be over-stated.

In section 11 are discussed various models (other than the HPEs) which allow gravity waves but not acoustic waves. Section 12 gives a brief survey of issues in numerical modelling for weather forecasting and climate simulation, and offers some concluding remarks.

The article is based on three lectures given during various phases of the Isaac Newton Institute programme on *Mathematics of Atmosphere and Ocean Dynamics* (December 1994, July 1996, December 1997). Its approach is elementary in so far as Hamiltonian methods are noted only in brief verbal summary; they are treated at proper length elsewhere in this volume. Much of the material is mainstream, and is covered in greater depth in the texts by Lorenz (1967), Phillips (1973), Haltiner and Williams (1981), Gill (1982), Pedlosky (1987), Lindzen (1990), Carlson (1991), Daley (1991), Holton (1992), Bluestein (1992), James (1994), Dutton (1995) and Green (1999), amongst others. Some new interpretations are presented, however, and later sections deal increasingly with developments which have not yet reverberated outside the research literature. Results that are thought to be new include: a bisection theorem relating the principal directions of curvature of the height field and the dilatation axis in geostrophic flow; a geometric solution of an acoustic/gravity wave dispersion relation; and a fresh perspective on the aptly-named 'omega equation' of QG1. [M.J. Sewell has demonstrated that the first of these results is an example of a general relationship between a certain pair of tensors associated with any 2-dimensional, solenoidal vector field; see section 7.2.] Sections 5.5 and 8.2 contain material covered in unpublished course notes by R.W. Riddaway and J.S.A. Green – notes to which I have been fortunate to have had access both as student and lecturer.

In mathematical respects, meteorological and oceanographic dynamics have much in common, and the atmosphere and oceans are closely-interacting systems, especially on climatological time-scales, but – in the interests of brevity – this article will refer only incidentally to oceanography and the oceans.

2 Fluid kinematics

Deformability is a key feature of a fluid: except in certain very simple flows, particles do not retain the fixed relative spatial relationships that are characteristic of a rigid body in motion. Our discussion in this section draws on the treatments given by Batchelor (1967), Wiin-Nielsen (1973), Ottino (1990) and Bluestein (1992).

Consider the motion of a fluid in two spatial dimensions relative to Cartesian axes Oxy; see Figure 1(a). Suppose that the velocity field $\mathbf{v} = \mathbf{v}(x, y, t) =$

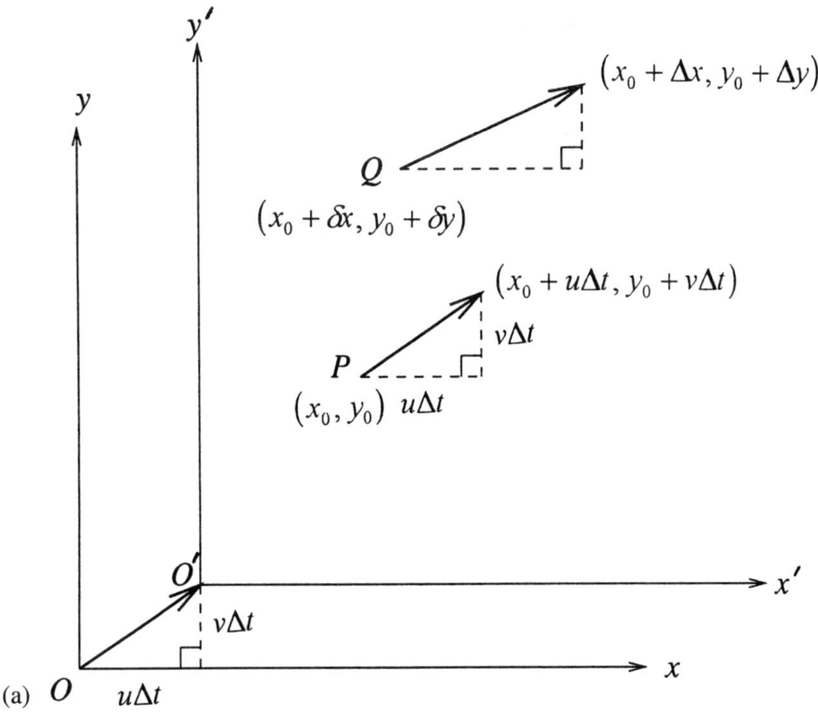

Figure 1: (a) Displacement in time Δt of fluid particles that are in the neighbourhood of the point $P = (x_0, y_0)$ at time $t = t_0$. To leading order, the fluid particle which is at P at $t = t_0$ is displaced to $(x_0 + u\Delta t, y_0 + v\Delta t)$ at $t = t_0 + \Delta t$, where u and v (the components of the flow in the x and y directions) are evaluated at (x_0, y_0, t_0). Also to leading order, a fluid particle which is at $Q = (x_0 + \delta x, y_0 + \delta y)$ at $t = t_0$ is displaced to $(x_0 + \Delta x, y_0 + \Delta y)$ at $t = t_0 + \Delta t$, where Δx and Δy are related to u, v and the spatial derivatives u_x, u_y, v_x, v_y at (x_0, y_0, t_0) according to (2.1). As well as the coordinate system Oxy relative to which u and v are measured, the diagram shows (at $t = t_0 + \Delta t$) the coordinate system $O'x'y'$ which moves with the flow velocity at (x_0, y_0, t_0) and is coincident with the Oxy system at $t = t_0$.

$(u(x, y, t), v(x, y, t))$ varies smoothly in space and time, so that the derivatives u_x, u_y, v_x, v_y are well defined, at least in the neighbourhood of a chosen point $P = (x_0, y_0)$ and time t_0. If a particle which is at point $Q = (x_0 + \delta x, y_0 + \delta y)$ at t_0 is at $(x_0 + \Delta x, y_0 + \Delta y)$ a short time Δt later, then it follows (from the definition of velocity as rate of change of position) that:

$$\begin{pmatrix} \Delta x \\ \Delta y \end{pmatrix} = \begin{pmatrix} \delta x \\ \delta y \end{pmatrix} + \begin{pmatrix} u \\ v \end{pmatrix} \Delta t + \begin{pmatrix} u_x & u_y \\ v_x & v_y \end{pmatrix} \begin{pmatrix} \delta x \\ \delta y \end{pmatrix} \Delta t. \tag{2.1}$$

Here u, v and their first derivatives are evaluated at (x_0, y_0, t_0), and higher-order terms in the Taylor expansion of $\mathbf{v}$ about (x_0, y_0, t_0), have been neglected. The second term on the right side of (2.1) represents translation with the flow at point $P = (x_0, y_0)$. Measuring position $(\delta x', \delta y')$ in a Cartesian system

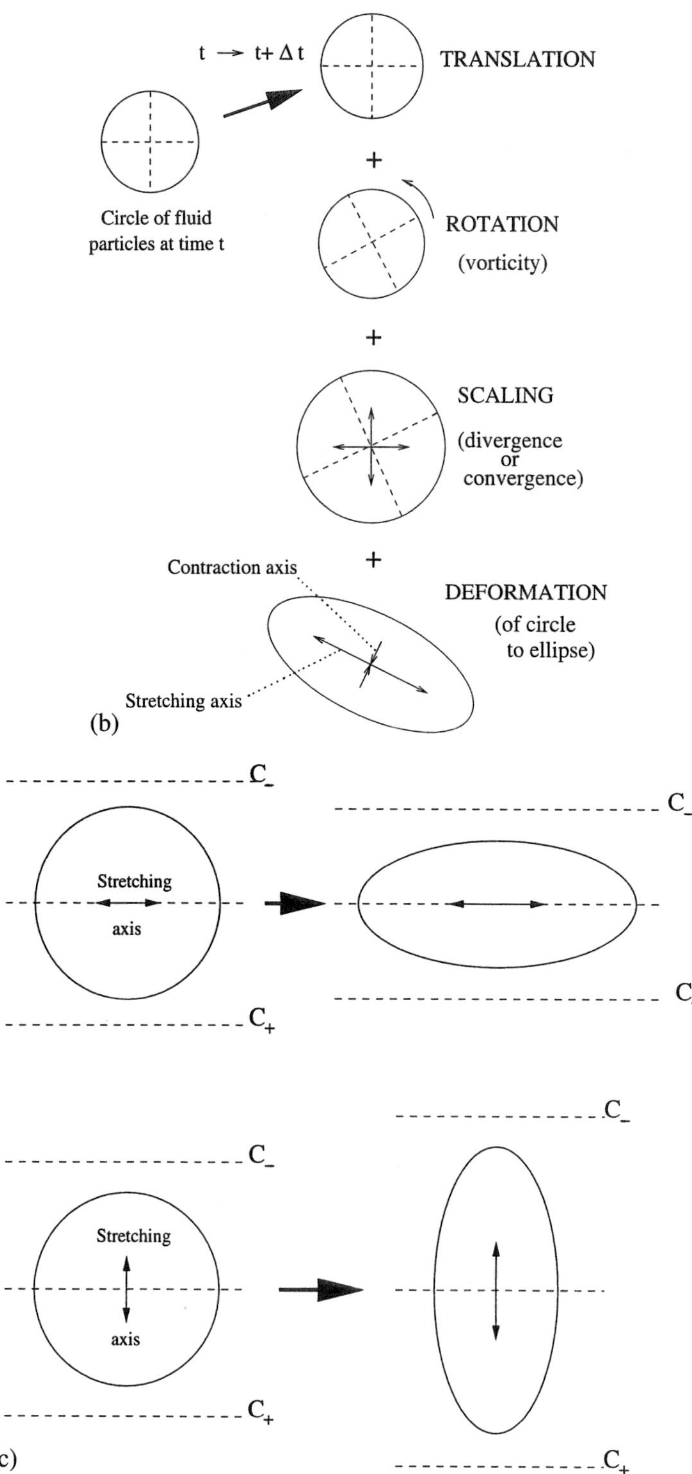

Figure 1: (b) Illustrating that the evolution of an initial circle of fluid particles in a short time Δt is the sum of a translation, a rotation, a scaling and a deformation. (c) Showing the effects of deformation on pre-existing gradients of a conserved scalar C when the stretching axis is respectively perpendicular to and parallel to the gradient of C.

$O'x'y'$ (Figure 1(a)) moving with this translation velocity (i.e. $\delta x' = \Delta x - u\Delta t$, $\delta y' = \Delta y - v\Delta t$)) gives

$$\delta \mathbf{x}' = \mathbf{A}\delta \mathbf{x},$$

where

$$\mathbf{A} = \begin{pmatrix} 1 + u_x\Delta t & u_y\Delta t \\ v_x\Delta t & 1 + v_y\Delta t \end{pmatrix} \tag{2.2}$$

and

$$\delta \mathbf{x} = (\delta x, \delta y)\,; \quad \delta \mathbf{x}' = (\delta x', \delta y')\,.$$

Define divergence δ, vorticity ζ and deformation components as

$$\begin{aligned} \delta &= u_x + v_y; \quad & \zeta &= v_x - u_y; \\ D_1 &= u_x - v_y; \quad & D_2 &= v_x + u_y. \end{aligned} \tag{2.3}$$

From (2.2) and (2.3) we get

$$\mathbf{A} = \mathbf{I} + (\mathbf{R} + \mathbf{S} + \mathbf{D})\Delta t, \tag{2.4}$$

where $\mathbf{I}$ is the unit diagonal matrix, and

$$2\mathbf{R} = \begin{pmatrix} 0 & -\zeta \\ \zeta & 0 \end{pmatrix}, \quad 2\mathbf{S} = \mathbf{I}\delta, \quad 2\mathbf{D} = \begin{pmatrix} D_1 & D_2 \\ D_2 & -D_1 \end{pmatrix}. \tag{2.5}$$

Also, to first order in Δt, $\mathbf{A}$ can be expressed as the product of three matrices:

$$\mathbf{A} = (\mathbf{I} + \mathbf{R}\Delta t)(\mathbf{I} + \mathbf{S}\Delta t)(\mathbf{I} + \mathbf{D}\Delta t) + O(\Delta t^2) = \hat{\mathbf{R}}\hat{\mathbf{S}}\hat{\mathbf{D}} + O(\Delta t^2)$$

with

$$\begin{aligned} \hat{\mathbf{R}} &= \begin{pmatrix} 1 & -\frac{1}{2}\zeta\Delta t \\ \frac{1}{2}\zeta\Delta t & 1 \end{pmatrix}; \hat{\mathbf{S}} = \left(1 + \tfrac{1}{2}\delta\Delta t\right)\mathbf{I}; \\ \hat{\mathbf{D}} &= \begin{pmatrix} 1 + \frac{1}{2}D_1\Delta t & \frac{1}{2}D_2\Delta t \\ \frac{1}{2}D_2\Delta t & 1 - \frac{1}{2}D_1\Delta t \end{pmatrix}. \end{aligned} \tag{2.6}$$

Consider particles which formed a *circle* centred on (x_0, y_0) at time t_0; see Figure 1(b). It is readily shown that the matrices $\hat{\mathbf{R}}$, $\hat{\mathbf{S}}$ and $\hat{\mathbf{D}}$ correspond respectively to (infinitesimal) *rotation*, *scaling* and *deformation* of the circle of particles over the time interval $[t_0, t_0 + \Delta t]$.

The rotation ($\hat{\mathbf{R}}$) is associated with vorticity (ζ), and corresponds to a turning of the initial circle through an angle $\frac{1}{2}\zeta\Delta t$ counterclockwise. The scaling ($\hat{\mathbf{S}}$) is associated with divergence (δ); it represents an isotropic change of size (a uniform magnification or minification) in which the radius of the circle changes by a factor $(1 + \frac{1}{2}\delta\Delta t)$.

The deformation ($\hat{\mathbf{D}}$) corresponds to a change of shape: the initial circle becomes an ellipse. The major axis of the ellipse (the stretching or dilatation axis) is inclined to the x axis at an angle $\frac{1}{2}\tan^{-1}(D_2/D_1)$. If the initial radius of the circle is chosen as the unit of distance, the semi-major axis of the ellipse is $1+\frac{1}{2}D\Delta t$, where $D^2 = D_1^2 + D_2^2$ is the square of the total deformation; and the semi-minor axis of the ellipse, the contraction axis of the initial circle, is $(1-\frac{1}{2}D\Delta t)$. The magnitudes and directions of the major and minor axes are given by the eigenvalues and eigenvectors of $\hat{\mathbf{D}}$. (The eigenvalues of $\mathbf{D}$ are $\pm\frac{1}{2}D\Delta t$. Its eigenvectors too are parallel to the axes of stretching and contraction.) Area is preserved, to order Δt, during the deformation.

As illustrated in Figure 1(b), the evolution of the initial circle of particles is (for small Δt) a combination of (i) translation, (ii) rotation, (iii) scaling, and (iv) deformation (to an ellipse).

In analytical terms, particle locations in the neighbourhood of (x_0, y_0) are transformed into locations in the neighbourhood of $(x_0 + u\Delta t, y_0 + v\Delta t)$ according to an infinitesimal general (non-conformal, non-isometric) mapping; see Klein (1938), p. 105. The details of the mapping are determined by the first derivatives of u and v in the neighbourhood of (x_0, y_0).

The matrices $\mathbf{R}$, $\mathbf{S}$ and $\mathbf{D}$ defined by (2.3) and (2.5) together constitute a decomposition of the 2D velocity gradient tensor $\mathbf{T}$:

$$\mathbf{T} = \begin{pmatrix} u_x & u_y \\ v_x & v_y \end{pmatrix} \quad (= \mathrm{grad}_2\,\mathbf{v}). \tag{2.7}$$

The quantity $\mathbf{R}$ (sometimes called the *body spin matrix*) is the skew-symmetric part of $\mathbf{T}$; $\mathbf{S}+\mathbf{D}$ is the symmetric part of $\mathbf{T}$ (the *Eulerian rate of strain matrix*). Vorticity (associated with $\mathbf{R}$) is seen to be essentially a rigid body property. Deformation (associated with $\mathbf{D}$) is essentially a non-rigid body property; the same statement could be made about the divergence (associated with $\mathbf{S}$), and some authors treat divergence as a special kind of deformation.

For 3D flow $\mathbf{u} = \mathbf{u}(x,y,z,t) = (u(x,y,z,t), v(x,y,z,t), w(x,y,z,t))$, the treatment may be repeated for an initial *sphere* of particles and 3D velocity gradient tensor $\mathbf{T}$:

$$\mathbf{T} = \begin{pmatrix} u_x & u_y & u_z \\ v_x & v_y & v_z \\ w_x & w_y & w_z \end{pmatrix} \quad (= \mathrm{grad}\,\mathbf{u}). \tag{2.8}$$

The results are similar to those of the 2D case, though more complicated in analytical terms. The sphere undergoes a translation, a rotation, a scaling and a deformation to an ellipsoid. The rotation is through an angle $\frac{1}{2}|\mathbf{Z}|\Delta t$ about the direction of $\mathbf{Z} = \mathrm{curl}\,\mathbf{u}$ (the vorticity vector), and the scaling is $1+\frac{1}{2}\Delta t\,\mathrm{div}\,\mathbf{u}$. The deformation is specified by the orientation and magnitude of the principal axes of the ellipsoid. In general, there is a stretching axis, a contraction axis and an intermediate axis, which may be an axis of contraction or stretching; degenerate cases may occur. The components of the deformation (not given here) determine the orientation and size of the principal axes and the extents of the stretching and contraction. The spatial relationship of the velocity and vorticity vectors to the principal axes of the deformation ellipsoid will be of general kinematic and dynamic importance.

Tensor considerations obviously enter fluid dynamics at a pre-Newtonian level. The tensorial character of flow kinematics is evident also on direct physical grounds from a consideration of the effect of a deformation on a pre-existing gradient of some conserved scalar field. Figure 1(c) (representing a 2D case) shows that a pre-existing gradient perpendicular to the stretching axis increases as a consequence of the deformation, whereas a gradient parallel to the stretching axis decreases. A pre-existing gradient at 45° to the stretching axis remains unchanged in magnitude. These effects are important in the formation of *fronts* – regions of large horizontal gradients of temperature and other properties – in the atmosphere and oceans [see Hoskins (1982) and Hewson (1998)]. Tensor considerations also play an important role in the proper representation of viscous effects, in the analysis of interactions between eddies and mean flows, and in the parametrization of subgridscale Reynolds stresses in numerical models; see Williams (1972), Hoskins *et al.* (1983) and Adcroft and Marshall (1998). It turns out, however, that vorticity – a vector quantity – figures more prominently than deformation in the dynamics of meteorological flows. Although we shall refer again in this article to deformation, the bulk of the treatment will involve nothing more complicated than vector analysis and the manipulation of vector differential operators.

3 Fluid dynamics and thermodynamics

This section gives an elementary account of those equations of thermodynamics and fluid dynamics from which the future state of the atmosphere may be forecast, given its present state.

3.1 Local and total time derivatives; advection

Consider some meteorological field $\mathfrak{I}$. This field might be a scalar quantity, such as temperature; or a vector, such as the flow velocity $\mathbf{u}$. Assume that $\mathfrak{I}$ is a function of time t and position $\mathbf{r}$ in some chosen coordinate frame:

$$\mathfrak{I} = \mathfrak{I}(\mathbf{r}, t).$$

Assume also that $\mathfrak{I}(\mathbf{r}, t)$ is differentiable with respect to each argument. Then first-order Taylor expansion of $\mathfrak{I}$ about $\mathfrak{I}(\mathbf{r}, t)$ gives

$$\delta\mathfrak{I} \equiv \mathfrak{I}(\mathbf{r}+\delta\mathbf{r}, t+\delta t) - \mathfrak{I}(\mathbf{r}, t) = (\delta\mathbf{r}.\,\mathrm{grad})\mathfrak{I} + (\partial\mathfrak{I}/\partial t)\delta t. \tag{3.1}$$

Equation (3.1) applies to any (infinitesimal) choice of $\delta\mathbf{r}$, δt. Choose $\delta\mathbf{r}$ to be the displacement in time δt corresponding to the velocity $\mathbf{u}$ of the air currently at position $\mathbf{r}$. Then $\delta\mathbf{r}/\delta t = \mathbf{u}$, and (3.1) becomes

$$\frac{D\mathfrak{I}}{Dt} \equiv \frac{\delta\mathfrak{I}}{\delta t} = (\mathbf{u}.\,\mathrm{grad})\mathfrak{I} + \frac{\partial\mathfrak{I}}{\partial t}. \tag{3.2}$$

The term $D\mathfrak{I}/Dt$ is the rate of change of $\mathfrak{I}$ following a parcel of air; it is known as the *total* (or material, or substantial, or individual, or Lagrangian) time derivative of $\mathfrak{I}$. The term $\partial\mathfrak{I}/\partial t$ is the *local* (or Eulerian) time derivative of $\mathfrak{I}$; it is the rate of change of $\mathfrak{I}$ at a point fixed in the chosen coordinate frame.

Some important physical laws (such as Newton's second law of motion) give information about material time derivatives. The users of weather forecasts are usually – not always – interested in the consequences of the local rate of change of $\mathfrak{I}$. A Grampian farmer in the highlands of Scotland may wish to know what the temperature of the air in the neighbourhood of the farm will be tomorrow, but is unlikely to want to know what the temperature of the air which is at the farm now will be tomorrow; that body of air may be over the North Sea by then. Hence the trivial re-expression of (3.2) as

$$\frac{\partial\mathfrak{I}}{\partial t} = \frac{D\mathfrak{I}}{Dt} - (\mathbf{u}.\,\mathrm{grad})\mathfrak{I} \tag{3.3}$$

is of fundamental importance in meteorology. Within its generality, (3.3) expresses the key physical notion that when $\mathfrak{I}$ is conserved on fluid particles ($D\mathfrak{I}/Dt = 0$) the value of $\mathfrak{I}$ at a fixed point in our coordinate frame will nevertheless be changing ($\partial\mathfrak{I}/\partial t \neq 0$) if fluid having a different value of $\mathfrak{I}$ is being brought in, or advected, by the flow ($-(\mathbf{u}.\,\mathrm{grad})\mathfrak{I} \neq 0$). The term $-(\mathbf{u}.\,\mathrm{grad})\mathfrak{I}$ represents the (rate of) advection of $\mathfrak{I}$. A vexed issue of terminology will be side-stepped in this article by using the expression 'advection term' to describe both $-(\mathbf{u}.\,\mathrm{grad})\mathfrak{I}$ (as in (3.3)) and $+(\mathbf{u}.\,\mathrm{grad})\mathfrak{I}$ (as in (3.2)).

We now consider how various choices of $\mathfrak{I}$, and the application of various physical laws, lead to expressions for the local rates of change of meteorological fields. With the needs of our Grampian farmer in mind, we begin by choosing $\mathfrak{I} = T$, the temperature.

3.2 First law of thermodynamics

Suppose that a parcel of air having unit mass, temperature T and (specific) volume α undergoes a change of (specific) entropy δs. According to the first law of thermodynamics, the concomitant changes δT and $\delta\alpha$ of T and α are related by

$$c_v\delta T + p\delta\alpha = T\delta s. \tag{3.4}$$

Here c_v is the specific heat at constant volume and p is the pressure of the parcel of air. Since the first law of thermodynamics applies to the parcel of air as it moves, it follows from (3.4) that

$$c_v\frac{DT}{Dt} + p\frac{D\alpha}{Dt} = T\frac{Ds}{Dt} \equiv Q. \tag{3.5}$$

In meteorology, Q is usually thought of as the total heating rate per unit mass; strictly, it is the heating rate that would achieve, by reversible processes, the same rates of change of T and α as those occurring in the actual irreversible system (Lorenz 1967, p.14). Equation (3.5) can be written in terms of density $\rho = 1/\alpha$ as

$$c_v \frac{DT}{Dt} - \frac{p}{\rho^2}\frac{D\rho}{Dt} = Q. \tag{3.6}$$

In either form, however, the first law of thermodynamics gives only a relationship between the material derivatives of T and a density variable.

3.3 Mass continuity

Information about the material derivative of density, $D\rho/Dt$ (see (3.6)), may be obtained from mass conservation. The mass within a volume τ (fixed relative to the chosen coordinate frame) changes only to the extent that there is net inflow or outflow of mass at the boundary S of the volume. Hence

$$\frac{\partial}{\partial t}\int_\tau \rho\, d\tau = -\int_S \rho\mathbf{u}.\,\mathbf{dS} = -\int_\tau \operatorname{div} \rho\mathbf{u}\, d\tau \tag{3.7}$$

by the divergence theorem. Equation (3.7) applies to any volume τ, so the local equality

$$\frac{\partial \rho}{\partial t} + \operatorname{div} \rho\mathbf{u} = 0 \tag{3.8}$$

must hold. Equation (3.8) is a form of the (mass) continuity equation. By using (3.3), we may deduce an alternative form:

$$\frac{D\rho}{Dt} + \rho \operatorname{div} \mathbf{u} = 0. \tag{3.9}$$

3.4 Perfect gas law

Equations (3.6) and (3.9), taken together with (3.3) in the form

$$\frac{\partial T}{\partial t} = \frac{DT}{Dt} - (\mathbf{u}.\operatorname{grad})T,$$

enable us to evaluate the local rate of change $\partial T/\partial t$ so long as we know the current values of Q, p, ρ and the flow vector $\mathbf{u}$. The current value of ρ can be found from observations of p and T by using the perfect gas law in the form

$$p = \rho RT \tag{3.10}$$

where R is the gas constant per unit mass. Equation (3.10) has no time derivatives. In meteorological parlance, it is a *diagnostic* equation; equations involving time derivatives are called *prognostic*. We have now set up the apparatus to evaluate, and hence (knowing the current values of T, p and $\mathbf{u}$) to calculate T at our chosen location at a later time $t + \delta t$. If we were content to take δt

to equal 24 hours, we could calculate an expected value of T at the chosen location tomorrow. The calculated value would probably be very inaccurate, and for this reason (and others) the calculation of a 24-hour temperature forecast proceeds in practice by performing a number of *time steps* δt which are much shorter than 24 hours. (Typically, the time steps are of the order of 10 minutes.) This process requires values of Q, p, ρ and $\mathbf{u}$ at each time step. Hence we require a prognostic equation for the flow $\mathbf{u}$; in general, we cannot forecast the temperature accurately for more than (say) an hour ahead without forecasting the flow too. In any case, many users of weather forecasts – including our Scottish farmer, if there are new lambs on the hill – will want to know what tomorrow's wind speed and direction are likely to be.

3.5 Newton's second law

Newton's second law of motion relates the inertial acceleration of an element of air to the net force acting on it. Contributory forces include the pressure gradient force, gravity, and friction. If (as is usually convenient) velocities and accelerations are measured relative to the rotating frame of the solid Earth, Coriolis and centrifugal 'forces' must be introduced to allow for the transformation from inertial to accelerating (rotating) frame; see Stommel and Moore (1989) and Persson (1998) for discussion.

The Lagrangian rate of change of the velocity $\mathbf{u}$ of an element of air, relative to the rotating Earth, is then given by

$$\frac{D\mathbf{u}}{DT} = \underbrace{-2\boldsymbol{\Omega}\times\mathbf{u}}_{\text{Coriolis}} - \underbrace{\alpha\,\mathrm{grad}\,p}_{\substack{\text{Pressure}\\ \text{gradient}}} - \underbrace{\mathrm{grad}\,\Phi}_{\substack{\text{Apparent}\\ \text{gravity}}} + \underbrace{\mathbf{F}}_{\substack{\text{Friction and all}\\ \text{other forces}}} . \tag{3.11}$$

Equation (3.11) is the Navier–Stokes equation for motion and acceleration relative to the Earth, whose rotation vector is $\boldsymbol{\Omega}$. 'Apparent gravity', with potential function Φ, consists of the contribution (dominant in the atmosphere) of true Newtonian gravity and the contribution of the centrifugal force $-\boldsymbol{\Omega}\times(\boldsymbol{\Omega}\times\mathbf{r})$; here $\mathbf{r}$ is position vector relative to a frame rotating with the Earth, and having its origin at the centre of the Earth – see Figure 2. In (3.11) all forces are expressed per unit mass of air.

Equation (3.11) may be used in conjunction with

$$\frac{\partial\mathbf{u}}{\partial t} = \frac{D\mathbf{u}}{Dt} - (\mathbf{u}.\,\mathrm{grad})\mathbf{u}, \tag{3.12}$$

(the appropriate form of (3.3)) to give an expression for the local rate of change of $\mathbf{u}$, i.e. $\partial\mathbf{u}/\partial t$. The advection term is nonlinear in $\mathbf{u}$; the pressure gradient term, $\alpha\,\mathrm{grad}\,p$, is also in a certain sense nonlinear (as are the advection terms which arise from the first law of thermodynamics and the continuity equation).

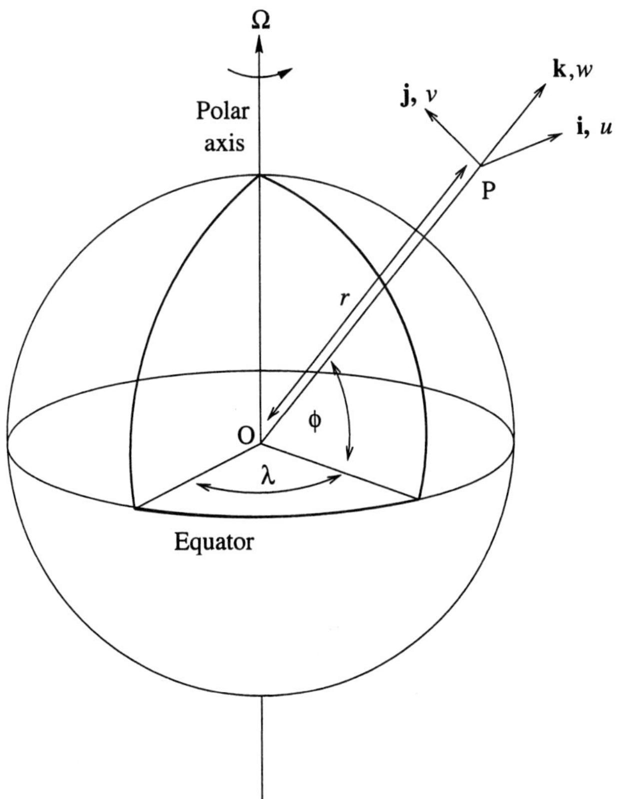

Figure 2: The (λ, ϕ, r) spherical polar system whose origin O is at the centre of the Earth and which co-rotates with angular velocity $\mathbf{\Omega}$. The unit vector triad $(\mathbf{i}, \mathbf{j}, \mathbf{k})$ at the generic point $P = (\lambda, \phi, r)$ is also indicated, as are the corresponding zonal, meridional and radial velocity components u, v and w.

3.6 The full set of forecasting equations

An audit of (3.6), (3.9), (3.10) and (3.11), together with appropriate forms of (3.3), shows that we have six equations from which p, ρ, T and the three components of $\mathbf{u}$ may be forecast by repeated time-stepping, so long as friction $\mathbf{F}$ and heating rate Q are known. For convenience and future reference we gather together the relevant equations:

$$\frac{\partial \mathbf{u}}{\partial t} = -(\mathbf{u}.\,\mathrm{grad})\mathbf{u} - 2\mathbf{\Omega} \times \mathbf{u} - \alpha\,\mathrm{grad}\,p - \mathrm{grad}\,\Phi + \mathbf{F} \qquad (3.13)$$

$$c_v \frac{\partial T}{\partial t} = -c_v(\mathbf{u}.\,\mathrm{grad})T - \frac{p}{\rho}\,\mathrm{div}\,\mathbf{u} + Q \qquad (3.14)$$

$$\frac{\partial \rho}{\partial t} = -(\mathbf{u}.\,\mathrm{grad})\rho - \rho\,\mathrm{div}\,\mathbf{u} \qquad (3.15)$$

$$p = \rho R T. \qquad (3.16)$$

To obtain (3.14) we have used the continuity equation (3.15) for $D\rho/Dt$.

Numerical techniques are needed for the practical time integration of (3.13)–(3.16). Hence only finite spatial and temporal resolution is possible. This means that Q and $\mathbf{F}$ include the effects of unresolved motions as well as physical processes such as radiative flux convergence, latent heat release/uptake and friction. The difficulties thus introduced are various and profound; see section 12 and Cullen (2002) for further discussion.

In addition to (3.14)–(3.16) and approximations to the components of (3.13), climate simulation models and many weather prediction models include prognostic equations for the local concentration of water substance in some or all of its phases. Water substance is a key quantity in practice – not only because humidity, cloud and precipitation are important meteorologically and climatologically – but because its distribution has a central effect on the distribution of the heating rate Q. We shall not discuss water conservation equations further. Neither shall we treat the variations in gas constant R and principal specific heats c_p and c_v which accompany variations in the amount of water substance present; Gill (1982) gives a concise account.

The equations (3.13)–(3-16) may be written in many alternative forms by using either other equations of the set, or various thermodynamic relations. One of the most important is an alternative form of the thermodynamic equation involving the potential temperature θ defined by

$$\theta = T\left(\frac{p_{\text{ref}}}{p}\right)^{R/c_p}. \tag{3.17}$$

Here p_{ref} is a reference pressure (conventionally 1000hPa), c_p is the specific heat at constant pressure and θ is the temperature that an element of air would have if it were to be brought adiabatically and reversibly to pressure p_{ref}. In terms of θ, (3.14) takes the simpler form

$$\frac{D\theta}{Dt} = \left(\frac{\theta}{Tc_p}\right)Q, \tag{3.18}$$

upon use of (3.15), (3.16) and the relation $c_p - c_v = R$. From (3.18) it is clear that θ remains constant following an element of air if the motion is adiabatic ($Q = 0$). The temperature θ is related to the specific entropy s by $\ln\theta = s/c_p$.

Potential temperature, a thermodynamic quantity, is conserved in adiabatic flow. A dynamic/thermodynamic quantity that is conserved in adiabatic, frictionless flow is potential vorticity, which is of central importance in meteorology. We discuss potential vorticity in the next section.

A useful alternative form of (3.17) arises if (3.16) is used to eliminate T:

$$\ln\theta = \ln T + \frac{R}{c_p}\ln\left(\frac{p_{\text{ref}}}{p}\right) = \frac{1}{\gamma}\ln p - \ln\rho + \text{constant}. \tag{3.19}$$

Here $\gamma = c_p/c_v$, and the relation $c_p - c_v = R$ has again been used.

Equation (3.13) provides three prognostic equations (for the three components of $\mathbf{u}$). It is usual to define the vertical direction by $\nabla\Phi$, the gradient of apparent geopotential; this direction, which is known as *apparent vertical*, is the direction indicated by a plumbline hanging at rest relative to the Earth. [Since both apparent gravity and the direction of apparent vertical depend on the rotation rate of the coordinate frame, which we have chosen to be that of the Earth, they are both frame-dependent quantities.] Also, the slightly spheroidal geopotential surfaces are customarily represented by spheres – an approximation which is amply justified by the smallness (for terrestrial parameter values) of the centrifugal contribution to apparent gravity; see Gill (1982) and White (1982). Convenient horizontal coordinates are then latitude ϕ and longitude λ; see Figure 2. Isolating the three components of (3.13) is not straightforward because the unit vectors change direction over the sphere and so metric (curvature) terms arise. The results are well-known (see Phillips 1973), but we postpone presentation of them until section 4, where conservation properties will be used to provide a rationalisation.

4 Conservation properties

Equations (3.13), (3.14) and (3.15) express conservation of momentum, thermodynamic energy and mass. Other quantities obey other conservation laws, and all such laws appear in various forms expressing, for example, the budget of a quantity in a fixed finite or infinitesimal volume (Eulerian form) or in an identifiable mass of fluid (Lagrangian form). When approximate versions of the governing equations are being set up, the fate of the conservation properties is naturally of interest and importance.

In this section we consider mass, total energy and axial angular momentum conservation, and obtain the components of (3.13) by using conservation arguments. We then derive the material conservation law for potential vorticity – which is implied by (3.13)–(3.16) but is by no means obvious. A Hamiltonian treatment which unifies the conservation laws is noted in conclusion.

4.1 Mass conservation

Equation (3.8) is a mass conservation law of Eulerian form. Equation (3.9) is of Lagrangian form, relating the material derivative of density to the divergence of the flow $\mathbf{u}$; it can be obtained directly by considering conservation of the mass $\rho\,\delta\tau$ of a parcel of air, upon noting that $D(\delta\tau)/Dt = \operatorname{div}\mathbf{u}$. A global mass conservation law can be obtained from (3.7) by taking τ to be the entire volume of the atmosphere:

$$\frac{\partial}{\partial t}\int_{\substack{\text{whole}\\ \text{atmosphere}}} \rho\, d\tau = -\int_{\text{boundaries}} \rho\,\mathbf{u}.\,\mathbf{dS} = 0. \tag{4.1}$$

The second equality assumes there is no net mass transfer into or out of the atmosphere.

4.2 Total energy conservation

By taking the scalar product of $\mathbf{u}$ with (3.13), and using (3.14), one readily obtains a Lagrangian conservation law for the total energy E per unit mass ($E = \frac{1}{2}\mathbf{u}^2 + \Phi + c_v T$ is the sum of the specific kinetic, potential and internal energy):

$$\rho\frac{DE}{Dt} = -\operatorname{div}(p\mathbf{u}) + \rho(Q + \mathbf{u}.\,\mathbf{F}). \tag{4.2}$$

Hence

$$\frac{\partial}{\partial t}(\rho E) = -\operatorname{div}[(\rho E + p)\mathbf{u}] + \rho(Q + \mathbf{u}.\,\mathbf{F}), \tag{4.3}$$

which is the Eulerian version of (4.2). Since it acts at right angles to $\mathbf{u}$, the Coriolis force in (3.13) does not figure directly in the energetics. Equation (4.3) may be regarded as a statement of the conservation of energy; for the case $\mathbf{F} = 0$, Holton (1992) derives (3.6) from (4.3).

Atmospheric energetics is a large subject; White (1978a) gives an elementary account. An important issue is the extent to which potential and internal energy may be converted into flow kinetic energy ($\frac{1}{2}\mathbf{u}^2$ per unit mass). Availability in this sense is the subject of continuing study – see Shepherd (1993), Marquet (1993), Kucharski (1997) and references in these papers.

4.3 Axial angular momentum conservation

The components of (3.13) in the zonal, meridional and vertical directions may be derived by considering the rates of change of unit vectors over the sphere. One finds (see Phillips (1973))

$$\frac{Du}{Dt} = 2\Omega v\sin\phi - 2\Omega w\cos\phi + \frac{uv\tan\phi}{r} - \frac{uw}{r} - \frac{1}{\rho r\cos\phi}\frac{\partial p}{\partial\lambda} + F_\lambda \tag{4.4}$$

$$\frac{Dv}{Dt} = -2\Omega u\sin\phi - \frac{u^2\tan\phi}{r} - \frac{vw}{r} - \frac{1}{\rho r}\frac{\partial p}{\partial\phi} + F_\phi \tag{4.5}$$

$$\frac{Dw}{Dt} = +2\Omega u\cos\phi + \frac{(u^2+v^2)}{r} - \frac{1}{\rho}\frac{\partial p}{\partial r} + F_r - g. \tag{4.6}$$

The arrangement of the terms has a purpose, as will be seen in section 4.4. By multiplying (4.4) by $r\cos\phi$, and noting that $u = r\cos\phi D\lambda/Dt$, $v = rD\phi/Dt$ and $w = Dr/Dt$, it follows that

$$\rho\frac{D}{Dt}\left[(\Omega r\cos\phi + u)r\cos\phi\right] = -\frac{\partial p}{\partial\lambda} + \rho F_\lambda r\cos\phi. \qquad (4.7)$$

Equation (4.7) relates the rate of change of the axial component of absolute angular momentum (per unit mass of air) to the axial components of the torques acting (see Figure 3(a)); it is a Lagrangian conservation law for axial angular momentum. Local and global versions are readily derived. The total axial angular momentum of the atmosphere is by no means constant. Changes of day-length of milliseconds over a few days are detectable by astronomical methods and reflect exchange of axial angular momentum between atmosphere and solid Earth – see Hide *et al.* (1997). Small changes of the direction of the Earth's rotation vector also occur; Barnes *et al.* (1983) give an account of the vectorial angular momentum dynamics involved. A notable aspect of angular momentum conservation is that it determines the frame invariance of the energy conservation laws (White 1989a).

4.4 Spherical polar components of the equation of motion – a derivation via conservation

Perhaps the most direct way of obtaining the three spherical polar components of (3.13) reverses the above argument by using (4.7) to derive (4.4), and then notes that the Coriolis and metric terms in the components of (3.13) must disappear when a kinetic energy equation is formed (see (4.2)). We outline the reasoning. By expanding the material derivative on the left side of (4.7) and multiplying by $1/\rho r\cos\phi$, one readily obtains (4.4). Multiplication by u then gives

$$u\frac{Du}{Dt} = 2\Omega uv\sin\phi - 2\Omega uw\cos\phi + \frac{u^2 v\tan\phi}{r} - \frac{u^2 w}{r} - \frac{u}{\rho r\cos\phi}\frac{\partial p}{\partial\lambda} + uF_\lambda. \quad (4.8)$$

Equation (4.8) contains two Coriolis and two metric terms which must cancel with corresponding terms in the expressions for vDv/Dt and wDw/Dt. Hence the meridional (ϕ) component of (3.13) must contain a Coriolis term $-2\Omega u\sin\phi$, and the radial (r) component a Coriolis term $+2\Omega u\cos\phi$; also, the meridional component must contain a metric term $-(u^2/r)\tan\phi$ to ensure cancellation with $+(u^2 v/r)\tan\phi$ in (4.8). The remaining metric term in (4.8), $-u^2 w/r$, must cancel with a term in the expression for wDw/Dt, so the radial component must contain a term $+u^2/r$. To ensure isotropy with respect to horizontal flow direction, a term $+v^2/r$ must accompany $+u^2/r$ in the radial component. A term $-vw/r$ must then appear in the meridional component. This reasoning reproduces all the Coriolis and metric terms seen in (4.4)–(4.6).

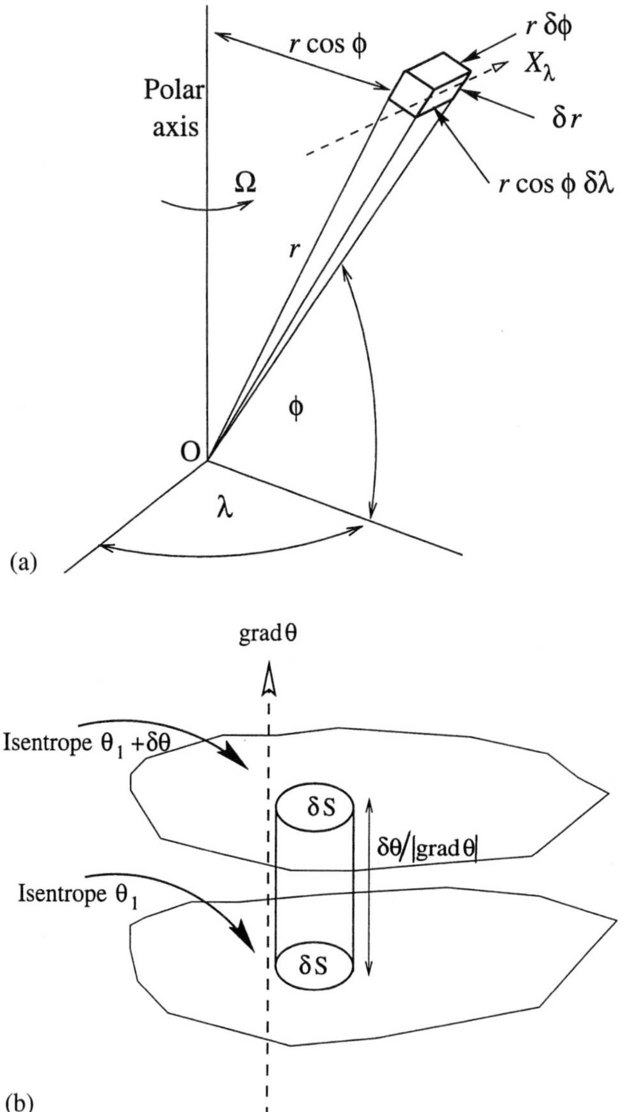

Figure 3: (a) An element of air of mass $\rho\delta\tau = \rho r^2 \cos\phi\,\delta\lambda\,\delta\phi\,\delta r$ centred (at some time t) at longitude λ, latitude ϕ and distance r from the centre of the Earth. If the net zonal force acting on the element is X_λ, then the net torque about the polar axis is $X_\lambda r\cos\phi$. The axial component of the absolute angular momentum of the element is $\delta A = \rho\delta\tau(\Omega r\cos\phi + u)r\cos\phi$, where u is the zonal component of its velocity relative to the Earth. Since the mass $\rho\delta\tau$ of the element is by definition constant, equating the rate of change of δA to the net torque gives

$$\rho\delta\tau\frac{D}{Dt}\{(\Omega r\cos\phi + u)r\cos\phi\} = X_\lambda\cos\phi.$$

Equation (4.7) then results when X_λ is appropriately expressed as the sum of contributions from the pressure gradient force and the force **F** (see (3.11)).
(b) A small cylinder has bases δS which lie within isentropes θ_1 and $\theta_1 + \delta\theta$, and generators parallel to $\mathrm{grad}\,\theta$. In the case considered (see text) the motion is assumed adiabatic, and the cylinder is a material volume.

4.5 Potential vorticity conservation

Equations (4.2) and (4.7) show that neither the total energy nor the axial angular momentum is generally conserved following the flow, even if it is frictionless and adiabatic. Axial angular momentum is conserved in this sense in frictionless flow if the pressure field is independent of longitude, but such axisymmetric flow is rather special. (It can be engineered in laboratory systems – see Hide and Mason (1975).) Even for axial angular momentum, then, the Lagrangian conservation law might more accurately be called a non-conservation law.

Since (3.13) contains two gradient terms (albeit one of them multiplied by α) a reasonable strategy for deriving a Lagrangian conserved quantity is to take the curl of (3.13). By using

$$(\mathbf{u}.\,\mathrm{grad})\mathbf{u} = \mathrm{grad}(\mathbf{u}^2/2) - \mathbf{u} \times \mathrm{curl}\,\mathbf{u} \tag{4.9}$$

and various other vector differential identities, one obtains from (3.13):

$$\frac{D}{Dt}\{\mathbf{Z}+2\mathbf{\Omega}\} = -(\mathbf{Z}+2\mathbf{\Omega})\,\mathrm{div}\,\mathbf{u} + [(\mathbf{Z}+2\mathbf{\Omega}).\,\mathrm{grad}]\,\mathbf{u} + \frac{1}{\rho^2}\,\mathrm{grad}\,\rho \times \mathrm{grad}\,p + \mathrm{curl}\,\mathbf{F}. \tag{4.10}$$

Here $\mathbf{Z} \equiv \mathrm{curl}\,\mathbf{u}$ is the relative vorticity, and $(\mathbf{Z}+2\mathbf{\Omega})$ is the absolute vorticity. Equation (4.10) is the vorticity equation. In spite of its complexity, it is an important equation, and we have not space to do it justice here; see Batchelor (1967) and Pedlosky (1987) for detailed treatments.

Suppose there is no motion ($\mathbf{u} = 0$ everywhere) at some instant. If $\mathrm{curl}\,\mathbf{F}$ vanishes when $\mathbf{u} = 0$, which will be the case if $\mathbf{F}$ consists entirely of the contribution of (Newtonian) friction, then (4.10) shows that motion will develop ($D\mathbf{Z}/Dt \neq 0$) if the surfaces of constant density and constant pressure do not coincide. Fluids obeying $\rho = \rho(p)$ are called barotropic; their surfaces of constant density and constant pressure coincide. Fluids not obeying $\rho = \rho(p)$ are called baroclinic; their constant density and constant pressure surfaces intersect. We deduce Jeffreys' theorem (see Hide (1977)): motion must develop, or already be present, in a baroclinic fluid.

From our perspective of wishing to derive a Lagrangian conserved quantity, (4.10) might seem to represent several steps backwards. However, if we:

(i) multiply (4.10) by $1/\rho$ and apply the continuity equation in the form (3.9);

(ii) use the vector identity, valid for any vector $\mathbf{A}$ and scalar S,

$$\mathbf{A}.\,\frac{D}{Dt}(\mathrm{grad}\,S) = \mathbf{A}.\,\mathrm{grad}\left(\frac{DS}{Dt}\right) - \mathrm{grad}\,S.\,[\mathbf{A}.\,\mathrm{grad}]\mathbf{u};$$

(iii) note that ρ can be expressed (via (3.10) and (3.17)) as a function of p and θ;

then we find that

$$\rho\frac{D}{Dt}\left[\frac{(\mathbf{Z}+2\mathbf{\Omega}).\operatorname{grad}\theta}{\rho}\right]=\operatorname{div}\left[(\mathbf{Z}+2\mathbf{\Omega})\frac{D\theta}{Dt}+\theta\operatorname{curl}\mathbf{F}\right]. \tag{4.11}$$

Hence if (but not only if) the motion is frictionless ($\mathbf{F}=0$) and adiabatic ($D\theta/Dt=0$) then

$$\frac{DP}{Dt}=0, \quad \text{where } P\equiv\frac{(\mathbf{Z}+2\mathbf{\Omega}).\operatorname{grad}\theta}{\rho}. \tag{4.12}$$

The quantity P, which is called Ertel's potential vorticity (Ertel 1942) or simply EPV, *is materially conserved in frictionless, adiabatic flow.* The form of (4.11) implies that any local creation of EPV by heating and friction will tend to be balanced by destruction elsewhere. Equation (4.11) is a central result in fluid dynamics, especially rotating fluid dynamics; see Hoskins *et al.* (1985), Haynes and McIntyre (1987), Hoskins (1991), Lait (1995) and Viúdez (1999).

Result (4.12) may be obtained by applying Kelvin's circulation theorem in isentropic surfaces (surfaces of constant potential temperature, θ). Kelvin's theorem takes the form

$$\begin{aligned}\frac{D}{Dt}\oint_C[\mathbf{u}+\mathbf{\Omega}\times\mathbf{r}].\mathbf{dl} &= \frac{D}{Dt}\oint_S\operatorname{curl}[\mathbf{u}+\mathbf{\Omega}\times\mathbf{r}].\mathbf{dS}\\ &= \frac{D}{Dt}\oint_S[\operatorname{curl}\mathbf{u}+2\mathbf{\Omega}].\mathbf{dS}\\ &= -\oint_C\frac{dp}{\rho}+\oint_C\mathbf{F}.\mathbf{dl}.\end{aligned} \tag{4.13}$$

Here C is any closed loop of material particles and S is any surface bounded by C. If C lies in an isentropic surface, then ρ is a function of p on C. Hence, if the motion is frictionless and adiabatic, one can apply (4.13) to a small material area δS within an isentrope $\theta=\theta_1$ to obtain

$$\frac{D}{Dt}\left\{\frac{([\operatorname{curl}\mathbf{u}+2\mathbf{\Omega}].\operatorname{grad}\theta)\,\delta S}{|\operatorname{grad}\theta|}\right\}=0. \tag{4.14}$$

Also, the quantity $\delta M\equiv\rho\delta S\delta\theta/|\operatorname{grad}\theta|$ – which is the mass within a right cylinder having bases δS on isentropes θ_1 and $\theta_1+\delta\theta$ (see Figure 3(b)) – remains constant. So $\delta M/\delta\theta$ may be taken outside the material derivative in (4.14), and (4.12) is revealed.

4.6 Lagrangian symmetries and conservation properties

In the analytical dynamics of rigid bodies it is well known that conservation laws correspond to symmetries of the Hamiltonian functional that appears in the variational formulation (Noether's theorem). For example, conservation of energy corresponds to time-parametrization invariance. Fluid dynamics is a more complicated problem, partly because of the choice between Eulerian and Lagrangian descriptions, but the theoretical position is now understood. Potential vorticity conservation (see (4.12)) corresponds to the symmetry whereby the Hamiltonian is invariant to the coordinates used to label particles (Ripa 1981, Salmon 1982). Noether's theorem offers a systematic method for deriving consistent approximate models: one approximates the Hamiltonian (whilst preserving its symmetries) and is then assured that the implied evolution equations reproduce the various conservation laws; see Salmon (1983), (1988) and Shepherd (1990). Some applications of the method are noted in section 9.5. Mobbs (1982), Wang (1984) and Sewell (1990) discuss other key aspects of variational formulations of fluid dynamics.

5 The hydrostatic approximation, the hydrostatic primitive equations and the shallow water equations

Jeffreys' theorem (see section 4.6) shows that motion must occur if the pressure and density surfaces in a fluid are not parallel, and the occurrence of motion in the atmosphere is evident even to the most casual observer. Nevertheless, on a wide range of time and space scales, the vertical component of the momentum equation is dominated by the contributions of gravity and the pressure gradient force; the atmosphere is close to *hydrostatic balance*. [The adjective *aerostatic* would seem more appropriate than *hydrostatic*, but the latter is irreversibly established in meteorological usage.] We begin this section by examining the relationships which exist between the thermodynamic fields when hydrostatic balance is precise. Having noted elementary static stability criteria, we then consider how, and under what conditions, we may construct equations describing the motion of an atmosphere that is close to hydrostatic balance. We present and discuss the hydrostatic primitive equations (HPEs), which are widely used in numerical weather prediction and climate simulation, and note the shallow water equations (SWEs), which are widely used as a testbed in both theory and numerical practice.

5.1 Hydrostatic atmospheres

In the absence of motion and of forcing, the governing equations (3.14)–(3.16) and (4.4)–(4.6) are satisfied so long as

$$g + \frac{1}{\rho}\frac{\partial p}{\partial z} = 0 \tag{5.1}$$

and there are no horizontal variations of pressure. Here z is height above mean sea level; see section 5.4. Equation (5.1) is the *hydrostatic equation.* Integration with respect to height gives (since $p\to 0$ as $z\to\infty$)

$$p(z) = \int_z^\infty \rho g \, dz. \tag{5.2}$$

The pressure at height z is equal to the 'weight' of the air above unit area. By using the perfect gas law (3.16) it also follows from (5.1) that

$$p(z) = p(z_s) \exp\left[-\int_{z_s}^{z} (g/RT)\, dz'\right], \tag{5.3}$$

where z_s is the height of the Earth's surface above mean sea level. In a hydrostatic atmosphere, the pressure field is determined by the variation of temperature with height, and temperature must vary only with height. (Spatial variations of g are neglected in this simple treatment.)

Knowing $T(z)$, one can find $p(z)$ from (5.3), $\rho(z)$ from the perfect gas equation (3.16), and $\theta(z)$ from (3.17). For illustration and later reference we list the results obtained in the special case of an isothermal atmosphere ($T = T_0$), assuming $z_s = 0$, uniform g and $p(0) = p_{\text{ref}}$ (see (3.17)):

$$p(z) = p(0)\exp\{-z/H_0\}; \quad H_0 = RT_0/g \tag{5.4}$$

$$\rho(z) = \rho(0)\exp\{-z/H_0\}; \quad \rho(0) = p(0)/RT_0 \tag{5.5}$$

$$\theta(z) = \theta(0)\exp\{+gz/c_pT_0\}; \quad \theta(0) = T_0 \tag{5.6}$$

$$\Rightarrow N^2 \equiv \frac{g}{\theta}\frac{d\theta}{dz} = \frac{g^2}{c_pT_0}. \tag{5.7}$$

The quantity $H_0 = RT_0/g$ is called the scale height of the isothermal atmosphere; it is the height over which the pressure and density decrease by a factor of e. If temperature varies with height – as it does, of course, in the real atmosphere (see Figure 4(a)) – then (5.4)–(5.7) are not valid. Nevertheless, substituting an appropriate mean temperature gives a useful measure of the rate of decrease of pressure and density with height: taking $T_0 = 250$K gives $H_0 = 7.4$km.

5.2 Static stability and the buoyancy frequency

If a parcel of air is displaced vertically (see Figure 5) it will experience a change of hydrostatic pressure, with consequent (adiabatic) changes of temperature and density. From the first law of thermodynamics in the form (3.4), the perfect gas law (3.16) and the relation $c_p - c_v = R$, it follows that the changes and δp in temperature and pressure of the parcel will be related by

$$c_p\delta T - \alpha\delta p = 0 \tag{5.8}$$

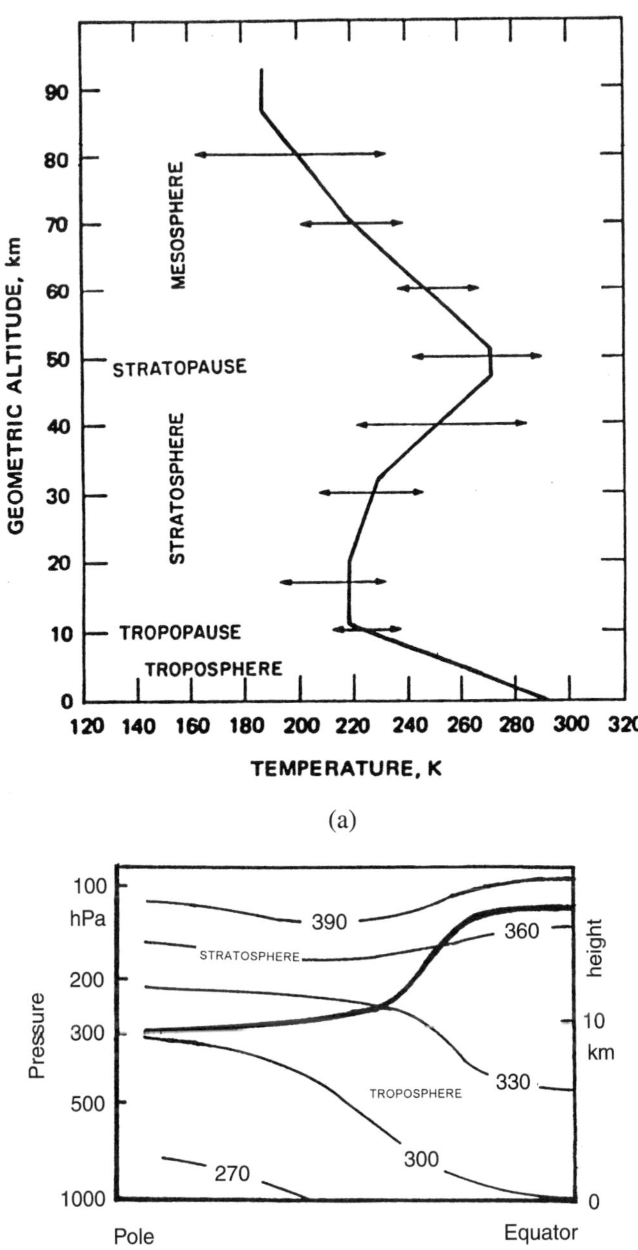

Figure 4: (a) Temperature variation with height to 90km in the US Standard Atmosphere. Profile consists of straight-line segments, as shown. Arrows span the lowest and highest mean monthly temperatures obtained for any location, and so indicate the spatial and temporal variability of monthly means about the standard profile. After NOAA/NASA/USAF (1976) and Gill (1982).
(b) A broad-brush view of the Northern Hemisphere potential temperature field $\theta(\phi, z)$, temporally and longitudinally averaged. Isentropes (contours of constant θ) are marked every 30K from 270–390K by thin lines; the thick line indicates the tropopause. After Hoskins (1991).

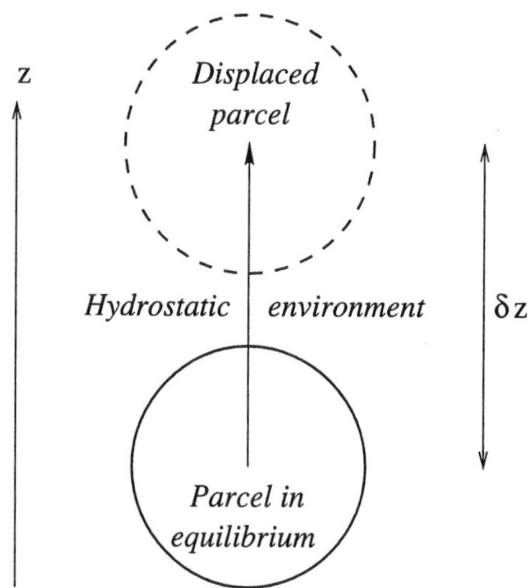

Figure 5: A parcel of air displaced vertically a distance δz from its equilibrium position within a hydrostatic environment.

in adiabatic displacement. Assuming that δp is equal to the change in hydrostatic pressure of the surrounding air over the distance δz (i.e. $\alpha\delta p = -g\delta z$), (5.8) becomes

$$c_p\delta T + g\delta z = 0. \tag{5.9}$$

The rate (with respect to height) at which the temperature of a parcel of air decreases on upward displacement or increases on downward displacement is therefore g/c_p (a quantity known as the dry adiabatic lapse rate). An atmosphere at rest will be stable to vertical displacements of parcels if its temperature $T(z)$ decreases less rapidly with respect to height than g/c_p, i.e. if $dT/dz \geq -g/c_p$. (A parcel of air displaced upwards will then become cooler than, and hence more dense than, its environment; and a parcel of air displaced downwards will become warmer than, and hence less dense than, its environment.) An atmosphere at rest will be unstable to vertical displacements if its temperature decreases more rapidly than g/c_p, i.e. if $dT/dz \leq -g/c_p$.

Air saturated with water vapour suffers a decrease of temperature smaller than $g\delta z/c_p$ upon upward displacement because the inevitable cooling brings about condensation and the release of latent heat (so long as condensation nuclei are present and prevent supersaturation). We shall not discuss further this important effect, which is one of the major complications and fascinations of meteorology; see Gill (1982) and Emanuel (1994) for clear discussion.

The conditions for stability to vertical displacement of unsaturated air are most easily expressed in terms of the vertical gradient of potential temperature. Use of (3.17), (5.1) and (5.9) shows that: if $\partial\theta/\partial z > 0$, unsaturated air is stable to vertical displacement; if $\partial\theta/\partial z < 0$ it is unstable; if $\partial\theta/\partial z = 0$, it is neutrally stable. Large positive values of $\partial\theta/\partial z$ tend to inhibit vertical motion.

In the Earth's atmosphere, well away from the surface, the vertical gradient of potential temperature is generally much greater numerically than the horizontal gradient; the atmosphere is said to be *stratified.* (Horizontal gradients are not, of course, negligible; the difference of potential temperature between different horizontal locations is a key driving agency of the circulation, as Jeffreys' theorem suggests.) Well away from the Earth's surface, values of $\partial\theta/\partial z$ are typically $4 \times 10^{-3}\text{Km}^{-1}$ in the troposphere, and about a factor of 4 greater in the stratosphere; see the schematic climatological section shown in Figure 4(b). Considerable spatial and temporal variations occur, however, especially in the troposphere. The transition region between the troposphere and the stratosphere – the tropopause – across which $\partial\theta/\partial z$ and potential vorticity both change markedly (see Thuburn and Craig (2000)) tends to act as a quasi-horizontal lid to motions beneath. Locally, the tropopause exhibits major variations in height associated with the passage of weather systems (see Keyser and Shapiro (1986) and Browning and Reynolds (1994)) and a general decrease with latitude is evident in Figure 4(b), but a typical value is 10km.

The physical significance of the quantity $\partial\theta/\partial z$ is further illuminated by considering the vertical displacement of a parcel of air in dynamic terms (Figure 5). Upon neglecting the (small) metric and Coriolis terms in (4.6) (terms which vanish if the motion is purely vertical), and assuming again that the displaced parcel experiences the pressure field $\overline{p}(z)$ of its surroundings, we find

$$\frac{Dw}{Dt} + g + \frac{1}{\rho}\frac{d\overline{p}}{dz} = 0. \tag{5.10}$$

Since $d\overline{p}/dz = -\overline{\rho}g$, the second and third terms in (5.10) combine to give $(\rho - \overline{\rho})g/\rho$. Given $p = \overline{p}(z)$ and small displacements δz, we have (from (3.16) and (3.17)):

$$\left(\frac{\rho - \overline{\rho}}{\rho}\right) = \left(\frac{\overline{T} - T}{\overline{T}}\right) = \left(\frac{\overline{\theta} - \theta}{\overline{\theta}}\right) \approx \frac{1}{\overline{\theta}}\frac{d\overline{\theta}}{dz}\delta z. \tag{5.11}$$

But $w = D/Dt(\delta z)$, so (5.10) becomes

$$\frac{D^2}{Dt^2}\delta z + N^2\delta z = 0, \tag{5.12}$$

where $N^2 = (g/\overline{\theta})\, d\overline{\theta}/dz$ is the *buoyancy frequency* (also known as the Brunt–Väisälla frequency). The period of small vertical oscillations in a stable atmosphere is thus $2\pi/N$. If $N^2 < 0$, small vertical displacements amplify with time as $\exp(Nt)$; this is consistent with our earlier identification of $\partial\theta/\partial z < 0$ as the condition for instability to vertical displacements.

5.3 The hydrostatic approximation for an atmosphere in motion

Vertical accelerations in air motions are typically much less than g. Even in the most violent cumulonimbus circulations one might find ascent rates of at most 10ms^{-1} being attained in times of order 1000s; then $Dw/Dt \sim 10^{-2}$ ms^{-2}, compared with $g \approx 10$ ms^{-2}. In a *diagnostic* sense, therefore, the hydrostatic approximation – of assuming hydrostatic balance – is very good indeed.

But this is a naïve view. Our deduced hydrostatic balance mainly reflects the contributions of pressure and density fields varying only with height, which are not associated with motion; our analysis in section 5.1 considered the behaviour of such precisely hydrostatic states. We enquire to what extent and under what conditions the hydrostatic approximation applies to the deviations of all fields from a state of precise static balance: is the hydrostatic approximation valid when we have subtracted out some background static balance? To achieve this we write

$$\begin{aligned} p &= p_0(z) + p'(\lambda, \phi, z, t) \\ \rho &= \rho_0(z) + \rho'(\lambda, \phi, z, t) \end{aligned} \qquad \text{with} \quad \frac{dp_0}{dz} = -\rho_0 g. \tag{5.13}$$

(The 'background' state could be defined by horizontal and temporal averages of density ρ at each height and of mean sea level pressure.) Given the decomposition (5.13), the horizontal components (4.4), (4.5) of the momentum equation change only in that p' replaces p. The vertical component (4.6) becomes

$$\frac{Dw}{Dt} - 2\Omega u \cos\phi - \left(\frac{u^2 + v^2}{r}\right) + g\frac{\rho'}{\rho} + \frac{1}{\rho}\frac{\partial p'}{\partial z} = 0.$$

Our criterion for the validity of the hydrostatic approximation is therefore that Dw/Dt be small compared with $g\rho'/\rho$ or $(1/\rho)\partial p'/\partial z$ (we neglect the other terms in this simple treatment; see section 11.3 for a formulation that includes them). This is a much more testing condition than we had earlier, since typically $|\rho'/\rho| \ll 1$. Supposing the motion to be adiabatic, the thermodynamic equation gives

$$\frac{D\theta'}{Dt} + w\frac{d\theta_0}{dz} = 0.$$

Here $\theta_0 = \theta_0(z)$ is the potential temperature variation implied by $p_0(z)$ and $\rho_0(z)$, and $\theta' = \theta - \theta_0$. Since, *to order of magnitude*, $(\theta'/\theta) \sim (\rho'/\rho)$ we obtain the criterion

$$\tau^2 \gg \frac{1}{N^2} \tag{5.14}$$

if we assume $D/Dt \approx 1/\tau$, where τ is a Lagrangian time scale. Perhaps not surprisingly, our condition (5.14) is that the time-scale of the motion should be much longer than that of vertical buoyancy oscillations (which are essentially non-hydrostatic). The hydrostatic approximation is clearly not dynamically appropriate for an atmosphere that is neutrally stratified ($N = 0$).

The argument leading to (5.12) is readily repeated for parcel oscillations constrained to lie in a plane making an angle α to the horizontal; the resulting frequency is $N \sin\alpha$. Small values of α are characteristic of motion having its horizontal space scale L much larger than its vertical space scale H. The often-quoted condition $H \ll L$ for the applicability of the hydrostatic approximation is thus consistent with condition (5.14). However, α may be small even when L and H are comparable (see Hide and Mason 1975); (5.14) is considered to be the more fundamental condition for the applicability of the hydrostatic approximation.

5.4 The traditional approximation, the shallow atmosphere approximation and the hydrostatic primitive equations

Compared to the dimensions of the Earth (mean radius 6360km) the atmosphere is shallow: consistent with our conclusions in section 5.1, 90% of its mass lies below 17km. Shallow in this sense it is, but two caveats should be noted. First, nearly every field varies with height as well as with horizontal location. For example, winds at 10km are usually markedly different from those found near the surface, as regards both speed and direction. Second, the atmosphere is generally not shallow in relation to the Earth's topography. Mountains locally attain heights of about 8km above mean sea level, and they certainly influence the motion and behaviour of the atmosphere to an important extent, but there is little tendency for the atmosphere to be divided up into 'basins' in the way that the continents divide the Earth's seas into ocean basins, or in the way that high mountains effectively divide the Martian atmosphere (see Hide 1976). [Some local phenomena, for example the East African Jet (Findlater 1969) and coastal lows in Southern Africa (Gill 1977), do depend on mountain ranges acting as lateral 'walls'. The effects of mountains on air flow generally depend on the stratification of the air – as measured by $\partial\theta/\partial z$ – as well as on the flow itself and the elevation of the mountains. See Baines (1995).]

With these caveats in mind, it is reasonable to seek a simplification of the equations of motion which exploits the fact that the atmosphere's depth is a small fraction of the Earth's radius – a *shallow atmosphere* approximation. We aim to replace the variable radius r by a mean value a, whilst retaining differentiations with respect to height as $\partial/\partial z$, where z is height above mean sea level. An implication of this strategy is clear if we re-consider the derivation of the components of the momentum equation given in section 4.4. The Coriolis and metric terms $-2\Omega w \cos\phi$ and $-uw/r$ in (4.4) are lost if we re-define absolute axial angular momentum per unit mass as

$$(u + \Omega a \cos\phi) a \cos\phi \tag{5.15}$$

and the material derivative as

$$\frac{D}{Dt} = \frac{\partial}{\partial t} + \frac{u}{a\cos\phi}\frac{\partial}{\partial\lambda} + \frac{v}{a}\frac{\partial}{\partial\phi} + w\frac{\partial}{\partial z}. \tag{5.16}$$

Pursuit of the argument given in section 4.4 shows that the terms $2\Omega u\cos\phi$ and $(u^2+v^2)/r$ in the vertical component (4.6) and $-vw/r$ in the meridional component (4.5) must then be neglected for consistent energetics. The neglect of the $\cos\phi$ Coriolis terms – known as the *traditional approximation* (Eckart 1960) – is less comfortable than neglect of the quadratic metric terms not involving $\tan\phi$, although for many purposes it turns out to be a good approximation; we shall return to this issue in section 11.3 during a discussion of acoustically-filtered global models. Accepting that the stated omissions should accompany the shallow atmosphere approximation, we obtain the *hydrostatic primitive equations* as

$$\frac{Du}{Dt} = 2\Omega v\sin\phi + \frac{uv\tan\phi}{a} - \frac{1}{\rho a\cos\phi}\frac{\partial p}{\partial\lambda} + F_\lambda \tag{5.17}$$

$$\frac{Dv}{Dt} = -2\Omega u\sin\phi - \frac{u^2\tan\phi}{a} - \frac{1}{\rho a}\frac{\partial p}{\partial\phi} + F_\phi \tag{5.18}$$

$$g + \frac{1}{\rho}\frac{\partial p}{\partial z} = 0. \tag{5.19}$$

The thermodynamic and continuity equations remain

$$\frac{D\theta}{Dt} = \left(\frac{\theta}{c_p T}\right)Q \tag{5.20}$$

and

$$\frac{D\rho}{Dt} = -\rho\nabla.\mathbf{u} \tag{5.21}$$

but, as in (5.17)–(5.19), D/Dt is defined by (5.16), and

$$\nabla.\mathbf{u} = \frac{1}{a\cos\phi}\left[\frac{\partial u}{\partial\lambda} + \frac{\partial}{\partial\phi}(v\cos\phi)\right] + \frac{\partial w}{\partial z}. \tag{5.22}$$

In (5.19), g is properly considered constant. The quantity $2\Omega\sin\phi$ that appears in (5.17) and (5.18) is usually referred to as the Coriolis parameter and accredited the symbol f. The other Coriolis parameter, $2\Omega\cos\phi$, which is absent from the HPEs, has no universally accepted title.

The axial angular momentum conservation law of the HPEs, readily derived from (5.17), is

$$\rho\frac{D}{Dt}\{(u + \Omega a\cos\phi)\,a\cos\phi\} = \rho F_\lambda a\cos\phi - \frac{\partial p}{\partial\lambda}. \tag{5.23}$$

The energy conservation law (Lagrangian form) is

$$\rho\frac{D}{Dt}\left[\frac{1}{2}\mathbf{v}^2 + gz + c_v T\right] = -\nabla.(p\mathbf{u}) + \rho(Q + \mathbf{v}.\mathbf{F}_h). \tag{5.24}$$

Here $\mathbf{v}$ is the *horizontal* flow (the 'wind'). Equation (5.24) shows that the vertical motion w does not contribute to the kinetic energy in the HPEs, but it appears in the pressure–work convergence term $-\nabla.(p\mathbf{u})$ and in the definition of D/Dt (see (5.16)). We consider in 5.5 what determines w.

The HPEs' analogue of potential vorticity conservation is

$$\rho\frac{D}{Dt}\left\{\frac{\boldsymbol{\xi}.\nabla\theta}{\rho}\right\} = \boldsymbol{\xi}.\nabla\left(\frac{D\theta}{Dt}\right) + \nabla\theta.\nabla\times\mathbf{F}_h, \tag{5.25}$$

where

$$\nabla\theta = \left(\frac{1}{a\cos\phi}\frac{\partial\theta}{\partial\lambda}, \frac{1}{a}\frac{\partial\theta}{\partial\phi}, \frac{\partial\theta}{\partial z}\right) \equiv \left(\nabla_z\theta, \frac{\partial\theta}{\partial z}\right), \tag{5.26}$$

and

$$\boldsymbol{\xi} = 2\Omega\mathbf{k}\sin\phi + \nabla\times\mathbf{v}. \tag{5.27}$$

Here $\mathbf{k}$ is unit vector in the (upward) vertical direction, and

$$\nabla\times\mathbf{v} \equiv \left(-\frac{\partial v}{\partial z}, \frac{\partial u}{\partial z}, \frac{1}{a\cos\phi}\left(\frac{\partial v}{\partial\lambda} - \frac{\partial}{\partial\phi}(u\cos\phi)\right)\right). \tag{5.28}$$

The horizontal component equations (5.17), (5.18) of the HPEs may be written in vector form as

$$\frac{\partial\mathbf{v}}{\partial t} + \nabla_z\left(\frac{\mathbf{v}^2}{2}\right) + \zeta\mathbf{k}\times\mathbf{v} + w\frac{\partial\mathbf{v}}{\partial z} = -f\mathbf{k}\times\mathbf{v} - \frac{1}{\rho}\nabla_z p + \mathbf{F}_h. \tag{5.29}$$

Here ∇_z is the horizontal part of ∇, as defined in (5.26), $\zeta \equiv \mathbf{k}.\nabla\times\mathbf{v}$ is the vertical component of the relative vorticity, and $\mathbf{F}_h \equiv (F_\lambda, F_\phi)$. From (5.29) may be derived prognostic equations for ζ and $\nabla_z.\mathbf{v}$, the divergence of the horizontal flow. Such *vorticity, divergence* forms are used in some HPE numerical models, particularly Eulerian spectral models (see Hoskins and Simmons (1975)). The second and third terms on the left side of (5.29) are not precisely equivalent to $(\mathbf{v}.\nabla_z)\mathbf{v}$, which contains a small vertical component when ∇_z is defined as in (5.26); see Côté (1988), Ritchie (1988) and Bates *et al.* (1990).

5.5 Richardson's equation

A by-product of the hydrostatic approximation is that the prognostic equation for w is lost. The implication is *not* that $w = 0$, or that w does not vary with time; rather, w takes that spatial form which maintains hydrostatic equilibrium as the thermodynamic and horizontal flow fields evolve. A diagnostic equation for w may be derived in several ways. We use a route which gives physical insight and an explicit expression for $\partial w/\partial z$, and then note a second-order differential equation that w obeys. Our treatment follows that of unpublished Met Office College lecture notes (1981) by R.W. Riddaway; see also Wiin-Nielsen (1968) and Dutton (1995).

By writing the HPE continuity equation (5.21) in terms of $\partial\rho/\partial t$ and using the hydrostatic approximation (5.1) one obtains the important relation

$$\nabla_z.(\rho\mathbf{v}) + \frac{\partial}{\partial z}\left(\rho w - \frac{1}{g}\frac{\partial p}{\partial t}\right) = 0. \tag{5.30}$$

Here, once again, the flow $\mathbf{u}$ has been separated into its horizontal part $\mathbf{v}$ and vertical part $w\mathbf{k}$, and $\nabla_z.$ indicates the horizontal part of the divergence:

$$\nabla_z.(\rho\mathbf{v}) = \frac{1}{a\cos\phi}\left[\frac{\partial(\rho u)}{\partial\lambda} + \frac{\partial}{\partial\phi}(\rho v\cos\phi)\right]. \tag{5.31}$$

Integrating (5.30) over the interval $[z,\infty]$ gives

$$\frac{\partial p}{\partial t} = \rho g w - g\int_z^\infty \nabla_z.(\rho\mathbf{v})\,dz'. \tag{5.32}$$

Equation (5.32) states simply that the time rate of change of pressure at height z is equal to the product of g with the rate of convergence of mass into the column (of unit horizontal cross-section) above z.

In addition to (5.32) we have another equation for $\partial p/\partial t$. Using (5.21) and (3.10), the thermodynamic equation (3.14) can be written as

$$\frac{Dp}{Dt} = -\gamma p\nabla.\mathbf{u} + \frac{\rho RQ}{c_v}. \tag{5.33}$$

Hence, using (5.19),

$$\frac{\partial p}{\partial t} = -\mathbf{v}.\nabla_z p + \rho w g - \gamma p\nabla.\mathbf{u} + \frac{\rho RQ}{c_v}. \tag{5.34}$$

The right sides of (5.32) and (5.34) must be equal; we find that

$$\gamma p\frac{\partial w}{\partial z} = \gamma p\left[\frac{Q}{Tc_p} - \nabla_z.\mathbf{v}\right] - \mathbf{v}.\nabla_z p + g\int_z^\infty \nabla_z.(\rho\mathbf{v})\,dz'. \tag{5.35}$$

Equation (5.35) determines $\partial w/\partial z$ at height z in terms of p, ρ, $\mathbf{v}$ and Q at z' and ρ and $\mathbf{v}$ at greater heights; $w(z)$ itself may be obtained by integrating (5.35) from $z = z_s$ upwards, assuming a reasonable lower boundary condition (such as $w = 0$ at a flat lower boundary). The explicit expression for $w(z)$ so obtained is known as Richardson's equation from its use in the first numerical weather prediction experiment (Richardson 1922). A different treatment is necessary if an upper boundary condition is applied at a finite height (Kasahara and Washington 1967).

Differentiation of (5.35) leads to a form that does not contain a vertical integral:

$$\gamma\frac{\partial}{\partial z}\left\{p\left[\frac{\partial w}{\partial z} + \nabla_z.\mathbf{v} - \frac{Q}{Tc_p}\right]\right\} = \frac{\partial p}{\partial z}\nabla_z.\mathbf{v} - \frac{\partial\mathbf{v}}{\partial z}.\nabla_z p; \tag{5.36}$$

[Tapp and White (1976) give an equivalent form in the case $Q = 0$]. Equation (5.36), which is unchanged if any upper boundary condition is applied at a finite height, can be obtained more directly by writing (5.19) as $\rho g + \partial p/\partial z = 0$, differentiating locally with respect to t, then using (5.21) and (5.33) to substitute for $\partial \rho/\partial t$ and $\partial p/\partial t$, and finally applying (5.19) for ρ.

5.6 The shallow water equations

The HPEs describe the motion of a compressible atmosphere, and allow height variation of all fields (within the shallow atmosphere approximation, and criterion (5.14)). For both theoretical and numerical testing it is often convenient to have recourse to a set of equations which does not involve height variations or compressibility. The shallow water equations (SWEs) are such a set.

Waves on the surface of a *non*-rotating, incompressible, homogeneous liquid of mean depth d under the influence of gravity behave differently in the long and short wave limits (see, for example, Lighthill (1978)). If the wavelength λ of the surface waves obeys $\lambda \ll d$, then we are close to the deep limit: the waves are dispersive, particle paths are circles (of exponentially decreasing radius as one goes deeper into the fluid) and the motion is essentially non-hydrostatic. But if $\lambda \gg d$, then we are close to the shallow limit: the waves are non-dispersive, particle paths are horizontal, amplitude is independent of depth, and the motion is essentially hydrostatic.

With this background, consider how the HPE momentum and continuity equations may be applied to a *rotating* incompressible, homogeneous fluid of density $\hat{\rho}$ bounded by a rigid horizontal surface at $z = 0$ and having a free surface at $z = h(\lambda, \phi, t)$. In the shallow limit, the pressure is (plausibly) hydrostatic, its horizontal gradient is $\hat{\rho}g$ multiplied by the free surface gradient, and (5.17), (5.18) become

$$\frac{Du}{Dt} = 2\Omega v \sin\phi + \frac{uv\tan\phi}{a} - \frac{g}{a\cos\phi}\frac{\partial h}{\partial\lambda} + F_\lambda \tag{5.37}$$

$$\frac{Dv}{Dt} = -2\Omega u \sin\phi - \frac{u^2\tan\phi}{a} - \frac{g}{a}\frac{\partial h}{\partial\phi} + F_\phi \tag{5.38}$$

with

$$\frac{D}{Dt} \equiv \frac{\partial}{\partial t} + \frac{u}{a\cos\phi}\frac{\partial}{\partial\lambda} + \frac{v}{a}\frac{\partial}{\partial\phi}. \tag{5.39}$$

If the horizontal velocity components u, v are initially independent of depth, they will remain so, since the pressure gradient is independent of depth. The time evolution of h may then be obtained by integrating the continuity equation (5.21) over the depth h and noting that $w(h) = Dh/Dt$ (and $\rho = \hat{\rho}$ is constant):

$$\int_0^h \nabla_z.\mathbf{v}\,dz + w(h) = 0 \Rightarrow \frac{Dh}{Dt} + h\nabla_z.\mathbf{v} = 0. \tag{5.40}$$

Equations (5.37)–(5.40) are the shallow water equations (SWEs); we shall use them in our account of approximately geostrophic models (section 9). [To avoid ambiguity, we retain the symbol ∇_z for the horizontal part of the ∇ operator – see (5.26) and (5.31) – although for the SWEs ∇ reduces to ∇_z anyway, since no height variations of u, v or h are involved.]

The SWEs are closed for u, v or h, and have the following conservation properties (see Salmon (1983) for a Hamiltonian treatment):

Axial angular momentum:
$$\frac{D}{Dt}\{(u+\Omega a\cos\phi)a\cos\phi\} = F_\lambda a\cos\phi - g\frac{\partial h}{\partial\lambda} \tag{5.41}$$

Energy:
$$h\frac{D}{Dt}\left(\frac{1}{2}\mathbf{v}^2\right) = -gh\mathbf{v}.\nabla_z h + h\mathbf{v}.\mathbf{F}_h \tag{5.42}$$

Potential vorticity:
$$h\frac{D}{Dt}\left(\frac{\zeta+2\Omega\sin\phi}{h}\right) = \mathbf{k}.\nabla_z\times\mathbf{F}_h. \tag{5.43}$$

Here, $\mathbf{F}_h \equiv (F_\lambda, F_\phi)$ and ζ is the SWE relative vorticity:

$$\zeta = \frac{1}{a\cos\phi}\left(\frac{\partial v}{\partial\lambda} - \frac{\partial}{\partial\phi}(u\cos\phi)\right). \tag{5.44}$$

An important limiting case of the SWEs occurs when variations of the depth h are negligible (we examine in section 8 the conditions under which this occurs). Then (5.40) becomes

$$\nabla_z.\mathbf{v} = 0 \tag{5.45}$$

and (5.43) reduces to

$$\frac{D}{Dt}(\zeta + 2\Omega\sin\phi) = \mathbf{k}.\nabla_z\times\mathbf{F}_h \tag{5.46}$$

which is the *barotropic vorticity equation.* The material derivative in (5.46) is given by (5.39), with $\mathbf{v} = (u, v)$ satisfying the non-divergence condition (5.45). A streamfunction ψ may be introduced for $\mathbf{v}$, whereupon (5.46) becomes a prognostic equation for $\zeta = \nabla_z^2\psi$ in terms of the advection of the absolute vorticity $\nabla_z^2\psi + 2\Omega\sin\phi$ by the flow $\mathbf{v} = \mathbf{k}\times\nabla_z\psi$:

$$\frac{\partial}{\partial t}(\nabla_z^2\psi) = -(\mathbf{k}\times\nabla_z\psi).\nabla_z(\nabla_z^2\psi + 2\Omega\sin\phi) + \mathbf{k}.\nabla_z\times\mathbf{F}_h. \tag{5.47}$$

Equation (5.47) determines the time evolution of ψ, given appropriate boundary conditions and a specification of $\mathbf{F}_h$. Studies of (5.47) and close variants (some of them Cartesian – see section 6.3) have given insight into Rossby waves (section 8.3), steady flow structures and geostrophic turbulence in rotating fluids; see, for example, Platzman (1968), Hoskins (1973), Rhines (1975), Baines (1976), Held (1983), Shutts (1983a), McWilliams (1984), Marshall (1984), White (1990) and Verkley (1993).

6 Vertical coordinate systems; the 'f-plane' and the 'β-plane'

This section deals with some further aspects of formulation and approximation that are characteristic of meteorological dynamics. The use of pressure as vertical coordinate is first discussed, and other choices are noted. The use of Cartesian geometry and approximate treatments of the spatial variation of the Coriolis parameter are then briefly considered.

6.1 Use of pressure as vertical coordinate

Any quantity that bears a one-to-one relation to height z may be used as a vertical coordinate. If the hydrostatic approximation is made, then pressure is certainly such a quantity, since $\rho > 0$ ensures that $\partial p/\partial z = -\rho g$ is everywhere negative. In 'pressure coordinates' the independent variables are (λ, ϕ, p, t) instead of (λ, ϕ, z, t), and z becomes a dependent variable. The material derivative is

$$\frac{D}{Dt} = \frac{\partial}{\partial t} + \frac{u}{a\cos\phi}\frac{\partial}{\partial\lambda} + \frac{v}{a}\frac{\partial}{\partial\phi} + \omega\frac{\partial}{\partial p}. \tag{6.1}$$

Here $\omega \equiv Dp/Dt$. The horizontal and local time derivatives in (6.1) are taken at constant pressure (but with distances measured on constant height surfaces); u, v are the velocity components in constant height surfaces (*not* the components in constant pressure surfaces); and $\partial/\partial p$ is taken at constant λ, ϕ, t. Equation (6.1) may be derived either from first principles, or from (5.21) by using the following rules, valid for any well-behaved $Q = Q(\lambda, \phi, z, t)$, with $X = t$, λ, ϕ and then $Q = p$:

$$\frac{\partial Q}{\partial z} = \frac{\partial p}{\partial z}\frac{\partial Q}{\partial p} = -\rho g\frac{\partial Q}{\partial p}; \quad \left.\frac{\partial Q}{\partial X}\right|_z = \left.\frac{\partial Q}{\partial X}\right|_p - \left.\frac{\partial Q}{\partial z}\frac{\partial z}{\partial X}\right|_p; \tag{6.2}$$

see Figure 6. The quantity $\omega \equiv Dp/Dt$ is often referred to as the pressure-coordinate 'vertical velocity', although, as the material derivative of a scalar, it is frame-invariant.

The hydrostatic relation (5.19) may be written as

$$g\frac{\partial z}{\partial p} = -\frac{1}{\rho} = -\frac{RT}{p}. \tag{6.3}$$

The pressure-coordinate versions of (5.17) and (5.18) are

$$\frac{Du}{Dt} = \left(2\Omega + \frac{u}{a\cos\phi}\right)v\sin\phi - \frac{g}{a\cos\phi}\frac{\partial z}{\partial\lambda} + F_\lambda \tag{6.4}$$

$$\frac{Dv}{Dt} = -\left(2\Omega + \frac{u}{a\cos\phi}\right)u\sin\phi - \frac{g}{a}\frac{\partial z}{\partial\phi} + F_\phi, \tag{6.5}$$

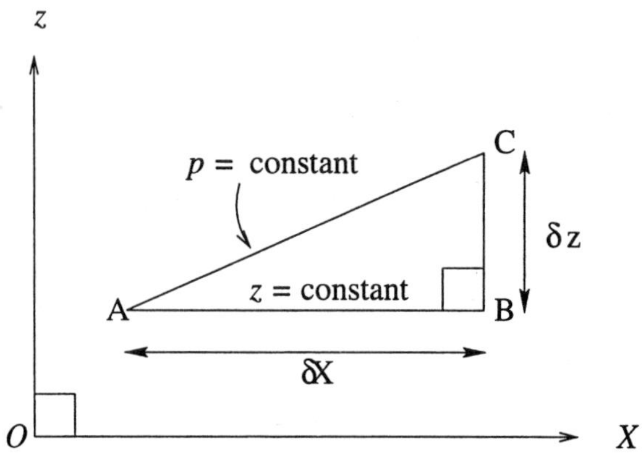

Figure 6: Transformation of horizontal and local time derivatives between height and pressure coordinates; $X = \lambda$, ϕ or t. Line AB (parallel to OX) has length δX, line BC (parallel to Oz) has length δz; pressure p is constant on AC. Let δQ_{RS} denote the difference in $Q = Q(\lambda, \phi, z, t)$ between any points R and S. Then

$$\delta Q_{\mathrm{AC}} = \delta Q_{\mathrm{AB}} + \delta Q_{\mathrm{BC}}.$$

Thus

$$\left.\frac{\partial Q}{\partial X}\right|_p \delta X = \left.\frac{\partial Q}{\partial X}\right|_z \delta X + \frac{\partial Q}{\partial z}\delta z$$

i.e.

$$\left.\frac{\partial Q}{\partial X}\right|_p = \left.\frac{\partial Q}{\partial X}\right|_z + \left.\frac{\partial Q}{\partial z}\frac{\partial z}{\partial X}\right|_p$$

in which the nonlinear pressure gradient terms in (5.17), (5.18) have become linear in the horizontal gradient components (on pressure surfaces) of z.

The thermodynamic equation remains as (5.20), but the material derivative is expressed as (6.1). A major simplification occurs in the continuity equation (5.21) [see Sutcliffe (1947) and Eliassen (1949)]. The form (5.30) shows that hydrostatic balance reduces the continuity equation to non-divergence even in height coordinates (accepting a suitably redefined vertical velocity). Use of

$$w = \frac{Dz}{Dt} = \frac{\partial z}{\partial t} + \frac{u}{a\cos\phi}\frac{\partial z}{\partial \lambda} + \frac{v}{a}\frac{\partial z}{\partial \phi} + \omega\frac{\partial z}{\partial p} \tag{6.6}$$

in (5.29), along with (6.2) and the hydrostatic relation, shows that (5.21) becomes simply

$$\nabla_p . \mathbf{v} + \frac{\partial \omega}{\partial p} = 0 \tag{6.7}$$

in which

$$\nabla_p . \mathbf{v} \equiv \frac{1}{a\cos\phi}\left(\frac{\partial u}{\partial \lambda} + \frac{\partial}{\partial \phi}(v\cos\phi)\right) \tag{6.8}$$

Equations (6.3)–(6.7) [with (5.20)] are exact transforms of the height coordinate HPEs. Pressure-coordinate forms of the various conservation laws are readily derived, but will not be given here.

Similar transformations of the height coordinate HPEs may be made using any suitable function of pressure as vertical coordinate. Frequent choices include p^{R/c_p} (see Hoskins and Bretherton 1972) and $\ln p$ (see Holton (1975)); these coordinates are often given the symbols z or Z, and so it is easy to lose sight of the fact that they are pressured-based coordinates.

6.2 Other choices of vertical coordinate

Given the hydrostatic approximation, pressure coordinates offer at once a simplification and a complication. From the continuity equation (6.7) one can readily derive a diagnostic equation for the 'vertical velocity', ω, in the pressure system:

$$\omega(p) = -\int_0^p \nabla_p . \mathbf{v}\, dp. \tag{6.9}$$

Equation (6.9) is simpler than Richardson's equation for the usual height-coordinate vertical velocity, $w = Dz/Dt$; see section 5.5. The complication is that the Earth's surface is generally not a coordinate surface in the pressure system – even in the absence of topography. The local rate of change of surface pressure $p_s = p_s(\lambda, \phi, t)$ can be calculated from (6.9) with $p = p_s$:

$$\frac{\partial p_s}{\partial t} = \frac{Dp_s}{Dt} - \mathbf{v}. \nabla p_s = -\int_0^{p_s} \nabla_p . \mathbf{v} dp - \mathbf{v}. \nabla p_s. \tag{6.10}$$

The quantity $\partial p_s/\partial t$ is known as the surface pressure *tendency*.

The boundary condition $w = 0$ at a horizontal surface ($z = 0$, say) becomes (from (6.3) and (6.6))

$$\frac{\partial z}{\partial t} + \frac{u}{a \cos\phi} \frac{\partial z}{\partial \lambda} + \frac{v}{a} \frac{\partial z}{\partial \phi} - \frac{\omega}{\rho g} = 0 \tag{6.11}$$

on $p = p_s$. In theoretical analyses, approximations to (6.11) are often resorted to, and should be carefully justified in each case (see, for example, Haynes and Shepherd (1989)). A common procedure is to apply $\omega = 0$ on $p = p_0$, where p_0 is a horizontal average surface pressure; a more accurate approximation under certain conditions (see section 10.1) is to retain the local time derivative in (6.11) and to apply $\omega = -\rho g \partial z/\partial t$ at $p = p_0$.

In numerical weather forecasting and climate simulation models it is usual to adopt a vertical coordinate for which the Earth's surface is a coordinate surface. The prototype choice is the sigma coordinate $\sigma = p/p_s$ (Phillips 1957), for which $\sigma = 1$ at the Earth's surface whether or not topography is present. The continuity equation (6.8) becomes

$$\frac{\partial p_s}{\partial t} + \nabla_\sigma . (p_s \mathbf{v}) + \frac{\partial \dot{\sigma}}{\partial \sigma} = 0 \tag{6.12}$$

where $\dot{\sigma} \equiv D\sigma/Dt$ is the σ-coordinate 'vertical velocity'. Thus (since $\dot{\sigma} = 0$ at $\sigma = 0$ and $\sigma = 1$):

$$\frac{\partial p_s}{\partial t} = -\int_0^1 \nabla_\sigma . (p_s \mathbf{v})\, d\sigma. \tag{6.13}$$

The quantity $\dot{\sigma}$ may be found by eliminating $\partial p_s/\partial t$ from (6.12), (6.13), and integrating over σ:

$$\dot{\sigma} = -\int_0^\sigma \nabla_\sigma . (p_s \mathbf{v})\, d\sigma' - \sigma \int_0^1 \nabla_\sigma . (p_s \mathbf{v})\, d\sigma'; \tag{6.14}$$

Haltiner and Williams (1981) give further details.

A vertical coordinate that has particular manipulative and conceptual advantages is potential temperature, θ (Starr 1945; see also Eliassen 1987). It is a permissible choice so long as no regions of neutrality or static instability exist (i.e. so long as $\partial\theta/\partial z > 0$). The material derivative in θ-coordinates is

$$\frac{D}{Dt} = \frac{\partial}{\partial t} + \frac{u}{a\cos\phi}\frac{\partial}{\partial\lambda} + \frac{v}{a}\frac{\partial}{\partial\phi} + \dot{\theta}\frac{\partial}{\partial\theta} \tag{6.15}$$

(with the t, λ and ϕ derivatives taken at constant θ). In the case of adiabatic motion ($\dot{\theta} = 0$) the $\partial/\partial\theta$ contribution to (6.15) vanishes and advection is purely 2-dimensional (on surfaces of constant θ). The continuity equation also takes a quasi–2-dimensional form, and (whether or not the motion is adiabatic) the pressure gradient terms in (5.18) and (5.19) become linear in the horizontal gradients of the quantity $M \equiv gz + c_pT$ (known as the Montgomery potential). A partial similarity to the shallow water equations may be noted, although the fields described by the SWEs have no vertical variation – see section 5.6. Against these (and other) considerable advantages must be weighed the disadvantage that the Earth's surface is not a constant θ surface. The difficulty is not insuperable, however; see Bleck (1984) and Hsu and Arakawa (1990).

Kasahara (1974) derived forms of the HPEs using a generalised vertical coordinate s (such that $\partial s/\partial z \neq 0$), and some numerical weather prediction and climate simulation models use so-called hybrid coordinates which behave like σ near the Earth's surface but like pressure at high levels (Simmons and Burridge 1981, Simmons and Strüfing 1983). Hybrid coordinates have been used that behave like σ near the Earth's surface, like θ at intermediate levels and like p at high levels (Zhu *et al.* 1992, Thuburn 1993).

As we shall see in section 11, the use of pressure and sigma coordinates is not limited to models in which the hydrostatic approximation is applied. Also, a vertical coordinate equivalent to *hydrostatic pressure* has been successfully used in fully non-hydrostatic models; see Laprise (1992), Bubnová *et al.* (1995) and Geleyn and Bubnová (1997).

6.3 Further geometric and Coriolis approximations

Geometric and Coriolis approximations beyond the shallow atmosphere and 'traditional' approximations of section 5.4 are common in meteorology, especially in theoretical treatments.

For accurate modelling of the global atmosphere it is essential to represent the sphericity of the Earth and the latitude variation of the Coriolis parameter $f = 2\Omega \sin\phi$. In the study of sub-planetary scale phenomena, especially when quantitatively accurate conclusions are not required, the use of simplified geometries and coarse treatment of the Coriolis parameter are convenient and justifiable. For example, if one wishes to model the circulation of a cumulus cloud, for which time-scales are typically tens of minutes and space scales a few kilometres, the use of local Cartesian geometry and neglect of Coriolis effects are entirely reasonable simplifications.

For weather systems having a horizontal space scale of 1000km and a time-scale of a few days – the so-called synoptic scale – the Coriolis force must be accounted for, but the latitude variation of f, and spherical geometry, may be considered unimportant. The use of Cartesian geometry with a constant Coriolis parameter is a scheme known as the 'f-plane'. Often, Cartesian geometry is used, but in differentiated terms the Coriolis parameter is allowed a linear variation $f = f_0 + \beta y$, where f_0 and β are constants and y is northward distance from the latitude at which $f = f_0$. This scheme is known as a 'β-plane': if $f_0 \cong \pm 10^{-4}\ \mathrm{s}^{-1}$, it is a 'mid-latitude β-plane'; if $f_0 = 0$, it is an 'equatorial β-plane'. [The latitude variation of the Coriolis parameter is itself often referred to as 'the β-effect'.] These approximations are often introduced in a rather *ad hoc* fashion in theoretical analyses (though with due regard to the latitudinal scale of the motion being studied); for critical discussion see Pedlosky (1987), and for a Hamiltonian approach to the issue, Ripa (1997) and Graef (1998). We shall treat a particular case in section 8.3.

For illustration and later use (see sections 7–10) we note here how the height-coordinate HPEs (5.17)–(5.22) are modified in Cartesian 'β-plane' form. The material derivative becomes

$$\frac{D}{Dt} \equiv \frac{\partial}{\partial t} + u\frac{\partial}{\partial x} + v\frac{\partial}{\partial y} + w\frac{\partial}{\partial z}. \tag{6.16}$$

Equations (5.17) and (5.18) are written in vector form as

$$\frac{D\mathbf{v}}{Dt} = -f\mathbf{k} \times \mathbf{v} - \frac{1}{\rho}\nabla_z p + \mathbf{F}_h \tag{6.17}$$

with $\nabla_z \equiv (\partial/\partial x, \partial/\partial y)$, $\mathbf{F}_h \equiv (F_x, F_y)$ and the metric terms in (5.17), (5.18) neglected. The 3-dimensional divergence term in the continuity equation (5.21) is expressed as

$$\nabla.\mathbf{u} = \frac{\partial u}{\partial x} + \frac{\partial v}{\partial y} + \frac{\partial w}{\partial z}. \tag{6.18}$$

The Cartesian coordinates x and y may be regarded as $x = a\lambda\cos\phi_0$, $y = a(\phi - \phi_0)$, where ϕ_0 is the central latitude of the 'β-plane'; we also have $f = f_0 + \beta y = 2\Omega\sin\phi_0 + (2\Omega/a)y\cos\phi_0$. On an '$f$-plane', $\beta = 0$ and the orientation of Oxy in the horizontal is immaterial.

7 The geostrophic approximation

During our discussion of hydrostatic balance, the buoyancy frequency $N \equiv ((g/\theta)\partial\theta/\partial z)^{1/2}$ emerged in section 5.2 as a key inverse time-scale in a stratified atmosphere. Another important inverse time-scale, but one having a much more systematic spatial variation, is the inertial frequency $f = 2\Omega\sin\phi$. This is the frequency with which parcels of air may circulate in the horizontal under the action only of the horizontal component of the Coriolis force. If friction and horizontal pressure gradients are absent, and the $\tan\phi$ metric terms and the latitude variation of f are neglected, then (5.18) and (5.19) give

$$\frac{D^2u}{Dt^2} + f^2u = 0.$$

The period of these inertial oscillations, $2\pi/f = \pi/\Omega\sin\phi$, is half the local pendulum day – i.e. half the period with which a Foucault pendulum will circulate about the local vertical at latitude ϕ. See Paldor and Killworth (1988) and Stommel and Moore (1989) for detailed discussion.

Large-scale motion in the extra-tropical atmosphere, on length scales of 1000km and more and time scales of a day and more (the 'synoptic scale' – as noted in section 6.3), is typified by a quite different balance: the $\sin\phi$ part of the Coriolis force is nearly balanced by the horizontal pressure gradient force. In geostrophic flow, this balance is precise (see (Figure 7(a)), and (6.17) becomes

$$-f\mathbf{k}\times\mathbf{v} - \frac{1}{\rho}\nabla_z p = 0. \tag{7.1}$$

Consistent with (7.1), the geostrophic wind $\mathbf{v}_G$, is *defined* as

$$\mathbf{v}_G \equiv \frac{1}{\rho f}\mathbf{k}\times\nabla_z p. \tag{7.2}$$

Other definitions of geostrophic wind are sometimes useful (Blackburn 1985), and one of them will be used extensively in sections 9 and 10. A definition which combines geostrophic and hydrostatic balance, and involves the $\cos\phi$ parts of the Coriolis force as well as the $\sin\phi$ parts, has been used by Hide (1971) and others; see also Shutts (1989).

In (7.1) and (7.2) (as in section 5) $\mathbf{k}$ is a unit vector in the upward vertical direction. The criterion for validity of the geostrophic *approximation*, $\mathbf{v} \approx \mathbf{v}_G$, is that the acceleration term $D\mathbf{v}/Dt$ in (6.17) should be negligible compared

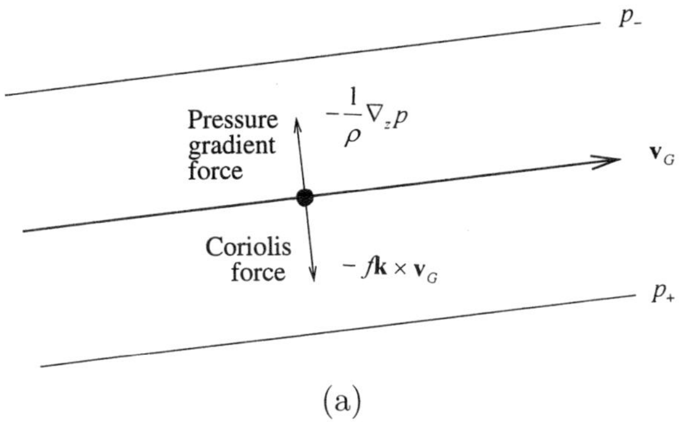

(a)

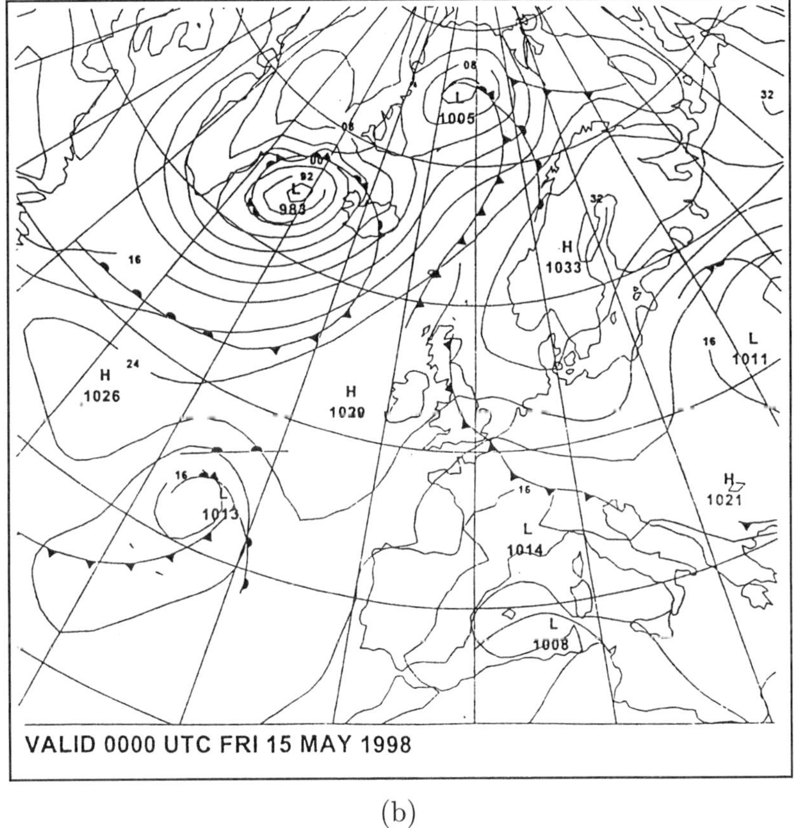

(b)

Figure 7: (a) Illustrating the balance between the horizontal components of the Coriolis and pressure gradient forces acting on unit mass of air in (horizontal) geostrophic flow $\mathbf{v}_G$. The diagram is drawn assuming $f > 0$ (Northern Hemisphere). If $f < 0$ (Southern Hemisphere) $\mathbf{v}_G$ would be oppositely directed, given the horizontal pressure gradient $\nabla_z p$ shown. The quantity $\mathbf{k}$ is a unit vector in the upward vertical direction (perpendicular to the plane of the diagram).
(b) A typical map of mean sea-level pressure; analysed locations of warm, cold and occluded fronts are also shown. Contour interval 4hPa. From the UK Met. Office's *Daily Weather Summary.* [In land regions the pressure at mean sea-level has been obtained by a standard extrapolation based on the hydrostatic approximation and knowledge of the atmosphere's temperature structure.]

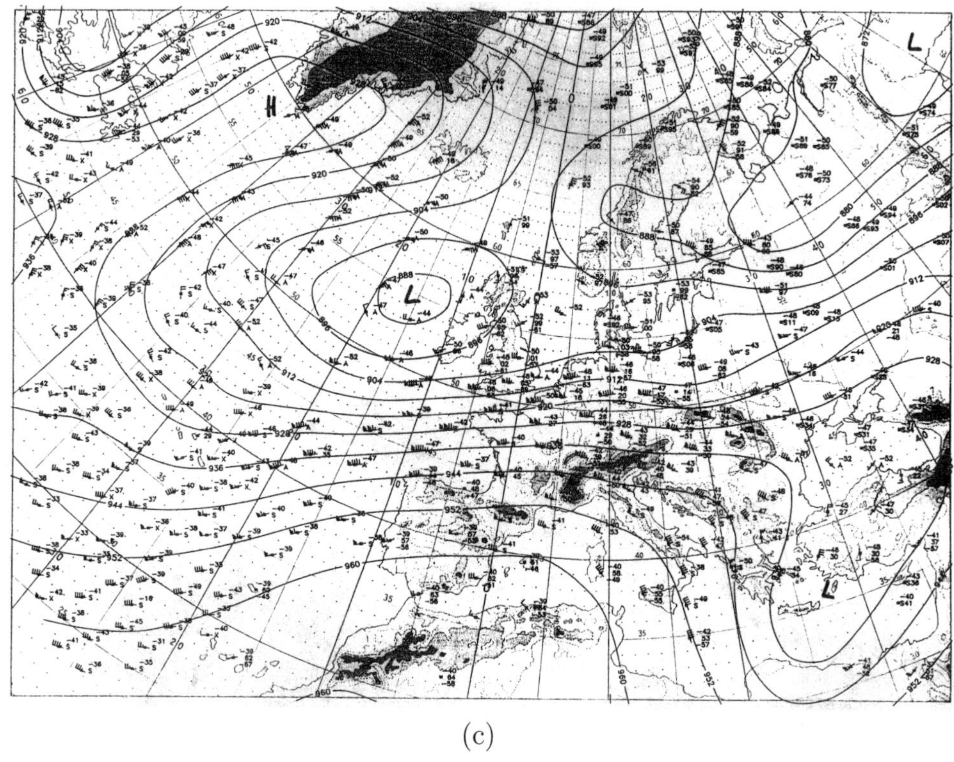

(c)

Figure 7: (c) A typical 300hPa height map, valid 1200 UTC 11 May 1999. Contour interval 8 decametres. Data entries indicate the observation density (and use a standard code). From the *European Meteorological Bulletin* of the Deutscher Wetterdienst, Offenbach, by permission.

with the Coriolis term $-f\mathbf{k}\times\mathbf{v}$. Assuming a horizontal space scale of variation L, and a horizontal velocity scale V (i.e. the horizontal flow varies by V over horizontal distance L), then $\mathbf{v} \approx \mathbf{v}_G$ according to a simple scale analysis if

$$\mathrm{Ro} \equiv \frac{V}{fL} \ll 1.$$

Here Ro is a Rossby number, and it has been assumed that $D/Dt \sim V/L$. Putting $V \sim 10$ ms^{-1}, $f \sim 10^{-4}$ s^{-1} and $L \sim 10^6$ m gives Ro $\sim 10^{-1}$; this is a typical value for synoptic-scale weather systems in middle and high latitudes.

We consider in this section various aspects of the geostrophic wind, and the interesting consequences of combining the geostrophic and hydrostatic approximations – which together account in a diagnostic sense for many synoptic-scale features of the extra-tropical atmosphere.

7.1 Pressure and height signatures

A third possible balance in the horizontal components of the momentum equation is between the acceleration and the pressure gradient force:

$$\frac{D\mathbf{v}}{Dt} = -\frac{1}{\rho}\nabla_z p. \tag{7.3}$$

This balance may be achieved in motion systems having a much shorter time scale than the pendulum day. By applying a scale analysis to (7.3), and assuming again that $D/Dt \sim V/L$, we find that such systems will be characterised by horizontal pressure fluctuations Δp of magnitude ρV^2 (independent of horizontal scale). On the other hand, *geostrophically balanced* flow, according to (7.2), will be characterised by pressure fluctuations Δp_{G} of magnitude $\rho f V L$. Hence

$$\frac{\Delta p}{\Delta p_{\mathrm{G}}} \sim \frac{V}{fL} \equiv \mathrm{Ro}\,. \tag{7.4}$$

The Rossby number, Ro, therefore measures the magnitude of pressure fluctuations due to circulations characterised by (7.3) compared with pressure fluctuations due to geostrophically-balanced circulations characterised by (7.1) and similar flow speeds V. In other words, the *pressure signature* of nearly-geostrophic flows is an order of magnitude greater than that of flows (of similar strength) characterised by (7.3). To the extent that $\mathrm{Ro} \ll 1$, a map of (say) pressure at sea-level, will be dominated by the contributions of geostrophically balanced flows. Taking $\rho \sim 1\mathrm{kg\,m^{-3}}$ and $V \sim 10\ \mathrm{ms^{-1}}$, we find $\Delta p \sim 10^2\mathrm{Pa}$ $= 1\mathrm{hPa}$ for short time-scale circulations. Taking $L \sim 10^6$ m for synoptic-scale flow gives $\sim 10^3\mathrm{Pa} = 10\mathrm{hPa}$. Maps of sea-level pressure are therefore expected to show fluctuations of order 10hPa about a spatial mean, and such fluctuations are indeed observed: see Figure 7(b), which shows a typical sea-level pressure map.

By use of (6.2), and assuming hydrostatic balance, the definition (7.2) of geostrophic wind can be written in terms of the gradient of the height h of pressure surfaces as

$$\mathbf{v}_{\mathrm{G}} \equiv \frac{g}{f}\mathbf{k} \times \nabla_p h. \tag{7.5}$$

Height variations Δh of a pressure surface associated with geostrophic flow are thus of order $fLV_{\mathrm{G}}/g \sim 10^2$ m (given $g \approx 10\ \mathrm{ms^{-2}}$ and other values as quoted earlier). Maps of the height of a pressure surface are widely used in meteorology. Figure 7(c) shows a typical map of the height of the 300hPa (= 300mb) surface. This surface is roughly 9km above the Earth's surface, and Figure 7(c) shows variations of about $\pm 5 \times 10^2$ m in its local height; the flow at 300hPa attains values substantially greater than $10\mathrm{ms^{-1}}$. Even with height fluctuations of this magnitude, the 300hPa surface is very gently sloping: $\Delta h/L \ll 10^{-3}$.

The geostrophic wind $\mathbf{v}_\mathrm{G}$ is horizontally divergent on pressure surfaces only to the extent that the latitude variation of the Coriolis parameter f contributes:

$$\nabla_p.\, \mathbf{v}_\mathrm{G} = -\frac{\beta v_\mathrm{G}}{f},$$

where (as in section 6.3) $\beta = (2\Omega/a)\cos\phi$ is the rate at which f increases with distance northward. Thus $\nabla_p.\, \mathbf{v}_\mathrm{G}$ is much smaller than its two constituent terms if $\beta L/f \ll 1$. In middle and high latitudes this condition reduces to $L/a \ll 1$, which is reasonably well satisfied by motion having a horizontal space scale of 10^6 m. In extra-tropical latitudes, the geostrophic wind is therefore nearly non-divergent on pressure surfaces (given $L/a \ll 1$).

7.2 The differential geometry of the height field

According to (7.5), $\mathbf{v}_\mathrm{G}$ is directed parallel to the height contours $h =$ constant and has magnitude $(g/f)|\nabla_p h|$. Other differential geometric properties of the height field are related to other properties of the geostrophic wind field. The vertical component of the vorticity of $\mathbf{v}_\mathrm{G}$ is

$$\mathbf{k}.\, \nabla_p \times \mathbf{v}_\mathrm{G} = \frac{g}{f}\nabla_p^2 h + \frac{\beta u_\mathrm{G}}{f},$$

which is dominated by the $\nabla_p^2 h$ term so long as $L/a \ll 1$. Thus $(g/f)\nabla_p^2 h$ is a good approximation to the vertical component of the vorticity of the geostrophic wind if $L/a \ll 1$.

A less well-known property of the height field is a relationship between its principal directions of curvature and the stretching and contraction axes of the geostrophic flow. Because the height field typically has a slope much less than 10^{-3} (see section 7.1), classical expressions for the principal directions of curvature may be simplified, to a very good approximation.

Consider the height of a pressure surface as a function of horizontal Cartesian coordinates on an f-plane: $z = h(x, y)$ with $f = f_0 =$ constant. Assume that h and its first and second derivatives h_x, h_y, h_{xx}, h_{xy} and h_{yy} are continuous. The projections on the (x, y)-plane of the principal directions of curvature of a surface specified in Monge form $z = h(x, y)$ are lines having dy/dx given by

$$\left(\frac{dy}{dx}\right)^2 \left\{h_{xy}\left(1 + h_y^2\right) - h_{yy}h_x h_y\right\} - \left(\frac{dy}{dx}\right)\left\{h_{yy}\left(1 + h_x^2\right) - h_{xx}\left(1 + h_y^2\right)\right\} + h_{xx}h_x h_y - h_{xy}\left(1 + h_x^2\right) = 0; \tag{7.6}$$

see, for example, Bell (1912), p.338. If the second derivatives h_{xx}, h_{xy}, h_{yy} are of similar order of magnitude and the slopes h_x, h_y are very small ($\ll 1$), then (7.6) reduces to

$$\left(\frac{dy}{dx}\right)^2 h_{xy} - \left(\frac{dy}{dx}\right)\{h_{yy} - h_{xx}\} - h_{xy} = 0. \tag{7.7}$$

In this approximation of small slope, the horizontal projections of the principal directions are perpendicular to one another. For 2D flow, the angle θ between the dilatation axis and the x-axis is given (see section 2) by

$$\tan 2\theta = \frac{D_2}{D_1} = \frac{(v_x + u_y)}{(u_x - v_y)}. \tag{7.8}$$

In geostrophic flow on an f-plane, $u = u_G = -(g/f_0)h_y$ and $v = v_G = (g/f_0)h_x$, and (7.8) becomes

$$\tan 2\theta \equiv \tan 2\theta_G = -\frac{(h_{xx} - h_{yy})}{2h_{xy}}. \tag{7.9}$$

Choose the axes Oxy such that x lies along the dilatation axis. In this system $\theta_G = 0$, and (from (7.9)) $h_{xx} = h_{yy}$; substitution into (7.7) now shows that $(dy/dx)^2 = 1$. Hence (given that the height field is characterised by very small slopes) *the dilatation axes and contraction axes of the geostrophic flow on an f-plane bisect the principal directions of curvature of the corresponding height field.* M.J. Sewell (private communication, 1998) has shown that this result is an example of a general bisection relationship between the principal axes of the following two tensors associated with any 2-dimensional, solenoidal vector field: the symmetric part of its gradient; and the second derivative of its scalar potential.

7.3 The thermal wind equation

An important result follows by combining the geostrophic relation (7.5) with the hydrostatic relation (6.3):

$$-\frac{\partial \mathbf{v}_G}{\partial p} = -\frac{g}{f}\mathbf{k} \times \nabla_p \left(\frac{\partial h}{\partial p}\right) = \frac{R}{fp}\mathbf{k} \times \nabla_p T. \tag{7.10}$$

Hence

$$\frac{\partial \mathbf{v}_G}{\partial z} = \frac{\partial p}{\partial z}\frac{\partial \mathbf{v}_G}{\partial p} = \frac{g}{fT}\mathbf{k} \times \nabla_p T = \frac{g}{f\theta}\mathbf{k} \times \nabla_p \theta. \tag{7.11}$$

Thus the vertical *shear* of the geostrophic wind is at right angles to the temperature gradient on pressure surfaces; see Figure 8(a). Equation (7.10) is the differential form of the *thermal wind* equation. Hydrostatic and geostrophic balance tie the wind and thermodynamic fields together in a specific way that is one of the key features of synoptic-scale meteorology.

A useful height-integrated form of (7.10) is readily obtainable in terms of the vertical distance Δz_{12} between two pressure surfaces p_2 and $p_1 < p_2$. From the hydrostatic approximation (5.1) and the perfect gas law (3.16):

$$\Delta z_{12} = \frac{R}{g}\int_{p_1}^{p_2} T\, d(\ln p). \tag{7.12}$$

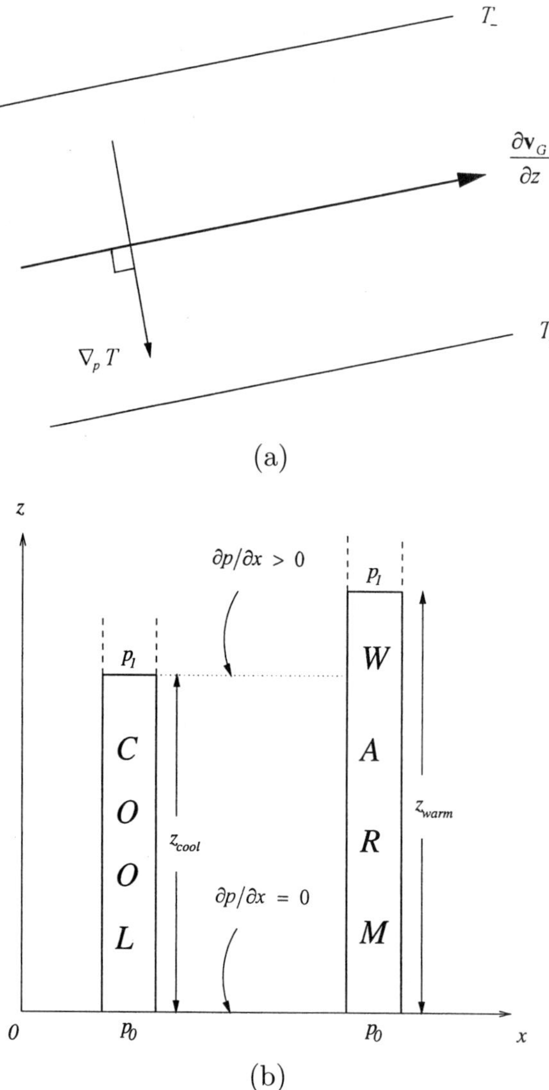

Figure 8: (a) Illustrating the relative orientation (in the horizontal plane) of the shear of the geostrophic wind in the vertical ($\partial \mathbf{v}_G/\partial z$) and the temperature gradient ($\nabla_p T$) on pressure surfaces in the Northern Hemisphere. (In the Southern Hemisphere, the geostrophic wind shear vector ($\partial \mathbf{v}_G/\partial z$) would be oppositely directed, given the temperature gradient shown.)
(b) Two columns of air having different mean temperatures. Suppose that the hydrostatic approximation is applicable and that the pressure at level $z = 0$ is p_0 in each column, so that the component of geostrophic wind $\mathbf{v}_G$ perpendicular to the plane of the paper is zero there. Define the top of each column as the level at which pressure is p_1 (which is less than p_0). The top of the warm column, thus defined, is at a level z_{warm} greater than that of the top of the cool column z_{cool} (see (7.12)); the 'thickness' of the warm column is greater than the 'thickness' of the cool column. At height z_{cool}, the pressure in the warm column is therefore greater than p_1; hence, at this level, a horizontal pressure gradient exists and the geostrophic wind component perpendicular to the plane of the diagram is non-zero.

The quantity Δz_{12} is known as the 'thickness' of the layer between pressures p_2 and p_1.

Equation (7.12) shows that the thickness is a measure of the mean temperature of the layer. Charts of thickness, often of the layer between 1000 and 500hPa, are part of the stock-in-trade of the synoptic meteorologist. Along with appropriate height charts, they show at a glance where warm and cold air are being advected by the geostrophic wind. From (7.10) and (7.12) we find

$$\mathbf{v}_G(p_1) - \mathbf{v}_G(p_2) = \frac{g}{f}\mathbf{k} \times \nabla(\Delta z_{12}). \tag{7.13}$$

Thus the vector *difference* in the geostrophic flow between two pressure surfaces (at the same horizontal location) bears the same relation to the thickness contours as the geostrophic wind does to the pressure or height field. This is readily understood in physical terms; see Figure 8(b).

It is worth noting that the geostrophic wind generally changes its direction as well as its magnitude with height. From (7.5) and (7.11),

$$\mathbf{k}.\left(\mathbf{v}_G \times \frac{\partial \mathbf{v}_G}{\partial z}\right) = \frac{g^2}{f^2 T}\mathbf{k}.\,(\nabla_p h \times \nabla_p T). \tag{7.14}$$

Hence the geostrophic wind shear $\partial \mathbf{v}_G/\partial z$ is parallel or anti-parallel to the geostrophic wind $\mathbf{v}_G$ only if the height gradient $\nabla_p h$ is parallel or anti-parallel to the temperature gradient $\nabla_p T$.

In this brief account we have been able to mention only a few of the diagnostic results which may be obtained by combining the hydrostatic and geostrophic relations. This is currently a rather underplayed area of meteorology, but it is well described in older textbooks (see Saucier (1955)). For forecasting or for the elucidation of forecasts generated numerically (using, say, the HPEs), a time-dependent picture is required; we consider in section 9 some models which answer this need.

7.4 Other steady, balanced flows

In strictly geostrophic flow, particle accelerations and friction are absent; Coriolis and pressure-gradient forces are in precise balance and the flow is rectilinear. Balanced *circular* motion under the influence of the Coriolis and pressure gradient forces is readily analysed, and gives what is known as *gradient flow*. The balanced circular flow around a centre of low pressure is weaker than the geostrophic flow implied by the pressure or height field (the contours of which are circular in this case): the excess of the pressure gradient force over the Coriolis force supplies the acceleration that is necessary to maintain circular motion. The balanced circular flow around a centre of high pressure is larger than the geostrophic flow implied by the pressure or height field: the excess of the Coriolis force over the pressure gradient force now supplies the acceleration necessary to maintain circular motion. A possibly less expected aspect of the problem is that the supergeostrophic flow around a centre of high pressure has an upper bound. In plausibility terms, one may argue that the required acceleration in circular motion of radius r is $|\mathbf{v}|^2/r$, but the Coriolis force varies only as $|\mathbf{v}|$; hence it is reasonable that a limit exists to the extent to which the acceleration can be supplied by the Coriolis force. See Holton (1992).

Straight flow in the presence of friction may be analysed by assuming a tractable relation between the flow and the friction. The customary example is Ekman's classical treatment of the case in which $\mathbf{F} = k\partial^2\mathbf{v}/\partial z^2$; this is covered in textbooks such as Holton (1992) and Gill (1982). [The assumed force balance between frictional, pressure-gradient and Coriolis forces is sometimes referred to as *geotriptic*; see Bannon (1998) and references therein.] The essential physics may be exposed by considering the case in which friction is assumed to oppose the flow according to a simple linear law (first used, according to Eliassen (1984), by Guldberg and Mohn in 1876):

$$f\mathbf{k} \times \mathbf{v} - f\mathbf{k} \times \mathbf{v}_{\mathrm{G}} = -C\mathbf{v}. \tag{7.15}$$

Hence

$$f\mathbf{k} \times \mathbf{v}_{\mathrm{AG}} = -C\mathbf{v}, \tag{7.16}$$

where $\mathbf{v}_{\mathrm{AG}} = \mathbf{v} - \mathbf{v}_{\mathrm{G}}$ is the *ageostrophic wind*, and $\mathbf{v}_{\mathrm{AG}}$ is perpendicular to $\mathbf{v}$. If $\mathbf{v}_{\mathrm{G}}$ is plotted along the diameter of a circle, $\mathbf{v}$ (the actual horizontal wind) and $\mathbf{v}_{\mathrm{AG}}$ will meet one another on the circle (see Figure 9); also, $|\mathbf{v}| < |\mathbf{v}_{\mathrm{G}}|$ and $|\mathbf{v}_{\mathrm{AG}}| < |\mathbf{v}_{\mathrm{G}}|$ (given $C > 0$). A simple calculation gives

$$\mathbf{v}^2 = \frac{\mathbf{v}_{\mathrm{G}}^2}{\left(1 + \frac{C^2}{f^2}\right)} \quad \text{and} \quad \tan\alpha \equiv \frac{C}{f}, \tag{7.17}$$

where α is the angle between $\mathbf{v}_{\mathrm{G}}$ and $\mathbf{v}$. Carrying the analysis through for a quadratic friction law is straightforward. In physical terms, friction reduces the flow below the geostrophic value (which is not a general property of the Ekman solution) and directs it towards lower pressure.

8 Atmospheric waves

The nonlinearity of the advection terms in the equations of motion cannot safely be ignored in quantitative forecasting or simulation of the atmosphere's motion. Nevertheless, a knowledge of the small amplitude oscillations and waves that are possible in a compressible, stratified, rotating atmosphere is fundamental to an appreciation of meteorological dynamics. The nonlinear terms sometimes turn out to be less important than a crude analysis might suggest (see, for example, White (1990)) and they can in any case be regarded as a forcing agency for the linearised dynamics and thermodynamics (along with diabatic and frictional sources and sinks). The properties of the possible wave motions determine how, and how quickly, local disturbances may

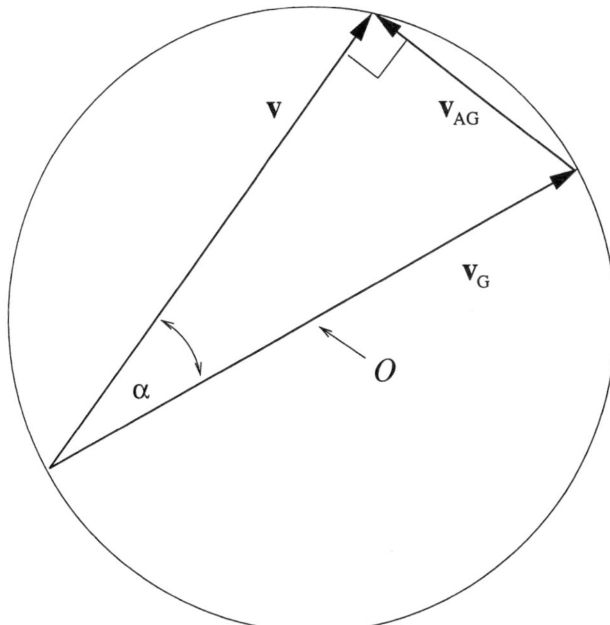

Figure 9: The geostrophic wind $\mathbf{v}_G$ plotted as the diameter of a circle (centre O). By definition, $\mathbf{v} = \mathbf{v}_\mathrm{G} + \mathbf{v}_\mathrm{AG}$, where $\mathbf{v}$ is the total (horizontal) wind and $\mathbf{v}_\mathrm{AG}$ is the ageostrophic wind. If $\mathbf{v}$ and $\mathbf{v}_G$ are steady and spatially uniform, and a friction law $\mathbf{F} = -c\mathbf{v}$ holds, then $\mathbf{v}.\,\mathbf{v}_\mathrm{AG} = 0$, and $\mathbf{v}$ and $\mathbf{v}_\mathrm{AG}$ meet on the circle of which $\mathbf{v}_\mathrm{G}$ is a diameter. Equation (7.17) specifies the angle α between $\mathbf{v}_\mathrm{G}$ and $\mathbf{v}$. The diagram has been drawn assuming $f > 0$ (Northern Hemisphere) so that high pressure lies to the right of $\mathbf{v}_\mathrm{G}$, and $\mathbf{v}$ to the left of $\mathbf{v}_\mathrm{G}$. If $f < 0$ (Southern Hemisphere) high pressure lies to the left of $\mathbf{v}_\mathrm{G}$, and $\mathbf{v}$ would lie to the right of $\mathbf{v}_\mathrm{G}$.

influence distant regions (Lighthill 1978). Their properties also affect the application of numerical schemes in weather forecasting and climate simulation models (Haltiner and Williams 1981). Perhaps most important in our context, an appreciation of the possible wave motions illuminates the various approximate formulations – which do not support all modes of oscillation.

In this section we consider small oscillations of a frictionless, adiabatic, perfect gas atmosphere about an isothermal state of rest relative to the rotating Earth. Analytical results are readily obtained by use of the f-plane and β-plane approximations (see section 6.3). Only neutral waves are found because there is no energy available apart from that initially present in the perturbations. Cases in which the initial state has available energy because of velocity or horizontal temperature gradients will not be addressed, although they have played a key role in setting up the conceptual furniture of meteorological dynamics. For discussion of relevant instability problems see Drazin and Reid (1981), Gill (1982), Held (1985), Farrell (1989) and Holton (1992), for example.

The profiles of pressure, density, potential temperature and buoyancy frequency in an isothermal atmosphere were given in section 5 (equations (5.4)–(5.7)). Pressure and density decrease exponentially with height, potential temperature increases exponentially with height, and the buoyancy frequency N is constant. The classical adiabatic sound speed, c_0, given by

$$c_0 = \sqrt{\gamma R T_0}, \tag{8.1}$$

(where $\gamma = c_p/c_v$), is also independent of height, as is the scale height $H_0 = RT_0/g$. From (3.19), N, H_0 and c_0 obey a relation that will be used repeatedly in this and later sections:

$$\frac{N^2 H_0}{g} + \frac{g H_0}{c_0^2} = 1. \tag{8.2}$$

8.1 Oscillations of an isothermal atmosphere: f-plane case

We begin with some comments on notation. In linearised analyses it is usual to indicate perturbations from the chosen basic state by primes: $\mathfrak{I} = \mathfrak{I}_0 + \mathfrak{I}'$, where $\mathfrak{I}$ is a generic field and $\mathfrak{I}_0$ its value in the basic state. The equations obtained after linearization involve only $\mathfrak{I}_0$ and $\mathfrak{I}'$, and the use of primes to indicate perturbations becomes tedious and redundant. We shall drop the primes in the linearised equations: in this section, u, v, w, p, ρ and θ are to be understood as the perturbation velocity components and thermodynamic quantities. Various combinations of the thermodynamic variables feature in the linearised equations: ρ/ρ_0, p/ρ_0 and θ/θ_0 (to exercise the notation just introduced). It is tempting to introduce new symbols for all or some of these quantities, but we shall resist the temptation: it would lead us into slightly tidier equations, but their physical content might be obscured. A final issue is the choice of symbol for the angular frequency of a wave. We shall follow common usage in mathematical physics, and denote this quantity by ω; our choice is not to be confused with the use of ω in sections 6 and 9–11 to represent Dp/Dt (which is common usage in meteorology).

Linearisation of the adiabatic, frictionless, f-plane equations about an isothermal rest state gives:

$$\frac{\partial \mathbf{v}}{\partial t} + f_0 \mathbf{k} \times \mathbf{v} + \nabla_z \left(\frac{p}{\rho_0}\right) = 0 \tag{8.3}$$

$$\frac{\partial w}{\partial t} - g\left(\frac{\theta}{\theta_0}\right) + \left(\frac{\partial}{\partial z} - \frac{N^2}{g}\right)\left(\frac{p}{\rho_0}\right) = 0 \tag{8.4}$$

$$\frac{\partial}{\partial t}\left(\frac{\rho}{\rho_0}\right) + \nabla_z . \mathbf{v} + \frac{1}{\rho_0}\frac{\partial}{\partial z}(\rho_0 w) = 0 \tag{8.5}$$

$$\frac{\partial}{\partial t}\left(\frac{\theta}{\theta_0}\right) + \frac{N^2}{g} w = 0 \tag{8.6}$$

$$\frac{\theta}{\theta_0} - \frac{1}{c^2}\left(\frac{p}{\rho_0}\right) + \frac{\rho}{\rho_0} = 0. \tag{8.7}$$

Here $\nabla_z \equiv \left(\frac{\partial}{\partial z}, \frac{\partial}{\partial y}\right)$. Equation (8.3) is the linearised, frictionless form of the horizontal momentum equation (6.17). Equations (8.5) and (8.6) are respectively the linearised continuity and (adiabatic) thermodynamic equations. Equation (8.7) may be obtained by linearising (3.19) and then using (3.10) and (8.1). Equation (8.4) is the linearised vertical component of the momentum equation [with the shallow atmosphere approximation, use of Cartesian geometry, z as vertical coordinate and neglect of the Coriolis and metric terms in (4.6)]; (8.7) has been used to eliminate the perturbation density, and (8.2) applied.

Elimination of θ/θ_0 between (8.4) and (8.6) gives

$$\left(\frac{\partial^2}{\partial t^2} + N^2\right) w + \left(\frac{\partial}{\partial z} - \frac{N^2}{g}\right) \frac{\partial}{\partial t}\left(\frac{p}{\rho_0}\right) = 0. \tag{8.8}$$

Another relation between w and p/ρ_0, obtainable from (8.3) and (8.5)–(8.7), is

$$c_0^2\left(\frac{\partial^2}{\partial t^2} + f_0^2\right)\left(\frac{\partial}{\partial z} - \frac{g}{c_0^2}\right) w + \left(\frac{\partial^2}{\partial t^2} + f_0^2 - c_0^2\nabla_z^2\right)\frac{\partial}{\partial t}\left(\frac{p}{\rho_0}\right) = 0. \tag{8.9}$$

An important special solution of (8.8) has $w = 0$ everywhere, and hence

$$\frac{p}{\rho_0} \propto \exp\left[\frac{N^2 z}{g}\right].$$

If a wave-like form $\exp\{i(kx + ly - \omega t)\}$ is assumed, (8.9) then requires that the angular frequency ω should obey

$$\omega^2 = c_0^2(k^2 + l^2) + f_0^2. \tag{8.10}$$

These horizontally-propagating waves are known as *Lamb waves* (see Lamb 1932). With them are associated fluctuations of pressure, density and horizontal velocity, but not potential temperature (or vertical velocity). They are anisotropic in character, being in hydrostatic balance in the vertical, but having the structure of classical sound waves as regards their horizontal field variations. Apart from the effect of rotation ($f_0 \neq 0$) they are non-dispersive and have the phase speed of classical sound waves.

Other modes permitted by (8.8) and (8.9) obey a partial differential equation obtained by eliminating w:

$$\left[\left(\frac{\partial^2}{\partial t^2} + N^2\right)\nabla_z^2 + \left(\frac{\partial^2}{\partial t^2} + f_0^2\right)\left(\frac{\partial^2}{\partial z^2} - \frac{1}{H_0}\frac{\partial}{\partial z} - \frac{1}{c_0^2}\frac{\partial^2}{\partial t^2}\right)\right]\frac{\partial}{\partial t}\left(\frac{p}{\rho_0}\right) = 0. \tag{8.11}$$

Wave-like solutions of (8.11) exist in the form

$$\frac{p}{\rho_0} \propto \exp\left(\frac{z}{2H_0}\right)\exp\{i(kx + ly + mz - \omega t)\}. \tag{8.12}$$

These solutions have angular frequencies ω that obey the dispersion equation

$$\omega\left(\omega^4 - \Omega_a^2\omega^2 + \Omega_a^2\Omega_g^2\right) = 0 \tag{8.13}$$

in which the parameters Ω_a^2 and Ω_g^2 are given by

$$\Omega_a^2 \equiv c_0^2\left(k^2 + l^2 + m^2 + \frac{1}{4H_0^2}\right) + f_0^2 \tag{8.14}$$

and

$$\Omega_a^2\Omega_g^2 \equiv N^2c_0^2(k^2 + l^2) + f_0^2c_0^2\left(m^2 + \frac{1}{4H_0^2}\right). \tag{8.15}$$

Equation (8.13), which is obtained by substituting (8.12) into (8.11), has five solutions. Corresponding to $\omega = 0$ is a *geostrophic mode* having $f\mathbf{v} = \mathbf{k} \times \nabla(p/\rho_0)$, with u, v, p/ρ_0 proportional to $\exp(N^2z/g)$, $\rho \propto \exp(-gz/c_0^2)$ and $w = \theta = 0$. The other four solutions consist of two pairs. One pair, having high frequencies, has

$$\omega^2 \approx \Omega_a^2 = c_0^2\left(k^2 + l^2 + m^2 + \frac{1}{4H_0^2}\right) + f_0^2. \tag{8.16}$$

These are acoustic waves modified by rotation ($f_0 \neq 0$) and static compressibility ($1/H_0 \neq 0$). Even horizontally-propagating waves ($m = 0$) of this type are distinct from the Lamb waves. They are weakly dispersive through the terms in (8.16) in f_0^2 and $1/4H_0^2$. The second pair of solutions, having lower frequencies (see Gill 1982, p174), has

$$\omega^2 \approx \Omega_g^2 = \frac{N^2(k^2 + l^2) + f_0^2\left(m^2 + \frac{1}{4H_0^2}\right)}{\frac{f_0^2}{c_0^2} + k^2 + l^2 + m^2 + \frac{1}{4H_0^2}}. \tag{8.17}$$

These are buoyancy, or *gravity* waves, modified by rotation and static compressibility; they are often called *inertio-gravity* waves. Even if $f_0 = 0$, they are dispersive.

The approximations (8.16) and (8.17) are generally very good in terrestrial parameter ranges, and are sufficiently accurate for many purposes. Exact solutions may be obtained by noting that $\Omega_a^2 > 4\Omega_g^2$ and writing

$$\frac{2\Omega_g}{\Omega_a} = \sin 2\psi. \tag{8.18}$$

Then

$$\omega = \omega_n = \Omega_a \sin\left(\psi + \frac{n\pi}{2}\right), \quad n = 0, 1, 2, 3, \tag{8.19}$$

and the solutions may be represented graphically, as in Figure 10. From the quartic bracket of (8.13) it follows that the exact solutions $\pm\omega_a$, $\pm\omega_g$ obey

$$\omega_a^2 + \omega_g^2 = \Omega_a^2; \quad \omega_a^2\omega_g^2 = \Omega_a^2\Omega_g^2. \tag{8.20}$$

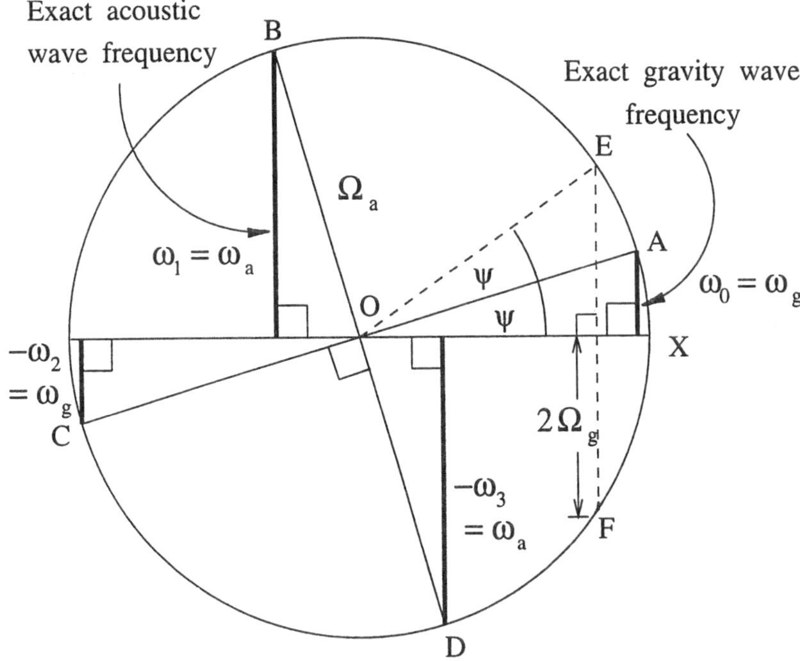

Figure 10: Showing the four non-zero solutions of the acoustic/gravity wave dispersion equation (8.13). The radius of the circle is the approximate acoustic wave frequency Ω_a given by (8.16), and the angle ψ, which is less than $\pi/4$, is given in terms of Ω_a and the approximate gravity wave frequency Ω_g (see (8.17)) by (8.18). The quantities ω_0 and ω_2 are the exact gravity wave frequencies $\pm\omega_g$; ω_1 and ω_3 are the exact acoustic wave frequencies $\pm\omega_a$. The straight lines COA and DOB are perpendicular diameters of the circle, and OA subtends ψ with OX. The angle ψ may be constructed by plotting a chord EF $= 4\Omega_g$ perpendicular to OX, joining OE, and bisecting the angle EOX $= 2\psi$ (see (8.18)).

The approximation (8.16) thus overestimates the true sound wave frequency, while (8.17) underestimates the true gravity wave frequency.

Apart from the introduction of the geostrophic mode having $\omega = 0$, the presence of rotation does not lead to any new modes of motion. If $f_0 = 0$, a pair of sound waves and a pair of gravity waves are still found; rotation only serves to modify their frequencies (and generally to increase dispersion). In particular, it is noticeable that there are no modes corresponding to inertial oscillations – see section 7 – whereas the (inertio-) gravity waves can be identified with buoyancy oscillations (modified by rotation). This reflects the fact that the pressure field plays a key role in gravity waves but not in pure inertial oscillations.

8.2 Filtering approximations in the f-plane problem

The consequences of various approximations and modifications of the equations of motion may be explored by repeating the above analysis with various terms omitted. An apt way of doing this [J.S.A. Green, unpublished lecture notes, Imperial College, 1970; G.J. Haltiner (1971)] is to attach multiplicative tracer parameters n_i to the target terms; then $n_i = 0$ or 1 according as the associated term is omitted or retained. Our treatment in this section closely follows Green's.

Of particular interest are the *hydrostatic approximation*, in which $\partial w/\partial t$ is omitted from (8.4), and the *anelastic approximation*, in which $\partial\rho/\partial t$ is omitted from (8.5). Our freedom to omit these terms is not complete, as an examination of the energy equation implied by (8.3)–(8.7) readily shows. Consider (8.4) and (8.5) in the forms

$$n_1\frac{\partial w}{\partial t} - g\frac{\theta}{\theta_0} + \left(\frac{\partial}{\partial z} - n_0\frac{N^2}{g}\right)\frac{p}{\rho_0} = 0 \tag{8.4a}$$

$$n_2\frac{\partial}{\partial t}\left(\frac{\rho}{\rho_0}\right) + \nabla_z.\mathbf{v} + \frac{1}{\rho_0}\frac{\partial}{\partial z}(\rho_0 w) = 0. \tag{8.5a}$$

A tracer n_0 for $-(N^2/g)(p/\rho_0)$ has been placed in (8.4a). The local energy conservation law is

$$\frac{\partial}{\partial t}\left(\frac{1}{2}\rho_0\left(\mathbf{v}^2 + n_1 w^2 + \frac{g^2}{N^2}\left(\frac{\theta}{\theta_0}\right)^2 + \frac{n_2}{c_0^2}\left(\frac{p}{\rho_0}\right)^2\right)\right)$$
$$= -\nabla_z.(p\mathbf{v}) - \frac{\partial}{\partial z}(pw) + \frac{N^2 pw}{g}(n_0 - n_2). \tag{8.21}$$

When $n_0 = n_1 = n_2 = 1$, this reduces to a familiar form (see Gill (1982), p.170); in particular, the term in $N^2 w$ vanishes. To ensure that we do not introduce a spurious energy source, we therefore require that n_0 take the same value as n_2. In place of (8.4a) we use

$$n_1\frac{\partial w}{\partial t} - g\frac{\theta}{\theta_0} + \left(\frac{\partial}{\partial z} - n_2\frac{N^2}{g}\right)\frac{p}{\rho_0} = 0 \tag{8.4b}$$

Equations (8.8) and (8.9) become

$$\left(n_1\frac{\partial^2}{\partial t^2} + N^2\right)w + \left(\frac{\partial}{\partial z} - n_2\frac{N^2}{g}\right)\frac{\partial}{\partial t}\left(\frac{p}{\rho_0}\right) = 0. \tag{8.8a}$$

and

$$c_0^2\left(\frac{\partial^2}{\partial t^2} + f_0^2\right)\left(\frac{\partial}{\partial z} + n_2\frac{N^2}{g} - \frac{1}{H_0}\right)w$$
$$+\left(n_2\left(\frac{\partial^2}{\partial t^2} + f_0^2\right) - c_0^2\nabla_z^2\right)\frac{\partial}{\partial t}\left(\frac{p}{\rho_0}\right) = 0. \tag{8.9a}$$

The solution having $w = 0$, $\left(\frac{p}{\rho_0}\right) \neq 0$ has vertical structure $\exp(N^2 z/g)$ if $n_2 = 1$, but [from (8.9a)] $\partial/\partial t(p/\rho_0) = 0$ if $n_2 = 0$. The Lamb wave is thus absent if the term $\partial\rho/\partial t$ is omitted from the continuity equation (*anelastic* approximation). The Lamb wave is still present (given $n_2 = 1$) if $n_1 = 0$ (*hydrostatic* approximation).

Elimination of w between (8.8a) and (8.9a) gives

$$\left[\left(n_1 \frac{\partial^2}{\partial t^2} + N^2\right) \nabla_z^2 + \left(\frac{\partial^2}{\partial t^2} + f_0^2\right)\left(\frac{\partial^2}{\partial z^2} - \frac{1}{H_0}\frac{\partial}{\partial z} - \frac{n_1 n_2}{c_0^2}\frac{\partial^2}{\partial t^2}\right)\right] \frac{\partial}{\partial t}\left(\frac{p}{\rho_0}\right) = 0, \tag{8.11a}$$

which reduces to (8.11) if $n_1 = n_2 = 1$. We examine the fate of wave-like solutions of the form (8.12) in the remaining cases.

$n_1 = 0$, $n_2 = 0$ (*Hydrostatic, anelastic*)

In this case, (8.11a) gives $\omega = 0$ (geostrophic mode) or

$$\omega^2 = \frac{N^2(k^2 + l^2) + f_0^2\left(m^2 + \frac{1}{4H_0^2}\right)}{\left(m^2 + \frac{1}{4H_0^2}\right)}. \tag{8.22}$$

There are no sound waves (or Lamb waves) in this case. Equation (8.22) represents a pair of gravity waves, whose frequencies differ even from those of the approximate solution (8.17) (section 8.1).

$n_1 - 0$, $n_2 - 1$ (*Hydrostatic, elastic*) This gives the same results as the previous case, except that Lamb waves remain.

$n_1 = 1$, $n_2 = 0$ (*Non-hydrostatic, anelastic*)

Now (8.11a) gives $\omega = 0$ (geostrophic mode), or

$$\omega^2 = \frac{N^2(k^2 + l^2) + f_0^2\left(m^2 + \frac{1}{4H_0^2}\right)}{\left(k^2 + l^2 + m^2 + \frac{1}{4H_0^2}\right)}. \tag{8.23}$$

There are no sound waves (or Lamb waves). Equation (8.23) represents a pair of gravity waves; in the absence of rotation ($f_0 = 0$), their phase speeds are as given by Ω_g (see (8.17)).

In summary, the *anelastic* approximation removes sound waves and Lamb waves and leaves the gravity wave frequencies almost intact. The *hydrostatic* approximation removes sound waves but not Lamb waves, and the frequencies of the remaining gravity waves are noticeably modified.

8.3 Hydrostatic waves in an isothermal atmosphere: mid-latitude β-plane

The treatment given in sections 8.1 and 8.2 assumed a constant value f_0 of the Coriolis parameter f (as well as Cartesian geometry). Allowing f to vary with latitude (y) opens up new possibilities and brings new problems too. These problems are typical of those that arise when one seeks to approximate the equations for motion subject to gravity on a rotating sphere. We treat the variable-f linearised case because it offers a vignette of more complicated cases as well as revealing an important new type of wave motion. The hydrostatic approximation will be applied, thus limiting attention to motion having a frequency much less than the buoyancy frequency N; wave motion having a horizontal scale large compared with its vertical scale is of this type.

We consider the linearised equations of motion with, initially, $f = f_0 + \beta y$, where f_0 and β are constants. In place of (8.3) and (8.4b), we have

$$\frac{\partial \mathbf{v}}{\partial t} + f\mathbf{k} \times \mathbf{v} + \nabla_z \left(\frac{p}{\rho_0}\right) = 0 \tag{8.24}$$

$$-g\frac{\theta}{\theta_0} + \left(\frac{\partial}{\partial z} - n_2\frac{N^2}{g}\right)\frac{p}{\rho_0} = 0. \tag{8.25}$$

The linearised continuity equation (8.5a) – with tracer parameter – remains unchanged, as does the linearised thermodynamic equation (8.6) and the linearised relation (8.7).

For the *equatorial* β-plane ($f = \beta y$), analysis of (8.5a), (8.6), (8.7), (8.24) and (8.25) can be carried through without further approximation (Gill 1982); we shall refer to this case in section 8.5. Analysis of the mid-latitude β-plane case ($f = f_0 + \beta y$; $f_0 \neq 0$) can also be pursued without further approximation, but unwieldy latitude structure functions arise. Instead of following this route, we seek to replace $f = f_0 + \beta y$ by constant values, *wherever possible in a consistent way.*

From (8.24) we form equations for the time evolution of the (vertical) relative vorticity $\zeta \equiv \partial v/\partial x - \partial u/\partial y$ and the (horizontal) divergence $\delta \equiv \partial u/\partial x + \partial v/\partial y$:

$$\frac{\partial \zeta}{\partial t} + f\delta + \beta v = 0 \tag{8.26}$$

$$\frac{\partial \delta}{\partial t} - f\zeta + \beta u + \nabla_z^2\left(\frac{p}{\rho_0}\right) = 0. \tag{8.27}$$

With a Helmholtz decomposition of $\mathbf{v} = (u, v)$ into rotational/non-divergent, and divergent/irrotational parts, i.e., $\mathbf{v} = \mathbf{k} \times \nabla_z \psi + \nabla_z \chi$ where ψ and χ are streamfunction and velocity potential, (8.26) and (8.27) become

$$\frac{\partial}{\partial t}\nabla_z^2\psi + f\nabla_z^2\chi + \beta\frac{\partial \psi}{\partial x} + \beta\frac{\partial \chi}{\partial y} = 0 \tag{8.28}$$

$$\frac{\partial}{\partial t}\nabla_z^2\chi - f\nabla_z^2\psi + \beta\frac{\partial \chi}{\partial x} - \beta\frac{\partial \psi}{\partial y} + \nabla_z^2\left(\frac{p}{\rho_0}\right) = 0. \tag{8.29}$$

A naïve application of the β-plane approximation would involve setting $f = f_0$ in (8.28) and (8.29), and then proceeding with f_0 (as well as β) constant thereafter. Grimshaw (1975) noted that the β-plane approximation, in this guise, is ill-posed because it does not commute with other operations such as differentiation with respect to latitude. In the present case we reason instead that, if we set $f = f_0$ in (8.28) and (8.29), we should also omit the terms $\beta\partial\chi/\partial y$ and $-\beta\partial\psi/\partial y$ to ensure that the resulting forms

$$\frac{\partial}{\partial t}\nabla_z^2\psi + f\nabla_z^2\chi + \beta\frac{\partial\psi}{\partial x} = 0 \tag{8.30}$$

and

$$\frac{\partial}{\partial t}\nabla_z^2\chi - f\nabla_z^2\psi + \beta\frac{\partial\chi}{\partial x} + \nabla_z^2\left(\frac{p}{\rho_0}\right) = 0 \tag{8.31}$$

imply an acceptable kinetic energy equation. [To obtain a kinetic energy equation, multiply (8.28) by ψ, multiply (8.29) by χ and add the results. The term $f_0(\psi\nabla_z^2\chi - \chi\nabla_z^2\psi)$ can be written in divergence form as $\nabla_z\cdot(f_0\psi\nabla_z\chi - \chi\nabla_z\psi)$. The term $\beta(\psi\partial\chi/\partial y - \chi\partial\psi/\partial y)$, which arises if $\beta\partial\chi/\partial y$ in (8.28) and $-\beta\partial\psi/\partial y$ in (8.29) are retained, is not of the required divergence form.] The omissions can also be justified by scale analysis, as follows. We wish to represent the latitude variation of f in some WKB sense; thus the scale L_y of latitude variation of the motion must be much less than that of f – i.e. the planetary scale a which equals the radius of the Earth. Hence (for wave-like motion which is not evanescent in the horizontal), $\beta\partial\chi/\partial y$ in (8.28) must be much less than $f\nabla_z^2\chi$ in numerical terms, since $\beta \sim f/a$. Similarly, $\beta\partial\psi/\partial y$ in (8.29) must be much less than $f\nabla_z^2\psi$. [Some published accounts achieve these omissions by assuming $\partial/\partial y = 0$, which is not the appropriate limit.]

For reasons that will soon be clear, we attach a single tracer parameter (n_3) to both the first and third terms in (8.31):

$$n_3\left(\frac{\partial}{\partial t}\nabla_z^2\chi + \beta\frac{\partial\chi}{\partial x}\right) - f_0\nabla_z^2\psi + \nabla_z^2\left(\frac{p}{\rho_0}\right) = 0. \tag{8.32}$$

From (8.5a), (8.6), (8.7) and (8.25) we obtain, after a lengthy calculation assuming that w does not vanish everywhere,

$$N^2\nabla_z^2\chi - \left[\frac{\partial}{\partial z}\left(\frac{\partial}{\partial z} - \frac{1}{H_0}\right) + n_2(1-n_2)\frac{N^4}{g^2}\right]\frac{\partial}{\partial t}\left(\frac{p}{\rho_0}\right) = 0. \tag{8.33}$$

The term in N^4/g^2 in (8.33) vanishes whether $n_2 = 0$ or 1. From (8.30), (8.32) and (8.33):

$$\left\{\left[n_3\left(\frac{\partial}{\partial t}\nabla_z^2 + \beta\frac{\partial}{\partial x}\right)^2 + f_0^2\nabla_z^4\right]\left[\frac{\partial}{\partial z}\left(\frac{\partial}{\partial z} - \frac{1}{H_0}\right)\right]\frac{\partial}{\partial t}\right.$$
$$\left. + N^2\left(\frac{\partial}{\partial t}\nabla_z^2 + \beta\frac{\partial}{\partial x}\right)\nabla_z^4\right\}\left(\frac{p}{\rho_0}\right) = 0. \tag{8.34}$$

Solutions of (8.34) of the form (8.12) obey the dispersion equation

$$\left[f_0^2K^4 - n_3(\beta k + \omega K^2)^2\right]\omega + (\beta k + \omega K^2)N^2\left(\frac{K^4}{M^2}\right) = 0 \tag{8.35}$$

$$\text{in which} \quad K^2 \equiv k^2 + l^2 \tag{8.36}$$

$$\text{and} \quad M^2 \equiv m^2 + \frac{1}{4H_0^2}. \tag{8.37}$$

Equation (8.35) is a cubic in ω. We shall not give a detailed analysis of the general case ($n_3 = 1$): two of the solutions are a pair of (inertio-)gravity waves modified by the β-effect; the third solution is a lower frequency solution, a *Rossby* or *planetary wave*. When $n_3 = 0$, (8.35) becomes linear in ω; the gravity waves disappear, but the Rossby wave remains:

$$\omega = -\frac{\beta k}{\left[K^2 + (f_0^2/N^2)M^2\right]}. \tag{8.38}$$

The westward propagation of Rossby waves arises because of the latitude variation of the Coriolis parameter (the β-effect). For our present purposes, the key aspect is that gravity waves are 'filtered' by omitting the term $\partial/\partial t\left(\nabla_z^2\chi\right) = \partial\delta/\partial t$ from the divergence equation (8.32) [i.e. $n_3 = 0$], but Rossby waves remain (Thompson 1956). [Putting $n_3 = 0$ in (8.32) also implies omission of $\beta\partial\chi/\partial x$. Separate treatment of this term unproductively complicates the analysis.]

Our derivation and discussion has assumed that $w \neq 0$. What about Lamb waves? If $w = 0$ everywhere, then (from (8.6)) $\theta = 0$ also, and use of (8.7) shows that (8.5a) becomes

$$\frac{n_2}{c_0^2}\frac{\partial}{\partial t}\left(\frac{p}{\rho_0}\right) + \delta = 0. \tag{8.39}$$

The corresponding vorticity and divergence equations are the same as before ((8.30) and (8.32)). We find, instead of (8.34),

$$\left\{\left[n_3\left(\frac{\partial}{\partial t}\nabla_z^2 + \beta\frac{\partial}{\partial x}\right)^2 + f_0^2\nabla_z^4\right]\frac{n_2}{c_0^2}\frac{\partial}{\partial t} - \left(\frac{\partial}{\partial t}\nabla_z^2 + \beta\frac{\partial}{\partial x}\right)\nabla_z^4\right\}\left(\frac{p}{\rho_0}\right) = 0. \tag{8.40}$$

Once again, we obtain a cubic dispersion relation (if $(p/\rho_0) \propto \exp[i(kx + ly - \omega t)]$ is assumed). If $n_3 = 0$ we find

$$\omega = -\frac{\beta k}{\left[K^2 + n_2(f_0^2/c_0^2)\right]}. \tag{8.41}$$

Setting $n_3 = 0$ removes the two (paired) hydrostatic Lamb waves, but leaves a Rossby mode – known as the Rossby–Lamb or external Rossby mode. This mode's frequency (see (8.41)) is then dependent on whether horizontal divergence is retained by setting $n_2 = 1$ or non-divergence is enforced by setting $n_2 = 0$; see (8.39).

8.4 Waves on shallow water: mid-latitude β-plane

The above analysis is readily repeated for the shallow water equations on a mid-latitude β-plane. Appropriate linearization of the β-plane versions of (5.36), (5.37) and (5.39) gives

$$\frac{\partial \mathbf{v}}{\partial t} + f\mathbf{k}\times\mathbf{v} + \nabla_z(gh) = 0 \tag{8.42}$$

$$\frac{\partial h}{\partial t} + h_0\nabla_z.\mathbf{v} = 0. \tag{8.43}$$

Equations (8.42) is of the same form as (8.24), with gh replacing p/ρ_0. With the same approximations and tracer scheme as before, we obtain (8.30) and (8.32), with gh replacing p/ρ_0. Equation (8.43) is much simpler than (8.33). In place of (8.34) we find

$$\left\{\left[n_3\left(\frac{\partial}{\partial t}\nabla_z^2 + \beta\frac{\partial}{\partial x}\right)^2 + f_0^2\nabla_z^4\right]\frac{\partial}{\partial t} - gh_0\left(\frac{\partial}{\partial t}\nabla_z^2 + \beta\frac{\partial}{\partial x}\right)\nabla_z^4\right\}h = 0. \tag{8.44}$$

Solutions of (8.44) of the form $\exp\{i(kx+ly-\omega t)\}$ have angular frequency ω which obeys the cubic

$$\{f_0^2K^4 - n_3(\beta k + \omega K^2)^2\}\,\omega + (\beta k + \omega K^2)gh_0K^4 = 0. \tag{8.45}$$

Equation (8.45) is the same as (8.35), except gh_0 that replaces $N^2/M^2 = N^2/\left(m^2 + 1/(4H_0^2)\right)$. The quantity $d_\mathrm{E} \equiv N^2/g\left(m^2 + 1/(4H_0^2)\right)$ is called the *equivalent depth*. Every Rossby wave or (inertio-) gravity wave in an isothermal atmosphere at rest has the same dispersion relation as a counterpart Rossby or (inertio-) gravity wave on a shallow layer of incompressible fluid having mean depth d_E. By comparison of (8.40) and (8.44), it is clear that Rossby–Lamb waves have equivalent depth $c_0^2/g = \gamma RT_0/g = \gamma H_0$.

Putting $n_3 = 0$ in (8.45) reduces it to an explicit linear expression for ω:

$$\omega = -\frac{\beta k}{\left[K^2 + (f_0^2/gh_0)\right]}. \tag{8.46}$$

Gravity waves have been removed by setting $n_3 = 0$, and (8.46) gives the angular frequency of the remaining shallow water Rossby wave (which is an approximation to the corresponding root of the cubic dispersion equation (8.45) with $n_3 = 1$). If we omit the term $\partial h/\partial t$ from (8.43) then we oblige the flow to be non-divergent, and the vorticity equation derived from (8.42) is simply

$$\left(\frac{\partial}{\partial t}\nabla_z^2 + \beta\frac{\partial}{\partial x}\right)\psi = 0 \tag{8.47}$$

with $\psi = gh/f_0$. Waves of the form $\exp\{i(kx+ly-\omega t)\}$ have

$$\omega = -\frac{\beta k}{K^2}. \tag{8.48}$$

(Comparison with (8.46) shows that the imposition of non-divergence is valid if $K^2 \gg f_0^2/gh_0$.) These are prototypical Rossby waves – non-divergent, barotropic waves (Rossby 1939). See Hoskins *et al.* (1985), Durran (1988) and Holton (1992) for discussion of their mechanism.

The shallow-water equations in spherical polar geometry have been the vehicle of analyses of tidal motion dating back to Laplace (see Lamb (1932) and Gill (1982)); and the linearised free-wave problem, which subsumes both equatorial (section 8.5) and mid-latitude cases, was thoroughly studied by Longuet-Higgins (1968). The mid-latitude β-plane analysis given in this section provides a straightforward illustration of the key result that omission of the $\partial\delta/\partial t$ term from the divergence equation leads to the removal, or 'filtering' of gravity waves, and we have already seen (section 8.3) that the result extends to the case of an isothermal, compressible atmosphere.

8.5 Tropical modes

If $f = \beta y$ – the equatorial β-plane case – the linearised problems of section 8.3 and 8.4 can be completed without approximation; Gill (1982) gives a full account. Equatorially trapped modes, which propagate in the equatorial plane, are found. As well as gravity waves and Rossby waves, two other types occur: equatorial *Kelvin waves*, and *mixed Rossby-gravity waves*. We discuss the shallow water case (in which the waves propagate in the zonal direction). Equatorial Kelvin waves are non-dispersive, eastward propagating, and similar in many ways to the classical Kelvin waves which are permitted in middle latitudes in the presence of a vertical boundary. They are hybrid, anisotropic modes, being in geostrophic balance in the meridional direction (perpendicular to the equator), but having the character of gravity waves as regards the force balance in the zonal direction (parallel to the equator). Mixed Rossby-gravity waves behave like Rossby waves in their westward propagating branch, but like gravity waves in their eastward-propagating branch. Behaviour in the case of an isothermal, compressible atmosphere (with the hydrostatic approximation and a basic state of no motion) is similar, but with the possibility of vertical propagation.

The consequences of omitting the term $\partial\delta/\partial t$ in the divergence equation are not obvious *a priori* because of the special character of some of the tropical modes. Results depend on which other terms are omitted from the divergence equation (Gent and McWilliams 1983). Kelvin modes are absent, but one branch of mixed Rossby-gravity waves remains, and spurious high frequency modes occur if the term $\beta\partial\chi/\partial x$ is retained (cf. the pairing of this term with $\partial\delta/\partial t$ by the tracer parameter n_3 in (8.32)). Such spurious modes are also found on the sphere and on a mid-latitude β-plane if $\beta\partial\chi/\partial x$ is retained but $\partial\delta/\partial t$ omitted (Moura 1976, Allen *et al.* 1990b).

9 Approximately geostrophic models

There are many dynamical models that are intermediate in accuracy between the HPEs (section 5) and the diagnostic geostrophic approximation (section 7) and from which inertio-gravity waves have been filtered. Wide-ranging accounts are given by McWilliams and Gent (1980) and Allen and Newberger (1993); see also Phillips (1963) and Eliassen (1984). In this section we aim not to review, but to indicate the major types of model and the guiding principles. We use the shallow water equations (5.37)–(5.40) as a simple vehicle for discussion of each of the major types except the balance class (section 9.6), for which the HPEs in pressure coordinates are more appropriate. For simplicity we shall ignore both heating and friction.

9.1 Planetary geostrophic equations (QG2)

The planetary geostrophic equations were first discussed by Burger (1958), and are known as QG2 (following Phillips (1963)). In their shallow-water guise, they replace the horizontal momentum equations (5.37), (5.38) by the diagnostic geostrophic approximation, and retain time evolution only in the continuity equation (5.40); $\mathbf{v}$ is replaced by the geostrophic wind $\mathbf{v}_G$ in the material derivative, and spherical geometry is retained – as is the latitude variation of $f = 2\Omega \sin\phi$:

$$\mathbf{v} = \mathbf{v}_G \equiv \frac{g}{f}\mathbf{k} \times \nabla_z h \tag{9.1}$$

$$\frac{Dh}{Dt_G} + h\nabla_z . \mathbf{v}_G = 0 \tag{9.2}$$

where

$$\frac{D}{Dt_G} \equiv \frac{\partial}{\partial t} + \mathbf{v}_G . \nabla_z = \frac{\partial}{\partial t} + \frac{u_G}{a\cos\phi}\frac{\partial}{\partial \lambda} + \frac{v_G}{a}\frac{\partial}{\partial \phi}. \tag{9.3}$$

Gravity waves are absent because the implied divergence equation lacks the term $\partial\delta/\partial t$. The vorticity equation is also necessarily diagnostic, and (in the terminology of vorticity dynamics) represents a balance between planetary vorticity advection and vortex stretching/compression:

$$\nabla_z . (f\mathbf{v}_G) = \frac{v_G}{a}\frac{df}{d\phi} + f\nabla_z . \mathbf{v}_G = 0. \tag{9.4}$$

By using the continuity equation (9.2), (9.4) can be written

$$\frac{D}{Dt_G}\left\{\frac{2\Omega a \sin\phi}{h}\right\} = 0. \tag{9.5}$$

Equation (9.5) is a form of the potential vorticity equation in which the contribution of relative vorticity is completely neglected. This is an extreme approximation, valid to the extent that the omission in (9.4) of relative vorticity advection is justified: $V/fL \ll L/a$. Since Ro $\equiv V/fL \ll 1$ is the condition for geostrophic motion, it is required that $L \sim a$: the horizontal scale of the (nearly geostrophic) motion must be comparable with the radius of the Earth.

The energy equation of QG2 is

$$\frac{\partial}{\partial t}\left(\frac{1}{2}gh^2\right) = -\nabla_z.(gh^2\mathbf{v}_G). \tag{9.6}$$

Angular momentum conservation is reflected in the meridional component of (9.1) in the form

$$\frac{D}{Dt_G}(\Omega a^2 \cos^2\phi) = -g\frac{\partial h}{\partial \lambda}. \tag{9.7}$$

In the context of the shallow water equations, QG2 is of interest mainly in theoretical rather than practical terms. It is a compact model that exhibits analogues of the main conservation properties, and is in this respect a fully consistent approximation; see also section 9.5.

9.2 Quasi-geostrophic model (QG1)

QG1 originated in attempts by various meteorologists in the 1930s and 1940s to derive equations describing the time-evolution of extra-tropical weather systems having a horizontal space scale, L, of about 1000km (the 'synoptic scale') and typified by horizontal flow speeds, V, of order 10 ms^{-1}. The term *quasi-geostrophic* was suggested by Sutcliffe (1938). For such systems the Rossby number is of order 10^{-1}, the β-plane approximation is applicable since $L \ll a$, and the use of Cartesian geometry is justified. A 3D version of this important model will be discussed in section 10. Here we give an outline derivation of the shallow water version, and describe how it defines both the geostrophic and ageostrophic parts of the flow.

Suppose that the fluid exhibits variations h' about its mean depth h_0:

$$h = h(x, y, t) = h_0 + h'(x, y, t). \tag{9.8}$$

Define the geostrophic flow in terms of a *mean* Coriolis parameter, f_0, as

$$\mathbf{v}_g \equiv \frac{g}{f_0}\mathbf{k}\times\nabla_z h = \frac{g}{f_0}\mathbf{k}\times\nabla_z h' = \mathbf{k}\times\nabla_z\left(\frac{gh'}{f_0}\right). \tag{9.9}$$

[In (9.9), and throughout this section, ∇_z is the Cartesian operator $(\partial/\partial x, \partial/\partial y)$.] The use of f_0, rather than the variable f, in (9.9) is a key simplifying feature in the subsequent analysis; note that $\nabla_z.\mathbf{v}_g = 0$, so that the divergent part of the flow is contained in the ageostrophic flow $\mathbf{v}_a \equiv \mathbf{v} - \mathbf{v}_g$. (The ageostrophic flow also has a rotational part, as we shall see.) The choice (9.9) of $\mathbf{v}_g$ is a good approximation to $\mathbf{v}_G$ (see (9.1)), given $L \ll a$. From (9.9), the streamfunction, ψ, of the geostrophic flow is

$$\psi = \frac{gh'}{f_0}. \tag{9.10}$$

The horizontal components of the momentum equation (the SWE form of (6.17)) may now be written in vector form as

$$\frac{D\mathbf{v}}{Dt} + f\mathbf{k} \times \mathbf{v}_a + \beta y\mathbf{k} \times \mathbf{v}_g = 0. \tag{9.11}$$

Since $\mathbf{v} \approx \mathbf{v}_g$ to the extent that the Rossby number is small, it is reasonable to replace $D\mathbf{v}/Dt$ in (9.9) by the geostrophically-approximated (but still nonlinear) quantity

$$\frac{D\mathbf{v}_g}{Dt_g} \equiv \left(\frac{\partial}{\partial t} + \mathbf{v}_g.\nabla_z\right)\mathbf{v}_g. \tag{9.12}$$

Note that $\mathbf{v}$ has been replaced by $\mathbf{v}_g$ in both the advecting and the advected flow. The replacement of the advected flow by $\mathbf{v}_g$ (and the non-divergence of $\mathbf{v}_g$) ensures the absence of gravity waves.

Given $L \ll a$, the Coriolis term in (9.11) depending on the ageostrophic flow may be approximated by $f_0\mathbf{k} \times \mathbf{v}_a$, so that the latitude variation of the Coriolis parameter enters only via the term βy associated with the (much larger) geostrophic flow. Equation (9.11) then becomes

$$\frac{D\mathbf{v}_g}{Dt_g} + f_0\mathbf{k} \times \mathbf{v}_a + \beta y\mathbf{k} \times \mathbf{v}_g = 0. \tag{9.13}$$

An equation for the time-evolution of the geostrophic vorticity $\zeta_q \equiv (\partial v_q/\partial x - \partial u_g/\partial y) = \nabla_z^2\psi$ may be formed from (9.13). Noting the non-divergence of $\mathbf{v}_g$, we find the simple result

$$\frac{D}{Dt_g}\left(\nabla_z^2\psi + \beta y\right) = -f_0\nabla_z.\mathbf{v}_a. \tag{9.14}$$

This is the shallow-water QG1 vorticity equation.

Consider the shallow-water continuity equation (5.40) in the Cartesian form

$$\frac{Dh}{Dt} + h\nabla_z.\mathbf{v}_a = 0. \tag{9.15}$$

Equation (9.15) is replaced by

$$\frac{Dh'}{Dt_g} + h_0\nabla_z.\mathbf{v}_a = 0. \tag{9.16}$$

This step involves the same approximation of the material derivative as that made in the momentum equation to reach (9.13). Also, the term $h\nabla_z.\mathbf{v}_a$ in (9.15) has been approximated by $h_0\nabla_z.\mathbf{v}_a$, which requires that fluctuations h' about the mean depth h_0 be small, i.e. $|h'|/h_0 \ll 1$; see (9.18). Elimination of $\nabla_z.\mathbf{v}_a$ between (9.14) and (9.16), and use of (9.10), then gives

$$\frac{D}{Dt_g}\left\{\nabla_z^2\psi + \beta y - \frac{f_0^2}{gh_0}\psi\right\} = 0. \tag{9.17}$$

Since D/Dt_g (see (9.12)) involves only $\partial/\partial t$, ∇_z and $\mathbf{v}_g$, (9.17) *defines the time-evolution of the geostrophic streamfunction* ψ (given suitable initial and spatial boundary conditions).

Equation (9.17) is the shallow-water QG1 potential vorticity equation. The advected quantity is readily seen to be an approximation to $h_0(\zeta+f)/h$, valid in the case of small Rossby number and small height deviation $|h'|/h_0 \ll 1$. The criterion for the latter may be deduced by simple scale analysis:

$$h' \sim \frac{f_0 VL}{g} \Rightarrow \frac{h'}{h_0} \sim \frac{f_0 VL}{gh_0}.$$

Hence we require $gh_0/f_0VL \gg 1$, which is equivalent to

$$\mathfrak{R} \equiv \frac{gh_0}{V^2} \gg \frac{f_0 L}{V} = \mathrm{Ro}^{-1}, \tag{9.18}$$

the applicability of which depends on the mean depth h_0 as well as quantities already discussed. Taking $V = 10$ ms^{-1} and $h_0 = 10$km gives $\mathfrak{R} = 10^3$, while $h_0 = 1$km gives $\mathfrak{R} = 10^2$; so for a wide range of choices of h_0 (9.18) is obeyed if Ro $= 10^{-1}$ or greater. Indeed, (9.16) shows that the dynamics reduces to that of the barotropic vorticity equation (see section 5.6) if $gh_0/f_0^2L^2 \gg 1$ (a result also noted in section 8.4).

The derivation of Equation (9.17) depends on f_0 being a constant. If f_0 had been a function of y, a conservation law would not have resulted. If $h\nabla_z.\mathbf{v}_a$ in (9.15) had not been replaced by $h_0\nabla_z.\mathbf{v}_a$ in (9.16), a conservation law would have resulted, but not in terms of a quantity linear in ψ.

Finding the height deviation h' and geostrophic flow from the streamfunction ψ, via (9.9) and (9.10), is just a matter of multiplication and spatial differentiation. The determination of the ageostrophic flow $\mathbf{v}_a$ is more subtle. Rather than eliminating $\nabla_z.\mathbf{v}_a$ between (9.14) and (9.16) we may eliminate the local time derivatives (noting (9.10)). The result is a diagnostic partial differential equation for $\nabla_z\mathbf{v}_a$:

$$\left(\nabla_z^2 - \frac{f_0^2}{gh_0}\right)\nabla_z.\mathbf{v}_a = \frac{f_0}{gh_0}\left\{(\mathbf{v}_g.\nabla_z)\nabla_z^2\psi + \beta v_g\right\}. \tag{9.19}$$

The RHS term is known if the streamfunction is known, so (9.19) determines the irrotational part of $\mathbf{v}_a$ (given appropriate boundary conditions). Equation (9.19) may also be obtained from (9.17) by algebraic application of (9.10) and use of (9.16).

The rotational part of $\mathbf{v}_a$ may be determined from an elliptic PDE obtained by taking the divergence of (9.13):

$$f_0\nabla_z^2\psi_a = \nabla_z.\left[(\mathbf{v}_g.\nabla_z)\mathbf{v}_g\right] + \beta u_g - \beta y\nabla_z^2\psi, \tag{9.20}$$

where ψ_a is the streamfunction of the ageostrophic flow.

Thus the ageostrophic flow is completely defined in QG1; it is that flow which is required to maintain geostrophic balance between the geostrophic flow and height fields as the time-evolution occurs.

An energy equation is readily derived from (9.17):

$$\frac{\partial}{\partial t}\left\{\frac{1}{2}\left[(\nabla_z^2\psi)^2 + \frac{f_0^2}{gh_0}\psi^2\right]\right\} = -\nabla_z.\left\{\frac{1}{2}\left[(\nabla_z^2\psi)^2 + \frac{f_0^2}{gh_0}\psi^2\right]\mathbf{v}_g + f_0\psi\mathbf{v}_a\right\}. \tag{9.21}$$

The axial angular momentum balance is governed by the zonal component of (9.13), which – upon restoring the terms representing geostrophic balance – may be written as

$$\frac{Du_g}{Dt_g} - f_0 v_a - \beta y v_g - f_0 v_g + g\frac{\partial h}{\partial x} = 0.$$

Hence

$$\frac{D}{Dt_g}\left\{u_g - \int f dy'\right\} - f_0 v_a = -g\frac{\partial h}{\partial x}. \tag{9.22}$$

This form allows for the fact that the zonal (x) average of the meridional geostrophic flow vanishes, so the contribution of the ageostrophic flow must be represented.

9.3 Models based on formal considerations of accuracy

The derivation of QG1 given in section 9.2 may be formalised by a truncated Rossby number expansion of the velocity field $\mathbf{v}$. We sketch the procedure, and note that it can be taken to higher order to generate models of higher formal accuracy than QG1. For simplicity we consider the f-plane case ($\beta = 0$), in which (9.11) may be written as

$$\mathbf{v} = \mathbf{v}_g + \mathbf{v}_a = \frac{g}{f_0}\mathbf{k}\times\nabla_z h' + \frac{1}{f_0}\mathbf{k}\times\frac{D\mathbf{v}}{Dt}. \tag{9.23}$$

Making $\mathbf{v}$, h', ∇_z and $\partial/\partial t$ dimensionless by extracting factors of V, f_0VL/g, $1/L$ and V/L gives:

$$\mathbf{v} = V\hat{\mathbf{v}};\ h' = f_0\frac{VL}{g}\hat{h};\ \nabla_z = \frac{1}{L}\hat{\nabla}_z;\ \frac{\partial}{\partial t} = \frac{V}{L}\frac{\partial}{\partial\hat{t}}. \tag{9.24}$$

The dimensionless velocity $\hat{\mathbf{v}}$, depth deviation $\hat{h}$ and operators $\hat{\nabla}_z$ and $\frac{\partial}{\partial\hat{t}}$ are each assumed to have magnitude of order unity. Equation (9.23) becomes

$$\hat{\mathbf{v}} = \mathbf{k}\times\hat{\nabla}_z h' + \mathrm{Ro}\,\mathbf{k}\times\left(\frac{\partial}{\partial\hat{t}} + \hat{\mathbf{v}}.\hat{\nabla}_z\right)\hat{\mathbf{v}}, \tag{9.25}$$

where $\mathrm{Ro} \equiv V/f_0L$. Equation (9.25) formally expresses the horizontal flow as the sum of the geostrophic contribution and an ageostrophic flow that is one order of magnitude smaller in Rossby number terms. The continuity equation (9.15) becomes

$$\left(\frac{\partial}{\partial \hat{t}} + \hat{\mathbf{v}}.\hat{\nabla}_z\right)\hat{h} + \left(\frac{\mathrm{B}}{\mathrm{Ro}} + \hat{h}\right)\hat{\nabla}_z.\hat{\mathbf{v}} = 0, \tag{9.26}$$

where

$$\mathrm{B} \equiv \frac{g h_0}{f_0^2 L^2} = \mathcal{R}\,\mathrm{Ro}^2\,. \tag{9.27}$$

From (9.25), the zeroth-order approximation to $\hat{\mathbf{v}}$ is $\mathbf{k} \times \hat{\nabla}_z h'$, which is simply the dimensionless geostrophic flow $\hat{\mathbf{v}}_g$. If B is of order unity, or greater, the leading order balance in (9.26) is simply $\mathrm{B}\,\mathrm{Ro}^{-1}\,\hat{\nabla}_z.\hat{\mathbf{v}} = 0$; this is consistent with $\hat{\mathbf{v}} = \hat{\mathbf{v}}_g$, since the geostrophic flow is non-divergent.

To find the next order approximation, put

$$\hat{\mathbf{v}} = \hat{\mathbf{v}}_g + \mathrm{Ro}\,\hat{\mathbf{v}}_1 \tag{9.28}$$

and isolate the coefficient of Ro in (9.25) and the coefficient of Ro^0 in (9.26) – assuming that $\mathrm{B} = O(1)$:

$$\hat{\mathbf{v}}_1 = \mathbf{k} \times \left(\frac{\partial}{\partial \hat{t}} + \hat{\mathbf{v}}_g.\hat{\nabla}_z\right)\hat{\mathbf{v}}_g \tag{9.29}$$

$$\left(\frac{\partial}{\partial \hat{t}} + \hat{\mathbf{v}}_g.\hat{\nabla}_z\right)\hat{h} + \mathrm{B}\hat{\nabla}_z.\hat{\mathbf{v}}_1 = 0. \tag{9.30}$$

Equations (9.29) and (9.30) are dimensionless forms of the f-plane versions of (9.13) and (9.16); elimination of $\hat{\mathbf{v}}_1$ gives

$$\left(\frac{\partial}{\partial \hat{t}} + \hat{\mathbf{v}}_g.\hat{\nabla}_z\right)\left\{\nabla_z^2\hat{h} - \frac{1}{\mathrm{B}}\hat{h}\right\} = 0. \tag{9.31}$$

Equation (9.31) is a dimensionless, f-plane form of the QG1 potential vorticity equation (9.17).

A second-order approximation may be obtained by putting

$$\hat{\mathbf{v}} = \hat{\mathbf{v}}_g + \mathrm{Ro}\,\hat{\mathbf{v}}_1 + \mathrm{Ro}^2\,\hat{\mathbf{v}}_2 \tag{9.32}$$

and isolating the coefficient of Ro^2 in (9.25) and the coefficient of Ro in (9.26). Higher-order approximations may be obtained. A broadly similar procedure has been used by Allen (1993) to obtain a hierarchy of increasingly accurate 'iterated geostrophic models' of 3D stratified flow; Allen and Newberger (1993) found that the third member of the hierarchy performed very well in numerical simulations against the (Cartesian) hydrostatic primitive equations.

Power series expansions are a useful way of systematising the derivation of approximately geostrophic models which happen to conform to a single truncation of the assumed series, and of giving a critical perspective on those that do not. Such expansion methods may be suspected of lending a cosmetic veneer to what is rather crude and restricted scale analysis. In the present case (which is typical) it has been assumed that the local time-scale is of order L/V, and that a single velocity scale (V) describes spatial and temporal variations of the flow. The method may lead to lengthy equations, especially at higher orders of accuracy; these may be amenable to numerical solution but not necessarily to analysis aimed at developing insight into the physical processes involved. The method is not guaranteed to deliver equations that reproduce conservation properties at any chosen truncation (although the order Ro truncation in the above case gives the QG1 model, which does have good conservation properties). A more subtle aspect of our chosen example is that the deviation height field h' has been given special status (Muraki *et al.* 1999); it has not been expanded in powers of Ro. Other fields may equally well be granted special status: in derivations of some of the PV-balance models noted briefly in section 9.6 the potential vorticity field is considered as central to the dynamics, and not expanded in powers of the Rossby number. Pedlosky (1987), section 3.12, expands *all* variable fields as powers of Ro; see also Pedlosky (1964).

9.4 Semi-geostrophic model: SG

In QG1 the advecting flow is replaced by the geostrophic flow $\mathbf{v}_g$ wherever it occurs. QG1 requires for its validity the replacement of f by f_0 (i.e. $L \ll a$) and only small deviations of height h from a mean value (as well as small Rossby number). In order to remove gravity waves, only the advected flow need be replaced by $\mathbf{v}_g$ (or some other non-divergent flow) in the horizontal momentum equation. The semi-geostrophic model (SG) takes advantage of this situation by retaining advection by the total flow throughout. The shallow water equations in SG form are

$$\frac{D\mathbf{v}_g}{Dt} + f_0\mathbf{k}\times\mathbf{v}_a = 0 \qquad (9.33)$$

$$\frac{Dh}{Dt} + f\nabla_z.\mathbf{v}_a = 0, \qquad (9.34)$$

with

$$\frac{D}{Dt} \equiv \left(\frac{\partial}{\partial t} + \mathbf{v}.\nabla_z\right) \quad \text{and} \quad \mathbf{v}_g = \frac{g}{f_0}\mathbf{k}\times\nabla_z h. \qquad (9.35)$$

The f-plane approximation is made, and Cartesian geometry assumed. Within this framework, the *only* approximation made in SG is the neglect of the term $D\mathbf{v}_a/Dt$ in the horizontal momentum equation. (This implies retention of some terms or order Ro^2 but neglect of others; see Fjørtoft (1962), p.158, and Craig (1993a).) There is no restriction on the depth h, and no need to divide it into mean and deviation parts. The definition (9.35) of geostrophic flow $\mathbf{v}_g$ is the same as in QG1.

Equation (9.33) implies the axial angular momentum principle

$$\frac{D}{Dt}\{u_g - f_0 y\} = -g\frac{\partial h}{\partial x}. \tag{9.36}$$

An energy equation exists in the form

$$h\frac{D}{Dt}\{\mathbf{v}_g^2 + gh\} = -\nabla_z.(gh^2\mathbf{v}_a). \tag{9.37}$$

The prospects for the existence of a potential vorticity conservation analogue in SG do not at first look bright, since (9.33) involves $(\mathbf{v}.\nabla)\mathbf{v}_g$, and such mixed vector advection terms are notoriously difficult to handle by the usual differential operator methods. However, following Allen *et al.* (1990a), consider the components of (9.33) as linear algebraic expressions for u and v:

$$\begin{aligned} u\left(\frac{\partial u_g}{\partial x}\right) - v\left(f_0 - \frac{\partial u_g}{\partial y}\right) &= -\frac{\partial u_g}{\partial t} - f_0 v_g \\ u\left(f_0 + \frac{\partial v_g}{\partial x}\right) + v\left(\frac{\partial v_g}{\partial y}\right) &= -\frac{\partial v_g}{\partial t} + f_0 u_g. \end{aligned} \tag{9.38}$$

'Solving' (9.38) for u and v gives

$$u = \frac{1}{f_0\xi_{\mathrm{SG}}}\left[\left(f_0 u_g - \frac{\partial v_g}{\partial t}\right)\left(f_0 - \frac{\partial u_g}{\partial y}\right) - \left(f_0 v_g + \frac{\partial u_g}{\partial t}\right)\frac{\partial v_g}{\partial y}\right] \tag{9.39}$$

$$v = \frac{1}{f_0\xi_{\mathrm{SG}}}\left[\left(f_0 v_g + \frac{\partial u_g}{\partial t}\right)\left(f_0 + \frac{\partial v_g}{\partial x}\right) + \left(f_0 u_g - \frac{\partial v_g}{\partial t}\right)\frac{\partial u_g}{\partial x}\right], \tag{9.40}$$

in which

$$f_0\xi_{\mathrm{SG}} \equiv \left(f_0 + \frac{\partial v_g}{\partial x}\right)\left(f_0 - \frac{\partial u_g}{\partial y}\right) + \frac{\partial u_g}{\partial x}\frac{\partial v_g}{\partial y} \tag{9.41}$$

Now form $\nabla_z.(\xi_{\mathrm{SG}}\mathbf{v})$ from (9.39)–(9.41). After some easy algebra and a few exhilarating cancellations we find that

$$\frac{\partial}{\partial x}(u\xi_{\mathrm{SG}}) + \frac{\partial}{\partial y}(v\xi_{\mathrm{SG}}) = -\frac{\partial \xi_{\mathrm{SG}}}{\partial t}. \tag{9.42}$$

Hence

$$\frac{D}{Dt}\xi_{\mathrm{SG}} + \xi_{\mathrm{SG}}\left(\frac{\partial u}{\partial x} + \frac{\partial v}{\partial y}\right) = 0. \tag{9.43}$$

From (9.34) and (9.43) the SG potential vorticity equation follows:

$$\frac{D}{Dt}\left(\frac{\xi_{\mathrm{SG}}}{h}\right) = 0. \tag{9.44}$$

The quantity ξ_{SG} is the SG absolute vorticity. Its definition (9.41) may be rewritten as

$$\xi_{\mathrm{SG}} = f_0 + \frac{\partial v_g}{\partial x} - \frac{\partial u_g}{\partial y} + \frac{1}{f_0}\frac{\partial(u_g, v_g)}{\partial(x,y)} = \xi_g + \frac{1}{f_0}\frac{\partial(u_g, v_g)}{\partial(x,y)}. \tag{9.45}$$

Thus ξ_{SG} is the usual geostrophic absolute vorticity, ξ_g, augmented by a Jacobian term which is small in comparison if the Rossby number is small.

From (9.33) the SG divergence equation may be derived as

$$\frac{\partial \mathbf{v}}{\partial x}.\nabla_z u_g + \frac{\partial \mathbf{v}}{\partial y}.\nabla_z v_g - f_0\left(\frac{\partial v}{\partial x} - \frac{\partial u}{\partial y}\right) + g\nabla_z^2 h = 0.$$

To a leading-order approximation ($\mathbf{v} \to \mathbf{v}_g$) of its nonlinear terms, this implies

$$\xi \equiv f_0 + \frac{\partial v}{\partial x} - \frac{\partial u}{\partial y} = f_0 + \frac{\partial v_g}{\partial x} - \frac{\partial u_g}{\partial y} - \frac{2}{f_0}\frac{\partial(u_g, v_g)}{\partial(x, y)}. \tag{9.46}$$

From (9.45) and (9.46), we see that ξ_{SG} is a worse approximation to the absolute vorticity ξ than might have been hoped; ξ_g is in fact a better approximation to ξ.

Hoskins (1975) made notable advances in a 3D version of SG by transforming from spatial to *geostrophic coordinates*:

$$X = x + \frac{v_g}{f_0} = x + \frac{g}{f_0^2}\frac{\partial z}{\partial x}; \quad Y = y - \frac{u_g}{f_0} = y + \frac{g}{f_0^2}\frac{\partial z}{\partial y}; \quad Z = z. \tag{9.47}$$

Then:

$$\frac{DX}{Dt} = u + \frac{1}{f_0}\frac{Dv_g}{Dt} = u_g \text{ and } \frac{DY}{Dt} = v - \frac{1}{f_0}\frac{Du_g}{Dt} = v_g,$$

and it is readily shown that the Jacobian of the transformation from physical to geostrophic space is none other than the (3D) SG absolute vorticity (divided by f_0). In our 2D context of the shallow water equations, a similar result follows for the transformation $(x, y) \to (X, Y)$:

$$\frac{\partial(X, Y)}{\partial(x, y)} = \frac{\xi_{SG}}{f_0}. \tag{9.48}$$

Further, Hoskins showed that the SG potential vorticity equation can be written in terms of derivatives of an augmented potential function with respect to the geostrophic coordinates in a form nearly isomorphic to the QG1 potential vorticity equation in its usual space-coordinate form.

The SG model has given important insights into the dynamics of weather systems (in particular, the formation of fronts) and into the status of QG1. It has also excited interest in other ways, prompting various questions. We have space only to juxtapose some of the questions and some of the studies that have addressed them. What is the mathematical significance of the geostrophic coordinate transformation? [Blumen 1981, Roulstone and Sewell 1997.] Can a version of SG having a more satisfactory definition of ξ_{SG} be derived? [McIntyre and Roulstone 2002.] Can SG be extended to the case of variable Coriolis parameter? [Shutts 1980, Magnusdottir and Schubert 1991.]

9.5 Hamiltonian models

Of the nearly geostrophic models so far presented, QG2 is the only one that succeeds in retaining the conservation properties of the SWEs whilst allowing latitude variation of both the Coriolis parameter and the mean depth h_0. QG2, however, is applicable only to motion on planetary scales; it is not appropriate for motion on the synoptic scale of extra-tropical weather systems. Useful extensions of the SG model have been proposed, but neither of those cited at the end of section 9.4 represents the true latitude variation of the Coriolis parameter whilst retaining SG's accuracy.

QG1, QG2 and SG have each been proposed or derived as sets of approximate equations that represent more or less heavily approximated versions of the SWEs. Conservation properties have then been investigated as a sort of health check. A requirement of good conservation properties is useful in limiting the vast number of conceivable approximations of the SWEs (or HPEs) which present themselves, but it is unhelpful if none of the candidate models passes muster.

Salmon (1983), (1988) proposed a systematic method of deriving consistent approximate models; see also Allen and Holm (1996). As noted in section 4.6, the unapproximated equations are equivalent to a variational statement, and by Noether's theorem, the symmetries of the Hamiltonian functional in that variational statement are associated with the conservation properties of the system. Making the desired approximations in the Hamiltonian then ensures that the implied (approximate) equations have consistent conservation properties.

Salmon (1983) applied this method to the shallow water equations. The coarsest level of approximation, involving the complete neglect of the velocities u, v in the Hamiltonian, delivers the planetary geostrophic equations QG2. The next level, in which u and v are replaced by their geostrophic values, leads to forms reminiscent of the SG equations, but with further terms. For the f-plane case, Salmon's approximate momentum equation reduces to

$$\left(\frac{\partial}{\partial t} + \mathbf{v}.\nabla_z\right)\mathbf{v}_g + f_0\mathbf{k}\times\mathbf{v}_a = -(\mathbf{v}_g.\nabla_z)\mathbf{v}_a - \frac{g}{h}\nabla_z\left(\frac{h^2}{f_0}\zeta_a\right), \qquad (9.49)$$

where ζ_a is the relative vorticity of the ageostrophic flow. The right-hand terms in (9.49), both of which are absent in SG, are of order Ro smaller than the left-hand terms. Their presence is consistent with the following potential vorticity conservation and global energy conservation laws:

$$\frac{D}{Dt}\left\{\frac{1}{h}\left(f_0 + \frac{\partial v_g}{\partial x} - \frac{\partial u_g}{\partial y}\right)\right\} = 0 \qquad (9.50)$$

and

$$\frac{d}{dt}\iint(\mathbf{v}_g^2 + gh)h\,dx\,dy = 0. \qquad (9.51)$$

Allen *et al.* (1990a) give details of the derivation of (9.50) and (9.51) from (9.49).

Salmon's method can be relied upon to give consistent equations, but in the present case they are not simple or familiar ones. Salmon (1985) showed that the SG model (section 9.4) may be obtained from an augmented Hamiltonian whose extra terms are compatible with the formal accuracy of the model. This demonstration of Hamiltonian structure enabled SG *per se* to be generalised to the case of variable f; see also Purser (1993), (1999).

The variational method has been successfully applied in the derivation of a number of approximate models of rotating flows having vertical structure: see, for example, Shutts (1989), Craig (1993b), Roulstone and Brice (1995), Holm (1996) and Ripa (1997).

9.6 Balance equations

Lorenz (1960) considered how the vorticity and divergence forms of the HPEs (see section 5.4) might be approximated so as to preserve a global energy invariant. There is little point in illustrating this important technique in a shallow water model because applying it delivers only QG1 and SG (Gent and McWilliams 1982). The reason for this perhaps surprising result is that the energy integrand in the shallow water system is essentially a cubic quantity ($h\mathbf{v}^2$); in the stratified flow case considered by Lorenz (1960) the integrand in pressure coordinates is quadratic ($\mathbf{v}^2/2$). Perhaps less surprisingly, the retention of potential vorticity conservation in approximated forms of the vorticity equation is far easier to achieve in the shallow water case than in the stratified flow case (in which the potential vorticity is a scalar product of vectors rather than the absolute vorticity divided by the depth of the fluid). The SWEs thus exhibit nearly the opposite properties to the HPEs written in pressure coordinates.

Lorenz's method depends on dividing the horizontal flow into its rotational (solenoidal), non-divergent part $\mathbf{v}_\psi$ and its divergent, irrotational part $\mathbf{v}_\chi$:

$$\mathbf{v} = \mathbf{v}_\psi + \mathbf{v}_\chi = \mathbf{k} \times \nabla_p \psi + \nabla_p \chi. \tag{9.52}$$

This Helmholtz decomposition differs from geostrophic/ageostrophic decompositions of $\mathbf{v}$, since the geostrophic flow has a non-zero divergence if the latitude variation of the Coriolis parameter is taken into account (see section 7.1), and the ageostrophic flow in QG1 has a rotational part (see sections 9.2 and 10.1).

Vorticity and divergence equations are obtained by taking $\mathbf{k}.\,\nabla_p\times$ and $\nabla_p\cdot$ of the p-coordinate version of the HPE horizontal momentum equation (5.29):

$$\frac{\partial \mathbf{v}}{\partial t} + \nabla_p(\mathbf{v}^2/2) + \zeta \mathbf{k} \times \mathbf{v} + \omega \frac{\partial \mathbf{v}}{\partial p} = -f\mathbf{k} \times \mathbf{v} - g\nabla_p z. \tag{9.53}$$

Here ∇_p is the (spherical polar) horizontal gradient operator on pressure surfaces of height $z = z(\lambda, \phi, p, t)$; $\zeta = \nabla_p^2 \psi$ is the relative vorticity and $\delta = \nabla_p^2 \chi$ is the divergence (both defined in p-coordinate terms). Multiplication of the two resulting equations respectively by ψ and χ, and use of the identity

$$F\frac{\partial}{\partial t}\nabla_p^2 F = \nabla_p.\left\{F\nabla_p\frac{\partial F}{\partial t}\right\} - \frac{\partial}{\partial t}\left\{\frac{(\nabla_p F)^2}{2}\right\},$$

(with $F = \psi$ or χ) then gives equations for the time-evolution of the rotational and divergent flow specific kinetic energies $\mathbf{v}_\psi^2/2$ and $\mathbf{v}_\chi^2/2$. The thermodynamic and continuity equations are then applied to produce a total energy equation. Associations between groups of terms in the vorticity and divergence equations which retain total energy conservation are sought. One consistent approximation of the vorticity and divergence equations which is recognised in this way (see Haltiner and Williams (1981)) is the pair

$$\frac{\partial \zeta}{\partial t} = -\mathbf{v}_\psi.\nabla_p(\zeta+f) - \nabla_p.(f\mathbf{v}_\chi) - \mathbf{v}_\chi.\nabla_p\zeta - \zeta\delta - \omega\frac{\partial \zeta}{\partial p} - \nabla_p\omega.\nabla_p\frac{\partial \psi}{\partial p}, \tag{9.54}$$

$$\nabla_p.\left[(\mathbf{v}_\psi.\nabla_p)\mathbf{v}_\psi\right] - \nabla_p.(f\nabla_p\psi) + g\nabla_p^2 z = 0. \tag{9.55}$$

Equation (9.54) is a nearly complete form of the vorticity equation. Equation (9.55) is a form of the divergence equation known as the Charney balance equation. It neglects the term $\partial\delta/\partial t$ (so gravity waves are absent), several elements of $\nabla_p.\left\{(\mathbf{v}.\nabla_p)\mathbf{v} + \omega\partial\mathbf{v}/\partial p\right\}$, and $\nabla_p.(f\mathbf{k}\times\nabla_p\chi)$.

A further energetically-consistent pair is obtained if the last four terms on the right side of (9.54) and the first on the left side of (9.55) are omitted:

$$\frac{\partial \zeta}{\partial t} = -\mathbf{v}_\psi.\nabla_p(\zeta+f) - \nabla_p.(f\mathbf{v}_\chi), \tag{9.56}$$

$$-\nabla_p.(f\nabla_p\psi) + g\nabla_p^2 z = 0. \tag{9.57}$$

Equation (9.57) is known as the linear balance equation. The resemblance of (9.56) and (9.57) to QG1, if the f-plane or β-plane approximations *are* applied, is noticeable. However, (9.56) and (9.57) are an energetically consistent pair (as are (9.54) and (9.55)) even when the latitude variation of f is fully represented (though potential vorticity conservation is then lost). Energy consistency requires in each case the use of the complete thermodynamic equation, and the definition of kinetic energy includes only the contribution of the rotational flow. The latter aspect shows that the filtering of gravity waves by omission of the term $\partial\delta/\partial t$ from the divergence equation is intimately related to the absence of divergent flow kinetic energy from the prognostic energy equation. The same link occurs between the kinetic energy of vertical motion and the filtering of vertically-propagating sound waves via the hydrostatic approximation (see equation (5.24) and section 8.2).

A variant of the vorticity/divergence equation approach that is more tractable in many respects is the use of separate momentum equations for the rotational flow and for the divergent flow. The divergent flow equation is rendered in diagnostic form in order to eliminate gravity waves. Such a momentum form of the balance equations which conserves both energy and potential vorticity has been proposed by Allen (1991). This model implies spurious high frequency modes similar to those noted in section 8.5; they may be controlled by choosing initial conditions and time integration schemes carefully.

Other workers, especially in recent years, have used what may be called PV-balance models. These use the PV equation, perhaps in complete (HPE) form, as a forecasting equation, in conjunction with the Charney balance equation (9.45), the linear balance equation (9.46) or some variant. Energy conservation is generally not reproduced, but another quadratic quantity – the potential enstrophy (PV^2) – is conserved in the global average; see Gent and McWilliams (1984). Models of this type have been constructed and used by Lynch (1989), Raymond (1992), Warn *et al.* (1995) and Vallis (1996); an earlier example is that of Charney (1962). The same rationale underlies the static PV inversions (see section 10.4) carried out by Davis and Emanuel (1991), Demirtas and Thorpe (1999) and others.

10 The 3D quasi-geostrophic model QG1

The shallow-water version of QG1 was discussed in section 9.2. Here we focus on a version applicable to the synoptic-scale, quasi-geostrophic evolution of a 3-dimensional, perfect gas atmosphere. We begin with an outline derivation of the model in pressure coordinates, and then note a height-coordinate version that illuminates various issues, including the status of the so-called *omega equation*. Conditions for the applicability of QG1 are then summarised. In conclusion, we note the frequent occurrence in QG1 of variants of Poisson's differential equation, and discuss the application of well-known properties of these equations, with particular regard to various forms of 'PV inversion'. Cartesian geometry will be assumed throughout this section, and, for simplicity, friction and diabatic forcing will be neglected.

10.1 Pressure-coordinate development of QG1

Central to the development of QG1 in pressure coordinates is a hydrostatic reference state (of no motion) in which all thermodynamic variables, and height z, are functions of pressure only. The fields themselves are expressed as deviations from the reference state values. For example:

$$T = T_s(p) + T'(x, y, p, t) \tag{10.1}$$

$$\theta = \theta_s(p) + \theta'(x, y, p, t) \tag{10.2}$$

$$z = z_s(p) + z'(x, y, p, t). \tag{10.3}$$

From (6.3), hydrostatic balance of the reference state is expressed by

$$g\frac{dz_s}{dp} + \frac{RT_s}{p} = 0. \tag{10.4}$$

From (6.3), (10.1) and (10.3), the deviations z' and T' obey a similar relation:

$$g\frac{\partial z'}{\partial p} + \frac{RT'}{p} = 0. \tag{10.5}$$

The geostrophic wind, $\mathbf{v}_g$, is defined as

$$\mathbf{v}_g \equiv \frac{g}{f_0}\mathbf{k} \times \nabla_p z' = \mathbf{k} \times \nabla_p \psi; \quad \psi \equiv \frac{gz'}{f_0}, \ \nabla_p \equiv \left(\left.\frac{\partial}{\partial x}\right|_p, \left.\frac{\partial}{\partial y}\right|_p \right). \tag{10.6}$$

As in the SWE case (section 9.2), $|\mathbf{v}_g| \approx |\mathbf{v}_G|$ if $L \ll a$ (recall $\mathbf{v}_G$ is defined by (7.2)). From (10.5) and (3.17), the streamfunction $\psi = \psi(x, y, p, t)$ defined in (10.6) obeys

$$-\frac{\partial \psi}{\partial p} = \frac{RT'}{f_0 p} = \left(\frac{RT_s}{f_0 p \theta_s} \right) \theta'. \tag{10.7}$$

Differentiating ψ with respect to p thus gives the temperature and potential temperature deviations multiplied by functions of pressure; horizontal differentiation gives $\mathbf{v}_g$ via (10.6). In terms of the ageostrophic wind $\mathbf{v}_a \equiv \mathbf{v} - \mathbf{v}_g$, the continuity equation becomes

$$\nabla_p . \mathbf{v}_a + \frac{\partial \omega}{\partial p} = 0. \tag{10.8}$$

Apart from the use of Cartesian geometry, no approximation of the HPE forms of section 6.1 has been made so far. Approximations *are* made in the HPE horizontal momentum and thermodynamic equations. Extraction of (10.6) from (6.17) (in which $f = f_0 + \beta y$) gives

$$\frac{D\mathbf{v}}{Dt} + f\mathbf{k} \times \mathbf{v}_a + \beta y \mathbf{k} \times \mathbf{v}_g = 0. \tag{10.9}$$

As in the shallow water case, consistent with $\mathrm{Ro} \ll 1$ and $L \ll a$, we replace the horizontal flow $\mathbf{v}$ by the geostrophic value $\mathbf{v}_g$ in the material derivative term in (10.9), and $f\mathbf{k} \times \mathbf{v}_a$ by $f_0\mathbf{k} \times \mathbf{v}_a$. In addition, we neglect the vertical advection term $\omega \partial \mathbf{v}/\partial p$ by comparison with $(\mathbf{v}.\nabla)\mathbf{v}$ in $D\mathbf{v}/Dt$. This is justified if $\hat{\omega}/\hat{p} \ll V/L$ (where $\hat{\omega}$ is a typical magnitude of ω and $\hat{p}$ is a scale of pressure variation in the vertical), which requires $\mathrm{Ri}\,\mathrm{Ro} \gg 1$ – see section 10.3. Hence

$$\frac{D\mathbf{v}}{Dt} = \left(\frac{\partial}{\partial t} + \mathbf{v}.\nabla_p + \omega \frac{\partial}{\partial p} \right) \mathbf{v} \rightarrow \left(\frac{\partial}{\partial t} + \mathbf{v}_g .\nabla_p \right) \mathbf{v}_g \equiv \frac{D\mathbf{v}_g}{Dt_g}, \tag{10.10}$$

and (10.9) is replaced by

$$\frac{D\mathbf{v}_g}{Dt_g} + f_0\mathbf{k} \times \mathbf{v}_a + \beta y \mathbf{k} \times \mathbf{v}_g = 0. \tag{10.11}$$

Equation (10.11) is nonlinear through the geostrophic self-advection term $(\mathbf{v}_g.\nabla_p)\mathbf{v}_g$; see (10.10). By taking $\mathbf{k}.\nabla_p\times(10.11)$ and using (10.6) and (10.8), one obtains, without further approximation:

$$\frac{D}{Dt_g}\left(\nabla_p^2\psi+\beta y\right)=f_0\frac{\partial\omega}{\partial p}, \tag{10.12}$$

which is the p-coordinate, QG1 vorticity equation.

The p-coordinate form of the thermodynamic equation (5.20) may be written (in Cartesian geometry and with $Q=0$) as

$$\left(\frac{\partial}{\partial t}+\mathbf{v}.\nabla_p\right)\theta'+\omega\frac{\partial}{\partial p}(\theta_s+\theta')=0. \tag{10.13}$$

Approximations are now made in (10.13) that parallel those made in (10.9), and are justified under the same conditions: $\omega\partial\theta'/\partial p$ is neglected compared with $\mathbf{v}.\nabla\theta'$, and $\mathbf{v}$ is replaced by $\mathbf{v}_g$. Upon use of (10.7), the QG1 thermodynamic equation is obtained as

$$\frac{D}{Dt_g}\left\{\frac{f_0^2}{S}\frac{\partial\psi}{\partial p}\right\}+f_0\omega=0, \tag{10.14}$$

in which

$$S=S(p)\equiv-\frac{RT_s}{p\theta_s}\frac{d\theta_s}{dp}. \tag{10.15}$$

Elimination of ω between (10.12) and (10.14) gives the QG1 potential vorticity equation:

$$\frac{D}{Dt_g}\{QGPV\}=0 \tag{10.16}$$

$$QGPV\equiv\nabla_p^2\psi+\beta y+f_0^2\frac{\partial}{\partial p}\left(\frac{1}{S}\frac{\partial\psi}{\partial p}\right); \tag{10.17}$$

QGPV is the (p-coordinate) quasi-geostrophic potential vorticity.

Equation (10.16) (with (10.17)) is an approximation to the conservation of Ertel's potential vorticity (Bretherton 1966, Green 1970, Kuo 1972). The analogy is between the conservation laws and not the conserved quantities; Ertel's PV is conserved under D/Dt, but QGPV under D/Dt_g, although vertical motion is allowed for in QG1. To indicate this, QGPV is sometimes called the quasi-geostrophic *pseudo* potential vorticity (Charney 1971).

Given appropriate initial and spatial boundary conditions, Equation (10.16) determines the time evolution of the streamfunction, ψ; it is the central prognostic equation of QG1, the 'signal accomplishment' of quasi-geostrophic theory (Dutton 1974).

Elimination of the local time derivatives between (10.14) and (10.11) or (10.12), leads to a diagnostic equation for ω:

$$\frac{S}{f_0}\nabla_p^2\omega+f_0\frac{\partial^2\omega}{\partial p^2}=G. \tag{10.18}$$

The source function G in (10.18), which involves ψ and its spatial derivatives, may be expressed in many different forms; see Hoskins *et al.* (1985), Sanders and Hoskins (1990), Xu (1992), Carroll (1995) and Martin (1999). One of the most useful is the Q-vector form of Hoskins *et al.* (1978):

$$G = -2\nabla . \mathbf{Q} + \frac{f_0}{S}\beta\frac{\partial v_g}{\partial p}, \tag{10.19}$$

with

$$\mathbf{Q} \equiv \frac{f_0}{S}\left(\frac{\partial \mathbf{v}_g}{\partial p} . \nabla_p\right)\nabla_p \psi.$$

At any time t, we may determine ω from (10.18) and appropriate boundary conditions. Knowledge of ω enables the divergent part of $\mathbf{v}_a$ to be found from (10.8) [and appropriate boundary conditions]. The rotational part of $\mathbf{v}_a$ may be found from the result of taking the divergence of (10.11):

$$f_0 \nabla_p^2 \psi_a = \nabla_p . [(\mathbf{v}_g . \nabla)\mathbf{v}_g] + \beta u_g - \beta y \nabla_p^2 \psi. \tag{10.20}$$

Here $\psi_a = \psi_a(x, y, p, t)$ is the steamfunction of the ageostrophic flow (cf. (9.20)).

A knowledge of the geostrophic streamfunction $\psi = \psi(x, y, p, t)$ at some time t thus enables all other variables of the model to be determined at that time (given appropriate boundary conditions). The relevant equations are: (10.6) [for $\mathbf{v}_g$]; (10.7) [for T' and θ']; (10.18) [for ω]; (10.8) and (10.20) [for the divergent and rotational parts of $\mathbf{v}_a$].

An energy equation may be formed by multiplying (10.14) by $\partial\psi/\partial p$, adding the result to the dot product of $\nabla_p \psi$ and (10.11), and using (10.8):

$$\frac{D}{Dt_g}\left\{\frac{1}{2}\left(\mathbf{v}_g^2 + \frac{f_0^2}{S}\left(\frac{\partial\psi}{\partial p}\right)^2\right)\right\} + f_0\left\{\nabla_p . (\mathbf{v}_a \psi) + \frac{\partial}{\partial p}(\omega\psi)\right\} = 0. \tag{10.21}$$

The boundary condition $\omega = 0$ on $p = p_0$ is often applied in this model – and, from (10.14), determines a boundary condition on $\partial/\partial t(\partial\psi/\partial p)$. A more accurate choice is $\omega = -(f_0 p_0/RT_0)\partial\psi/\partial t$ on $p = p_0$, which introduces various interesting features, both to the time evolution and to the energetics; see White (1978b) and Rõõm (1996).

10.2 QG1 in height coordinates

It is revealing to compare the analysis and results given in section 10.1 with the development of QG1 in ordinary height coordinates. In this case, all thermodynamic variables (including pressure) are represented as deviations from a hydrostatic reference state that is a function of height z only:

$$q = q(x, y, z, t) = q_0(z) + q'(x, y, z, t), \tag{10.22}$$

where $q = p$, ρ, θ or T and

$$\frac{dp_0}{dz} = -\rho_0 g; \; p_0 = \rho_0 R T_0; \; \theta_0 = T_0 \left(\frac{p_{\text{ref}}}{p_o}\right)^{R/c_p}. \tag{10.23}$$

Approximations are made in the hydrostatic and continuity equations as well as in the horizontal momentum and thermodynamic equations. Pedlosky (1964) gives a power series derivation assuming Ro $\ll 1$, $B \equiv N^2H^2/f^2L^2 \sim 1$, $N^2H/g \ll 1$, where

$$N^2 \equiv \frac{g}{\theta_0}\frac{d\theta_0}{dz}, \tag{10.24}$$

and L and H are, as usual, horizontal and vertical length scales of the motion. The third of Pedlosky's conditions is readily relaxed to $N^2H/g \sim 1$ in his derivation; the result is an extended QG1 z-coordinate model which includes terms that are negligible if $N^2H/g \ll 1$. It has been referred to variously as the 'modified', 'non-Doppler' or 'deep' QG1 model (Blumen 1978, White 1982, Bannon 1989) and is very similar to the formulation originally proposed by Charney (1948). The geostrophic flow $\mathbf{v}_g$ is defined as

$$\mathbf{v}_g \equiv \frac{1}{\rho_0 f_0}\mathbf{k} \times \nabla_z p' = \mathbf{k} \times \nabla_z \psi; \; \psi \equiv \frac{p'}{\rho_0 f_0}, \; \nabla_z \equiv \left(\left.\frac{\partial}{\partial x}\right|_z, \left.\frac{\partial}{\partial y}\right|_z\right). \tag{10.25}$$

The horizontal momentum, continuity and thermodynamic equations of the model, as obtained by White (1977), may be written in the forms

$$\frac{D\mathbf{v}_g}{Dt_g} + f_0\mathbf{k} \times \hat{\mathbf{v}}_a + \beta y \mathbf{k} \times \mathbf{v}_g = 0 \tag{10.26}$$

$$\rho_0 \nabla_z . \hat{\mathbf{v}}_a + \frac{\partial}{\partial z}(\rho_0 \hat{w}) = 0 \tag{10.27}$$

$$\frac{D}{Dt_g}\left(\frac{\partial \psi}{\partial z}\right) + \frac{N^2}{f_0}\hat{w} = 0, \tag{10.28}$$

in which

$$\frac{D}{Dt_g} \equiv \frac{\partial}{\partial t} + \mathbf{v}_g . \nabla_z \tag{10.29}$$

and

$$\hat{\mathbf{v}}_a \equiv \mathbf{v}_a + \frac{\rho'}{\rho_0}\mathbf{v}_g; \quad \hat{w} \equiv w - \frac{f_0}{g}\frac{\partial \psi}{\partial t}; \tag{10.30}$$

thus $\hat{\mathbf{v}}_a$ is an extended ageostrophic flow and $\hat{w}$ an extended vertical velocity.

Given appropriate boundary conditions, (10.26)–(10.30) imply a global energy equation having a quadratic integrand (Blumen 1978) and Hamiltonian structure may be demonstrated (Holm and Zeitlin 1998). Equations (10.26)–(10.30) imply a prognostic equation for (height-coordinate) QGPV and a diagnostic equation for the extended vertical velocity $\hat{w}$:

$$\frac{D}{Dt_g}\{QGPV\} = 0 \tag{10.31}$$

$$QGPV \equiv \nabla_z^2\psi + \beta y + \frac{f_0^2}{\rho_0}\frac{\partial}{\partial z}\left(\frac{\rho_0}{N^2}\frac{\partial\psi}{\partial z}\right) \tag{10.32}$$

$$\frac{N^2}{f_0}\nabla_z^2\hat{w} + f_0\frac{\partial}{\partial z}\left\{\frac{1}{\rho_0}\frac{\partial}{\partial z}(\rho_0\hat{w})\right\}$$
$$= -2\nabla_z.\left[\left(\frac{\partial\mathbf{v}_g}{\partial z}.\nabla_z\right)\mathbf{k}\times\mathbf{v}_g\right] + \beta\frac{\partial v_g}{\partial z}. \tag{10.33}$$

The QGPV equation (10.31) [with (10.32)] is the same in the cases $N^2H/g \ll 1$ (Pedlosky 1964) and $N^2H/g \sim 1$ (White 1977). Equation (10.33) is also the same in both cases, but when $N^2H/g \ll 1$ the local time derivative term in the definition (10.30) of $\hat{w}$ becomes negligible, so that $\hat{w}$ reduces to w; and $\hat{\mathbf{v}}_a$ also reduces to $\mathbf{v}_a$ when $N^2H/g \ll 1$.

Various aspects of this 'non-Doppler' QG1 model are of interest.

(i) From (10.28) and (10.30), the condition on ψ at a rigid horizontal boundary ($w = 0$) is

$$\frac{D}{Dt_g}\left(\frac{\partial\psi}{\partial z}\right) - \frac{N^2}{g}\frac{\partial\psi}{\partial t} = 0. \tag{10.34}$$

The term in $\partial\psi/\partial t$ (which is negligible if $N^2H/g \ll 1$) allows for the change of apparent vertical – see section 3.6 – that accompanies a steady zonal frame translation (Betts and McIlveen 1969, White 1982). Its presence means that the effect of adding a constant U_0 to the zonal flow is not simply to shift the evolution by U_0; hence the epithet *non-Doppler* (Lindzen 1968). The same effect is seen in the pressure-coordinate QG1 model if the boundary condition $\omega = -(f_0p_0/RT_0)\partial\psi/\partial t$ is applied at $p = p_0$ (see section 10.1).

(ii) In terms of w and $\hat{\mathbf{v}}_a$, rather than $\hat{w}$ and $\hat{\mathbf{v}}_a$, the continuity equation (10.27) becomes

$$\frac{\partial\rho'}{\partial t} + (\mathbf{v}_g.\nabla_z)\,\rho' + \rho_0\nabla_z.\mathbf{v}_a + \frac{\partial}{\partial z}(\rho_0 w) = 0. \tag{10.35}$$

Equation (10.35), which is equivalent to the continuity equation used by Charney (1948), is *not* of anelastic form (see sections 8 and 11). When $N^2H/g \ll 1$, the terms in ρ' are negligible, and (10.35) reduces to the anelastic form of Pedlosky's (1964) model. The two models give widely different external Rossby wave phase speeds at planetary scales (White 1978b); those predicted by the non-Doppler model are in better accord with observation.

(iii) Equation (10.33) is diagnostic for the extended vertical velocity $\hat{w}$, and for the usual vertical velocity w only in the case $N^2H/g \ll 1$, when $\hat{w}$ reduces to w. One might have expected that development of QG1 in height coordinates would lead to a diagnostic equation for w that was in some way a constrained version of Richardson's equation (see section 5.5), but this is not the case. Further, from (10.23), (10.25) and (10.30) we have:

$$\hat{w} \equiv w - \frac{f_0}{g}\frac{\partial\psi}{\partial t} = w - \frac{1}{\rho_0 g}\frac{\partial p'}{\partial t} = -\frac{1}{\rho_0 g}\left(\frac{\partial p'}{\partial t} + w\frac{\partial p_o}{\partial z}\right) \approx -\frac{\omega}{\rho_0 g}, \tag{10.36}$$

since $\mathbf{v}_g . \nabla_z p' = 0$. Hence $-\rho_0 g\hat{w}$ is an approximation to the pressure-coordinate 'vertical velocity' $\omega \equiv Dp/Dt$. Clearly, (10.33) is an omega equation, although it has emerged from a height-coordinate analysis. This result suggests (as one would hope, though perhaps not expect) that the development of QG1 is essentially independent of the vertical coordinate used. [See Berrisford *et al.* (1993) for a development of QG1 in θ-coordinates.]

(iv) From (ii), (iii) and Equation (10.32), we see that compressibility may be taken into account to varying degrees in QG1 models. By using the p-coordinate form (section 10.1) we achieve the most complete treatment as regards the interior equations, but at the expense (in practice) of an approximate treatment of the boundary conditions at quasi-horizontal surfaces. Within a z-coordinate framework, formally the same interior accuracy can be achieved by using the non-Doppler model; and boundary conditions are more clearly defined. Both models represent the effect of dynamic compressibility in the continuity equation: in the p-coordinate development the full HPE form is used, whilst in the non-Doppler model the term $D\rho'/Dt$ is represented by $\partial\rho'/\partial t + (\mathbf{v}_g . \nabla_z)\rho'$ (see (10.35)). In addition to dynamic compressibility, there is also a static compressibility effect (Green 1960): the variation with height of the reference state density $\rho_0(z)$. If this is neglected in (10.32), and the buoyancy frequency N is assumed independent of height also, then the pseudo-potential vorticity reduces to

$$QGPV = \nabla_z^2 \psi + \beta y + \frac{f_0^2}{N^2} \frac{\partial^2 \psi}{\partial z^2}. \tag{10.37}$$

In this Boussinesq limit, $QGPV - \beta y$ is simply the 3-dimensional Laplacian of the stream-function, ψ, if z is scaled by N/f_0. A similar simplification occurs on the left side of (10.33).

10.3 Conditions for validity and application of QG1

A summary of the assumptions made in deriving QG1 may be timely. We consider the z-coordinate case examined in section 10.2. As before, L and H are horizontal and vertical length scales over which $|\mathbf{v}|$ and $|w|$ change by the characteristic values V and W respectively.

(a) The central condition is that the Rossby number be small: $\text{Ro} \equiv V/fL \ll 1$.

(b) The Lagrangian time-scale is assumed to be of order L/V, so that $|D\mathbf{v}/Dt| \sim V^2/L$. Since $|(\mathbf{v}.\nabla)\mathbf{v}| \sim V^2/L$, the *local* time scale is assumed to be of order, or greater than, L/V.

(c) $L/a \leq \text{Ro}$ ensures that $\mathbf{v}_g$ is a good approximation to $\mathbf{v}_G$.

(d) The neglect of vertical advection of momentum and deviation potential temperature in comparison with the horizontal parts of the advection requires $W/H \ll V/L$. By noting that fractional variations of potential temperature and pressure in the horizontal are of the same order, one obtains from scale analysis of the thermodynamic equation (and previous assumptions) that $W/H \sim (V/L)(\mathrm{Ri\,Ro})^{-1}$, where $\mathrm{Ri} \equiv N^2H^2/V^2$ is a Richardson number. Hence it is required that $\mathrm{Ri\,Ro} \gg 1$; it is sufficient that $\mathrm{Ri\,Ro} \sim \mathrm{Ro}^{-1}$, i.e. that the Burger number $B \equiv \mathrm{Ri\,Ro}^2 \sim 1$. For synoptic-scale motion in mid-latitudes we have $H \sim 10^4$ m (the depth of the troposphere), $L \sim 10^6$ m (the synoptic horizontal scale), $N \sim 10^{-2}\ \mathrm{s}^{-1}$, $f \sim 10^{-4}\ \mathrm{s}^{-1}$; thus $B \sim 1$.

(e) Fractional variations in pressure in the horizontal are of order $fVL/gH = (V^2/gH)\,\mathrm{Ro}^{-1}$, and fractional variations of density in the horizontal will be of the same order. Hence we require that the Froude number $\mathrm{F} \equiv V^2/gH$ should obey $\mathrm{F} \ll \mathrm{Ro}$, in order that the neglect of horizontal variations of ρ in the definition of $\mathbf{v}_g$ is to be reasonable; the values quoted earlier give $\mathrm{F} \sim 10^{-3}$, $\mathrm{F Ro}^{-1} \sim 10^{-2}$.

(f) Notice that $\mathrm{Ri} \equiv N^2H^2/V^2 = (N^2H/g)\mathrm{F}^{-1}$. The importance of the quantity N^2H/g becomes clear from a scale analysis of the continuity equation using results already obtained: $\partial w/\partial z \sim (V/L)(\mathrm{Ri\,Ro})^{-1}$ and $(1/\rho)D\rho'/Dt_h \sim (V/L)\mathrm{F\,Ro}^{-1}$. Hence dynamic compressibility is important if $N^2H/g \sim 1$. The values quoted at (d) give $N^2H/g \sim 10^{-1}$, but motion having a height scale substantially greater than the depth of the troposphere will give a substantially larger value. Also, $N^2H_0/g = 2/7$ for an isothermal, diatomic, perfect gas atmosphere.

Assumptions (a)–(d) obviously appear also in the p-coordinate case. Assumptions (e)–(f) are not required for the interior equations in the p-coordinate case, but they are required for the validity of the usual boundary conditions. See White (1977) for further discussion of (a)–(f).

Derivation of the conservation properties of QG1 depends on f_0 being a constant, and on N being a function of height only. It is tempting to apply the model in contexts for which the ranges of variation of f and N are not small – to treat f_0 and N as functions of space and time, for example within the definition of QGPV (used, perhaps, as the prognostic variable in a numerical model). Such variations, particularly of f, are sometimes allowed on the understanding that they have small fractional variations over the horizontal space scale of the motion; see, for example, Kuo (1959), Charney and Stern (1962) and Pedlosky (1987).

The conservation properties of QG1 are retained if f_0 is held constant but spherical geometry is assumed and βy is replaced (in the definition of QGPV) by the true planetary vorticity $2\Omega \sin\phi$. Such a formulation has been used by many authors: see Baer (1970), Simons (1972), Baines and Frederiksen (1978), Shutts (1983b), Wu and White (1986) and Marshall and Molteni (1993). This spherical polar version of QG1 is analytically and numerically convenient but involves coarse approximation of f except in the planetary vorticity term.

If N is allowed to vary horizontally in QG1 in height coordinates (or S in a pressure coordinate version – see section 10.1), then the global potential temperature budget is disrupted (Haltiner and Williams 1981). Advection by the horizontally divergent flow should be retained in this case, with the consequence that the model ceases to be of QG1 type.

The desire to allow horizontal variations of f and N, and time variations of the latter, has been a stimulus to development of the more general nearly-geostrophic models discussed in section 9 (and their 3-dimensional relatives). Another stimulus has been a desire to remove gravity waves by less invasive surgery: to make minimal approximations in the momentum equation, and – ideally – to leave the other equations intact.

10.4 Equations of Poisson type in QG1

Although it retains the nonlinearity of advection by the geostrophic flow, QG1 yields a number of linear, elliptic partial differential equations. Two-dimensional Poisson equations arise in the determination of the ageostrophic flow; see, for example, (10.20). The omega equation ((10.18) in p-coordinates, (10.33) in z-coordinates) is a 3-dimensional elliptic PDE, the source function being a function of ψ and its spatial derivatives. If $QGPV - \beta y$ is regarded as known, then, in (10.17) and (10.32), it is the source function in another 3D Poisson-type equation, in this case for ψ. The QGPV equation itself ((10.16), (10.31)) can be written as yet another Poisson-type equation – for the streamfunction *tendency* $\partial\psi/\partial t$ (see, for example, Nielsen-Gammon and Lefevre 1996). Considering the z-coordinate case, we can write (10.31) as

$$\left[\nabla_z^2 + \frac{f_0^2}{\rho_0}\frac{\partial}{\partial z}\left(\frac{\rho_0}{N^2}\frac{\partial}{\partial z}\right)\right]\frac{\partial\psi}{\partial t} = -\mathbf{v}_g.\nabla_z\left[\nabla_z^2\psi + \beta y + \frac{f_0^2}{\rho_0}\frac{\partial}{\partial z}\left(\frac{\rho_0}{N^2}\frac{\partial\psi}{\partial z}\right)\right], \tag{10.38}$$

for which the boundary condition at rigid horizontal surfaces is, from (10.34) and (10.29), the mixed Dirichlet–Neumann form

$$\left[\frac{\partial}{\partial z} - \frac{N^2}{g}\right]\frac{\partial\psi}{\partial t} = -(\mathbf{v}_g.\nabla_z)\frac{\partial\psi}{\partial z}. \tag{10.39}$$

From classical treatments of Newtonian gravitation, electrostatics, magnetostatics, steady-state heat conduction and elastic membranes – and, indeed, fluid dynamics – equations of Poisson type are amongst the most extensively analysed and best understood in mathematical physics [see Eriksson *et al.* (1996), chapter 15, and Batchelor (1967), section 2.4]. All the insights gained can be used to rationalise the behaviour of the linear, elliptic QG1 problems.

For example, the total solution for ω or $\partial\psi/\partial t$ can be additively *attributed* to different regions or elements of the forcing or boundary conditions. This approach has been applied by Hoskins *et al.* (1985) and Clough *et al.* (1992) to the omega equation, and to the stream-function tendency equation by Hakim *et al.* (1996) and Nielsen-Gammon and Lefevre (1996); see also Räisänen (1997). Such methods offer a rational basis for identifying cause and effect links between fields of ω or $\partial\psi/\partial t$ (the effects) and the relevant source terms (the causes).

The problem in which (10.38) is inverted for $\partial\psi/\partial t$ is conveniently referred to as *prognostic PV inversion*, and that in which $QGPV - \beta y$ is inverted for ψ as *static PV inversion* (Hakim *et al.* 1996). Static PV inversion is of particular interest. Since $QGPV$ is conserved in the sense that $D/Dt_g(QGPV) = 0$ [in the absence of heat sources and friction, the effects of which can be taken into account if desired] $QGPV$ may be regarded – to use the language of gravitation or electrostatics – as a 'mass-like' or 'charge-like' quantity. To the extent that inverting $QGPV - \beta y$ for the streamfunction ψ may be achieved, and all other fields may be calculated from ψ, the analogy of $QGPV$ with mass or charge becomes even closer. Generalizations of this picture to EPV (with the hydrostatic approximation) subject to a balance condition such as the Charney balance equation (see section 9.6) offer a still more compelling view. A gap in the vision is that there is no unique specification of boundary conditions that can be justified by physical arguments; the boundary conditions in static PV inversion are ultimately a matter of choice (Bishop and Thorpe 1994). [In the current QG1 case, note that (10.39) gives a well-defined boundary condition on $\partial\psi/\partial t$, but no condition on ψ, at rigid horizontal boundaries.] Hakim *et al.* (1996) have noted that some choices of boundary condition violate regularity requirements; this observation is helpful in providing a constraint on the choice of boundary conditions, but it is a non-holonomic constraint in that it does not define a particular choice. Nevertheless, given awareness of the flexibility in choice of boundary conditions on horizontal surfaces, *static* PV inversion, as well as the clearly defined *prognostic* PV inversion, is useful in the development of well-founded conceptual models of weather systems and their behaviour. For a recent application of this type, involving an approximation to EPV, see Griffiths *et al.* (2000).

11 Acoustically-filtered models

The HPEs (section 5.4) are not the only 3D meteorological model that lacks vertically-propagating acoustic waves but supports gravity waves. Some other acoustically-filtered (or 'soundproofed') models are briefly addressed in this section. Anelastic models are discussed in section 11.1, in section 11.2 we describe a model which might be seen as an anelastic variant but uses pressure as vertical coordinate, and section 11.3 discusses application of its technique to represent the non-hydrostatic effect of the vertical component of the Coriolis force rather than the relative acceleration Dw/Dt. An important non-hydrostatic model which is *not* acoustically-filtered, but retains the shallow atmosphere approximation in spherical geometry, is considered in section 11.4.

11.1 Anelastic models

The linear mode analysis presented in section 8.2 suggests that gravity waves are more accurately treated when the continuity equation is written in incompressible form than when the hydrostatic approximation is applied. Using an incompressible form of the continuity equation to remove acoustic waves is therefore an attractive proposition – the more so because Lamb waves are removed as well as vertically-propagating acoustic waves. A typical *anelastic model* uses the continuity equation in the form

$$\frac{\partial u}{\partial x} + \frac{\partial v}{\partial y} + \frac{1}{\rho_0}\frac{\partial}{\partial z}(\rho_0 w) = 0, \tag{11.1}$$

in which $\rho_0 = \rho_0(z)$ is a fixed profile of mean density; see Ogura and Phillips (1962). The analysis given in section 8.2 also suggests (see (8.21)) that use of (11.1) should be accompanied by neglect of a certain term in the vertical component of the momentum equation, and this is usually done. Appropriate Boussinesq forms of the horizontal components and a form of the thermodynamic equation complete the model. Application of (11.1) to the three components of the momentum equation gives a diagnostic 3D elliptic equation for the pressure field. The formulation is then similar in many respects to the Navier–Stokes equations for incompressible flow (Williams 1969), and indeed becomes equivalent if the height variation $\rho_0(z)$ is neglected [as is appropriate if the vertical scale of the motion is much less than the scale height $H_0 = RT_0/g$ – see, for example, Mason and Brown (1999)].

Nonlinear conservation properties are good if $\rho_0(z)$ corresponds to certain simple thermodynamic states, but more general choices require specific investigation. Bannon (1995) gives a thorough discussion of this and related issues regarding a number of models of anelastic type.

The meteorological context of the anelastic equations is commonly that of cumulonimbus-scale convection; then the Coriolis terms are usually neglected and Cartesian geometry is used. If the hydrostatic approximation is applied, and Coriolis terms are included, the anelastic model becomes in the geostrophic limit the height-coordinate QG1 model that is valid when $N^2H/g \ll 1$; see section 10.2.

11.2 Non-hydrostatic convection models using pressure coordinates

Miller (1974) and Miller and Pearce (1974) first proposed and used a pressure coordinate model to describe *non*-hydrostatic motion of cumulonimbus scale. Their model incorporates a reference state in hydrostatic balance and deals with non-hydrostatic departures from this state.

The horizontal momentum equation is written

$$\frac{D\mathbf{v}}{Dt} + g\nabla_p z' = \mathbf{F}_{h'} \tag{11.2}$$

where z' is the deviation of the height z of a pressure surface from the reference state $z_s(p)$ which is associated hydrostatically with a temperature profile $T_s(p)$:

$$z(x, y, p, t) = z_s(p) + z'(x, y, p, t) \tag{11.3}$$

$$g\frac{dz_s}{dp} + \frac{RT_s}{p} = 0. \tag{11.4}$$

The continuity and thermodynamic equations are applied in the forms

$$\frac{\partial u}{\partial x} + \frac{\partial v}{\partial y} + \frac{\partial \omega}{\partial p} = 0 \tag{11.5}$$

$$\frac{DT}{Dt} - \frac{\omega RT}{pc_p} = \frac{Q}{c_p}. \tag{11.6}$$

Very small terms are neglected in writing (11.2) and (11.5) according to the criterion $g \gg Dw/Dt$. Non-hydrostatic effects are retained in the vertical component of the momentum equation by applying the approximation

$$w \approx -\frac{\omega}{g\rho_s(p)} = -\frac{\omega RT_s}{gp} \tag{11.7}$$

in the vertical acceleration term:

$$\frac{R}{g}\frac{D}{Dt}\left\{\frac{\omega T_s}{p}\right\} + g\frac{T'}{T_s} + \frac{g^2 p}{RT_s}\frac{\partial z'}{\partial p} = 0; \tag{11.8}$$

see the comment after (11.12), below. In (11.8), $T' = T - T_s(p)$. The material derivative is

$$\frac{D}{Dt} \equiv \frac{\partial}{\partial t} + u\frac{\partial}{\partial x} + v\frac{\partial}{\partial y} + \omega\frac{\partial}{\partial p}, \tag{11.9}$$

with $\omega \equiv Dp/Dt$, and differentiations with respect to t, x and y taken at constant p.

Miller (1974) justified the model by considering numerical magnitudes of the (small) terms omitted, and Miller and White (1984) obtained the same equations via power series expansion. The approximation (11.7), as applied in (11.8), has the effect of eliminating vertically-propagating acoustic waves, and Lamb waves may be eliminated by applying the lower boundary condition $\omega = 0$ at $p = p_0$. Integration proceeds by time-stepping in conjunction with solution of a 3D Poisson-like equation for the height deviation z'; it is obtained by taking ∇_p.(11.2), adding $(gp/RT_s)\partial/\partial p$(11.8), and applying (11.5). The model gives analogues of energy and potential vorticity conservation laws, and is virtually isomorphic to an anelastic model in height coordinates. It has been used in a range of numerical simulations of cumulonimbus and squall-line motion; see, for example, Miller (1978) and Brugge and Moncrieff (1985). Sigma-coordinate forms (which imply Lamb waves) have been used by Xue and Thorpe (1991) and Miranda and Valente (1997) to model flow over and around orography.

White (1989b) noted the dependence of the Miller–Pearce model on the reference state profiles $z_s(p)$, $T_s(p)$, and pointed out that the (Cartesian) vertical component of the momentum equation can be written, without approximation, as

$$\frac{Dw}{Dt} - \frac{gp}{RT}\left(1 + \frac{1}{g}\frac{Dw}{Dt}\right)\left(g\frac{\partial z'}{\partial p} + \frac{RT'}{p}\right) = 0. \tag{11.10}$$

Use of the uncritical approximation $g \gg Dw/Dt$, together with

$$w \approx -\frac{\omega}{\rho g} = -\frac{\omega RT}{gp}, \tag{11.11}$$

then replaces (11.8) by

$$\frac{R}{g}\frac{D}{Dt}\left\{\frac{\omega T}{p}\right\} + g\frac{T'}{T} + \frac{g^2 p}{RT}\frac{\partial z'}{\partial p} = 0, \tag{11.12}$$

which does not involve the reference state. (Setting $T = T_s(p)$ in (11.12) gives (11.8).)

Equations (11.2), (11.5) and (11.6), (11.12) constitute a modified Miller–Pearce model that retains analogues of energy and potential vorticity conservation, and implies a diagnostic, Poisson-like equation for z'; Salmon and Smith (1994) demonstrate its Hamiltonian form. Rõõm (1998) and Rõõm and Mannik (1999) describe related formulations and compare their linearised behaviour.

Economical time integration of the fully non-hydrostatic equations, with acoustic waves present, may be achieved by using a semi-implicit scheme (Tapp and White 1976). Such a formulation was the basis of a regional, mesoscale model used operationally by the UK Met. Office during the 1980s. The use of semi-implicit methods requires the solution of a 3D Helmholtz-type equation at each timestep. This illustrates a common situation: a diagnostic elliptic PDE has to be solved at each timestep whether special numerical methods are used to handle high frequency modes or whether these modes are filtered by approximating the governing equations. Lie (1999) gives a survey of non-hydrostatic models in the context of mesoscale weather forecasting.

11.3 Acoustically-filtered global models having a full representation of the Coriolis force

Miller and Pearce's method can be used to represent other non-hydrostatic terms in the vertical component of the momentum balance. White and Bromley (1995) noted that the term $-2\Omega w\cos\phi$ in the *zonal* component of the momentum equation (4.4) is not comfortably negligible in tropical, synoptic-scale flow systems in which diabatic heating is important. To include it requires the inclusion of other terms and effects, if conservation principles are to be respected: as discussed in section 5.4, the corresponding term in the vertical component (4.6) must be kept, the shallow-atmosphere approximation must be relaxed, and various metric terms retained. [The $2\Omega\cos\phi$ terms are negligible in *adiabatic* motion if $2\Omega \ll N$; see Gill (1982), p.449. This condition is obeyed given $2\Omega \sim 10^{-4}$ s^{-1} and $N \sim 10^{-2}$ s^{-1}, but it is clearly not satisfied as $N \to 0$.]

White and Bromley (1995) proposed a model based on a pseudo-radius defined as

$$r_s(p) = a + \int_p^{p_0} \frac{RT_s(p')}{p'}\,dp', \tag{11.13}$$

in which p' is a dummy variable and p_0 a mean sea-level pressure. Use of $r_s(p)$ as a vertical coordinate entails no approximation; interpreting r_s as distance from the centre of the Earth does. From (11.13),

$$\frac{Dr_s}{Dt} = -\frac{RT_s(p)}{gp}\omega \equiv \tilde{w}. \tag{11.14}$$

This is the approximation to the vertical velocity used in the Miller–Pearce model (section 11.2). The material derivative is written as

$$\frac{D}{Dt} \equiv \frac{\partial}{\partial t} + \mathbf{u}.\tilde{\nabla}, \tag{11.15}$$

with

$$\mathbf{u} = (u, v, \tilde{w}) \quad \text{and} \quad \tilde{\nabla} = \left(\frac{1}{r_s\cos\phi}\frac{\partial}{\partial\lambda}, \frac{1}{r_s}\frac{\partial}{\partial\phi}, \frac{\partial}{\partial r_s}\right). \tag{11.16}$$

The following pressure-coordinate equations were proposed:

$$\frac{Du}{Dt} - \left(2\Omega + \frac{u}{r_s\cos\phi}\right)(v\sin\phi - \tilde{w}\cos\phi) + \frac{g}{r_s\cos\phi}\frac{\partial z'}{\partial\lambda} = F_\lambda \tag{11.17}$$

$$\frac{Dv}{Dt} + \left(2\Omega + \frac{u}{r_s\cos\phi}\right)u\sin\phi + \frac{v\tilde{w}}{r_s} + \frac{g}{r_s}\frac{\partial z'}{\partial\phi} = F_\phi \tag{11.18}$$

$$-2\Omega u\cos\phi - \left(\frac{u^2+v^2}{r_s}\right) - g\frac{T'}{T_s} + g\frac{\partial z'}{\partial r_s} = 0 \tag{11.19}$$

$$\tilde{\nabla}_p.\mathbf{v} + \frac{1}{r_s^2}\frac{\partial}{\partial p}(r_s^2\omega) = 0 \tag{11.20}$$

$$\frac{D\theta}{Dt} = \left(\frac{\theta}{Tc_p}\right) Q. \tag{11.21}$$

In the continuity equation (11.20),

$$\tilde{\nabla}_p . \mathbf{v} = \frac{1}{r_s \cos\phi} \left\{ \frac{\partial u}{\partial \lambda} + \frac{\partial}{\partial \phi}(v \cos\phi) \right\}. \tag{11.22}$$

Some of the terms retained in these equations are typically very small, but they are needed for the delivery of the following conservation properties:

$$\frac{D}{Dt}\{(u + \Omega r_s \cos\phi)\, r_s \cos\phi\} = F_\lambda r_s \cos\phi - g\frac{\partial z'}{\partial \lambda} \tag{11.23}$$

$$\frac{D}{Dt}\left(\frac{1}{2}\mathbf{v}^2 + c_p T\right) + \tilde{\nabla}_p . (\mathbf{v}gz) + \frac{1}{r_s^2}\frac{\partial}{\partial p}(r_s^2 \omega g z) = Q + \mathbf{v}.\mathbf{F}_h \tag{11.24}$$

$$\rho_s \frac{D}{Dt}\left(\frac{\tilde{\mathbf{Z}}.\tilde{\nabla}\theta}{\rho_s}\right) = \tilde{\nabla}.\left[\theta\tilde{\nabla} \times \mathbf{F}_h + \tilde{\mathbf{Z}}\frac{D\theta}{Dt}\right]. \tag{11.25}$$

In (11.25),

$$\begin{aligned} \tilde{\mathbf{Z}} \;&\equiv\; \left[-\frac{1}{r_s}\frac{\partial(vr_s)}{\partial r}, 2\Omega\cos\phi + \frac{1}{r_s}\frac{\partial}{\partial r_s}(ur_s),\right. \\ &\qquad \left. 2\Omega\sin\phi + \frac{1}{r_s\cos\phi}\left(\frac{\partial v}{\partial \lambda} - \frac{\partial}{\partial \phi}(u\cos\phi)\right)\right] && (11.26) \\ &= 2\boldsymbol{\Omega} + \tilde{\nabla} \times \mathbf{u} && (11.27) \end{aligned}$$

and, for any vector $\mathbf{A} = (\mathbf{A}_h, A_r)$,

$$\tilde{\nabla}.\mathbf{A} = \tilde{\nabla}_p . \mathbf{A}_h + \frac{1}{r_s^2}\frac{\partial}{\partial r_s}(r_s^2 A_r). \tag{11.28}$$

Equations (11.23)–(11.25) are axial angular momentum, energy and potential vorticity conservation laws; their derivation is eased by noting an isomorphism with corresponding equations for the motion of an incompressible fluid. Only the term Dw/Dt (in the vertical component of the momentum equation) is unrepresented in (11.18)–(11.22). Its inclusion could be achieved by using the Miller–Pearce technique directly, but at the expense of having to solve an elliptic 3D PDE for z at each timestep; in the global model as set out above this is not necessary.

Roulstone and Brice (1995) demonstrated the Hamiltonian structure of isomorphs of (11.17)–(11.21), and White and Bromley (1995) derived σ-coordinate versions by direct transformation and described an integration strategy. Versions using another vertical coordinate system form the dynamical basis of the UK Met. Office's Unified Model (see Cullen (1993)) which is a gridpoint numerical model. The presence of the pseudo-radius $r_s(p)$ would complicate implementation of these equations in current spectral numerical models (see section 12).

11.4 A non-hydrostatic, global, shallow atmosphere model

As we noted in section 5.4, the HPEs omit the $\cos\phi$ Coriolis terms and various metric terms from the components of the momentum equation, and adopt the shallow-atmosphere approximation throughout. The non-hydrostatic, global, shallow-atmosphere model used by Tanguay *et al.* (1990) [see also Müller (1989)] consists of the HPEs augmented only by the term Dw/Dt in the vertical component of the momentum equation. In place of (5.19), the model therefore has

$$\frac{Dw}{Dt} + g + \frac{1}{\rho}\frac{\partial p}{\partial z} = 0, \tag{11.29}$$

with D/Dt given by the shallow atmosphere form (5.16). The retention of Dw/Dt in (11.28) appears unjustifiable for a range of mesoscale motion given that $-2\Omega u \cos\phi$ has been neglected (Draghici 1989), but the model is of theoretical interest because of its good conservation properties. The axial angular momentum conservation law is the HPE form (5.23), the energy conservation law is the HPE form (5.24) but with specific kinetic energy $\frac{1}{2}(\mathbf{v}^2 + w^2)$, and the PV law is of the HPE form (5.25) but with absolute vorticity $\boldsymbol{\xi}$ defined by

$$\boldsymbol{\xi} \equiv \left(\frac{1}{a}\frac{\partial w}{\partial\phi} - \frac{\partial v}{\partial z}, \frac{\partial u}{\partial z} - \frac{1}{a\cos\phi}\frac{\partial w}{\partial\lambda}, 2\Omega\sin\phi + \frac{1}{a\cos\phi}\left[\frac{\partial v}{\partial\lambda} - \frac{\partial}{\partial\phi}(u\cos\phi)\right]\right). \tag{11.30}$$

Roulstone and Brice (1995) showed that isomorphs of (11.17)–(11.21) arise when the functional form of the Hamiltonian is modified to exclude the contribution of the vertical motion to the kinetic energy. They showed too that the HPEs arise if the geometric factors in the Hamiltonian integral (its 'phase space') are also modified. It seems likely that the model of Tanguay *et al.* (1990) arises when the geometric factors in the Hamiltonian are modified but its functional form is left unchanged, and thus that there are two dynamically-consistent models intermediate in accuracy between the HPEs and the unapproximated equations – the model of Tanguay *et al.* (1990) and the model of White and Bromley (1995). These suggestions deserve further study.

12 Discussion: dynamical models, numerical weather prediction and climate simulation

This article has given an account of the basis and nature of many of the approximate models of meteorological dynamics. Here some remarks are offered on the approximation problem, on the applications of the approximate models, and on basic issues in the design of numerical models.

A theory of approximation for the equations governing meteorological flows is not yet fully developed or its rationale agreed upon. The retention of conservation properties during the approximation process is an attractive guiding principle, and if it is given exclusive priority, then the Hamiltonian technique pioneered by Salmon (1983) offers the best way forward. If a dynamical model having the desired conservation properties has been derived by other means, then the demonstration of Hamiltonian structure lends further credence. Prompted by evidence that Hamiltonian structure does not ensure superior performance in numerical practice (see, for example, Barth *et al.* (1990) and Allen and Newberger (1993)) some researchers consider that retention of all conservation properties should not be the priority. Few consider that conservation properties should be disregarded, but opinions differ on which should be favoured. Traditionally, global energy conservation has received most emphasis, but Lagrangian potential vorticity conservation is increasingly seen as paramount; at the time of writing, some striking results are emerging from studies of balanced, PV-conserving versions of the SWEs (McIntyre and Norton 2000).

What are the approximate models used for in meteorology? As we have noted, the HPEs are the foundation of most of the numerical weather prediction and climate simulation models run by operational and research centres worldwide, but the use of more accurate models is becoming more widespread. For example, the UK Met. Office's Unified Model is based on the acoustically-filtered equations discussed in section 11.3, and the use of virtually unapproximated forms is planned – the strategy being to use semi-implicit integration schemes to overcome the restriction to very short time steps which the presence of acoustic modes would otherwise impose; see Staniforth (2001). A trend towards the use of formulations more accurate than the HPEs is evident also in ocean modelling (Marshall *et al.* 1997). The utility of approximate models – especially those more heavily approximated than the HPEs – also lies in the development of a conceptual framework for the analysis of numerically-generated and observational data. Such a framework is necessary both in general scientific terms and to guide the development of better techniques for assimilating data into numerical models and effecting their time integration. We shall briefly discuss these aspects.

Analysing and understanding a simplified model is clearly easier than analysing and understanding a complicated one. Some uses of the barotropic vorticity equation in this respect were noted in section 5.6. In so far as it embodies notions of vorticity and temperature evolution and advection, the QG1 model systematises these concepts of the synoptic meteorologist and weather forecaster. A more modern view – not necessarily a competing view – is the PV perspective (Hoskins *et al.* 1985) which is embodied in QG1 and in some of the other balanced models discussed in section 9. The articles in Meteorological Applications (1997) elucidate the current interplay of ideas in this area.

Approximate models also play a major role as apparatus for thought experiments (which may be carried out either analytically or numerically). A particularly influential type of thought experiment is the stability analysis: a steady flow is subject to perturbations at $t = 0$ and the subsequent evolution determined by solving linearised forms of the governing equations. Eady's baroclinic stability problem (Eady 1949) can be solved analytically in the QG1 case, and its dominant eigenmodes resemble structures seen in developing mid-latitude weather systems. A large literature has grown up which extends Eady's analysis to more realistic initial steady flows and assumed external conditions, and explores development into the nonlinear stages using either analytical or numerical methods (in many cases using less heavily approximated dynamics); see Hart (1979) and Held and Hoskins (1985) for reviews. It could be argued that such stability problems, though illuminating, have been somewhat overemphasised, since one may reasonably enquire how the real atmosphere could ever reach the supercritical states which may be chosen for investigation. [Some recent work in this area has focussed on influences that stabilise flows, and on the hypothesis that the real atmosphere evolves close to a stability threshold; see Mole and James (1990), Stone and Nemet (1996), Dong and James (1997), Harnik and Lindzen (1998) and Nakamura (1999).] Also, the complete initial value problem is complicated by the presence of continuous spectrum instabilities as well as normal mode growth (or decay): many early analyses emphasised the latter at the expense of the former – see Farrell (1989). This important aspect of the stability problems reflects the non-self-adjointness of the relevant operators; Held (1985) gives a lucid account.

As reviewed by Errico (1997), adjoint operator theory has recently found practical application in ensemble forecasting and data assimilation. Ensemble methods (see Farrell 1990, Buizza and Palmer 1995, Buizza *et al.* 1997) aim to determine the sensitivity of a numerical forecast to its initial conditions. Since numerical integrations are time-consuming and the number of degrees of freedom is vast, ways must be found to identify patterns which capture the main instabilities of the initial flow and hence the sensitivity of the forecast. One way (of several) is to calculate singular vectors, having defined a suitable norm to gauge differences between integrations with slightly different initial conditions. The assimilation of observed data is a key part of the process of numerical weather prediction; see Daley (1991). The 4D variational technique (Talagrand and Courtier 1987) minimises a *cost function* that measures differences between evolving model values and observations over a chosen assimilation 'window' (typically a few hours). The minimisation is carried out with respect to fields at the beginning of the window period, and may be subject to constraints whose nature reflects knowledge of atmospheric behaviour developed from more heavily approximated dynamical models such as the semi-geostrophic and quasi-geostrophic forms.

Further examples of the use of knowledge gained from approximate models are noted in the following brief discussion of numerical model design.

Because of the nature of the governing equations, any reliable numerical model of the atmosphere must use a finite representation of its fields. That finite representation may involve field values at a number of chosen points (the gridpoint method) or fields specified in terms of amplitudes of a number of chosen functions (the Galerkin method). The Galerkin representation most frequently chosen for horizontal variations is a *spectral* representation in terms of surface spherical harmonics Y_n^m. (The viability of the technique depends on the use of finite Fourier transforms and Gaussian integration to handle product terms; see Hoskins and Simmons (1975), Côté and Staniforth (1988), Hortal and Simmons (1991) and Temperton *et al.* (2000).) Almost all models use the gridpoint method for vertical variations; thus, in global spectral models, the fields are represented by finite spherical harmonic expansions at a number of levels (typically 30 or more). There is advantage in using a staggered arrangement in which different fields are held at different levels. Many models hold the relevant vertical velocities at levels between those at which the horizontal velocity components and the potential temperature are held. This 'Lorenz' arrangement cannot give the most natural and accurate depiction of thermal wind balance (see section 7.3) and it also leads to the occurrence of spurious vertical modes (Schneider 1987). Thermal wind balance (Equation (7.11)) is better represented by holding potential temperature at the intermediate level – the 'Charney–Phillips' arrangement. The practical advantage of the Charney–Phillips arrangement over the Lorenz arrangement has yet to be demonstrated conclusively, but its theoretical advantage (at least for geostrophically-balanced motion) is partly an implication of the QG1 model. Horizontal grid staggering in gridpoint models is an issue of even greater variety; see Adcroft *et al.* (1999) for a recent discussion.

The spectral method of representing horizontal field variations has the considerable advantage of satisfactorily treating field variations close to coordinate poles. (Indeed, the 'triangular truncation' of the spherical harmonic series gives an isotropic representation which is independent of the location of the coordinate pole; see Hoskins and Simmons (1975).) Use of a gridpoint representation on points defined by the intersections of circles of latitude and longitude leads to numerous difficulties in the vicinity of the poles. Amongst various ways of coping with these, one of the most obvious and attractive is to use another distribution of points. Because of the existence of only 5 regular polyhedra in 3D space, a regular distribution of more than 20 points over the surface of a sphere is not possible. However, a quasi-regular distribution may be achieved by triangulating an icosahedron and centrally projecting the triangle vertices onto the circumscribing sphere (Sadourny *et al.* 1968, Thuburn 1997, Majewski 1998). Alternatively, projections of points on an inscribed cube may be used (Rančič *et al.* 1995, McGregor 1996). Another way of mitigating the pole problem is to use a subsidiary grid in the vicinity of the geographical poles. This is a particularly attractive option when it is used in conjunction with the semi-Lagrangian representation of material derivatives; see Staniforth and

Côté (1991) and references therein. Regional models can avoid the pole problem by using a rotated coordinate system whose poles are outside the domain; another strategy is to use a distribution of gridpoints that covers the sphere, but has their separation smoothly increasing away from the region of main interest – see Staniforth (2001).

The gridpoint method is reasonably expected to be better than the spectral method at representing near-discontinuities such as fronts in the atmosphere. It also has the advantage of allowing choice in the locations at which the various fields are held, and generally permitting more freedom – and thus scope for improvement – via the finite differencing. At present, however, their superior treatment of the poles makes spectral models at least competitive with gridpoint models.

Global gridpoint models nowadays have a grid interval of 50km or so in the horizontal; so systems having a wavelength of less than 100km are not resolved. Global spectral models are subject to broadly similar restrictions. Many important scales are therefore not explicitly represented, but their effects – in terms of heat, moisture and momentum transfers – must be allowed for. Especially in climate simulation, this problem of subgridscale *parametrization* is acute. An understanding of the fluxes carried by, for example, cumulonimbus systems, and their relation to the resolved flow is crucial for the development of appropriate parametrizations. Numerical simulation and theoretical analysis of motion on the relevant scales are the subjects of intense study, and the formulations described in sections 11.1 and 11.2 are frequently used for this purpose. Closely related is the problem of the scales that are barely resolved, and thus poorly treated, by the large-scale model, be it gridpoint or spectral. For an analysis of this key issue see Lander and Hoskins (1997).

In conclusion, it should be emphasised that meteorological dynamics is not solely concerned with the equations used for numerical weather forecasting and climate simulation. A glance at a text on satellite imagery (such as Bader *et al.* 1995) – or, indeed, out of a window during most daylight hours – serves to remind that the atmosphere is populated by flow structures and associated phenomena. These are naturally the concern of the users of weather forecasts, and could be said to be the weather itself. An appreciation of the structure of weather systems and phenomena, as well as of the structure of the governing equations, should guide the development of numerical models of the atmosphere and the appraisal of their performance.

Acknowledgements

I am grateful to Dr Ian Roulstone for his advice during the preparation of this article and for many useful scientific discussions, and to Dr Andrew Staniforth and Dr Sean Swarbrick for their valuable comments on a draft of the text. I acknowledge with gratitude the scientific influence of many other colleagues and former colleagues in the Met Office, at Reading University and at Imperial

College, London. In particular, I wish to thank Dr John Green, part of whose Imperial College lecture notes form the basis of section 8.2 of this article, and Dr Bob Riddaway, whose Met Office College lecture notes inspired section 5.5; both of these sets of notes have guided the development of my understanding of meteorological dynamics over many years.

References

Adcroft, A. and Marshall, D. 1998. 'How slippery are piecewise-constant coastlines in numerical ocean models?', *Tellus* **50A** 95–108.

Adcroft, A.J., Hill, C.N. and Marshall, J.C. 1999. 'A new treatment of the Coriolis terms in C-grid models at both high and low resolutions', *Mon. Weather Rev.* **127** 1928–1936.

Allen, J.S. 1991. 'Balance equations based on momentum equations with global invariants of potential enstrophy and energy', *J. Phys. Oceanog.* **21** 265–276.

Allen, J.S. 1993. 'Iterated geostrophic intermediate models', *J. Phys. Oceanog.* **23** 2447–2461.

Allen, J.S. and Holm, D.D. 1996. 'Extended-geostrophic Hamiltonian models for rotating shallow water motion', *Physica D* **98** 229–248.

Allen, J.S. and Newberger, P.A. 1993. 'On intermediate models for stratified flow', *J. Phys. Oceanog.* **23** 2462–2486.

Allen, J.S., Barth, J.A. and Newberger, P.A. 1990a. 'On intermediate models for barotropic continental shelf and slope flow fields. Part I: formulation and comparison of exact solutions', *J. Phys. Oceanog.* **20** 1017–1042.

Allen, J.S., Barth, J.A. and Newberger, P.A. 1990b. 'On intermediate models for barotropic continental shelf and slope flow fields. Part III: comparison of numerical model solutions in periodic channels', *J. Phys. Oceanog.* **20** 1949–1973.

Bader, M.J., Forbes, G.S., Grant, J.R., Lilley, R.B.E. and Waters, A.J. 1995. *Images in Weather Forecasting: a Practical Guide for Interpreting Satellite and Radar Imagery.* Cambridge University Press.

Baer, F. 1970. 'Analytical solution to low-order spectral systems', *Arch. Met. Geoph. Biokl., Ser. A,* **19** 255–282.

Baines, P.G. 1976. 'The stability of planetary waves on a sphere', *J. Fluid Mech.* **73** 193–213.

Baines, P.G. 1995. *Topographic effects in stratified flows.* Cambridge University Press.

Baines, P.G. and Frederiksen, J.S. 1978. 'Baroclinic instability on a sphere in two-layer models', *Q.J.R. Meteorol. Soc.* **104** 45–68.

Bannon, P. 1989. 'On deep quasi-geostrophic theory', *J. Atmos. Sci.* **22** 3457–3463.

Bannon, P. 1995. 'Potential vorticity conservation, hydrostatic adjustment, and the anelastic approximation', *J. Atmos. Sci.* **52** 2302–2312.

Bannon, P. 1998. 'A comparison of Ekman pumping in approximate models of the accelerating planetary boundary layer', *J. Atmos. Sci.* **55** 1446–1451.

Barnes, R.T.H., Hide, R., White, A.A. and Wilson, C.A. 1983. 'Atmospheric angular momentum fluctuations, length-of-day changes and polar motion', *Proc. R. Soc. Lond.* **A387** 31–73.

Barth, J.A., Allen, J.S. and Newberger, P.A. 1990. 'On intermediate models for barotropic continental shelf and slope flow fields. Part II: comparison of numerical model solutions in doubly periodic domains', *J. Phys. Oceanog.* **20** 1044–1076.

Batchelor, G.K. 1967. *An Introduction to Fluid Dynamics.* Cambridge University Press.

Bates, J.R., Semazzi, F.H.M., Higgins, R.W. and Barros, S.R.M. 1990. 'Integration of the shallow water equations on the sphere using a vector semi-Lagrangian scheme with a multi-grid solver', *Mon. Weather Rev.* **118** 1615–1627.

Bell, R.J.T. 1912. *An Elementary Treatise on Coordinate Geometry of Three Dimensions.* Macmillan.

Berrisford, P., Marshall, J.C. and White, A.A. 1993. 'Quasigoestrophic potential vorticity in isentropic coordinates', *J. Atmos. Sci.* **50** 778–782.

Betts, A.K. and McIlveen, J.F.R. 1969. 'The energy formula in a moving reference frame', *Q.J.R. Meteorol. Soc.* **95** 639–642.

Bishop, C.H. and Thorpe, A.J. 1994. 'Potential vorticity and the electrostatics analogy: quasi-geostrophic theory', *Q.J.R. Meteorol. Soc.* **120** 713–731.

Blackburn, M. 1985. 'Interpretation of ageostrophic winds and implications for jet stream maintenance', *J. Atmos. Sci.* **42** 2604–2620.

Bleck, R. 1984. 'An isentropic coordinate model suitable for lee cyclogenesis simulation', *Rivista di Meteorologia Aeronautica* **44** 189–194.

Bluestein, H.B. 1992. *Synoptic-Dynamic Meteorology in Midlatitudes.* Oxford University Press.

Blumen, W. 1978. 'A note on horizontal boundary conditions and stability of quasigeostrophic flow', *J. Atmos. Sci.* **35** 1314–1318.

Blumen, W. 1981. 'The geostrophic coordinate transformation', *J. Atmos. Sci.* **38** 1100–1105.

Bretherton, F.P. 1966. 'Critical layer instability in baroclinic flows', *Q.J.R. Meteorol. Soc.* **92** 325–334.

Browning, K.A. and Reynolds, R. 1994. 'Diagnostic study of a narrow cold-frontal rainband and severe winds associated with a stratospheric intrusion', *Q.J.R. Meteorol. Soc.* **120** 235–257.

Brugge, R, and Moncrieff, M.W. 1985. 'The effect of physical processes on numerical simulation of 2D cellular convection', *Contrib. Atmos. Phys.* **58** 417–440.

Bubnová, R., Hello, G., Bénard, P. and Geleyn, J.-F. 1995. 'Integration of the fully elastic equations cast in the hydrostatic pressure terrain-following coordinate in the framework of the ARPEGE/Aladin NWP system', *Mon. Weather Rev.* **123** 515–535.

Buizza, R. and Palmer, T.N. 1995. 'The singular vector structure of the atmospheric general circulation', *J. Atmos. Sci.* **52** 1434–1456.

Buizza, R., Gelaro, R., Molteni, F. and Palmer, T.N. 1997. 'The impact of increased resolution on predictability studies with singular vectors', *Q.J.R. Meteorol. Soc.* **123** 1007–1033.

Burger, A.P. 1958. 'Scale considerations of planetary motions of the atmosphere', *Tellus* **10** 195–205.

Carlson, T.N. 1991. *Mid-Latitude Weather Systems.* Harper Collins Academic.

Carroll, E.B. 1995. 'Practical subjective application of the omega equation and Sutcliffe development theory', *Meteorol. Appl.* **2** 71–81.

Charney, J.G. 1948. 'On the scale of atmospheric motions', *Geofys. Publ.* **17** 1–17.

Charney, J.G. 1962. 'Integration of the primitive and balance equations'. In *Proceedings of the International Symposium on Numerical Weather Prediction*, Tokyo, November 7–13, 1960 (Meteorological Society of Japan), 131–152.

Charney, J.G. 1971. 'Geostrophic turbulence', *J. Atmos. Sci.* **28** 1087–95.

Charney, J.G. and Stern, M.E. 1962. 'On the stability of internal baroclinic jets in a rotating atmosphere', *J. Met.* **19** 159–172

Clough, S.A., Davitt, C.S.A. and Thorpe, A.J. 1996. 'Attribution concepts applied to the omega equation', *Q.J.R. Meteorol. Soc.* **122** 1943–1962.

Côté, J. 1988. 'A Lagrange multiplier approach for the metric terms of semi-Lagrangian models on the sphere', *Q.J.R. Meteorol. Soc.* **114** 1347–1352.

Côté, J. and Staniforth, A. 1988. 'A two-time-level semi-Lagrangian semi-implicit scheme for spectral models', *Mon. Weather Rev.* **116** 2003–2012.

Craig, G.C. 1993a. 'A scaling for the three-dimensional semigeostrophic approximation', *J. Atmos. Sci.* **50** 3350–3355.

Craig, G.C. 1993b. 'A three-dimensional generalization of Eliassen's balanced vortex equations derived from Hamilton's principle', *Q.J.R. Meteorol. Soc.* **117** 435–448.

Cullen, M.J.P. 1993. 'The unified forecast/climate model', *Meteorol. Mag.* **122** 81–94.

Cullen, M.J.P. 2002. 'New mathematical developments in atmosphere and ocean dynamics, and their application to computer simulations'. This volume.

Daley, R. 1991. *Atmospheric Data Analysis.* Cambridge University Press.

Davis, C.A. and Emanuel, K.E. 1991. 'Potential vorticity diagnostics of cyclogenesis', *Mon. Weather Rev.* **119** 1929–1953.

Demirtas, M. and Thorpe, A.J. 1999. 'Sensitivity of short-range weather forecasts to local potential vorticity modifications', *Mon. Weather Rev.* **126** 922–939.

Dong, B. and James, I.N. 1997. 'The effect of barotropic shear on baroclinic instability', *Dyn. Atm. Oceans* **25** 143–190.

Draghici, I. 1989. 'The hypothesis of a marginally shallow atmosphere', *Meteorol. Hydrol.* **19** 13–27.

Drazin, P.G. and Reid, W.H. 1981. *Hydrodynamic Stability.* Cambridge University Press.

Durran, D.R. 1988. 'On a physical mechanism for Rossby wave propagation', *J. Atmos. Sci.* **45** 4020–4022.

Dutton, J.A. 1974. 'The nonlinear quasi-geostrophic equation: existence and uniqueness of solutions in a bounded domain', *J. Atmos. Sci.* **31** 422–433.

Dutton, J.A. 1995. *Dynamics of Atmospheric Motion.* Dover.

Eady, E.T. 1949. 'Long waves and cyclone waves', *Tellus* **1** 33–52.

Eckart, C. 1960. *Hydrodynamics of Oceans and Atmospheres.* Pergamon.

Eliassen, A. 1949. 'The quasi-static equations of motion with pressure as independent variable', *Geofys. Publ.* **17** 44pp.

Eliassen, A. 1984. 'Geostrophy', *Q.J.R. Meteorol. Soc.* **110** 1–12.

Eliassen, A. 1987. 'Entropy coordinates in atmospheric dynamics', *Zeitschrift für Meteorologie* **37** 1–11.

Emanuel, K.A. 1994. *Atmospheric Convection.* Oxford University Press.

Eriksson, K., Estep, D., Hansbo, P. and Johnson, C. 1996. *Computational Differential Equations.* Cambridge University Press.

Errico, R.M. 1997. 'What is an adjoint model?', *Bull. Amer. Met. Soc.* **78** 2577–2591.

Ertel 1942. 'Ein Neuer Hydrodynamischer Wirbelsatz', *Met. Z.* **59** 271–281.

Farrell, B.F. 1989. 'Optimal excitation of baroclinic waves', *J. Atmos. Sci.* **46** 1193–1206.

Farrell, B.F. 1990. 'Small error dynamics and the predictability of atmospheric flows', *J. Atmos. Sci.* **47** 2409–2416.

Findlater, J. 1969. 'Interhemispheric transport of air in the lower troposphere over the western Indian Ocean', *Q.J.R. Meteorol. Soc.* **95** 400–403.

Fjørtoft, R. 1962. 'On the integration of a system of geostrophically-balanced prognostic equations'. In *Proceedings of the International Symposium on Numerical Weather Prediction,* Tokyo, November 7–13, 1960 (Meteorological Society of Japan), 153–159.

Geleyn, J.-F. and Bubnová, R. 1997. 'The fully-elastic equations cast in hydrostatic pressure coordinates: accuracy and stability aspects of the scheme as implemented in ARPEGE/ALADIN'. In *Numerical Methods in Atmospheric and Oceanic Modelling,* C.A. Lin, R. Laprise, H. Ritchie (eds.), NRC Research Press.

Gent, P.R. and McWilliams, J.C. 1982. 'Intermediate model solutions to the Lorenz equations: strange attractors and other phenomena', *J. Atmos. Sci.* **39** 3–13.

Gent, P.R., and McWilliams, J.C. 1983. 'The equatorial waves of balanced models', *J. Phys. Oceanog.* **13** 1179–1192.

Gent, P.R., and McWilliams, J.C. 1984. 'Balanced models in isentropic coordinates and the shallow water equations', *Tellus* **36A** 166–171.

Gill, A.E. 1977. 'Coastally trapped waves in the atmosphere', *Q.J.R. Meteorol. Soc.* **103** 431–440.

Gill, A.E. 1982. *Atmosphere-Ocean Dynamics.* Academic Press.

Graef, F. 1998. 'On the westward translation of isolated eddies', *J. Phys. Oceanog.* **28** 740–745.

Green, J.S.A. 1960. 'A problem in baroclinic stability', *Q.J.R. Meteorol. Soc.* **86** 237–251.

Green, J.S.A. 1970. 'Transfer properties of the large-scale eddies and the general circulation of the atmosphere', *Q.J.R. Meteorol. Soc.* **96** 157–185.

Green, J.S.A. 1999. *Atmospheric Dynamics.* Cambridge University Press.

Griffiths, M., Thorpe, A.J. and Browning, K.A. 2000. 'Convective destabilization by a tropopause fold diagnosed using potential vorticity inversion', *Q.J.R. Meteorol. Soc.* **126** 125–144.

Grimshaw, R.H.J. 1975. 'A note on the β-plane approximation', *Tellus* **27** 351–357.

Hakim, G.J., Keyser, D. and Bosart, L.F. 1996. 'The Ohio Valley wave-merger cyclogenesis event of 25–26 January 1978. Part II: Diagnosis using quasigeostrophic potential vorticity inversion', *Mon. Weather Rev.* **124** 2176–2205.

Haltiner, G.J. 1971. *Numerical Weather Prediction.* Wiley.

Haltiner, G.J. and Williams, R.T. 1981. *Numerical Prediction and Dynamic Meteorology.* Wiley.

Harnik, N. and Lindzen, R.S. 1998. 'The effect of basic state PV gradients on the growth rate of baroclinic waves and the height of the tropopause', *J. Atmos. Sci.* **55** 344–360.

Hart, J.E. 1979. 'Finite amplitude baroclinic instability', *Ann. Rev. Fluid Mech.* **11** 147–172.

Haynes, P.H. and McIntyre, M.E. 1987. 'On the evolution of vorticity and potential vorticity in the presence of diabatic heating and frictional and other forces', *J. Atmos. Sci.* **44** 828–841.

Haynes, P.H. and Shepherd, T.G. 1989. 'The importance of surface pressure changes in the response of the atmosphere to zonally-symmetric thermal and mechanical forcing', *Q.J.R. Meteorol. Soc.* **115** 1181–1208.

Held, I.M. 1983. 'Stationary and quasi-stationary eddies in the extratropical troposphere'. In *Large scale Dynamical Processes in the Atmosphere*, B. Hoskins and R. Pearce (eds.), 127–168.

Held, I.M. 1985. 'Pseudomomentum and the orthogonality of modes in shear flow', *J. Atmos. Sci.* **42** 2280–2288.

Held, I.M. and Hoskins, B.J. 1985. 'Large-scale eddies and the general circulation of the troposphere', *Adv. Geophys.* **28A** 3–31.

Hewson, T.D. 1998. 'Objective fronts', *Meteorol. Appl.* **5** 37–65.

Hide, R. 1971. 'On geostrophic motion of a non-homogeneous fluid', *J. Fluid Mech.* **49** 745–751.

Hide, R. 1976. 'Motions in planetary atmospheres', *Q.J.R. Meteorol. Soc.* **102** 1–23.

Hide, R. 1977. 'Experiments with rotating fluids', *Q.J.R. Meteorol. Soc.* **103** 1–28.

Hide, R. and Mason, P.J. 1975. 'Sloping convection in a rotating fluid', *Adv. Geophys.* **24** 47–100.

Hide, R., Dickey, J.O., Marcus, S.L., Rosen, R.D. and Salstein, D.A. 1997. 'Atmospheric angular momentum fluctuations during 1979–1988 simulated by global circulation models', *J. Geophys. Res.* **102** 16423–16438.

Holm, D.D. 1996. 'Hamiltonian balance equations', *Physica D* **98** 379–414.

Holm, D.D. and Zeitlin, V. 1998. 'Hamilton's principle for quasigeostrophic motion', *Phys. Fluids* **10** 800–806.

Holton, J.R. 1975. *The Dynamic Meteorology of the Stratosphere and Mesosphere.* American Meteorological Society.

Holton, J.R. 1992. *An Introduction to Dynamic Meteorology.* Academic Press.

Hortal, M. and Simmons, A.J. 1991. 'Use of reduced Gaussian grids in spectral models', *Mon. Weather Rev.* **119** 1057–1074.

Hoskins, B.J. 1973. 'Stability of the Rossby–Haurwitz wave', *Q.J.R. Meteorol. Soc.* **99** 723–745.

Hoskins, B.J. 1975. 'The geostrophic momentum approximation and the semi-geostrophic equations', *J. Atmos. Sci.* **32** 233–242.

Hoskins, B.J. 1982. 'The mathematical theory of frontogenesis', *Ann. Rev. Fluid Mech.* **14** 131–151.

Hoskins, B.J. 1991. 'Towards a PV-θ view of the general circulation', *Tellus* **45AB** 27–35.

Hoskins, B.J. and Bretherton, F.P. 1972. 'Atmospheric frontogenesis models: mathematical formulation and solutions', *J. Atmos. Sci.* **29** 11–37.

Hoskins, B.J., Draghici, I. and Davies, H.C. 1978. 'A new look at the ω-equation', *Q.J.R. Meteorol. Soc.* **104** 31–38.

Hoskins, B.J., James, I.N. and White, G.H. 1983. 'The shape, propagation and mean-flow interaction of large-scale weather systems', *J. Atmos. Sci.* **40** 1595–1612.

Hoskins, B.J., McIntyre, M.E. and Robertson, A.W. 1985. 'On the use and significance of isentropic potential vorticity maps', *Q.J.R. Meteorol. Soc.* **111** 877–946.

Hoskins, B.J. and Simmons, A.J. 1975. 'A multi-layer spectral model and the semi-implicit method', *Q.J.R. Meteorol. Soc.* **101** 637–655.

Hsu, Y-J. G, and Arakawa, A. 1990. 'Numerical modeling of the atmosphere with an isentropic vertical coordinate', *Mon. Weather Rev.* **118** 1933–1959.

James, I.N. 1994. *Introduction to Circulating Atmospheres.* Cambridge University Press.

Kasahara, A. 1974. 'Various vertical coordinate systems used for numerical weather prediction', *Mon. Weather Rev.* **102** 509–522.

Kasahara, A. and Washington, W.E. 1967. 'NCAR global general circulation model of the atmosphere', *Mon. Weather Rev.* **95** 389–402.

Keyser, D. and Shapiro, M.A. 1986. 'A review of the structure and dynamics of upper-level frontal zones', *Mon. Weather Rev.* **114** 452–499.

Klein, F. 1938. *Elementary Mathematics from an Advanced Standpoint, Vol 2: Geometry.* English translation by E.R. Hedrick and C.A. Noble. Dover Publications.

Kucharski, F. 1997. 'On the concept of exergy and available potential energy', *Q.J.R. Meteorol. Soc.* **123** 2141–2156.

Kuo, H.-L. 1959. 'Finite amplitude three-dimensional harmonic waves on the spherical earth', *J. Met.* **16** 524–534.

Kuo, H.-L. 1972. 'On a generalized potential vorticity equation for quasi-geostrophic flow', *Pageoph.* **96** 171–175.

Lait, L. R. 1995. 'An alternative form for potential vorticity', *J. Atmos. Sci.* **51** 1754–1759.

Lamb, H. 1932. *Hydrodynamics.* Cambridge University Press.

Lander, J. and Hoskins, B.J. 1997. 'Believable scales and parameterizations in a spectral transform model', *Mon. Weather Rev.* **125** 292–303.

Laprise, R. 1992. 'The Euler equations of motion with hydrostatic pressure as an independent variable', *Mon. Weather Rev.* **120** 197–207.

Lie, I. 1999. 'Some aspects of non-hydrostatic models in the Hirlam perspective', Hirlam Tech. Rep. No. 41.

Lighthill, M.J. 1978. *Waves in Fluids.* Cambridge University Press.

Lindzen, R.S. 1968. 'Rossby waves with negative equivalent depth – comments on a note by G.A. Corby', *Q.J.R. Meteorol. Soc.* **94** 402–407.

Lindzen, R.S. 1990. *Dynamics in Atmospheric Physics.* Cambridge University Press.

Longuet-Higgins, M.S. 1968. 'The eigenfunctions of Laplace's tidal equations over a sphere', *Phil. Trans. Roy. Soc. London* **A262** 511–607.

Lorenz, E.N. 1960. 'Energy and numerical weather prediction', *Tellus* **12** 364–373.

Lorenz, E.N. 1967. 'The nature and theory of the general circulation of the atmosphere', WMO No. 218; TP 115.

Lynch, P. 1989. 'The slow equations', *Q.J.R. Meteorol. Soc.* **115** 201–219.

McGregor, J.L. 1996. 'Semi-Lagrangian advection on conformal-cubic grids', *Mon. Weather Rev.* **124** 1311–1322.

McIntyre M.E. and Norton, W.A. 2000. 'Potential vorticity inversion on a hemisphere', *J. Atmos. Sci.*, **57** 1214–1235.

McIntyre, M.E. and Roulstone, I. 2002. 'Are there higher-accuracy analogues of semi-geostrophic theory?' In *Large-Scale Atmosphere–Ocean Dynamics, II*, J. Norbury and I. Roulstone (eds.), Cambridge University Press, 301–364.

McWilliams, J.C. 1984. 'The emergence of isolated coherent vortices in turbulent flow', *J. Fluid Mech.* **146** 21–43.

McWilliams, J.C. and Gent, P.R. 1980. 'Intermediate models of planetary circulations in the atmosphere and ocean', *J. Atmos. Sci.* **37** 1657–1678.

Magnusdottir, G. and Schubert, W. 1991. 'Semi-geostrophic theory on the hemisphere', *J. Atmos. Sci.* **48** 1449–1456.

Majewski, D. 1998. 'Numerical weather prediction at the Deutscher Wetterdienst – from the third to the fourth generation', *Annalen der Meteorologie* **36** 39–63.

Marquet, P. 1993. 'Exergy in meteorology: definition and properties of moist available enthalpy', *Q.J.R. Meteorol. Soc.* **119** 567–590.

Marshall, J.C. 1984. 'Eddy-mean-flow interaction in a barotropic ocean model', *Q.J.R. Meteorol. Soc.* **110** 573–590.

Marshall, J.C. and Molteni, F. 1993. 'Toward a dynamical understanding of planetary-scale flow regimes', *J. Atmos. Sci.* **50** 1792–1818.

Marshall, J.C., Hill, C., Perelman, L. and Adcroft, A. 1997. 'Hydrostatic, quasi-hydrostatic and non-hydrostatic ocean modeling', *J. Geophys. Res.* **102** 5733–5752.

Martin, J.E. 1999. 'The separate roles of geostrophic vorticity and deformation in the midlatitude occlusion process', *Mon. Weather Rev.* **127** 2402–2418.

Mason, P.J. and Brown, A.R. 1999. 'On subgrid models and filter operations in large eddy simulations', *J. Atmos. Sci.* **56** 2101–2114.

Meteorological Applications 1997. 'Special issue on the 1996 Met. Office/Reading University summer study week on extratropical cyclones', *Meteorol. Appl.* **4** 291–382.

Miller, M.J. 1974. 'On the use of pressure as vertical coordinate in modelling convection', *Q.J.R. Meteorol. Soc.* **100** 155–162.

Miller, M.J. 1978. 'The Hampstead storm: a numerical simulation of a quasi-stationary cumulonimbus system', *Q.J.R. Meteorol. Soc.* **104** 413–427.

Miller, M.J. and Pearce, R.P. 1974. 'A three-dimensional primitive equation model of cumulonimbus convection', *Q.J.R. Meteorol. Soc.* **100** 133–154.

Miller, M.J. and White, A.A. 1984. 'On the non-hydrostatic equations in pressure and sigma coordinates', *Q.J.R. Meteorol. Soc.* **110** 515–533.

Miranda, P.M.A., and Valente, M.A. 1997. 'Critical-level resonance in three-dimensional flow past isolated mountains', *J. Atmos. Sci.* **54** 1574–1588.

Mobbs, S.D. 1982. 'Variational principles for perfect and dissipative fluid flows', *Proc. Roy. Soc. Lond.* **A381** 457–468.

Mole, N. and James, I.N. 1990. 'Baroclinic adjustment in a zonally varying flow', *Q.J.R. Meteorol. Soc.* **110** 247–268.

Moura, A.D. 1976. 'The eigensolutions of the linearized balance equations over sphere', *J. Atmos. Sci.* **33** 807–907.

Müller, R. 1989. 'A note on the relation between the "traditional approximation" and the metric of the primitive equations', *Tellus* **41A** 175–178.

Muraki, D.J., Snyder, C. and Rotunno, R. 1999. 'The next-order corrections to quasi-geostrophic theory', *J. Atmos. Sci.* **56** 1547–1560.

Nakamura, N. 1999. 'Baroclinic-barotropic adjustments in a meridionally wide domain', *J. Atmos. Sci.* **56** 2246–2260.

Nielsen-Gammon, J.W. and Lefevre, R.J. 1996. 'Piecewise tendency diagnosis of dynamical processes governing the development of an upper-tropospheric mobile trough', *J. Atmos. Sci.* **53** 3120–3142.

NOAA/NASA/USAF 1976. *US Standard Atmosphere*, 1976. Washington, D.C.

Ogura, Y. and Phillips, N.A. 1962. 'Scale analysis of deep and shallow convection in the atmosphere', *J. Met.* **19** 173–179.

Ottino, J.M. 1990. *The Kinematics of Mixing: Stretching, Chaos, and Transport.* Cambridge University Press.

Paldor, N. and Killworth, P.D. 1988. 'Inertial trajectories on a rotating Earth', *J. Atmos. Sci.* **45** 4013–4019.

Pedlosky, J. 1964. 'The stability of currents in the atmosphere and the ocean: Part I', *J. Atmos. Sci.* **53** 201–219.

Pedlosky, J. 1987. *Geophysical Fluid Dynamics.* Springer-Verlag.

Persson, A. 1998. 'How do we understand the Coriolis force?', *Bull. Amer. Meteorol. Soc.* **79** 1373–1385.

Phillips, N.A. 1957. 'A coordinate system having some special advantages for numerical forecasting', *J. Meteor.* **14** 184–185.

Phillips, N.A. 1963. 'Geostrophic motion', *Reviews of Geophysics* **1** 123–176.

Phillips, N.A. 1973. 'Principles of large scale numerical weather prediction'. In *Dynamic Meteorology*, P. Morel (ed.), Reidel, 3–96.

Platzman, G. 1968. 'The Rossby wave', *Q.J.R. Meteorol. Soc.* **94** 225–48.

Purser, R.J. 1993. 'Contact transformations and Hamiltonian dynamics in generalized semigeostrophic theories', *J. Atmos. Sci.* **50** 1449–1468.

Purser, R.J. 1999. 'Legendre-transformable semigeostrophic theories', *J. Atmos. Sci.* **56** 2522–2535.

Räisänen, J. 1997. 'Height tendency diagnostics using a generalized omega equation, the vorticity equation, and a nonlinear balance equation', *Mon. Weather Rev.* **125** 1577–1597.

Rančič, M., Purser, R.J. and Mesinger, F. 1996. 'A global shallow-water model using an expanded spherical cube: gnomonic versus conformal coordinates', *Q.J.R. Meteorol. Soc.* **122** 959–982.

Raymond, D.J. 1992. 'Nonlinear balance and potential-vorticity thinking at large Rossby number', *Q.J.R. Meteorol. Soc.* **118** 987–1015.

Rhines, P.B. 1975. 'Waves and turbulence on a beta-plane', *J. Fluid Mech.* **69** 417–443.

Richardson, L.F. 1922. *Weather Prediction by Numerical Process.* Cambridge University Press. (Reprinted by Dover Publications, 1965).

Ripa, P. 1981. 'Symmetries and conservation laws for internal gravity waves', *Am. Inst. Phys. Conf. Proc.* **76** 281–306.

Ripa, P. 1997. ' "Inertial" oscillations and the β-plane approximation(s)', *J. Phys. Oceanog.* **27** 633–647.

Ritchie, H. 1988. 'Application of the semi-Lagrangian method to a spectral model of the shallow water equations', *Mon. Weather Rev.* **116** 1587–1598.

Rõõm, R. 1996. 'Free and rigid boundary quasigeostrophic models in pressure coordinates', *J. Atmos. Sci.* **53** 1496–1501.

Rõõm, R. 1998. 'Acoustic filtering in non-hydrostatic pressure coordinate dynamics: a variational approach', *J. Atmos. Sci.* **55** 654–668.

Rõõm, R. and Männik, A. 1999. 'Responses of different non-hydrostatic, pressure-coordinate models to orographic forcing', *J. Atmos. Sci.* **55** 2553–2570.

Rossby, C.-G. 1939. 'Relation between variations in the intensity of the zonal circulation of the atmosphere and the displacements of the semi-permanent centers of action', *J. Marine Res.* **2** 38–55.

Roulstone, I. and Brice, S. 1995. 'On the Hamiltonian formulation of the quasi-hydrostatic equations', *Q.J.R. Meteorol. Soc.* **121** 927–936.

Roulstone, I. and Sewell, M.J. 1997. 'The mathematical structure of theories of semi-geostrophic type', *Phil. Trans. R. Soc., Lond.* **A355** 2489–2517.

Sadourny, R., Arakawa, A., and Mintz, Y. 1968. 'Integration of the nondivergent barotropic vorticity equation with an icosahedral-hexagonal grid for the sphere', *Mon. Weather Rev.* **96** 351–356.

Salmon, R. 1982. 'Hamilton's principle and Ertel's theorem', *Am. Inst. Phys. Proc.* **88** 127–135.

Salmon, R. 1983. 'Practical use of Hamilton's principle', *J. Fluid Mech.* **132** 431–444.

Salmon, R. 1985. 'New equations for nearly geostrophic flow', *J. Fluid Mech.* **153** 461–477.

Salmon, R. 1988. 'Hamiltonian fluid dynamics', *Ann. Rev. Fluid Mech.* **20** 225–256.

Salmon, R. and Smith, L.M. 1994. 'Hamiltonian derivation of the non-hydrostatic pressure co-ordinate model', *Q.J.R. Meteorol. Soc.* **120** 1409–1413.

Sanders, F., and Hoskins, B.J. 1990. 'An easy method for estimating Q-vectors from weather maps', *Weather and Forecasting* **5** 346–353.

Saucier, W.J. 1955. *Principles of Meteorological Analysis.* University of Chicago Press.

Schneider, E. 1987. 'An inconsistency in the vertical discretization in some atmospheric models', *Mon. Weather Rev.* **115** 2166–2169.

Sewell, M.J. 1990. *Maximum and Minimum Principles.* Cambridge University Press.

Shepherd, T.G. 1990. 'Symmetries, conservation laws and Hamiltonian structure in geophysical fluid dynamics', *Adv. Geophys.* **32** 287–338.

Shepherd, T.G. 1993. 'A unified theory of available potential energy', *Atmosphere-Ocean* **31** 1–26.

Shutts, G.J. 1980. 'Angular momentum coordinates and their use in zonal geostrophic motion in a hemisphere', *J. Atmos. Sci.* **37** 1126–1132.

Shutts, G.J. 1983a 'The propagation of eddies in diffluent jetstreams: eddy vorticity forcing of "blocking" flow fields', *Q.J.R. Meteorol. Soc.* **109** 737–761.

Shutts, G.J. 1983b Parameterization of travelling weather systems in a simple model of large-scale atmospheric flow', *Adv. Geophys.* **25** 117–172.

Shutts, G.J. 1989. 'Planetary semi-geostrophic equations derived from Hamilton's principle', *J. Fluid Mech.* **208** 545–573.

Simmons, A.J. and Burridge, D.M. 1981. 'An energy and angular-momentum conserving vertical finite-difference scheme and hybrid vertical coordinates', *Mon. Weather Rev.* **109** 758–766.

Simmons, A.J. and Strüfing, R.. 1983. 'Numerical forecasts of stratospheric warming events using a model with a hybrid vertical coordinate', *Q.J.R. Meteorol. Soc.* **119** 81–111.

Simons, T.J. 1972. 'The nonlinear dynamics of cyclone waves', *J. Atmos. Sci.* **29** 38–52.

Staniforth, A. 2001. 'Developing efficient unified non-hydrostatic models'. *Proceedings of the Commemorative Symposium, 50th Anniversary of Numerical Weather Prediction*, Potsdam, 9–10 March, 2000, Spekat, A. (ed), 185–200.

Staniforth, A. and Côté, J. 1991. 'Semi-Lagrangian integration schemes for atmospheric models – a review', *Mon. Weather Rev.* **119** 2206–2223.

Starr, V.P. 1945. 'A quasi-Lagrangian system of hydrodynamical equations', *J. Met.* **2** 227–237.

Stommel, H.M. and Moore, D.W. 1989. *An Introduction to the Coriolis Force.* Columbia University Press.

Stone, P.H. and Nemet, B. 1996. 'Baroclinic adjustment: a comparison between theory, observation and models', *J. Atmos. Sci.* **53** 1663–1674.

Sutcliffe, R.C. 1938. 'On development in the field of barometric pressure', *Q.J.R. Meteorol. Soc.* **64** 495–509.

Sutcliffe, R.C. 1947. 'A contribution to the problem of development', *Q.J.R. Meteorol. Soc.* **73** 370–383.

Talagrand, O. and Courtier, P. 1987. 'Variational assimilation of meteorological observations with the adjoint vorticity equation. Part 1: Theory', *Q.J.R. Meteorol. Soc.* **113** 1311–1328.

Tanguay, M., Robert, A. and Laprise, R. 1990. 'A semi implicit semi-Lagrangian fully compressible regional forecast model', *Mon. Weather Rev.* **118** 1970–1980.

Tapp, M.C. and White, P.W. 1976. 'A non-hydrostatic mesoscale model', *Q.J.R. Meteorol. Soc.* **102** 277–296.

Temperton, C., Hortal, M. and Simmons, A.J. 2000. 'A two-time-level semi-Lagrangian global spectral model', *Q.J.R. Meteorol. Soc.* **127** 111–127.

Thompson, P.D. 1956. 'A theory of large-scale disturbances in non-geostrophic flow', *J. Met.* **13** 251–261.

Thuburn, J. 1993. 'Baroclinic-wave life cycles, climate simulations and cross-isentrope mass flow in a hybrid isentropic coordinate GCM', *Q.J.R. Meteorol. Soc.* **119** 489–508.

Thuburn, J. 1997. 'A PV-based shallow-water model on a hexagonal-icosahedral grid', *Mon. Weather Rev.* **125** 2328–2347.

Thuburn, J. and Craig, G.C. 2000. 'Stratospheric influence on tropopause height: the radiative constraint', *J. Atmos. Sci.* **57** 17–28.

Vallis, G.K. 1996. 'Potential vorticity inversion and balanced equations of motion for rotating and stratified flows', *Q.J.R. Meteorol. Soc.* **122** 291–322.

Verkley, W.T.M. 1993. 'A numerical method for finding form-preserving free solutions of the barotropic vorticity equation on a sphere', *J. Atmos. Sci.* **50** 1488–1503.

Viúdez, A. 1999. 'On Ertel's potential vorticity. On the impermeability theorem for potential vorticity', *J. Atmos. Sci.* **56** 507–516.

Wang, P.K. 1984. 'A brief review of the Eulerian variational principle for atmospheric motions in rotating coordinates', *Atmosphere-Ocean* **22** 387–392.

Warn, T., Bokhove, O., Shepherd, T.G. and Vallis, G.K. 1995. 'Rossby number expansions, slaving principles, and balance dynamics', *Q.J.R. Meteorol. Soc.* **121** 723–739.

White, A.A. 1977. 'Modified quasi-geostrophic equations using geometric height as vertical coordinate', *Q.J.R. Meteorol. Soc.* **103** 383–396.

White, A.A. 1978a. 'Atmospheric energetics (1 and 2), *Weather* **33** 408–416; 446–457.

White, A.A. 1978b. 'A note on the horizontal boundary condition in quasi-geostrophic models', *J. Atmos. Sci.* **35** 735–740.

White, A.A. 1982. 'Zonal translation properties of two quasi-geostrophic systems of equations', *J. Atmos. Sci.* **39** 2107–2118.

White, A.A. 1989a. 'A relationship between energy and angular momentum conservation in dynamical models', *J. Atmos. Sci.* **46** 1855–1860.

White, A.A. 1989b. 'An extended version of a non-hydrostatic pressure co-ordinate model', *Q.J.R. Meteorol. Soc.* **115** 1243–1251.

White, A.A. 1990. 'Steady states in a turbulent atmosphere', *Meteorol. Mag.* **119** 1–9.

White, A.A. and Bromley, R.A. 1995. 'Dynamically consistent quasi-hydrostatic equations for global models with a complete representation of the Coriolis force', *Q.J.R. Meteorol. Soc.* **121** 399–418.

Wiin-Nielsen, A. 1968. 'On the intensity of the general circulation of the atmosphere', *Reviews of Geophysics* **6** 559–579.

Wiin-Nielsen, A. 1973. *Compendium of Meteorology, Vol I.*, WMO No.364.

Williams, G.. 1969. 'Numerical integration of the three-dimensional Navier–Stokes equations for incompressible flow', *J. Fluid Mech.* **37** 727–750.

Williams, G. P. 1972 'Friction term formulation and convective instability in a shallow atmosphere', *J. Atmos. Sci.* **29** 870–876.

Wu, G.-X. and White, A.A. 1986. 'A further study of the surface flow predicted by an eddy flux parametrization scheme', *Q.J.R. Met. Soc.* **112** 1041–1056.

Xu, Q. 1992. 'Ageostrophic pseudovorticity and geostrophic C-vector forcing – a new look at the Q-vector in three dimensions', *J. Atmos. Sci.* **49** 981–990.

Xue, M., and Thorpe, A.J. 1991. 'A mesoscale numerical model using the non-hydrostatic, pressure-based sigma-coordinate equations: model experiments with dry mountain flows', *Mon. Weather Rev.* **119** 1168–1185.

Zhu, Z., Thuburn, J., Hoskins, B.J. and Haynes, P.H. 1992. 'A vertical finite-difference scheme based on a hybrid σ-θ-p coordinate', *Mon. Weather Rev.* **120** 851–862.

Extended-Geostrophic Euler–Poincaré Models for Mesoscale Oceanographic Flow

J.S. Allen, Darryl D. Holm and P.A. Newberger

1 Introduction

We continue the study of intermediate models (McWilliams and Gent, 1980) for possible application to mesoscale oceanographic flow fields. Intermediate models are derived under the assumption that the Rossby number ϵ is small and filter out high-frequency gravity-inertial waves. Previous work has involved intermediate models for flows of homogeneous fluids governed by the f-plane shallow water equations (SWE) (Allen *et al.*, 1990a,b; Barth *et al.*, 1990; Allen and Holm, 1996) and for flows of continuously stratified fluids governed by the hydrostatic primitive equations (PE) (Allen, 1991; Allen, 1993; Allen and Newberger, 1993; Holm, 1996).

We use a traditional modelling approach of making approximations in Hamilton's principle. This approach was developed for geophysical fluid dynamics (GFD) and applied by Salmon (1983, 1985, 1996) to construct approximate balanced equations by substituting leading order balance relations and asymptotic expansions into Hamilton's principle before taking variations (see also Allen and Holm, 1996, and Holm, 1996). In the present paper, we use this approach to derive approximate intermediate models for mesoscale oceanographic flow. For this, we work in the framework of the Euler–Poincaré theorem for ideal continua with advected parameters (Holm, Marsden and Ratiu, 1998a). Euler–Poincaré systems are the Lagrangian analogue of Lie–Poisson Hamiltonian systems (Holm, Marsden, Ratiu, and Weinstein, 1985, and references therein). In this framework, the resulting Eulerian approximate GFD equations possess a Kelvin–Noether circulation theorem, conserve potential vorticity on fluid particles and conserve volume integrated energy. In addition, following the derivations we assess the accuracy of the model equations through numerical experiments.

Motivation for this study is provided by the seemingly great potential usefulness of approximate models derived from Hamilton's principle and the apparent soundness of this approach. On the other hand, results of numerical experiments assessing the accuracy of different intermediate models applied to the SWE (Allen *et al*, 1990a,b; Barth *et al.*, 1990) demonstrate clearly that, at moderate values of ϵ, Salmon's (1983) HP model and the geostrophic momentum (GM) approximation (Hoskins, 1975) provide disappointingly inaccurate solutions to the SWE, compared e.g., to those obtained from the balance equations (BE) (Gent and McWilliams, 1983). This is in spite of the fact that the

HP and GM models have Hamiltonian structure, whereas the BE for the SWE do not conserve energy. Thus, possession of Hamiltonian structure is not sufficient in itself to ensure an accurate approximate model. The question arises of how more accurate approximate models can be derived from Hamilton's principle. Allen and Holm (1996) formulated an approximate model for the SWE by extending Salmon's (1983) approach and utilizing higher order approximations in Hamilton's principle. Allen and Holm (1996) contended that this extended-geostrophic model should give more accurate solutions than the HP or GM models.

Here we apply the expansion procedure of Allen and Holm (1996) to derive extended-geostrophic models for continuously stratified flows governed by the PE. The initial step involves derivation of an approximate model following a strategy similar to that of Salmon (1983, 1985, 1996), but utilizing the methods of Holm, Marsden, and Ratiu (1998a) for deriving the Euler–Poincaré equations for fluids. We refer to the resulting approximate equations as the L1 model. A second step involves including consistent $O(\epsilon)$ higher order approximations. We refer to the resulting approximate equations in that case as the L2 model.

The derivation is followed by numerical experiments to assess the accuracy of L1 and L2 models compared to the PE. The idealized, mesoscale oceanographic problems utilized in Allen and Newberger (1993) to quantify the accuracy of different intermediate models are repeated. Thus, we find information not only on the absolute accuracy of the L1 and L2 models compared to PE solutions, but also on the relative accuracy compared to other intermediate models.

The outline of this paper is as follows. The Euler–Poincaré equations for fluids are summarized briefly in section 2 and the derivations of the L1 and L2 models equations are given in sections 3 and 4, respectively. The solution procedure for the L1 and L2 models is presented in section 5 and the numerical experiments are discussed in section 6. The details of the numerical methods are explained in the appendix. Brief summary comments are given in section 7.

2 Applications of the Euler–Poincaré Theorem in GFD

Here we recall from Holm *et al.* (1998a) the statements of the Euler–Poincaré equations and their associated Kelvin–Noether theorem in the context of continuum mechanics and approximate models in geophysical fluid dynamics.

The Euler–Poincaré equations for a GFD Lagrangian $L[\mathbf{u}, D, b]$ involve fluid velocity $\mathbf{u}$, specific entropy b, and density D as functions of three dimensional space with coordinates $\mathbf{x}$ and time t. In vector notation, these equations are expressed as (Holm *et al.*, 1998a,b; 2002; cf also Holm 1996),

$$\frac{d}{dt}\frac{1}{D}\frac{\delta L}{\delta \mathbf{u}} + \frac{1}{D}\frac{\delta L}{\delta u^j}\nabla u^j + \frac{1}{D}\frac{\delta L}{\delta b}\nabla b - \nabla\frac{\delta L}{\delta D} = 0, \tag{2.1}$$

or, equivalently, in 'curl form' as

$$\frac{\partial}{\partial t}\left(\frac{1}{D}\frac{\delta L}{\delta \mathbf{u}}\right) - \mathbf{u}\times \operatorname{curl}\left(\frac{1}{D}\frac{\delta L}{\delta \mathbf{u}}\right) + \nabla\left(\mathbf{u}\cdot\frac{1}{D}\frac{\delta L}{\delta \mathbf{u}} - \frac{\delta L}{\delta D}\right) + \frac{1}{D}\frac{\delta L}{\delta b}\nabla b = 0. \quad (2.2)$$

The Euler–Poincaré system is completed by including the auxiliary equations for advection of the specific entropy b,

$$\frac{\partial b}{\partial t} + \mathbf{u}\cdot\nabla b = 0, \quad (2.3)$$

and the continuity equation for the density D,

$$\frac{\partial D}{\partial t} + \nabla\cdot D\mathbf{u} = 0. \quad (2.4)$$

For incompressible flows, one sets $D = 1$ in the continuity equation, so that $\nabla\cdot\mathbf{u} = 0$. For anelastic flows, one sets $D = \rho_s(z)$ in the continuity equation with a prescribed stably stratified reference density profile $\rho_s(z)$, so that $\nabla\cdot(\rho_s(z)\mathbf{u}) = 0$.

The Euler–Poincaré motion equation in either form (2.1) or (2.2) results in the Kelvin–Noether circulation theorem,

$$\frac{d}{dt}\oint_{\gamma_t(\mathbf{u})}\frac{1}{D}\frac{\delta L}{\delta \mathbf{u}}\cdot d\mathbf{x} = -\oint_{\gamma_t(\mathbf{u})}\frac{1}{D}\frac{\delta L}{\delta b}\nabla b\cdot d\mathbf{x}, \quad (2.5)$$

where the curve $\gamma_t(\mathbf{u})$ moves with the fluid velocity $\mathbf{u}$. Then, by Stokes' theorem, the Euler–Poincaré equations generate circulation of the quantity $D^{-1}\delta L/\delta\mathbf{u}$ whenever the gradients ∇b and $\nabla(D^{-1}\delta L/\delta b)$ are not collinear.

Taking the curl of equation (2.2) and using advection of the specific entropy b and the continuity equation for the density D yields conservation of potential vorticity on fluid particles, as expressed by

$$\frac{\partial q}{\partial t} + \mathbf{u}\cdot\nabla q = 0, \quad \text{where} \quad q \equiv \frac{1}{D}\nabla b\cdot\operatorname{curl}\left(\frac{1}{D}\frac{\delta L}{\delta \mathbf{u}}\right). \quad (2.6)$$

Consequently, the following domain integrated quantities are conserved, for any function Φ,

$$C_\Phi = \int d^3x\, D\,\Phi(b,q)\,, \quad \text{for all } \Phi\,. \quad (2.7)$$

The absence of explicit time dependence in the Lagrangian $L[\mathbf{u},D,b]$ gives the conserved domain integrated energy, via Noether's theorem for time translation invariance. This energy is easily calculated using the Legendre transform to be

$$E[\mathbf{u},D,b] = \int d^3x\left(\mathbf{u}\cdot\frac{\delta L}{\delta \mathbf{u}}\right) - L[\mathbf{u},D,b]\,. \quad (2.8)$$

When the Legendre transform is completed to express $E[\mathbf{u},D,b]$ as $H[\mathbf{m},D,b]$ with $\mathbf{m}\equiv\delta L/\delta\mathbf{u}$ and $\delta H/\delta\mathbf{m} = \mathbf{u}$, the Euler–Poincaré system (2.1), (2.3) and (2.4) may be expressed in Hamiltonian form

$$\frac{\partial \mu}{\partial t} = \{\mu, H\}\,, \quad \text{with} \quad \mu\in[\mathbf{m},D,b]\,, \quad (2.9)$$

and Lie–Poisson bracket given in Euclidean component form by

$$
\begin{aligned}
\{F,H\}[\mathbf{m},D,b] = -\int d^3x \Big\{ \frac{\delta F}{\delta m_i}\Big[(\partial_j m_i + m_j \partial_i)\frac{\delta H}{\delta m_j} + (D\partial_i)\frac{\delta H}{\delta D} \\
- (b_{,i})\frac{\delta H}{\delta b}\Big] + \frac{\delta F}{\delta D}(\partial_j D)\frac{\delta H}{\delta m_j} + \frac{\delta F}{\delta b}(b_{,j})\frac{\delta H}{\delta m_j}\Big\}.
\end{aligned}
\tag{2.10}
$$

The conserved quantities C_Φ in (2.7) are then understood in the Lie–Poisson Hamiltonian formulation (2.9)–(2.10) of the Euler–Poincaré system (2.1)–(2.4) as Casimirs that commute under the Lie–Poisson bracket (2.10) with any functional of $[\mathbf{m}, D, b]$. The Casimirs also result via Noether's theorem from symmetry of Hamilton's principle for the Euler–Poincaré system under the 'particle relabelling transformations' that leave invariant the Lagrangian $L[\mathbf{u}, D, b]$. From the viewpoint of Noether's theorem, this particle relabelling symmetry corresponds to invariance of Hamilton's principle for the Euler–Poincaré equations under the transformation from the Lagrangian to the Eulerian fluid description, by pullback of the right action of the diffeomorphism group on the configuration space of the Lagrangian fluid parcel positions and their velocities. For full mathematical details, consult Marsden and Ratiu (1994), Holm *et al.* (1998a,b; 2002).

The four properties (2.5)–(2.8) and the Lie–Poisson Hamiltonian formulation (2.9)–(2.10) of the Euler–Poincaré equation (2.1) and its auxiliary equations (2.3) and (2.4) are desirable elements of approximate models for applications in geophysical fluid dynamics expressed in the variables $[\mathbf{u}, D, b]$. Thus, the Euler–Poincaré theory offers a unified framework in which to derive approximate GFD models that possess these properties: the Kelvin–Noether circulation theorem, conservation of potential vorticity on fluid particles, and the Lie–Poisson Hamiltonian formulation with its associated conserved Casimirs and conserved domain integrated energy. Previous work Holm *et al.* (1998a,b; 2002) has shown that many useful GFD approximations may be formulated as Euler–Poincaré equations, whose shared properties thus follow from this underlying common framework.

3 Derivation of the L1 model equations

We consider the motion of a rotating, continuously stratified fluid governed by the hydrostatic, Boussinesq, adiabatic, primitive equations and derive an approximate model for small Rossby number through the use of Hamilton's principle.

The primitive equations (PE) in dimensionless variables are

$$\nabla_{3D} \cdot \mathbf{u}_{3D} = 0\,, \tag{3.1a}$$

$$\epsilon \frac{D\mathbf{u}}{Dt} + f\hat{\mathbf{z}} \times \mathbf{u} = -\nabla p, \tag{3.1b}$$

$$0 = -p_z - \rho, \tag{3.1c}$$

$$\frac{D\rho}{Dt} = 0, \tag{3.1d}$$

where

$$\mathbf{x} = (x, y, z)\,, \tag{3.2a}$$

$$\mathbf{u}_{3D} = (u, v, \epsilon w)\,, \qquad \mathbf{u} = (u, v, 0)\,, \tag{3.2b, c}$$

$$\nabla_{3D} = \left(\frac{\partial}{\partial x}, \frac{\partial}{\partial y}, \frac{\partial}{\partial z}\right), \qquad \nabla = \left(\frac{\partial}{\partial x}, \frac{\partial}{\partial y}, 0\right), \tag{3.2d, e}$$

$$\frac{D}{Dt} = \left(\frac{\partial}{\partial t} + \mathbf{u}_{3D}\cdot\nabla_{3D}\right), \tag{3.2f}$$

and $\hat{\mathbf{z}}$ is the unit vector in the vertical z direction. In this case, ρ replaces b in (2.3) and we change notation so that ∇_{3D} replaces ∇ and D/Dt (3.2*f*) replaces d/dt.

Dimensionless variables are formed using the characteristic values (L, H_0, U_0, f_0) for, respectively, a horizontal length scale, vertical depth scale, horizontal velocity, and Coriolis parameter. The Rossby number

$$\epsilon = U_0/f_0 L\,. \tag{3.3}$$

With dimensional variables denoted by primes, we have

$$(x, y) = (x', y')/L\,, \qquad z = z'/H_0\,, \tag{3.4a, b}$$

$$(u, v) = (u', v')/U_0\,, \qquad \epsilon w = w' L/(U_0 H_0)\,, \tag{3.4c, d}$$

$$t = t' U_0/L\,, \qquad f = f'(x', y')/f_0\,, \tag{3.4e, f}$$

so that $(u, v, \epsilon w)$ are the dimensionless velocity components in the (x, y, z) directions, t is time, and $f = f(x, y)$ is the dimensionless Coriolis parameter.

The total dimensional density is given by

$$\rho_T' = \rho_0 + \overline{\rho}'(z') - \theta'(\mathbf{x}', t')\,, \tag{3.5}$$

where ρ_0 is a constant reference density, $\overline{\rho}'(z')$ the basic undisturbed z'-dependent field, and θ' the negative of the density fluctuation. We define

$$\overline{\rho}(z) = \overline{\rho}'(z')/\rho_C\,, \qquad \theta = \theta'/\rho_C\,, \tag{3.6a, b}$$

and

$$\rho = \overline{\rho}(z) - \theta\,, \tag{3.7}$$

where $\rho_C = p_C/(H_0 g)$, $p_C = \rho_0 U_0 f L$, and g is the acceleration of gravity. In addition, we write

$$\overline{\rho}_z = -S(z)/\epsilon\,, \qquad S(z) = N^2(z) H_0^2/(f_0^2 L^2)\,, \tag{3.8a, b}$$

so that

$$\rho_z = -S(z)/\epsilon - \theta_z, \tag{3.8c}$$

where

$$N^2(z) = -g\overline{\rho}'_{z'}/\rho_0\,, \tag{3.8d}$$

is the square of the basic Brunt-Väisälä frequency. Subscripts (x, y, z, t) denote partial differentiation. Pressure variables $p, \overline{p}$ and $\tilde{p}$ are defined by nondimensionalizing with p_C such that

$$p = \overline{p}(z) + \tilde{p}\,, \tag{3.9}$$

where

$$\overline{p}_z = -\overline{\rho}, \qquad \nabla p = \nabla\tilde{p}\,. \tag{3.10a, b}$$

We are interested in the limit of small Rossby number,

$$\epsilon \ll 1\,, \tag{3.11}$$

with $S = O(1)$.

From section 2, the PE are Euler–Poincaré equations, with action L given in dimensionless variables by

$$L = \int dt\,dx\,dy\,dz \left\{ D\left[\mathbf{u}\cdot(\mathbf{R}+\epsilon\mathbf{u}) - \frac{\epsilon}{2}|\mathbf{u}|^2 - \rho z \right] - p(D-1) \right\}, \tag{3.12}$$

where $\mathbf{R} = \mathbf{R}(x, y), \mathbf{R}\cdot\hat{\mathbf{z}} = 0$, and

$$\text{curl}_3\ \mathbf{R} = f(x, y)\hat{\mathbf{z}}\,. \tag{3.13}$$

This may be verified by direct substitution into equation (2.2).

We derive approximate equations for $\epsilon \ll 1$ using the Euler–Poincaré framework, by following a procedure similar to that applied by Salmon (1983, 1985, 1996) and Allen and Holm (1996). Thus, for the L1 model we define

$$\mathbf{u}_1 = f^{-1}\hat{\mathbf{z}} \times \nabla\tilde{\phi}\,, \tag{3.14a}$$

with

$$\tilde{\phi}(\mathbf{x}, t) = \phi_S(x, y, t) + \int_z^0 dz'\,\rho, \qquad \tilde{\phi}_z = -\rho\,, \tag{3.14b, c}$$

where $\phi_S(x, y, t)$ is a function to be determined, and utilize the following order $O(\epsilon)$ approximation for the action L,

$$L_1 = \int dt\,dx\,dy\,dz \left\{ D\left[\mathbf{u}\cdot(\mathbf{R}+\epsilon\mathbf{u}_1) - \frac{\epsilon}{2}|\mathbf{u}_1|^2 - \rho z \right] - p(D-1) \right\}. \tag{3.15}$$

The action L_1 has variational derivatives given by

$$\begin{aligned}\delta L_1 \;=\; & \int dt\,dx\,dy\,dz\Big\{D(\mathbf{R}+\epsilon\mathbf{u}_1)\cdot\delta\mathbf{u} \\ & +\Big[\mathbf{u}\cdot(\mathbf{R}+\epsilon\mathbf{u}_1)-\frac{\epsilon}{2}|\mathbf{u}_1|^2-\rho z-p\Big]\,\delta D \\ & +\epsilon D(\mathbf{u}-\mathbf{u}_1)\cdot\delta\mathbf{u}_1-Dz\delta\rho-(D-1)\delta p\Big\}\,.\end{aligned} \tag{3.16}$$

Defining

$$\mathbf{a}=\epsilon Df^{-1}(\mathbf{u}-\mathbf{u}_1), \tag{3.17a}$$

and using the relations

$$\int_{-H}^{0} dz\;\mathbf{a}\cdot\hat{\mathbf{z}}\times\nabla\delta\phi_S=-\delta\phi_S\,\hat{\mathbf{z}}\cdot\mathrm{curl}\Big(\int_{-H}^{0} dz\;\mathbf{a}\Big)+\mathrm{div}\,\Big[\Big(\int_{-H}^{0} dz\,\mathbf{a}\Big)\times\hat{\mathbf{z}}\;\delta\phi_S\Big], \tag{3.17b}$$

$$\mathbf{a}\cdot\hat{\mathbf{z}}\times\nabla\Big(\int_z^0 dz'\delta\rho\Big)=-\Big(\int_z^0 dz'\,\delta\rho\Big)\hat{\mathbf{z}}\cdot\mathrm{curl}\,\mathbf{a}+\mathrm{div}\Big[\mathbf{a}\times\hat{\mathbf{z}}\,\Big(\int_z^0 dz'\,\delta\rho\Big)\Big], \tag{3.17c}$$

and

$$\int_{-H}^{0} dz\,b(z)\left(\int_z^0 dz'\,c(z')\right)=\int_{-H}^{0} dz\,c(z)\left(\int_{-H}^{z} dz'b(z')\right)\,, \tag{3.18}$$

in (3.16) we obtain

$$\begin{aligned}\delta L_1 = & \int dt\,dx\,dy\,dz\,\Big\{D(\mathbf{R}+\epsilon\mathbf{u}_1)\cdot\delta\mathbf{u}+\Big[\mathbf{u}\cdot(\mathbf{R}+\epsilon\mathbf{u}_1)-\frac{\epsilon}{2}|\mathbf{u}_1|^2-\rho z-p\Big]\,\delta D \\ & -(D-1)\delta p-\Big[Dz+\int_{-H}^{z} dz'\,\hat{\mathbf{z}}\cdot\mathrm{curl}\,\epsilon Df^{-1}(\mathbf{u}-\mathbf{u}_1)\Big]\,\delta\rho\Big\} \\ & -\int dt\,dx\,dy\;\delta\phi_S\,\hat{\mathbf{z}}\cdot\mathrm{curl}\int_{-H}^{0} dz\;\;\epsilon Df^{-1}(\mathbf{u}-\mathbf{u}_1) \\ & +\int dt\oint_C ds\int_{-H}^{0} dz\,\left(\delta\phi_S+\int_z^0\delta\rho\,dz'\right)\,\epsilon Df^{-1}(\mathbf{u}-\mathbf{u}_1)\cdot\hat{\mathbf{s}} \\ & -\int dt\,dx\,dy\,\hat{\mathbf{z}}\times\nabla H\cdot\Big[\epsilon Df^{-1}(\mathbf{u}-\mathbf{u}_1)\Big|_{z=-H}\int_{-H}^{0} dz\delta\rho\Big]\,,\end{aligned} \tag{3.19}$$

where the integral over C is around a vertical wall along the outer boundary B of the domain in (x, y) and where $\hat{\mathbf{s}}=\hat{\mathbf{z}}\times\hat{\mathbf{n}}$ is the unit tangent vector and $\hat{\mathbf{n}}$ is the unit outward normal vector on that boundary. The boundary integral over C in (3.19) vanishes provided

$$(\mathbf{u}-\mathbf{u}_1)\cdot\hat{\mathbf{s}}=0 \qquad \text{on}\;\; B\,, \tag{3.20a}$$

i.e., that the tangential component of $\mathbf{u} - \mathbf{u}_1$ is zero on the outer boundary vertical wall. This condition is directly analogous to that derived in Allen *et al.* (1990a) and Allen and Holm (1996) for Salmon's (1983) model, which involves an L1 type approximation applied to the shallow water equations. The boundary integral over (x,y,t) at $z = -H$ in the last term in (3.19) vanishes provided

$$\hat{\mathbf{z}} \times \nabla H \cdot (\mathbf{u} - \mathbf{u}_1) = 0 \qquad \text{at } z = -H\,, \tag{3.20b}$$

which provides an additional boundary condition on $(\mathbf{u} - \mathbf{u}_1)$ at $z = -H$ if H is variable.

The δp and $\delta \phi_S$ variations in (3.19) yield, respectively,

$$D = 1\,, \tag{3.21}$$

and

$$\hat{\mathbf{z}} \cdot \mathrm{curl} \int_{-H}^{0} dz\, f^{-1}(\mathbf{u} - \mathbf{u}_1) = 0\,. \tag{3.22}$$

In this nondimensional notation, the Euler–Poincaré equation for the fluid motion generated by L_1 is, cf. (2.2),

$$\frac{\partial}{\partial t}\left(\frac{1}{D}\frac{\delta L_1}{\delta \mathbf{u}}\right) - \mathbf{u}_{3D} \times \mathrm{curl}_{3D}\left(\frac{1}{D}\frac{\delta L_1}{\delta \mathbf{u}}\right)$$

$$-\nabla_{3D}\left(\frac{\delta L_1}{\delta D} - \frac{1}{D}\frac{\delta L_1}{\delta \mathbf{u}} \cdot \mathbf{u}\right) + \frac{1}{D}\frac{\delta L_1}{\delta \rho}\nabla_{3D}\,\rho = 0\,. \tag{3.23}$$

The resulting equation, from (3.19) and (3.23), is

$$\epsilon\frac{\partial \mathbf{u}_1}{\partial t} - \mathbf{u}_{3D} \times \mathrm{curl}_{3D}\,(\mathbf{R} + \epsilon \mathbf{u}_1) + \nabla_{3D}\left(p + \frac{\epsilon}{2}|\mathbf{u}_1|^2\right)$$

$$+\rho\hat{\mathbf{z}} - \epsilon^2 I_1 \nabla_{3D}\,\rho = 0\,. \tag{3.24a}$$

where

$$I_1 = \epsilon^{-1}\int_{-H}^{z} dz'\,\hat{\mathbf{z}} \cdot \mathrm{curl}\, f^{-1}(\mathbf{u} - \mathbf{u}_1)\,. \tag{3.24b}$$

The approximate equations are given by (3.24), the constraint (3.22), the equation of continuity,

$$\nabla_{3D} \cdot \mathbf{u}_{3D} = 0, \tag{3.25}$$

obtained from the equation (2.4) for D and (3.21), and the equation for the density (2.3),

$$\frac{D\rho}{Dt} = \frac{\partial \rho}{\partial t} + \mathbf{u}_{3D} \cdot \nabla_{3D}\,\rho = 0\,. \tag{3.26}$$

We will refer to (3.24), (3.22), (3.25), and (3.26) as the L1 model equations.

By equation (2.5) for Euler–Poincaré systems the L1 model possesses the following Kelvin–Noether circulation theorem,

$$\frac{D}{Dt}\oint_{\gamma_t(\mathbf{u}_{3D})}(\mathbf{R}+\epsilon\mathbf{u}_1)\cdot d\mathbf{x} = -\oint_{\gamma_t(\mathbf{u}_{3D})}(z+\epsilon^2 I_1)\nabla_{3D}\rho\cdot d\mathbf{x}\,, \tag{3.27}$$

where $\gamma_t(\mathbf{u}_{3D})$ moves with the fluid velocity $\mathbf{u}_{3D}$. Then, by Stokes' theorem, the L1 equations generate circulation of $(\mathbf{R}+\epsilon\mathbf{u}_1)$ whenever $\nabla_{3D}\rho$ and $\nabla_{3D}(z+\epsilon^2 I_1)$ are not collinear.

The curl of equation (3.24) together with (3.25) and (3.26) gives conservation of potential vorticity Q_1 on fluid particles, cf. (2.6),

$$\frac{DQ_1}{Dt}=0\,, \tag{3.28}$$

where

$$\begin{aligned} Q_1 &= \text{curl}_{3D}(\mathbf{R}+\epsilon\mathbf{u}_1)\cdot\nabla_{3D}\rho\,, && (3.29a)\\ &= (f\hat{\mathbf{z}}+\epsilon\,\text{curl}_{3D}\ \mathbf{u}_1)\cdot\nabla_{3D}\rho\,. && (3.29b)\end{aligned}$$

The L1 model equations also conserve the volume integrated energy E_1, given in the general theory by (2.8), i.e.,

$$\frac{dE_1}{dt}=0\,, \tag{3.30}$$

$$E_1=\int dx\,dy\,dz\,\left[\frac{\epsilon}{2}|\mathbf{u}_1|^2+\rho z\right]. \tag{3.31}$$

4 Derivation of the L2 model equations

One important objective here is to assess the accuracy of the L1 and L2 model equations by obtaining numerical solutions to the idealized, mesoscale oceanographic problems utilized in Allen and Newberger (1993). Consequently, for simplicity in the derivation of the more complex L2 model, we restrict consideration to the idealized conditions utilized for the problems in that study, i.e., to an f-plane ($f=1$) with a rigid lid and flat bottom ($H=1$) and with the domain periodic in the horizontal (x,y) directions.

For the L2 model we follow Allen and Holm (1996) and include consistent $O(\epsilon)$ terms in the approximation $\mathbf{u}_2$. These $O(\epsilon)$ terms are obtained by iteration of the lowest order geostrophic balance in the PE momentum equations and involve quasi-geostrophic dynamics (Allen, 1993). The equation for $\mathbf{u}_2$ is

$$\mathbf{u}_2=\hat{\mathbf{z}}\times\nabla\tilde{\phi}+\alpha\epsilon\left\{-J(\tilde{\phi},\nabla\tilde{\phi})+\nabla\left[\mathcal{L}^{-1}\left(J(\tilde{\phi},\mathcal{L}\tilde{\phi})\right)\right]\right\}, \tag{4.1}$$

where $\tilde{\phi}$ is defined in (3.14b,c) and where

$$\mathcal{L}\tilde{\phi}_t=[\nabla^2+\partial_z(S^{-1}\partial_z)]\tilde{\phi}_t=-J(\tilde{\phi},\mathcal{L}\tilde{\phi}), \tag{4.2}$$

with inverse

$$\tilde{\phi}_t = -\mathcal{L}^{-1}[J(\tilde{\phi}, \mathcal{L}\tilde{\phi})]\,, \tag{4.3a}$$

obtained with boundary conditions

$$\tilde{\phi}_{zt} = -J(\tilde{\phi}, \tilde{\phi}_z) \qquad \text{at } z = 0, -1\,. \tag{4.3b}$$

The operator $J(a,b) = a_x b_y - a_y b_x$ is the Jacobian and $\mathcal{L}\tilde{\phi}$ is the quasi-geostrophic potential vorticity. In (4.1), α is simply an accounting parameter such that $\alpha = 1$ for the L2 model and $\alpha = 0$ for the L1 model.

We utilize the following approximation for the action L,

$$L_2 = \int dt\, dx\, dy\, dz \left\{ D[\mathbf{u}\cdot(\mathbf{R} + \epsilon\mathbf{u}_2) - \frac{\epsilon}{2}|\mathbf{u}_2|^2 - \rho z] - p(D-1) \right\}. \tag{4.4}$$

The action L_2 has variational derivatives given by

$$\begin{aligned} \delta L_2 \;=\; & \int dt\, dx\, dy\, dz \Big\{ D(\mathbf{R} + \epsilon\mathbf{u}_2)\cdot\delta\mathbf{u} \\ & + \left[\mathbf{u}\cdot(\mathbf{R} + \epsilon\mathbf{u}_2) - \frac{\epsilon}{2}|\mathbf{u}_2|^2 - \rho z - p\right]\delta D \\ & + \epsilon D(\mathbf{u} - \mathbf{u}_2)\cdot\delta\mathbf{u}_2 - Dz\delta\rho - (D-1)\delta p \Big\}. \end{aligned} \tag{4.5}$$

It is convenient to define

$$\epsilon\mathbf{u}_D = \mathbf{u} - \mathbf{u}_2, \quad \mathbf{u}_D = (u_D, v_D, 0)\,. \tag{4.6}$$

The variations associated with the term $\epsilon D(\mathbf{u} - \mathbf{u}_2)\cdot\delta\mathbf{u}_2$ are

$$\begin{aligned} \mathbf{u}_D\cdot\delta\mathbf{u}_2 \;=\; & \mathbf{u}_D\cdot\Big\{ \hat{\mathbf{z}}\times\nabla\delta\tilde{\phi} - \alpha\epsilon[J(\delta\tilde{\phi}, \nabla\tilde{\phi}) + J(\tilde{\phi}, \nabla\delta\tilde{\phi})] \\ & + \alpha\epsilon\nabla\Big[\mathcal{L}^{-1}[J(\delta\tilde{\phi}, \mathcal{L}\tilde{\phi}) + J(\tilde{\phi}, \mathcal{L}\delta\tilde{\phi})]\Big]\Big\}. \end{aligned} \tag{4.7}$$

Utilizing the relation (3.17) and integrating by parts we obtain, in a manner similar to that in section 3.2.2 of Allen and Holm (1996) except that the operator $\mathcal{L}$ is three-dimensional here,

$$\begin{aligned} \int dt\, dx\, dy\, dz\, \mathbf{u}_D\cdot\delta\mathbf{u}_2 \;=\; & \int dt\, dx\, dy\, dz\, \delta\tilde{\phi}\Big\{ -\hat{\mathbf{z}}\cdot\operatorname{curl}\mathbf{u}_D \\ & + \alpha\epsilon\Big[2J(u_D, \tilde{\phi}_x) + 2J(v_D, \tilde{\phi}_y) + J(\nabla\cdot\mathbf{u}_D, \tilde{\phi}) \\ & + \mathcal{L}[J(\tilde{\phi}, \mathcal{L}^{-1}\nabla\cdot\mathbf{u}_D)] - J(\mathcal{L}\tilde{\phi}, \mathcal{L}^{-1}\nabla\cdot\mathbf{u}_D)\Big]\Big\}, \end{aligned} \tag{4.8}$$

where the boundary integral terms in (x, y) vanish as a result of periodicity. The vanishing of the boundary integral terms at $z = 0, -1$ require

$$(\mathcal{L}^{-1}\nabla\cdot\mathbf{u}_D)_z = 0 \qquad \text{at } z = 0, -1\,, \tag{4.9a}$$

and imply that in the calculations of $\mathcal{L}[J(\tilde{\phi}, \mathcal{L}^{-1}\nabla\cdot\mathbf{u}_D)]$ and $\mathcal{L}\tilde{\phi}$,

$$J_z(\tilde{\phi}, \mathcal{L}^{-1}\nabla\cdot\mathbf{u}_D) = 0 \quad \text{and} \quad \tilde{\phi}_z = 0 \qquad \text{at } z = 0, -1\,, \tag{4.9b,c}$$

respectively. Continued evaluation of (4.8) involves use of (3.18).

The Euler–Poincarè equation for the fluid motion generated by L2 is

$$\begin{aligned}\frac{\partial}{\partial t}\left(\frac{1}{D}\frac{\delta L_2}{\delta\mathbf{u}}\right) - \mathbf{u}_{3D}\times\mathrm{curl}_{3D}\left(\frac{1}{D}\frac{\delta L_2}{\delta\mathbf{u}}\right)\\ -\nabla_{3D}\left(\frac{\delta L_2}{\delta D} - \frac{1}{D}\frac{\delta L_2}{\delta\mathbf{u}}\cdot\mathbf{u}\right) + \frac{1}{D}\frac{\delta L_2}{\delta\rho}\nabla_{3D}\,\rho = 0\,.\end{aligned} \tag{4.10}$$

The resulting equation, from (4.5) and (4.10), is

$$\begin{aligned}\epsilon\frac{\partial\mathbf{u}_2}{\partial t} - \mathbf{u}_{3D}\times\mathrm{curl}_{3D}(\mathbf{R}+\epsilon\mathbf{u}_2) + \nabla_{3D}\left(p + \frac{\epsilon}{2}|\mathbf{u}_2|^2\right)\\ +\rho\hat{\mathbf{z}} - \epsilon^2 I_2\nabla_{3D}\rho = 0,\end{aligned} \tag{4.11a}$$

where

$$\begin{aligned}I_2 \;=\; &\int_{-1}^{z} dz'\tilde{\mathbf{z}}\cdot\mathrm{curl}\,\mathbf{u}_D - \alpha\epsilon\int_{-1}^{z} dz'\Big\{2J(u_D,\tilde{\phi}_x) + 2J(v_D,\tilde{\phi}_y) + J(\nabla\cdot\mathbf{u}_D,\tilde{\phi})\\ &+\mathcal{L}[J(\tilde{\phi},\mathcal{L}^{-1}\nabla\cdot\mathbf{u}_D)] - J(\mathcal{L}\tilde{\phi},\mathcal{L}^{-1}\nabla\cdot\mathbf{u}_D)\Big\},\end{aligned} \tag{4.11b}$$

with constraint, implied by the $\delta\phi_S$ variation and analogous to (3.22),

$$I_2(z=0) = 0. \tag{4.12}$$

The L2 model equations are given by (4.11), (4.12), (3.25) and (3.26). These equations imply the conservation of potential vorticity Q_2 on fluid particles

$$\frac{DQ_2}{Dt} = 0, \tag{4.13}$$

where

$$Q_2 = (\hat{\mathbf{z}} + \epsilon\,\mathrm{curl}_{3D}\,\mathbf{u}_2)\cdot\nabla_{3D}\rho\,. \tag{4.14}$$

The L2 model equations also conserve the volume integrated energy E_2, i.e.,

$$\frac{d}{dt}E_2 = 0, \tag{4.15}$$

$$E_2 = \int dx\,dy\,dz\left[\frac{\epsilon}{2}|\mathbf{u}_2|^2 + \rho z\right]. \tag{4.16}$$

5 Solution procedure

We describe the solution procedure for the L2 model with $f = 1$ and $H = 1$. The procedure for the L1 model is the same and may be obtained by setting $\alpha = 0$. The L2 model equations (3.25), (3.26) and (4.11) are written in the form:

$$\nabla_{3D} \cdot \mathbf{u}_{3D} = 0 \,, \tag{5.1a}$$

$$\epsilon \frac{\partial \mathbf{u}_2}{\partial t} - \mathbf{u} \times (1 + \epsilon\zeta_2)\hat{\mathbf{z}} + \epsilon^2 w \mathbf{u}_{2z} + \nabla\left(\tilde{p} + \frac{\epsilon}{2} \,|\, \mathbf{u}_2 \,|^2\right) + \epsilon^2 I_2 \nabla \phi_z = -\epsilon\nu\nabla^4 \mathbf{u}_2 \,, \tag{5.1b}$$

$$\tilde{p}_z - \phi_z - \epsilon(\mathbf{u} - \mathbf{u}_2) \cdot \mathbf{u}_{2z} + \epsilon I_2(S + \epsilon\phi_{zz}) = 0, \tag{5.1c}$$

$$\frac{\partial}{\partial t}\phi_z + \nabla_{3D} \cdot (\mathbf{u}_{3D}\phi_z) + Sw = 0 \,, \tag{5.1d}$$

where the density and pressure fields have been decomposed as in (3.7) and (3.9) and where ϕ and ζ_2 are defined below.

With (3.7), $\mathbf{u}_2$ may be written

$$\begin{aligned} \mathbf{u}_2 \;\; = \;\; & \hat{\mathbf{z}} \times \nabla\phi + \alpha\epsilon\Big\{ -J(\phi, \nabla\phi) \\ & + \nabla\left[\mathcal{L}^{-1}\left(J(\phi, \mathcal{L}\phi) + \nu\nabla^6\phi\right)\right] - \nu\nabla^4\nabla\phi \Big\}, \end{aligned} \tag{5.2}$$

where

$$\phi = \phi_S(x, y, t) - \int_z^0 dz'\,\theta, \qquad \phi_z = \theta \,. \tag{5.3a, b}$$

We include biharmonic momentum diffusion in the horizontal momentum equations (5.1b) so that it may be used in the numerical finite-difference solutions to provide dissipation at high wave numbers in otherwise nearly inviscid flows. In addition, we add a consistent diffusion term in $\mathbf{u}_2$ (5.2).

We utilize

$$\mathbf{u} = \hat{\mathbf{z}} \times \nabla\psi + \epsilon\nabla\chi, \qquad \mathbf{u}_2 = \hat{\mathbf{z}} \times \nabla\psi_2 + \epsilon\nabla\chi_2 \,, \tag{5.4a, b}$$

$$\mathbf{u}_D = \hat{\mathbf{z}} \times \nabla\psi_D + \nabla\chi_D \,, \tag{5.4c}$$

so that

$$\psi = \psi_2 + \epsilon\psi_D, \quad \chi = \chi_2 + \chi_D \,. \tag{5.5a, b}$$

It follows from (5.2) that

$$\zeta_2 = \nabla^2\psi_2 = \nabla^2\phi - 2\alpha\epsilon J(\phi_x, \phi_y), \tag{5.6}$$

$$\nabla^2\chi_2 = \alpha\left\{ -J(\phi, \nabla^2\phi) + \nabla^2\left[\mathcal{L}^{-1}\left(J(\phi, \mathcal{L}\phi) + \nu\nabla^6\phi\right)\right] - \nu\nabla^6\phi \right\}, \tag{5.7}$$

where

$$J(\phi, \mathcal{L}\phi) = J(\phi, \nabla^2\phi) + \{S^{-1}J(\phi, \phi_z)\}_z. \tag{5.8}$$

In addition, (4.11*b*) may be written

$$I_2 = I_1 + \alpha\epsilon I_2' \,, \tag{5.9a}$$

where

$$I_1 = \int_{-1}^{z} dz' \, \nabla^2 \psi_D, \tag{5.9b}$$

$$I_2' = -\int_{-1}^{z} dz' \, \{2J(u_D, \phi_x) + 2J(v_D, \phi_y) + J(\nabla^2 \chi_D, \phi) + \nu\nabla^6 \chi_D$$
$$+\mathcal{L}[J(\phi, \mathcal{L}^{-1}\nabla^2 \chi_D)] - J(\mathcal{L}\phi, \mathcal{L}^{-1}\nabla^2 \chi_D)\} \,. \tag{5.9c}$$

The constraint (4.12) is

$$I_2(z = 0) = 0. \tag{5.10}$$

The horizontal momentum equations (5.1*b*) are replaced by vorticity and divergence equations formed, respectively, by the operations $\hat{\mathbf{z}} \cdot \nabla \times$ (5.1*b*) and $\nabla \cdot$ (5.1*b*):

$$\frac{\partial \zeta_2}{\partial t} + J(\psi, \zeta_2) + \nabla^2 \chi + \epsilon \nabla \cdot [w \nabla \psi_{2z} + \zeta_2 \nabla \chi] + \epsilon J(I_2, \phi_z)$$
$$+\epsilon^2 J(w, \chi_{2z}) + \nu \nabla^4 \zeta_2 = 0 \,, \tag{5.11}$$

$$\nabla^2 \psi = \nabla^2 \tilde{p} - \epsilon 2 J(\psi_{2x}, \psi_{2y}) + \epsilon^2 \Big[-J(\zeta_2, \chi) - J(w, \psi_{2z}) - \nabla \cdot (\zeta_2 \nabla \psi_D)$$
$$+\nabla \cdot (I_2 \nabla \phi_z) + \nabla^2 J(\psi_2, \chi_2) + \nabla^2 \chi_{2t} + \nu \nabla^6 \chi_2\Big]$$
$$+\epsilon^3 \Big[\nabla \cdot (w \nabla \chi_{2z}) + \frac{1}{2} \nabla^2 (|\nabla \chi_2|^2)\Big] \,. \tag{5.12}$$

We rewrite (5.1*d*) as

$$\frac{\partial \phi_z}{\partial t} + J(\psi, \phi_z) + S w + \epsilon \left[\nabla \cdot (\phi_z \nabla \chi) + (w \phi_z)_z\right] = 0 \,, \tag{5.13}$$

and (5.1*a*) as

$$\nabla^2 \chi + w_z = 0 \,. \tag{5.14}$$

The L2 model equations now consist of (5.9), (5.10), (5.11), (5.12), (5.13), (5.14), (5.6), (5.7) and (5.5*a,b*). The variables are $\psi, \chi, w, \tilde{p}, \phi, \psi_2, \chi_2, \psi_D, \chi_D$ and I_2. For the L1 model ($\alpha = 0$), $\mathbf{u}_2 = \mathbf{u}_1, \psi_2 = \phi, \chi_2 = 0$, and $I_2 = I_1$ (5.9*b*). Note that if we set $\epsilon = 0$ in the L2 equations, they reduce to the quasigeostrophic (QG) approximation (Pedlosky, 1987). To obtain numerical solutions, we follow a procedure similar to that developed and applied in Allen and Newberger (1993) for other intermediate models. The procedure is based on the assumption that $\epsilon \ll 1$ and essentially uses the solution of the QG approximation as the starting point for an iteration scheme. Accordingly, we form an equation for a linear approximation to the potential vorticity by eliminating $w_z = -\nabla^2\chi$ between (5.11) and (5.13) and using (5.6):

$$\frac{\partial}{\partial t}[\nabla^2\phi + (S^{-1}\phi_z)_z] = \alpha\epsilon 2\frac{\partial}{\partial t}J(\phi_x, \phi_y) - J(\psi, \zeta_2)$$
$$-\epsilon\nabla\cdot[w\nabla\psi_{2z} + \zeta_2\nabla\chi] - \epsilon J(I_2, \phi_z) - \epsilon^2 J(w, \chi_{2z})$$
$$-\nu\nabla^4\zeta_2 - \Big\{S^{-1}\Big[J(\psi, \phi_z) + \epsilon\left[\nabla\cdot(\phi_z\nabla\chi) + (w\phi_z)_z\right]\Big]\Big\}_z . \tag{5.15}$$

We also write (5.11) as

$$\nabla^2\chi = -\frac{\partial\nabla^2\phi}{\partial t} + \alpha\,\epsilon 2\frac{\partial}{\partial t}J(\phi_x, \phi_y) - J(\psi, \zeta_2) - \epsilon\nabla\cdot[w\nabla\psi_{2z} + \zeta_2\nabla\chi]$$
$$-\epsilon J(I_2, \phi_z) - \epsilon^2 J(w, \chi_{2z}) - \nu\nabla^4\zeta_2 . \tag{5.16}$$

We eliminate $\tilde{p}$ by taking the z derivative of (5.12) and substituting for $\tilde{p}_z$ from (5.1c). The resulting equation is

$$\nabla^2 G = -2\Big[J_z(\psi_{2x}, \psi_{2y}) - \alpha J_z(\phi_x, \phi_y)\Big] + \epsilon\Big\{-J_z(\zeta_2, \chi) - J_z(w, \psi_{2z})$$
$$-\nabla\cdot(\zeta_2\nabla\psi_D)_z + \nabla\cdot(I_2\nabla\phi_z)_z + \nabla^2\Big[\nabla\psi_D\cdot\nabla\psi_{2z} + J(\psi_{2z}, \chi_D)$$
$$-\phi_{zz}I_2 + J_z(\psi_2, \chi_2)\Big] + \nabla^2\chi_{2zt} + \nu\nabla^6\chi_{2z}\Big\} + \epsilon^2\Big\{\nabla\cdot(w\nabla\chi_{2z})_z$$
$$+\nabla^2\Big[\nabla\chi_D\cdot\nabla\chi_{2z} + J(\psi_D, \chi_{2z})\Big] + \frac{1}{2}\nabla^2(|\nabla\chi_2|^2)_z\Big\}, \tag{5.17a}$$

where

$$G = \psi_{Dz} + SI_2 . \tag{5.17b}$$

If, for $\epsilon \ll 1$, G is obtained from the solution of (5.17a), then ψ_D may be found from the solution to

$$\nabla^2\psi_D + [S^{-1}\psi_{Dz}]_z = [S^{-1}G - \alpha\,\epsilon\, I_2'\,]_z, \tag{5.18a}$$

where (5.9), (5.10), and (5.17b) imply

$$\psi_{Dz} = G \quad \text{at} \quad z = 0, -1 . \tag{5.18b}$$

In a domain periodic in (x, y), (5.17a) determines G up to an arbitrary function of z. That function has no effect on the horizontal derivatives of ψ_D (or ψ) needed in the other equations. Thus, to obtain a unique solution for G we require

$$\int dx\, dy\, G = 0. \tag{5.18c}$$

The L2 model equations we solve are finally (5.14), (5.15), (5.16), (5.17), (5.18), (5.6), (5.7) (5.9) and (5.5a,b). The variables are $\psi, \chi, w, \phi, \psi_2, \chi_2, \psi_D$,

χ_D, G, and I_2. The numerical finite difference method used to solve this equation set is described in the appendix. Boundary conditions are

$$w(z=0) = w(z=-1) = 0, \tag{5.19}$$

with periodicity of all variables over the domain in x and y. In the calculations of χ_2 (5.7) and of I_2 (5.9) that involve the operators $\mathcal{L}$ and $\mathcal{L}^{-1}$, boundary conditions at $z = 0, -1$ are obtained from (4.3b) and (4.9a,b,c).

6 Numerical experiments

Numerical solutions to finite-difference approximations to the L1 and L2 models are obtained for the problems utilized to investigate accuracy of different intermediate models in Allen and Newberger (1993). The numerical experiments involve initial-value problems for the time-dependent development of an unstable baroclinic jet on an f-plane. Initial conditions involve a uniform, vertically-sheared jet with small perturbations. The jet is unstable and develops finite amplitude meanders that grow in time and eventually pinch off to form detached eddies. The weak jet, basic case, and strong jet numerical experiments from Allen and Newberger (1993) are repeated here for the L1 and L2 models. The accuracy is assessed by comparison with solutions of the primitive equations (PE).

The domain is periodic in the horizontal directions (x, y) and is of constant depth in the vertical direction (z). The finite difference methods are discussed in the appendix and in Allen and Newberger (1993). All models use the same variables on the same grid.

Dimensional variables are used for the numerical experiments. The Coriolis parameter $f_0 = 9.20 \times 10^{-5}$ s^{-1}. The total depth $H_T = 3172$ m. The number of vertical grid cells is 6. The horizontal domain is

$$0 \leq x \leq L^{(x)}, \qquad 0 \leq y \leq L^{(y)}, \tag{6.1a,b}$$

where the initial jet flow is parallel to the x axis and toward positive x. For the weak jet experiment, $L^{(x)} = 250$ km, $L^{(y)} = 640$ km. For the basic case and strong jet experiments, $L^{(x)} = 250$ km, $L^{(y)} = 810$ m. The horizontal grid spacing is $\Delta x = \Delta y = 5$ km. The horizontal biharmonic diffusion coefficient $\nu = 8 \times 10^8$ m^4 s^{-1} is chosen to be small so that dissipative processes play a nearly negligible role in the time-dependent dynamics.

The initial stratification and jet structure are based on observed oceanographic values from the Coastal Transition Zone (CTZ) off Northern California (Kosro *et al.*, 1991; Pierce *et al.*, 1991; Walstad *et al.*, 1991). The Rossby radius for the first baroclinic vertical mode calculated from the initial stratification is $\delta_{R1} = 24.6$ km. The initial jet has a half-width of $L_J = 30$ km, comparable in magnitude to δ_{R1}. The experiments are characterized by different velocity magnitudes in the initial basic jet profiles. For the weak jet,

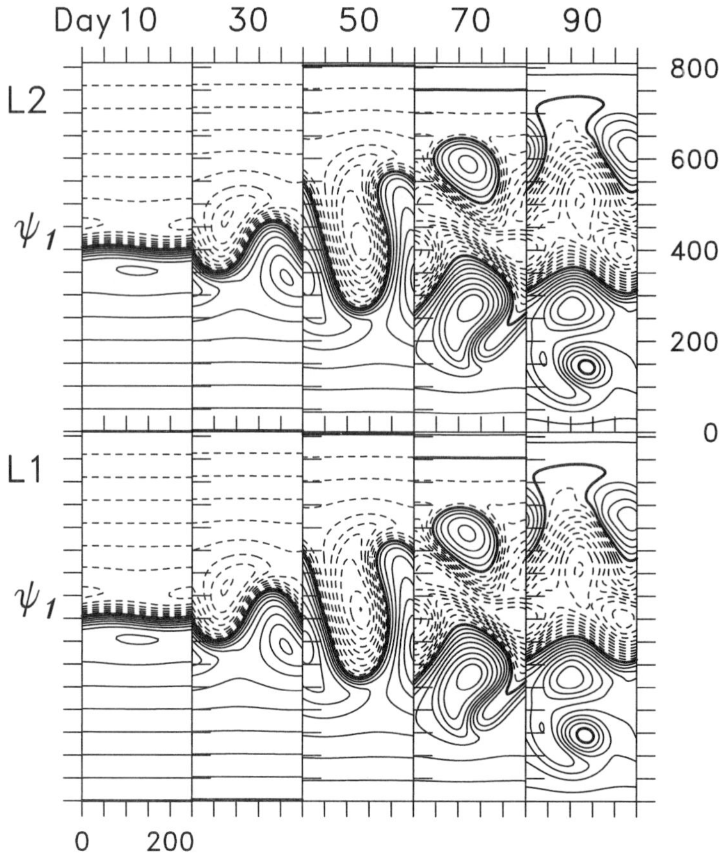

Figure 1: Contour plots of the ψ_1 fields from L2 and L1 as a function of (x, y) every 20 days from $t = 10$ to $t = 90$ days for the basic case experiment. The distance between tick marks on the axes is 50 km. Positive (negative) values are contoured by solid (dashed) lines with the zero contour a heavy solid line. The contour interval is 3000 m^2 s^{-1}.

basic case, and strong jet experiments, the maximum initial jet velocities are 0.52 m s^{-1}, 0.90 m s^{-1}, and 1.28 m s^{-1}, respectively. The vertical shear is such that the corresponding maximum initial jet velocities at 500 m depth are 0.23 m s^{-1}, 0.36 m s^{-1} and 0.56 m s^{-1}.

The function $|\zeta(x, y, z, t)|/f_0$, where $\zeta = v_x - u_y$, indicates the magnitude of the local, flow-determined Rossby number. For the weak jet, basic case, and strong jet experiments, the maximum initial values of $|\zeta|/f_0$ are 0.174, 0.287, and 0.404, respectively. The maximum values reached during the experiment are larger, 0.264, 0.555, and 1.025, respectively. Thus, the experiments cover a range of flow regimes characterized by maximum local Rossby numbers $|\zeta|/f_0$ that are moderately small (weak jet), moderate (basic case), and $O(1)$.

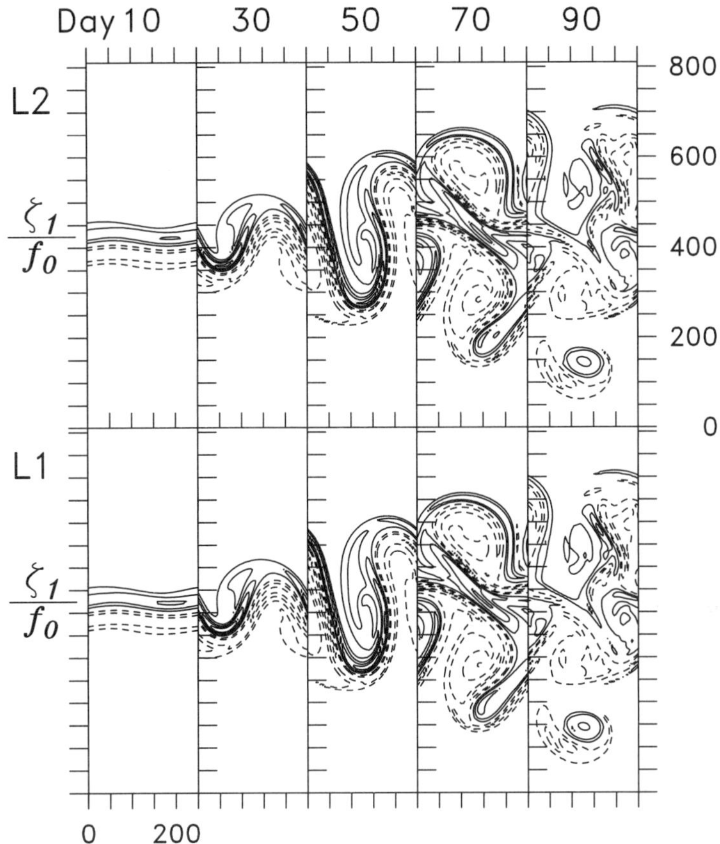

Figure 2: Contour plots of the ζ_1/f_0 fields from L2 and L1 every 20 days from $t = 10$ to $t = 90$ days for the basic case experiment. The contour interval is 0.1 and the zero contour line is omitted. Scaling of axes as in Figure 1.

Quantitative measures of the errors of the L1 and L2 model solutions, compared to PE, are found as a function of time by calculating normalized rms differences between the corresponding variables from the L1 or L2 model and from the PE solutions as described in Allen and Newberger (1993). Prior to comparison with L1 and L2 model results and calculation of errors, the PE solutions are averaged over an inertial period to eliminate high frequency variability.

The time-dependent development of the unstable jet flow field in the basic case experiment is illustrated in Fig. 1 by contour plots of the near-surface streamfunction fields $\psi(x, y, z_1, t) = \psi_1(x, y, t)$ where z_1 corresponds to the center of the top grid cell at 50 m depth. Fields from the solutions of the L2 and the L1 model are shown every 20 days from day 10 to day 90. The corresponding vorticity fields ζ_1/f_0 from L2 and L1 are shown in Fig. 2. There are no discernible visual differences between the L2 and PE fields (see, e.g.,

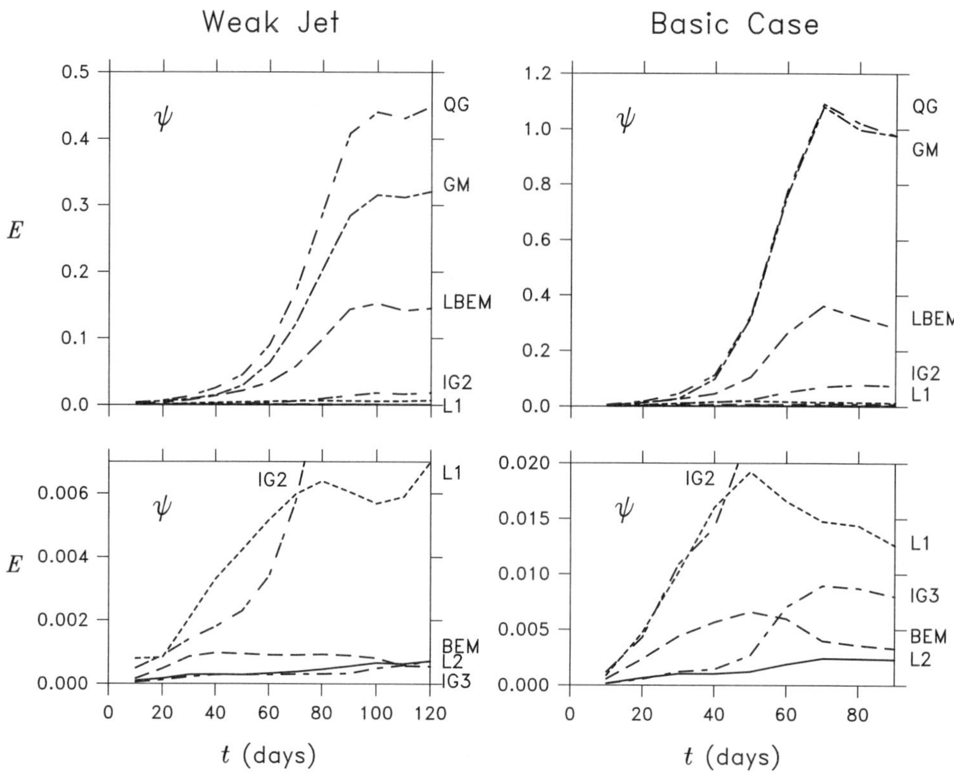

Figure 3: The normalized rms errors E for ψ as a function of time from the QG, GM, LBEM, IG2, L1, IG3, BEM, and L2 models compared to PE for the weak jet experiment (left) and for the basic case experiment (right). Note the different scales for the errors in the top and bottom panels. The type of line representing the errors from each model is consistent for both experiments.

Fig. 4). The close agreement of the ψ_1 and ζ_1/f_0 fields from the L1 model with the corresponding fields from the L2 model, and thus with the PE, is apparent.

The normalized errors for the streamfunction field $\psi(x, y, z, t)$ from the L1 and L2 models relative to PE for the weak jet and basic case experiments are shown in Fig. 3. Similar error characteristics are found for the other variables. For comparison, the errors from other intermediate model solutions as reported in Allen and Newberger (1993) are included. The other models are the quasigeostrophic (QG) equations (Pedlosky, 1987), the geostrophic momentum (GM) approximation (Hoskins, 1975), the linear BEM model (LBEM) (Allen and Newberger, 1993), the IG2 and IG3 iterated geostrophic models (Allen, 1993), and the balance equations based on momentum equations BEM model (Allen, 1991; Holm, 1996).

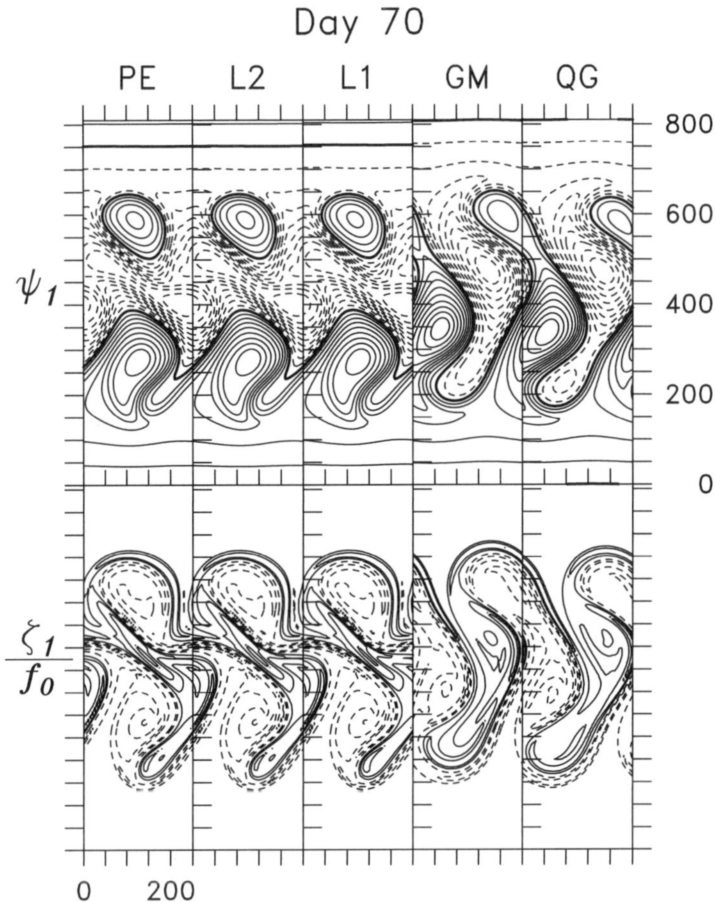

Figure 4: Contour plots of the ψ_1 and ζ_1/f_0 fields at day 70 of the basic case experiment from PE, L2, L1, GM and QG. Contour intervals and scaling of axes as in Figs. 1 and 2.

The errors are plotted with two different scales. An expanded scale plot is included at the bottom to show clearly the relative errors of the more accurate models. In both experiments, the errors from the L1 model are substantially lower than those from QG, GM or LBEM. That fact is further illustrated by a comparison of the ψ_1 and ζ_1/f_0 fields at day 70 of the basic case experiment from PE, L2, L1, GM and QG (Fig. 4). The L2, BEM, and IG3 models give approximate solutions with generally high accuracy. The balance equations BE (Gent and McWilliams, 1983) (errors not plotted) also give accurate solutions comparable to those of BEM. The errors from the new L1 model, however, are generally relatively small and remain so during the experiments. In the basic case experiment, the errors from L1 and IG2 are comparable for $t < 50$ days with IG2 slightly lower. After day 50, the errors for IG2 become considerably larger than those from L1. Similar qualitative behavior is found in the weak jet

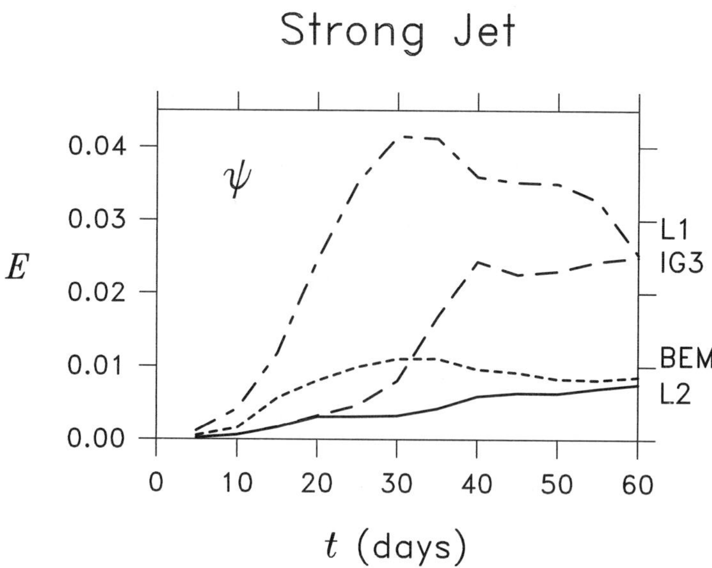

Figure 5: The normalized rms errors E for ψ as a function of time from the L1, IG3, BEM and L2 models compared to PE for the strong jet experiment. The type of line representing the errors from each model is consistent with Fig. 3.

experiment. The inviscid IG2 model does not have exact analogues of potential vorticity conservation on fluid particles or of conservation of volume integrals of energy (Allen, 1993). The L1 model, of course, does have analogues of these conservation equations, given in (3.28) and (3.30). It seems possible that the tendency of the IG2 model errors to increase at large time, while the L1 errors remain relatively constant, may be related to the differences in the models with regard to possession of analogue conservation equations for potential vorticity and energy.

In the basic case experiment the L2 model gives the most accurate approximate solution with errors that remain small for the duration (90 days) of the experiment. The same characteristic is found in the strong jet, $O(1)$ Rossby number, experiment (Fig. 5).

It is noteworthy that the L1 model has the smallest errors of any of the approximate models in Allen and Newberger (1993) that advect a vorticity of the form $\zeta_1 = \nabla^2\phi$. Those models include QG, LBEM, and the linear balance equations (LBE). All of the more accurate models advect a higher $O(\epsilon)$ approximation to the vorticity similar to that of L2 or, e.g., of IG2, where the advected vorticity ζ_2 is given by (5.6). From a comparison of the model errors in Allen and Newberger (1993), it appeared that advection of vorticity with $O(\epsilon)$ accuracy was a necessary property for an accurate model. The relatively small errors found with the L1 model do not seem to fit with that idea. In the L1 model, the vorticity of the advecting velocities

$$\zeta = \nabla^2 \psi = \nabla^2 \phi - \epsilon 2 J(\phi_x, \phi_y) + O(\epsilon^2)\,, \tag{6.2}$$

is accurate to $O(\epsilon)$ and that feature may contribute to the relatively small errors found with L1. Additionally, the requirement in L1 that

$$\int_{-1}^{0} dz\ \psi = \int_{-1}^{0} dz\ \phi\,, \tag{6.3}$$

which follows from the constraint (3.22), i.e., from (5.10) with $\alpha = 0$, may provide additional accuracy.

7 Summary comments

The L1 model produces generally accurate approximate solutions for the idealized, moderate Rossby number, mesoscale oceanographic flow problems examined in Allen and Newberger (1993). These solutions are not quite as accurate as those from the BEM or BE models, but are substantially more accurate than those from GM or QG, and are better for large time than IG2. The L2 model produces extremely accurate approximate solutions, with errors that are typically smaller than all of the other models evaluated. The relative complexity of the L2 model, however, may inhibit its general applicability. On the other hand, the results found for L1 are particularly encouraging. The fact that L1 is capable of producing accurate solutions for moderate Rossby number mesoscale flows, coupled with the capability of the general L1 model of section 3 to represent larger-scale flows with horizontally variable f, H, and N^2 on gyre scales, provides motivation for further development and application.

In summary, we note that the L1 and L2 models appear to realize some of the potential anticipated for approximate equations derived from Hamilton's principle. The accuracy of the L1 model solutions seems better than expected based on the asymptotics involved in its derivation. In addition, the errors from both L1 and L2 remain small as time increases, which may be the desired consequence of retaining analogue energy and potential vorticity conservation laws.

Acknowledgements For J.S.A. and P.A.N. this research was supported by the Office of Naval Research Coastal Dynamics Program under ONR Grant N00014-93-1-1301. Work of D.D.H. was supported by the U.S. Department of Energy under contracts W-7405-ENG-36 and the Applied Mathematical Sciences Program KC-07-01-01. The authors thank F. Beyer and C. Withrow for help in typing the manuscript.

8 Appendix — numerical methods

The numerical finite-difference approximations for the model equations in section 4 are discussed in this appendix. For consistency, the difference approximations are presented here in terms of the dimensionless variables of sections 3, 4, and 5 although the numerical solutions are obtained in corresponding dimensional variables. The finite difference grid for the variables (ψ, χ, w, ϕ) and the corresponding spatial difference operators are identical to those described in appendix A of Allen and Newberger (1993). Thus, those definitions are not repeated here.

The governing equations in difference form, corresponding to (5.14), (5.15), (5.16), (5.6), (5.7), (5.17) and (5.18) are

$$\delta_z w + \nabla^2 \chi = 0\,, \tag{A1}$$

$$\nabla^2 \phi_t + \delta_z (S^{-1} \delta_{\hat{z}} \phi_t) = RQ + \epsilon\alpha 2\{J(\delta_x \overline{\phi}^x, \delta_y \overline{\phi}^y)\}_t, \tag{A2}$$

$$\nabla^2 \chi = -\nabla^2 \phi_t + R\zeta + \epsilon\alpha 2\{J(\delta_x \overline{\phi}^x, \delta_y \overline{\phi}^y)\}_t, \tag{A3}$$

$$\nabla^2 \psi_2 = \nabla^2 \phi - 2\alpha\epsilon J(\delta_x \overline{\phi}^x, \delta_y \overline{\phi}^y), \tag{A4}$$

$$\nabla^2 \chi_2 = \alpha\{-J(\phi, \nabla^2 \phi) - \nu \nabla^6 \phi + \nabla^2 \mathcal{L}^{-1}(J(\phi, \mathcal{L}\phi) + \nu \nabla^6 \phi)\}\,, \tag{A5}$$

$$\nabla^2 G = RG\,, \tag{A6a}$$

$$\nabla^2 \psi_D + \delta_z (S^{-1} \delta_{\hat{z}} \psi_D) = \delta_z [(S^{-1} G) - \alpha\epsilon I_2']\,, \tag{A6b}$$

where (5.5) holds and where

$$RQ = R\zeta + RDZ\theta\,, \tag{A7a}$$

$$R\zeta = -J(\psi, \zeta_2) - \epsilon\Big[\nabla \cdot [\overline{w \nabla \delta_{\hat{z}} \psi_2}^z + \zeta_2 \nabla \chi]$$
$$+ \overline{J}^z (I_2, \delta_{\hat{z}} \phi)\Big] - \epsilon^2 \overline{J(w, \delta_{\hat{z}} \chi_2)}^z - \nu \nabla^4 \zeta_2), \tag{A7b}$$

$$RDZ\theta = -\delta_z \Big\{ S^{-1} \Big[J(\overline{\psi}^{\hat{z}}, \delta_{\hat{z}} \phi) + \epsilon[\nabla \cdot (\delta_{\hat{z}} \phi \overline{\nabla \chi}^{\hat{z}}) + \delta_{\hat{z}} (\overline{w \delta_{\hat{z}} \phi}^z)]\Big]\Big\}, \tag{A7c}$$

$$RG = -2[\delta_{\hat{z}} J(\delta_x \overline{\psi}_2^x, \delta_y \overline{\psi}_2^y) - \alpha \delta_{\hat{z}} J(\delta_x \overline{\phi}^x, \delta_y \overline{\phi}^y)] + \epsilon \Big\{ -\delta_{\hat{z}} J(\zeta_2, \chi)$$
$$- \delta_{\hat{z}} \overline{J}^z (w, \delta_{\hat{z}} \psi_2) - \delta_{\hat{z}} \nabla \cdot (\zeta_2 \nabla \psi_D) + \delta_{\hat{z}} \nabla \cdot \overline{(I_2 \nabla \delta_{\hat{z}} \phi)}^z$$
$$+ \nabla^2 \Big[\overline{\nabla \psi_D}^z \cdot \nabla \delta_{\hat{z}} \psi_2 + J(\delta_{\hat{z}} \psi_2, \overline{\chi}_D^z) - \delta_{\hat{z}} (\overline{\delta_{\hat{z}} \phi I_2}^z) + \delta_{\hat{z}} \overline{(I_2 \delta_{\hat{z}} \phi)}^z \Big]$$
$$+ \delta_{\hat{z}} J(\psi_2, \chi_2) + \nabla^2 \delta_{\hat{z}} \chi_{2t} + \nu \nabla^6 \delta_{\hat{z}} \chi_2 \Big\} + \epsilon^2 \Big\{ \delta_{\hat{z}} \nabla \cdot (\overline{w \nabla \delta_{\hat{z}} \chi_2}^z)$$
$$+ \nabla^2 \Big[\nabla \overline{\chi}_D^z \cdot \nabla \delta_{\hat{z}} \chi_2 + J(\overline{\psi}_D^z, \delta_{\hat{z}} \chi_2) \Big] + \frac{1}{2} \nabla^2 \delta_{\hat{z}} (|\nabla \chi_2|^2) \Big\}\,, \tag{A7d}$$

and the operator $\mathcal{L}$ (4.2) in (A5) is

$$\mathcal{L} \equiv \nabla^2 + \delta_z(S^{-1}\delta_{\hat{z}}) \,. \tag{A7e}$$

The function

$$I_2(z) = I_2(z_{k+(1/2)}) = \int_{-1}^{z} dz' \, (\nabla^2\psi_D + \alpha\epsilon\delta_z \, I_2') \,, \tag{A8a}$$

where the z integration is calculated as described for w in (A11) of Allen and Newberger (1993) and

$$\begin{aligned} \delta_z I_2' = -\{ & 2J(u_D, \delta_x\overline{\phi}^x) + 2J(v_D, \delta_y\overline{\phi}^y) \\ & + J(\nabla^2\chi_D, \phi) + \nu\nabla^6\chi_D \\ & + \mathcal{L}[J(\phi, \mathcal{L}^{-1}\nabla^2\chi_D)] - J(\mathcal{L}\phi, \mathcal{L}^{-1}\nabla^2\chi_D)\} \,, \end{aligned} \tag{A8b}$$

where

$$u_D = -\delta_y\overline{\psi}_D^y + \delta_x\overline{\chi}_D^x \,, \quad v_D = \delta_x\overline{\psi}_D^x + \delta_y\overline{\chi}_D^y \,. \tag{A8c, d}$$

We advance in time by using the implicit time difference scheme described for the balance equations (BE) in Allen and Newberger (1993). The equations (A2) and (A3) are time differenced as

$$\nabla^2\delta_t^{n+\frac{1}{2}}\phi + \delta_z(S^{-1}\delta_{\hat{z}}\delta_t^{n+\frac{1}{2}}\phi) = \overline{RQ}^{n+\frac{1}{2}} + RJT^{n+\frac{1}{2}} \,, \tag{A9a}$$

$$\nabla^2\overline{\chi}^{n+\frac{1}{2}} = -\nabla^2\delta_t^{n+\frac{1}{2}}\phi + \overline{R\zeta}^{n+\frac{1}{2}} + RJT^{n+\frac{1}{2}} \,, \tag{A9b}$$

where

$$RJT^{n+\frac{1}{2}} = \epsilon\alpha 2 \, \delta_t^{n+\frac{1}{2}} J(\delta_x\overline{\phi}^x, \delta_y\overline{\phi}^y) \,. \tag{A9c}$$

Equations (A1), (A4), (A5) and (A6*a*,*b*) are assumed to hold at each time level $t = n\Delta t$.

It follows from (A9*a*,*b*,*c*) that

$$\nabla^2\phi^{n+1} + \delta_z(S^{-1}\delta_{\hat{z}}\phi^{n+1}) = RIQ, \tag{A10a}$$

$$\nabla^2\chi^{n+1} = RI\zeta, \tag{A10b}$$

where

$$RIQ = \nabla^2\phi^n + \delta_z(S^{-1}\delta_{\hat{z}}\phi^n) + \Delta t \, \overline{RQ}^{n+\frac{1}{2}} + \Delta t \, RJT^{n+\frac{1}{2}} \,, \tag{A11a}$$

$$RI\zeta = -\nabla^2\chi^n + 2[-\nabla^2\delta_t^{n+\frac{1}{2}}\phi + \overline{R\zeta}^{n+\frac{1}{2}}] + 2\,RJT^{n+\frac{1}{2}} \,. \tag{A11b}$$

In solving (A10*a*) for ϕ^{n+1} and (A6*b*) for ψ_D^{n+1}, and calculating the $\mathcal{L}^{-1}$ terms in (A5) and (A8*b*), an expansion in terms of vertical linear normal modes is utilized as described in equations (A22) and (A23) in Allen and Newberger (1993).

With all variables known at $t = n\Delta t$ and at previous time levels, we solve (A10a,b), (A3), (A4) and (A6a,b) by iteration. Estimate all variables except ψ_2 and χ_2 at $t = (n+1)\Delta t$ by extrapolation, e.g., $\phi^{n+1} = 2\phi^n - \phi^{n-1}$. Solve (A4) and (A5) for ψ_2^{n+1} and χ_2^{n+1}. Use these estimates in the rhs of (A10a) and (A10b). Solve (A10a) for ϕ^{n+1}. Substitute the new value of ϕ^{n+1} in the time derivative term $\left(-\nabla^2 \delta_t^{n+\frac{1}{2}} \phi\right)$ on the rhs of (A10b) and solve (A10b) for χ^{n+1}. Calculate w^{n+1} from (A1). Using ϕ^{n+1}, calculate ψ_2^{n+1} from (A4), χ_2^{n+1} from (A5) and $\chi_D^{n+1} = \chi^{n+1} - \chi_2^{n+1}$. Approximate χ_{2t} in (A6a) as $(\chi_2^{n+1} - \chi_2^n)/\Delta t$. Substitute in the rhs of (A6a) and solve (A6a) for G^{n+1}. Calculate $\delta_z I_2'$ (A8a) and the rhs of (A6b) and solve (A6b) for ψ_D^{n+1}. Return to the step where (A10a) is solved for ϕ^{n+1} and substitute the latest values for the variables at $t = (n+1)\Delta t$ in the rhs. Repeat the cycle until convergence is obtained.

At $t = 0$, ϕ^0 is specified. Initial-values for $\psi_D^0, \chi^0, \psi_2^0, \chi_2^0$ and w^0 are found by an iterative procedure. Estimate $\psi_D^0 = \chi^0 = w^0 = 0$. Calculate ψ_2 (A4) and χ_2 (A5) which only depend on ϕ^0. Calculate ψ_D^0 from (A6a,b) with χ_{2t} set to zero, ϕ_t^0 from (A2), χ^0 from (A3), and w^0 from (A1). Repeat the calculations until convergence for ψ^0, χ^0, w^0, and ϕ_t^0 is obtained. For the first time step, estimate $\phi^1 = \phi^0 + \Delta t \phi_t^0, \psi_D^1 = \psi_D^0, \chi^1 = \chi^0, w^1 = w^0$ and calculate ψ_2^1 and χ_2^1 from (A4) and (A5) using the estimate for ϕ^1 and proceed with the general implicit time difference scheme.

References

Allen, J.S. (1991) 'Balance equations based on momentum equations with global invariants of potential enstrophy and energy', *J. Phys. Oceanogr.*, **21**, 265–276.

Allen, J.S. (1993) 'Iterated geostrophic intermediate models', *J. Phys. Oceanogr.*, **23**, 2447–2461.

Allen, J.S., Barth, J.A., Newberger, P.A. (1990a) 'On intermediate models for barotropic continental shelf and slope flow fields: Part I, Formulation and comparison of exact solutions', *J. Phys. Oceanogr.*, **20**, 1017–1042.

Allen, J.S., Barth, J.A., Newberger, P.A. (1990b) 'On intermediate models for barotropic continental shelf and slope flow fields: Part III, Comparison of numerical model solutions in periodic channels', *J. Phys. Oceanogr.*, **20**, 1949–1973.

Allen, J.S., Newberger, P.A. (1993) 'On intermediate models for stratified flow', *J. Phys. Oceanogr.*, **23**, 2462–2486.

Allen, J.S., Holm, D.D. (1996) 'Extended-geostrophic Hamiltonian models for rotating shallow water motion', *Physica D*, **98**, 229–248.

Barth, J.A., Allen, J.S., Newberger, P.A. (1990) 'On intermediate models for barotropic continental shelf and slope flow fields: Part II, Comparison of numerical model solutions in doubly periodic domains', *J. Phys. Oceanogr.*, **20**, 1044–1076.

Gent, P.R., McWilliams, J.C. (1983) 'Consistent balanced models in bounded and periodic domains', *Dyn. Atmos. Oceans*, **7**, 67–93.

Holm, D.D. (1996) 'Hamiltonian balance equations', *Physica D*, **98**, 379–414.

Holm, D.D., Marsden, J.E., Ratiu, T. (1998a) 'The Euler–Poincaré equations and semidirect products with applications to continuum theories', *Adv. in Math.*, **137**, 1–81.

Holm, D.D., Marsden, J.E., Ratiu, T. (1998b) 'Euler–Poincaré models of ideal fluids with nonlinear dispersion', *Phys. Rev. Lett.* **80**, 4173–4177.

Holm, D.D., Marsden, J.E., Ratiu, T. (2002) 'The Euler–Poincaré equations in geophysical fluid dynamics', in *Large-Scale Atmosphere–Ocean Dynamics, II*, J. Norbury and I. Roulstone (eds.), Cambridge University Press, 251–300.

Holm, D.D., Marsden, J.E., Ratiu, T., Weinstein, A. (1985) 'Nonlinear stability of fluid and plasma equilibria', *Physics Reports*, **123**, 1–116.

Hoskins, B.J. (1975) 'The geostrophic momentum approximation and the semigeostrophic equations', *J. Atmos Sci.*, **32**, 233–242.

Kosro, P.M., *et al.* (1991) 'The structure of the transition zone between coastal waters and the open ocean off northern California, winter and spring 1987', *J. Geophys. Res.*, **96**, 14707–14730.

Marsden, J.E., Ratiu, T.S. (1994) *Introduction to Mechanics and Symmetry*, Texts in Applied Mathematics, **17**, Springer-Verlag.

McWilliams, J.C., Gent, P.R. (1980) 'Intermediate models of planetary circulations in the atmosphere and ocean', *J. Atmos. Sci.*, **37**, 1657–1678.

Pedlosky, J. (1987) *Geophysical Fluid Dynamics*, Springer-Verlag, New York, 710 pp.

Pierce, S.D., Allen, J.S., Walstad, L.J. (1991) 'Dynamics of the coastal transition zone jet, 1. Linear stability analysis', *J. Geophys. Res.*, **96**, 14979–14993.

Salmon, R. (1983) 'Practical use of Hamilton's principle', *J. Fluid Mech.*, **132**, 431–444.

Salmon, R. (1985) 'New equations for nearly geostrophic flow', *J. Fluid Mech.*, **153**, 461–477.

Salmon, R. (1996) 'Large-scale semigeostrophic equations for use in ocean circulation models', *J. Fluid Mech.*, **318**, 85–105.

Walstad, L.J., Allen, J.S., Kosro, P.M., Huyer, A. (1991) 'Dynamics of the coastal transition zone through data assimilation studies', *J. Geophys. Res.*, **96**, 14959–14977.

Fast Singular Oscillating Limits of Stably-Stratified 3D Euler and Navier–Stokes Equations and Ageostrophic Wave Fronts

A. Babin, A. Mahalov, B. Nicolaenko

1 Introduction

Flows that are stably-stratified or are rotating have certain distinct characteristics which, unlike many flows, vary greatly in their form depending on how the flows are initiated. The characteristics also change as the flows move towards their respective equilibrium or quasi-equilibrium states. The initial effects of rotational and buoyancy forces with time scales $1/f_0$ and $1/N_0$, respectively, are to produce internal waves on those time scales and hence to exchange energy between distant points in the flow leading to significant changes in the form of the imposed flow. Here N_0 is the Brunt-Väisälä wave frequency and $f_0 = 2\Omega_0$ is the Coriolis parameter. The significance of the wave motion depends on the relative magnitude of the flow's time scale, T, to the rotational and buoyancy time scales. The length scales L and geometric shape (especially the ratio of the vertical to horizontal scale, H/L) of the initial disturbances are equally significant in determining the anisotropic form of the wave motion and the orientation of the constrained equilibrium forms, such as the 'Taylor' columns parallel to the rotation axis in rotation-dominated regimes, or the horizontal 'pancakes' or fronts characteristic of strong stable stratification in stratification-dominated regimes with rotation. The main objectives of a global mathematical study of these flows should be (cf. Cullen [25]):

(a) to identify, understand and quantify the key mechanisms, especially those dependent on nonlinear and frontal dynamics where numerical methods and linear analyses provide inaccurate and incomplete solution, and how they interact with others as the relative strengths of the rotational and buoyancy forces vary;

(b) the behaviour of the flows as the initial and boundary conditions vary, for example in the formation of fronts or multiple fronts, and whether or not they exhibit chaotic behaviour when averaged over timescales of interest ($1/f_0$ or $1/N_0$);

(c) to find efficient ways of calculating the flows using asymptotic and computational techniques;

(d) to assess how local and global errors might build up in approximate calculations.

The new mathematical approach reviewed here enables novel answers to be provided to the above questions; they are novel in that they are quite general and not specific to particular flows. The methods are based on the earlier theory of one-dimensional nonlinear oscillators (van der Pol, Bogoliubov) and of splitting the governing equations and the fields that satisfy them, in the case of three-dimensional unsteady inviscid rotating stratified flows, into quasi-geostrophic motions (zero horizontal divergence) and ageostrophic motions with non-zero horizontal divergence and a vertical velocity component that is a prognostic variable, a concept originated by Charney [20] and explained in detail by Gill [36] and Pedlosky [63]. The first step in this approach is to express the velocity as a modulation, varying on a 'slow' time scale T, of asymptotically much higher frequency oscillations, varying on time scales $1/f_0$ or $1/N_0$. It is then found that the nonlinear terms consist mainly of products of oscillatory functions with different periods. The second step is to average over time scales that are large compared to $1/f_0$ or $1/N_0$, which ensures that only the terms containing the products of resonant frequencies make a contribution. These initial disturbances are represented as Fourier series and it is found that only a restricted infinite series of Fourier coefficients need be evaluated to calculate these slow nonlinear terms and the 'slow envelope' variations of the flow (the restrictions are to resonant interactions of wave vectors). Thence many other aspects of the flow on this time scale can be derived, and many of the questions in (a)–(d) can be addressed. The third step is to separate regular terms in the averaged equations that depend continuously on parameters from the singular terms that depend discontinuously on the parameters; these two types should be treated in a different way.

This approach is quite different from the usual perturbation approach (e.g. geostrophic theory or rapid distortion theory) when the solution is expressed as a Taylor expansion or asymptotic expansion. Because these series typically are divergent with a finite 'radius of convergence' (which is usually very small) they become invalid after a short time. Therefore, unlike the van der Pol averaging method, they are not suitable methods for calculating flows over many time periods. Since these new methods show that errors in calculation on the slow time scales are controlled and do not grow, they indicate that errors in prediction grow more slowly than might be estimated from the initial value problem or from simple error estimates. A particularly critical problem of geophysical fluid dynamics that can be analysed using this approach, is the development of sharp interfaces in rotating stratified flows (see Section 9). These interfaces are where much of the 'weather' and other kinds of mixing events occur, such as the boundaries of the polar vortex. It is known that such interfaces can form from initial disturbances (e.g. Hoskins & Bretherton [41]) because of the variation of group speeds of waves of different wave lengths. But all the various different ways that fronts can form are not yet well established because the atmosphere is usually in a metastable state implying it has the capability of spawning large convective disturbances from smaller ones and small waves on fronts can grow into larger ones. These events will always be very difficult to predict, but the consequences of such growing disturbances in rotating stratified flows should be amenable to better understanding and analysis; that is the object of this review. One of the major difficulties encountered in understanding the dynamics of geophysical flows is the influence of the oscillations generated by rotation and stratification (buoyancy forces).

Very useful and thought provoking multi-scale analyses of rotating/stratified turbulence are presented in Riley *et al.* ([66]), Lilly ([49]), McWilliams ([56]). In particular, they argue that the velocity field of a rotating, stably-stratified fluid may be regarded as a superposition of waves which are modulated on a longer turbulence time scale. In our approach, the collective contribution to the dynamics made by waves is accounted for by rigorous estimates of wave resonances and quasi-resonances via small divisors analysis. Our theory handles rigorously all 3-wave resonances, but goes much deeper into the structure of quasi-3-wave resonances and their contributions. This mathematical approach in the context of geophysical flows was initiated in Babin, Mahalov & Nicolaenko (henceforth BMN) [5], Mahalov & Marcus [53]. In the context of symmetric hyperbolic systems, related singular limits have been investigated by Joly–Metivier–Rauch [43] and Schochet [69]. In Bartello ([16]), the relative physical importance of different resonances is discussed in depth. From the rigorous mathematics of such fast singular oscillating limits induced by fast inertio-gravity waves, we obtain a strong nonlinear interaction theory between potential vorticity dynamics and waves. Interactions between internal waves and the vortical (quasi-geostrophic) modes remained as one of the important questions to be addressed by strong interaction theory (Müller, Holloway *et al.*, [61]; Warn, [74]; Farge & Sadourny, [29]; Lelong & Riley, [47]).

The governing flow equations for 3D rotating stably-stratified fluids under the Boussinesq approximation are

$$\partial_t \mathbf{U} + \mathbf{U}\cdot\nabla\mathbf{U} + f_0 e_3 \times \mathbf{U} = -\nabla p + \rho_1\, e_3 + \nu_1 \Delta \mathbf{U} + \mathbf{F}, \qquad \nabla\cdot\mathbf{U} = 0, \tag{1.1}$$

$$\partial_t \rho_1 + \mathbf{U}\cdot\nabla\rho_1 = -N_0^2 U_3 + \nu_2 \Delta\rho_1 + F_4, \tag{1.2}$$

$$\mathbf{U}(t,x)|_{t=0} = \mathbf{U}(0,x), \qquad \rho_1(t,x)|_{t=0} = \rho_1(0,x) \tag{1.3}$$

where rotation and the mean stratification gradient are aligned parallel to the vertical axis x_3. Here $x = (x_1, x_2, x_3)$, $\mathbf{U} = (U_1, U_2, U_3)$ is the velocity field and ρ_1 is the buoyancy variable (relative density variation); N_0 is the Brunt-Väisälä wave frequency for constant stratification and Ω_0 is the frequency of background rotation, $f_0 = 2\Omega_0$, $\mathbf{F} = (F_1, F_2, F_3)$. Eqs. (1.1) and (1.2) are sometimes called the primitive (non-hydrostatic) equations of geophysical flows. In this chapter we consider the 3D initial value problem (1.1)–(1.3). We focus on both the inviscid situation and that with small uniform viscosities, $\nu_1 \geq 0, \nu_2 \geq 0$;

here ν_1 and ν_2 are the kinematic viscosity and the heat conductivity, respectively; the ratio $\mathrm{Pr} = \nu_1/\nu_2$ is known as the Prandtl number. We consider periodic boundary conditions in a parallepiped $[0, 2\pi a_1] \times [0, 2\pi a_2] \times [0, 2\pi a_3]$, as well as stress-free conditions $U_3 = 0$, $\partial U_1/\partial x_3 = \partial U_2/\partial x_3 = 0$ at $x_3 = 0$, $2\pi a_3$ (see [26]). For stress-free conditions one only needs to restrict the Fourier series to be even in x_3 for U_1, U_2 and odd in x_3 for U_3, ρ_1.

Let U_h be a characteristic horizontal velocity scale. Let H and L be the vertical and horizontal length scales and $a = H/L$ be the aspect ratio parameter. We define Froude numbers based on horizontal and vertical scales:

$$\mathrm{F}_h = U_h/(LN_0) \equiv 1/N, \; \mathrm{F}_v = U_h/(HN_0) = \mathrm{F}_h/a. \tag{1.4}$$

The classical Rossby and anisotropic Rossby numbers are defined as follows

$$\mathrm{Ro} = U_h/(f_0 L) \equiv 1/f, \qquad \mathrm{Ro}_a = a\mathrm{Ro}, \quad a = H/L. \tag{1.5}$$

In Eqs. (1.4)–(1.5) $f = \mathrm{Ro}^{-1}$ and $N = \mathrm{F}_h^{-1}$ are dimensionless rotation and stratification parameters. The Burger number characterises relative importance of the effects of rotation and stratification (e.g. McWilliams [56]):

$$\mathrm{Bu} = \mathrm{Ro}_a^2/\mathrm{F}_h^2 \equiv \mathrm{Ro}^2/\mathrm{F}_v^2 \equiv N^2 a^2/f^2 = N_0^2 a^2/f_0^2. \tag{1.6}$$

Flows with $\mathrm{Bu} \ll 1$ are rotation-dominated and $\mathrm{Bu} \gg 1$ corresponds to stratification-dominated flows. An equivalent measure of the relative importance of stratification and rotation is the internal radius of deformation Λ, which compares (stable) density stratification effects with respect to rotation. The internal (Rossby) radius of deformation Λ is defined as

$$\Lambda = N_0 H/f_0, \tag{1.7}$$

so that $\mathrm{Bu} = (\Lambda/L)^2$. Regimes where Bu is either much greater or much less than 1 are both important in the atmosphere ([25]). Processes which couple rotation and stratification, such as baroclinic instability, have $\mathrm{Bu} = O(1)$ as a natural case, [36].

Regimes of geophysical dynamics presenting the global picture for small Froude or Rossby numbers are shown in Figure 1. In [11] the asymptotic regimes of geophysical dynamics are described for different Burger number limits. As the Burger number increases from zero (rotation dominated flows) to infinity (stratification dominated flows), we demonstrate gradual unfreezing of energy cascades for ageostrophic (AG) dynamics. In the small Burger number regime, the process of geostrophic adjustment is inefficient and the inertio-gravity waves persist. On the other hand, in the $\mathrm{Bu} = O(1)$ and the large Burger number regimes, inertio-gravity waves efficiently cascade to small scales and are then destroyed by viscosity.

The case where there is strong rotation, but no stratification, is singular, (Figure 1, vertical axis). We prove in [6], [9] the generalised Taylor–Proudman theorem which controls the dynamics, and the flow is characterised by quasi-2D unsteady Taylor columns interacting with inertial waves ([6], [9]). In this regime energy cascades for the AG field are frozen in the vertical direction x_3 and the AG dynamics is pure *phase turbulence* ([6], [8]–[10]). In pure phase turbulence, the amplitudes of the AG modes remain approximately constant in absolute values; turbulent dynamics are restricted to the phases of the AG modes. This regime only occurs in limited regions of the atmosphere, though it can occur in neutrally stratified layers. A similar regime can occur in tropical cyclones, though here it is due to rapid system rotation.

Next is the regime of strong rotation and weak stratification as shown in Figure 1. In this regime $\mathrm{Bu} \ll 1$ and the horizontal length scale L is large compared with the Rossby radius Λ, $L \gg \Lambda$. This is the appropriate regime for large scale planetary waves in the extra-tropical atmosphere. In this regime energy cascades for the AG field are partially frozen in the vertical direction x_3 and the AG dynamics is predominantly phase turbulence. The total field consists of both geostrophic motion and inertio-gravity oscillations superimposed on the geostrophic motion. Geostrophic adjustment is inefficient in this regime. Observations of the internal structure of the atmosphere show that there are a lot of persistent 'quasi-inertia' oscillations with large horizontal and small vertical scale and they are also ubiquitous in the ocean ([70], [51]).

As the effects of stratification are increased (see Figure 1) AG cascades in the vertical x_3 become possible. If $\mathrm{Bu} = O(1)$, so that the horizontal length scale is comparable with the Rossby radius, we are considering the synoptic scale (e.g. [40]). In the limit of strong rotation and strong stratification, corresponding to $\mathrm{Ro}_a \to 0$, $\mathrm{Fr} \to 0$, but $\mathrm{Bu} = O(1) \Leftrightarrow L \approx \Lambda$, we established splitting between 3D QG (quasi-geostrophic) and the reduced AG field ([7]). Energy cascades are now allowed (unfrozen) for the AG field but they are restricted to anisotropic families of *rays* in Fourier space ([12]). Direct restricted energy cascades of the AG field along rays provide the mechanism for *nonlinear geostrophic adjustment.* This is fundamentally different from the rotation-dominated regimes where AG cascades in x_3 are frozen. This nonlinear geostrophic adjustment mechanism is indeed the capacity of the AG dynamics for transferring to smaller scales and eventually dissipating its inertio-gravitational energy (Sadourny, [68]). As shown by Farge & Sadourny, [29] in the context of rotating shallow-water equations, rotation inhibits nonlinear transfers and confines the inertio-gravity waves to scales larger than the Rossby deformation radius; therefore, geostrophic adjustment is possible only for scales smaller than the Rossby deformation radius.

The asymptotic regime $\mathrm{Bu} \gg 1$, $L \ll \Lambda$ holds for sub-synoptic horizontal scales, or most low latitude circulations. We expect a tendency towards horizontal density surfaces with gravity waves or gravity currents superposed on them. In the final sections of this chapter we analyse the intermediate asymptotic regime of strong stratification and weak rotation and we find the

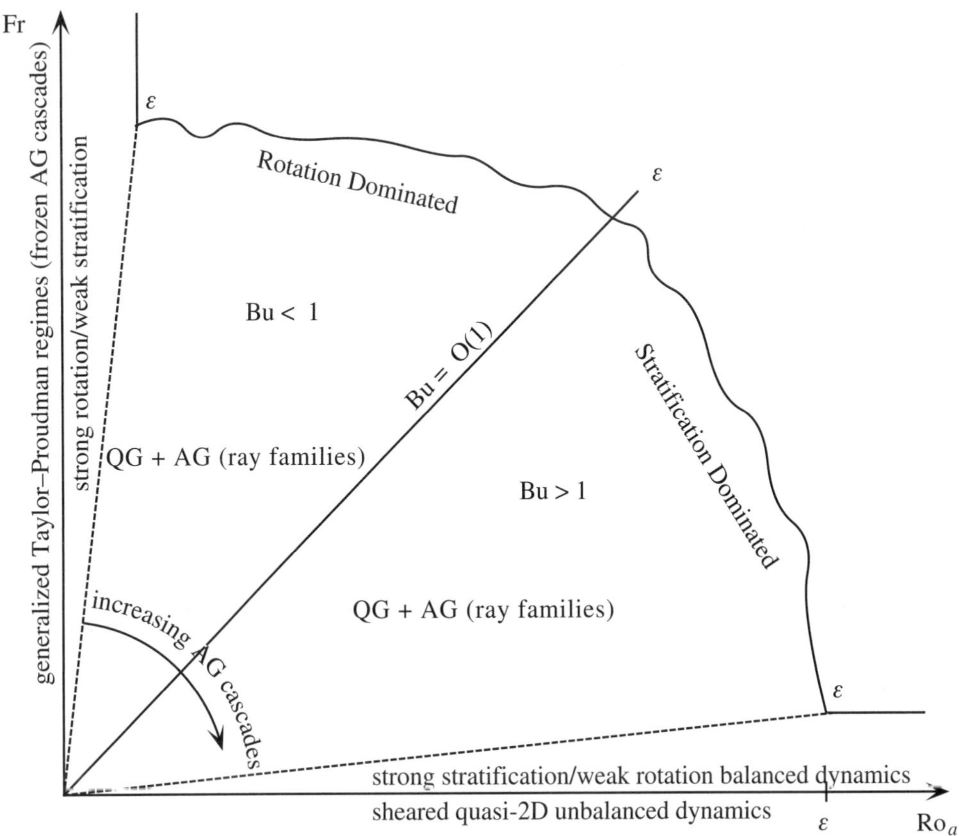

Figure 1: Geophysical Dynamics: the global picture for small Froude or small Rossby regimes.

principal term in asymptotics for sub-synoptic horizontal scales $L \approx \sqrt{\eta}\Lambda$, $\text{Bu} = (\Lambda/L)^2 = 1/\eta \gg 1$. Recalling that $\eta \approx 10^{-2}$ at mesoscale, our asymptotic analysis captures motions on horizontal scales which are about an order of magnitude smaller than the Rossby deformation radius.

In this chapter we present an in-depth mathematical investigation of the fast singular oscillating limits of Eqs. (1.1)–(1.3) as $f \to \infty$, $N \to \infty$, $\eta = f_0/N_0 = f/N$ fixed. In our approach, the collective contribution to the dynamics made by fast 'inertio-gravity' waves is accounted for by rigorous estimates of wave resonances and quasi-resonances via small divisors analysis. We briefly recall the principle of averaging Eqs. (1.1)–(1.2) over the fast time scales of inertio-gravity waves. The linear parts of inviscid Eqs. (1.1)–(1.2) written in dimensionless variables are

$$\partial_t \mathbf{U} + f e_3 \times \mathbf{U} - \rho_1 e_3 = -\nabla p, \quad \nabla \cdot \mathbf{U} = 0, \tag{1.8}$$

$$\partial_t \rho_1 + N^2 U_3 = 0. \tag{1.9}$$

The mathematical theory of Eqs. (1.8)–(1.9) and the corresponding non-homogeneous linear equations has attracted a considerable amount of attention. In the case of pure rotation Poincaré ([64]) reduced this linear system to one equation. Sobolev ([72]) studied the corresponding boundary problems for Eqs. (1.8)–(1.9) (see [2] for a historical review).

We denote by $E(Nt)$ the linear propagator solution to the initial value problem for (1.8)–(1.9); $E(Nt)$ is in fact a unitary group operator (preserves all Sobolev norms). The dispersion relation for inertio-gravity waves which are solutions of Eqs. (1.8)–(1.9) has the form

$$N^2\omega_n^2 = N^2\frac{|\check{n}'|^2}{|\check{n}|^2} + f^2\frac{\check{n}_3^2}{|\check{n}|^2} = N^2\left(\frac{|\check{n}'|^2}{|\check{n}|^2} + \eta^2\frac{\check{n}_3^2}{|\check{n}|^2}\right) \tag{1.10}$$

where $\check{n} = (n_1/a_1, n_2/a_2, n_3/a_3)$, $\check{n}' = (n_1/a_1, n_2/a_2, 0)$, $|\check{n}|^2 = n_1^2/a_1^2 + n_2^2/a_2^2 + n_3^2/a_3^2$, $|\check{n}'|^2 = n_1^2/a_1^2 + n_2^2/a_2^2$, $\eta = f/N = f_0/N_0$. Here a_1, a_2 and a_3 denote aspect ratios of the domain parallepiped. We note that all results in our work extend to boundary conditions periodic horizontally with zero flux in the vertical direction e_3 and no tangential stress on the boundary. One only needs to restrict Fourier series to be even in x_3 for U_1, U_2 and odd in x_3 for U_3, ρ_1. Such boundary conditions imply zero tangential stress on the vertical boundary (see [26]).

We have

$$\omega_n^2 = \frac{|\check{n}'|^2}{|\check{n}|^2} + \eta^2\frac{\check{n}_3^2}{|\check{n}|^2}. \tag{1.11}$$

It follows from (1.10) that the effects of rotation and stratification are *not uniform* on scales. In the case $|\check{n}'|/|\check{n}_3| \gg 1$ gravity waves are fast and inertial waves are slow. On the other hand, for scales satisfying $|\check{n}_3|/|\check{n}'| \gg 1$ gravity waves are slow and inertial waves are faster. This non-uniformity of the effects of rotation and stratification on different scales lies at the very heart of the nonlinear scale adjustment process described in Section 9. Control of resonances and quasi-resonances which resolves this non-uniformity can only be achieved through a careful analysis of small divisors in resonances ([5]–[12], [14], [15]).

After applying the Leray projection on divergence-free vector fields, we introduce the linear propagator directly into the nonlinearity in the dimensionless version of (1.1)–(1.2) using the change of variables (van der Pol transformation)

$$\mathbf{U}^\dagger(t) = E(-Nt)\mathbf{u}^\dagger(t), \tag{1.12}$$

where $\mathbf{U}^\dagger = (U_1, U_2, U_3, \rho_1)$ is the 'fast' field variable,; and $\mathbf{u}^\dagger$ the 'slow' Poincaré variable, after factorization via the fast oscillating ($N \gg 1$) propagator $\mathbf{E}(Nt)$. We define

$$\mathbf{B}(\mathbf{U}^\dagger, \mathbf{U}^\dagger) = (-\mathbf{P}(\mathbf{U}\cdot\nabla\mathbf{U}), -\mathbf{U}\cdot\nabla U_4), \qquad \mathbf{U}^\dagger = (\mathbf{U}, U_4) = (\mathbf{U}, \rho) \tag{1.13}$$

where $\mathbf{P}$ is the Leray projection on divergence-free vector fields. For $\nu_1 = \nu_2 = 0$, the 'primitive' Euler–Boussinesq equations (1.1)–(1.2), written in the Poincaré variables $\mathbf{u}^\dagger$, have the form

$$\begin{aligned} \partial_t \mathbf{u}^\dagger &= \mathbf{B}_p(Nt, \mathbf{u}^\dagger,\ \mathbf{u}^\dagger), \\ \mathbf{B}_p(Nt, \mathbf{u}^\dagger, \mathbf{u}^\dagger) &= \mathbf{E}(Nt)\mathbf{B}(\mathbf{E}(-Nt)\mathbf{u}^\dagger,\ \mathbf{E}(-Nt)\mathbf{u}^\dagger), \end{aligned} \tag{1.14}$$

where $\mathbf{B}_p$ is now an explicitly fast oscillating, non-autonomous operator in the 'slow' variable $\mathbf{u}^\dagger$. Equations (1.14) are explicitly time-dependent with rapidly varying coefficients. This is a problem of fast singular oscillating limits for a non-local hyperbolic system. Analogous problems are found in nonlinear geometric optics ([43]). The following equations describing asymptotic dynamics are associated with Eqs. (1.14) (BMN [7], [9], [14], [15]):

$$\partial_t \mathbf{w} = \widetilde{\mathbf{B}}(\mathbf{w},\ \mathbf{w}), \tag{1.15}$$

where the limit resonant operator $\widetilde{\mathbf{B}}$ in (1.15) is defined as

$$\begin{aligned} (\widetilde{\mathbf{B}}(\mathbf{u}, \mathbf{v}), \mathbf{z}) &= \lim_{T\to\infty} \frac{1}{T} \int_0^T (\mathbf{B}_p(Ns, \mathbf{u}, \mathbf{v}), \mathbf{z})\, ds \\ &= \lim_{N\to\infty} \frac{1}{T} \int_0^T (\mathbf{B}_p(Ns, \mathbf{u}, \mathbf{v}), \mathbf{z})\, ds \end{aligned} \tag{1.16}$$

where $\mathbf{u}$, $\mathbf{v}$ and $\mathbf{z}$ denote generic *time-independent* vector-functions; that is, we are averaging over the fast oscillations and keeping $\mathbf{u}$, $\mathbf{v}$ and $\mathbf{z}$ adiabatically frozen in Eqs. (1.16).

Clearly, when represented in Fourier modes in the limit $N \to +\infty$, $\eta = f/N$ fixed, the right-hand side of (1.15) will be determined by resonances $\pm\omega'_k \pm \omega'_m \pm \omega'_n = 0$ within terms of the type $\exp(iN(\pm\omega'_k \pm \omega'_m \pm \omega'_n)t)$, see (1.10)–(1.14). Here $\omega'_n = 0$ for QG modes and $\omega'_n = \omega_n$ is given by (1.11) for AG modes (similarly, ω_k and ω_m). With ω_n being the normalised spectral frequencies of inertio-gravity waves given by (1.11), the dependence of resonances

$$D_l(k, m, n) = \pm\omega'_k(a_1, a_2, a_3, \eta) \pm \omega'_m(a_1, a_2, a_3, \eta) \pm \omega'_n(a_1, a_2, a_3, \eta) = 0, \tag{1.17}$$

where $l = 1, \ldots, 8$ (eight combinations of $+$ and $-$ signs), and quasi-resonances

$$D_l(k, m, n) = \pm\omega'_k(a_1, a_2, a_3, \eta) \pm \omega'_m(a_1, a_2, a_3, \eta) \pm \omega'_n(a_1, a_2, a_3, \eta) = \delta, \tag{1.18}$$

on the parameters of the problem a_1, a_2, a_3 and η, and the algebraic geometry of this non-standard small divisor problem, are the basis of our analysis of fast singular oscillating limits for 3D primitive equations. In Eq. (1.18), $\delta = 0$ for exact resonances and is a small parameter for quasi-resonances ([1]). No resonances and quasi-resonances are neglected in our analysis but rather weights are assigned to them according to their importance. The concept of quasi-resonance goes back to Poincaré. In the limit $N \to +\infty$, $\eta = f/N$ fixed, our mathematical theory handles rigorously all resonances including 3-waves fast-fast-fast resonances and estimates the contribution of 3-wave quasi-resonances ([7]–[15]).

In earlier papers, an approach based on choosing special sets of 'prepared' initial data with infinite codimension was used in Bourgeois & Beale [18] and Chemin [21] to obtain long time existence, but it effectively filters out the nonlinear interactions between inertio-gravity waves and the QG (potential vorticity) fields. The first results on the global regularity of 3D Euler and Navier–Stokes systems (1.1)–(1.3) for general 'unprepared' initial data were obtained in BMN [5], [6], [9]. The first results on regularity in the context of geophysical flows with both rotation and stratification were obtained in BMN [7], [12]. Revealed in these papers was the crucial role of the parameters η, $\theta_2 = 1/a_2^2$ and $\theta_3 = 1/a_3^2$ for the properties of the dynamics (we set $a_1 = 1$ using a simple rescaling; in the general case one has to put $\theta_2 = a_1^2/a_2^2, \theta_3 = a_1^2/a_3^2$).

Our conditions on smoothness of initial data and forcing term were later relaxed in BMN [9], Avrin *et al.* [3] and Gallagher [31], [32], [33] (the '$2\frac{1}{2}$-dimensional' nonlinear limit equations for 3 wave resonances were not considered). In this chapter, we remove restrictions to non-resonant domains (non-resonant parameters θ_2, θ_3) as well as to the smoothness conditions of [31], [32], [33]. For the viscous case with both ν_1, and $\nu_2 > 0$, we demonstrate the global existence for infinite times for all values of η, a_1, a_2, a_3, without any restrictions (including all 3-wave resonances, [14] and [15]). We also relax (in the viscous case) conditions on the time behaviour of the forcing term $\mathbf{F}^\dagger(t)$; in contrast to [7], [9] and [31], [32], [33], no conditions are imposed here on $\partial_t \mathbf{F}^\dagger$. For the viscous case, our smoothness conditions for global regularity are now the same as those for local regularity theorems.

Indeed, in regularity theorems we impose only an integral regularity condition on the forcing term $\mathbf{F}^\dagger = (\mathbf{F}, F_4)$:

$$\sup_T \int_T^{T+1} \|\mathbf{F}^\dagger\|_{\alpha-1}^2 dt \le M_{\alpha F}^2, \tag{1.19}$$

with $\alpha \ge 3/4$. We do not impose (in contrast to Gallagher [31], [32], [33]) conditions on L_∞ norms of $\mathbf{F}^\dagger(t)$. Note also that Gallagher [31], [33] assumes that $\mathbf{F}^\dagger \in L_2(\mathbf{R}^+, H_0)$ which implies that $\int_T^\infty \|\mathbf{F}^\dagger\|_0 dt \to 0$ as $T \to \infty$; that is, $\mathbf{F}^\dagger(t)$ decays. We do not assume any decay of $\mathbf{F}^\dagger(t)$. We also consider the case when the kinematic viscosity $\nu_1 > 0$ but $\nu_2 = 0$; in this case we prove regularity on arbitrary large (but finite) interval when $\alpha \ge 1$.

In BMN [6], [9], [7] and here in Section 3 we obtain strong convergence results with uniform error estimates in η, θ_1, θ_2, θ_3 on parameter sets of full Lebesgue measure and with initial data being in the Sobolev space $H_{8\frac{1}{2}}$ uniformly in $\nu_1, \nu_2 \ge 0$. They cover the physically relevant case of huge atmospheric Reynolds numbers. This is in contrast with the work of Embid and

Majda [28] where, following general theorems of Schochet [69], they state a pointwise convergence theorem on a small time interval $[0,T]$ for every value of η, a_2, a_3 without explicit estimate of error; in fact, as is proven in BMN [9], [10], it is impossible to obtain explicit, uniform estimates if one does not delete almost-resonant sets of parameters. The limit equations are totally discontinuous in the Burger-like parameter η. No standard averaging theorems such as in Schochet [69] can handle our infinite time existence Theorems 1.1 and 1.2.

Following Métais and Herring [59] we introduce a change of variables $\rho_1 = N\rho$ and combine the velocity and buoyancy variables into one variable $\mathbf{U}^\dagger = (\mathbf{U}, \rho)$ after which Eqs. (1.1)–(1.2) written in dimensionless variables take the more symmetric form:

$$\begin{aligned} \partial_t \mathbf{U}^\dagger + \mathbf{U}\cdot\nabla\mathbf{U}^\dagger &= -\nabla^\dagger p - N\mathbf{M}\mathbf{U}^\dagger + \overline{\boldsymbol{\nu}}\Delta\mathbf{U}^\dagger + \mathbf{F}^\dagger, \quad \nabla\cdot\mathbf{U} = 0 \\ \mathbf{U}^\dagger(t,x)|_{t=0} &= \mathbf{U}^\dagger(0,x) \end{aligned} \tag{1.20}$$

where $\nabla^\dagger p = (\nabla p, 0), \mathbf{F}^\dagger = (\mathbf{F}, F_4)$ (where F_4 is rescaled),

$$\mathbf{M} = (\mathbf{S} + \eta\mathbf{R}),\ \eta = f/N,$$

$$\mathbf{R} = \begin{pmatrix} \mathbf{J} & \mathbf{0} \\ \mathbf{0} & \mathbf{0} \end{pmatrix},\ \mathbf{S} = \begin{pmatrix} \mathbf{0} & \mathbf{0} \\ \mathbf{0} & \mathbf{J} \end{pmatrix},\ \mathbf{J} = \begin{pmatrix} 0 & -1 \\ 1 & 0 \end{pmatrix}, \tag{1.21}$$

where $\overline{\boldsymbol{\nu}} = \mathrm{diag}(\nu_1, \nu_1, \nu_1, \nu_2)$ is the viscosity matrix, η being fixed.

There are three foremost issues with the analysis of (1.20) for large parameters N and f. First, the nature of the limit asymptotic equations as $N \to +\infty$ and the regularity of their solutions ('$2\frac{1}{2}$-dimensional' Navier–Stokes primitive equations). Second, the convergence of solutions of (1.20) to those of the limit equations; and, finally, bootstrapping from an analysis of the first two questions, the infinite time regularity of solutions of (1.20) for N large but finite.

The proof of the global regularity of the 3D primitive Navier–Stokes equations (1.20) for resonant domains presented in this chapter (Sections 6 and 7) relies on the global regularity of the '$2\frac{1}{2}$-dimensional' limit nonlinear 'primitive' Navier–Stokes equations and techniques for convergence theorems as $N \to \infty$ developed in [7], [9], [3], [13]. The technique of bootstrapping regularity of solutions of the 3D Navier–Stokes equations by perturbation from limit equations has been done in various contexts, for example, thin domains, [65], and helical flows, [52]. In these previous works, the limit equations are 2D Navier–Stokes equations for which global regularity is well known. In the present work, the limit equations are genuinely three-dimensional depending on all three variables x_1, x_2 and x_3 but with restricted wave-number interactions in the nonlinear term. The existence and regularity theory for those limit equations is non-trivial.

In [14], we demonstrated the global regularity of Eqs. (1.20) in the pure rotation case ($N = 0$, $\rho_1 = 0$) for large Coriolis parameters f including the case of 3-wave resonances with the '$2\frac{1}{2}$-dimensional' limit equations. In Sections 6 and 7 below we extend the results of [14] to the full primitive 3D equations (1.20) ($N \neq 0$). Our main mathematical result is the uniform existence in infinite time of regular strong solutions of Eqs. (1.20) for large but finite stratification parameters N. This result holds for all domain parameters a_1, a_2, a_3 including the case of domains with 3-wave resonances for inertio-gravity waves; such resonances yield strongly nonlinear '$2\frac{1}{2}$-dimensional' limit equations. The global existence is proven using techniques of Littlewood–Paley dyadic decomposition. The smoothness conditions we need are like those in standard local regularity theorems and do not include the technical smoothness conditions of BMN [7], [12]. All restrictions on the domain parameters are also removed.

In this chapter we prove the following main theorems: the Sobolev spaces H_α of periodic functions with zero mean are defined in Eqs. (2.1)–(2.2).

Theorem 1.1 *Let* $\eta = f/N$ *and the domain parameters* a_1, a_2, a_3 *be fixed but arbitrary. Let* $\nu_1, \nu_2 > 0$, $\nu = \min(\nu_1, \nu_2)$ *and the condition* (1.19) *on the force* $\mathbf{F}^\dagger(t, x)$ *be satisfied. Let* $\|\mathbf{U}^\dagger(0)\|_\alpha \leq M_\alpha$ *where* $\alpha > 3/4$. *Then for* $N \geq N_1(M_\alpha, M_{\alpha F}, \nu, a_1, a_2, a_3)$, *solutions of the 3D Navier–Stokes 'primitive' Eqs.* (1.20) *are regular for all* $t \geq 0$, *and* $\|\mathbf{U}^\dagger(t)\|_\alpha \leq M'_\alpha(M_\alpha, M_{\alpha F}, \nu, a_1, a_2, a_3)$ *for all* $t \geq 0$.

Theorem 1.2 *Let* $\eta = f/N$ *and the domain parameters* a_1, a_2, a_3 *be fixed but arbitrary. Let* ν_1, $\nu_2 > 0$, $\nu = \min(\nu_1, \nu_2)$, $\alpha > 3/4$ *and the condition* (1.19) *on the force be satisfied. Let* $||\mathbf{U}^\dagger(0)||_0 \leq M_0$, $\hat{T} = \hat{T}(M_0, M_{\alpha F}, \nu)$. *Then for every* $N \geq N'(a_1, a_2, a_3, \nu, M_{\alpha F})$, N' *independent of* M_0 *and for every weak solution* $\mathbf{U}^\dagger(t, x_1, x_2, x_3)$ *of the three-dimensional 'primitive' Navier–Stokes equations* (1.20) *defined on* $[0, \hat{T}]$ *which satisfies the classical energy estimates on* $[0, \hat{T}]$, *the following holds:* $\mathbf{U}^\dagger(t, x_1, x_2, x_3)$ *can be extended to* $0 < t < +\infty$ *and it is regular for every* $t : \hat{T} \leq t < +\infty$; $\mathbf{U}^\dagger(t, x_1, x_2, x_3)$ *belongs to* H_α *and* $||\mathbf{U}^\dagger(t, x_1, x_2, x_3)||_\alpha \leq C_1(a_1, a_2, a_3, M_{\alpha F}, \nu)$ *for every* $t \geq \hat{T}$. *If* $\mathbf{F}^\dagger$ *is independent of* t *then there exists a global attractor for the three-dimensional primitive Navier–Stokes equations* (1.20) *bounded in* H_α; *such an attractor has a finite fractal dimension and attracts every weak Leray solution as* $t \to +\infty$.

Remark 1.1 In the pure rotation case [14] only the condition $\alpha > 1/2$ is imposed on the force and the initial data. Here the condition $\alpha > 3/4$ is restricted only by the minimal regularity results for the viscous QG equations, see Section 4 below.

For the *inviscid* primitive Euler–Boussinesq Eqs. (1.20), we show in Section 8 that global regularity holds not simply for almost all values of the parameters η, a_2, a_3 (as is stated in [31], [32], [33]), but for triplets η, a_2, a_3 which do not belong to an explicitly described strictly resonant set. Specifically, (η, a_2, a_3) do not belong to $\Theta^* = \bigcup_{k,m} \Theta^*_{k,m}$ in the three-dimensional parameter space

$(\eta, \theta_2, \theta_3)$ where k, m are integer wave vectors and $\Theta^*_{k,m}$ is, for every k, m, a smooth analytic surface with equation $\eta = \eta^*_{k,m}(\theta_2, \theta_3)$ where $(\theta_2, \theta_3) = (1/a_2^2, 1/a_3^2)$. Thus global regularity holds for all a_2, a_3 (*all domains*) provided that $\eta \notin \Theta^*(\theta_2, \theta_3)$. The small divisor problem for the fast oscillating limits of Eqs. (1.1)–(1.2) is not of the simple type $\pm|k| \pm |m| \pm |n| = 0$, such as in [43] and [69].

Now we describe the structure of the limit resonant equations which will be derived in Sections 2 and 3. From now on we are going to restrict ourselves to η bounded, including $\eta \ll 1$. The case of strong rotation and weak stratification $\eta \gg 1$ must be treated separately and will be published elsewhere. The case $\eta = \infty$ ($f \to \infty$, $N = 0$) was the subject of our papers on pure fast rotating limit without stratification ([5]–[10], [14]).

For all parameters a_1, a_2 and a_3 and all values of the parameter $\eta = f/N$ in the limit resonant equations (1.15), the total velocity $\mathbf{w}$ splits into the QG field $\mathbf{w}_{\mathrm{QG}}(t)$ satisfying the 3D QG equations

$$\partial_t \mathbf{w}_{\mathrm{QG}} = \mathbf{B}_0(\mathbf{w}_{\mathrm{QG}}, \mathbf{w}_{\mathrm{QG}}) + \nu_{\mathrm{QG}} \Delta \mathbf{w}_{\mathrm{QG}} + \mathbf{F}_{\mathrm{QG}}, \tag{1.22}$$

and the AG components satisfying equations of the type:

$$\partial_t \mathbf{w}_{\mathrm{AG}} = \mathbf{B}_2(\mathbf{w}_{\mathrm{QG}}, \mathbf{w}_{\mathrm{AG}}) + \mathbf{B}_3(\mathbf{w}_{\mathrm{AG}}, \mathbf{w}_{\mathrm{AG}}) + \nu_{\mathrm{AG}} \Delta \mathbf{w}_{\mathrm{AG}} + \mathbf{F}_{\mathrm{AG}}; \tag{1.23}$$

here $\nu_{\mathrm{QG}}(\nu_1, \nu_2)$ and $\nu_{\mathrm{AG}}(\nu_1, \nu_2)$ are in general non-local zeroth-order pseudo-differential operators, whenever $\nu_1 \neq \nu_2$ (see [12]).

Eq. (1.22) results from the 'slow' $(\mathbf{w}_{\mathrm{QG}}, \mathbf{w}_{\mathrm{QG}}, \mathbf{w}_{\mathrm{QG}})$ triads as well as all resonant $(\mathbf{w}_{\mathrm{AG}}, \mathbf{w}_{\mathrm{AG}}, \mathbf{w}_{\mathrm{QG}})$ triads (the contribution of the latter is exactly zero in the limit, hence the operator splitting). The AG limit Eq. (1.23) is derived from both resonant $(\mathbf{w}_{\mathrm{AG}}, \mathbf{w}_{\mathrm{QG}}, \mathbf{w}_{\mathrm{AG}})$ and $(\mathbf{w}_{\mathrm{QG}}, \mathbf{w}_{\mathrm{AG}}, \mathbf{w}_{\mathrm{AG}})$ triads as well as the 3-wave resonances $(\mathbf{w}_{\mathrm{AG}}, \mathbf{w}_{\mathrm{AG}}, \mathbf{w}_{\mathrm{AG}})$. Notice that the slow-fast-slow $(\mathbf{w}_{\mathrm{QG}}, \mathbf{w}_{\mathrm{AG}}, \mathbf{w}_{\mathrm{QG}})$ triads are not resonant to the lowest order in $1/N$ and appear only at the next order in $1/N$ via 4-wave resonances (see also [16]). For any given parameters a_1, a_2 and a_3 there is only a rare non-dense discrete set $\{\eta_j\}_{j=1}^{\infty}$ such that only for $\eta = \eta_j$ there are 3-wave fast-fast-fast resonances; in particular, there is a whole interval of η centered at $\eta = f/N = 1$ where there are no 3-wave resonances. Even in the context of η equals resonant η_j, we demonstrate that the operator $\mathbf{B}_3$ in (1.23) generally induces only a finite-dimensional dynamical system (hence no energy cascades; energy cascades in (1.23) are controlled by $\mathbf{B}_2$).

The quasigeostrophic equations in the inviscid case have a global regular solution according to Bourgeois & Beale [18]; this is also so if $\nu_1 > 0$, according to a theorem we prove in Section 4. Note that the nonlinear operator $\mathbf{B}_3$ depends discontinuously on the parameters η, θ_2, θ_3; it is non-zero only on a set of measure zero, namely the set $\Theta^*(\theta_2, \theta_3)$ (see the proof in BMN [9] for the similar pure rotating case). In BMN [6], [7], [9] and in Section 3 below, it is shown that if one deletes a resonant set Θ^* of parameters η, θ_2, θ_3, then $\mathbf{B}_3(\mathbf{w}_{\mathrm{AG}}, \mathbf{w}_{\mathrm{AG}}) = 0$ and only 'catalytic' interactions described by the operator $\mathbf{B}_2(\mathbf{w}_{\mathrm{QG}}, \mathbf{w}_{\mathrm{AG}})$ linear in $\mathbf{w}_{\mathrm{AG}}$, rule AG dynamics:

$$\partial_t \mathbf{w}_{\mathrm{AG}} = \mathbf{B}_2(\mathbf{w}_{\mathrm{QG}}, \mathbf{w}_{\mathrm{AG}}) + \nu_{\mathrm{AG}} \Delta \mathbf{w}_{\mathrm{AG}} + \mathbf{F}_{\mathrm{AG}}, \tag{1.24}$$

where $\mathbf{w}_{\mathrm{QG}}(t)$ is a solution of the 3D QG equations. Here, we refine this result in proving that only the parameter η need not be resonant (Section 3). The reduction to (1.24) holds for all values of the parameter η except a non-dense set of discrete values $\{\eta_j\}_{j=1}^{\infty}$ $(\eta \neq \eta_j)$. We prove here that this linear system (1.24) has a global smooth solution even when $\nu_{\mathrm{AG}} = 0$, provided $\mathbf{w}_{\mathrm{QG}}$ is smooth enough (Section 5).

Further, for all η and all a_3 when a_2^2 is irrational, $\mathbf{B}_2(\mathbf{w}_{\mathrm{QG}}, \mathbf{w}_{\mathrm{AG}})$ splits in Fourier space into uncoupled, restricted interaction operators on 4-ray families in Fourier space, where λ is any rational number [11], [12]:

$$\begin{pmatrix} m_1 \\ m_2 \\ m_3 \end{pmatrix} = \lambda \begin{pmatrix} \pm 1 & 0 & 0 \\ 0 & \pm 1 & 0 \\ 0 & 0 & \pm 1 \end{pmatrix} \begin{pmatrix} n_1 \\ n_2 \\ n_3 \end{pmatrix}. \tag{1.25}$$

In Eq. (1.24) direct cascades of energy are allowed for $\mathbf{w}_{\mathrm{AG}}$ through $\mathbf{B}_2(\mathbf{w}_{\mathrm{QG}}, \mathbf{w}_{\mathrm{AG}})$. The fact that 'catalytic' fast-fast-slow interactions between the QG modes and two AG modes dominate the AG dynamics is confirmed by numerical simulations. Bartello, [16], shows that resonant 3-wave interactions are of secondary importance in the overall picture of interactions when both rotation and stratification are present. They do not lead to slow-fast energy exchange and are difficult to resonate. An interaction is 'catalytic' in that it does not influence slow modes, but serves to transfer fast AG energy downscale. In our work, non-resonant fast-slow-slow interactions appear at the next order in Ro_a or $1/N$ at $\mathrm{Bu} = O(1)$, and contribute to the feedback of the AG field onto the QG one. Our resonance theory lets us treat them in a systematic way as a next order term.

For such resonant values of the parameter η and/or the aspect ratios a_1, a_2, a_3 for which there do exist fast-fast-fast 3-wave resonances, we demonstrate in Section 6 that the limiting '$2\frac{1}{2}$-dimensional' Navier–Stokes equations (1.23) do have global regular solutions on infinite time intervals, through non-trivial estimates of $\mathbf{B}_3(\mathbf{w}_{\mathrm{AG}}, \mathbf{w}_{\mathrm{AG}})$ in Eq. (1.23). This implies the global existence on infinite time intervals of regular solutions to the 3D Navier–Stokes primitive equations of geophysics (1.1)–(1.2) for small Froude and/or Rossby numbers (Section 7).

In this chapter we use the Craya–Fourier cyclic basis for representing physical fields. It is convenient to change variables so that the primitive physical variables $\mathbf{U}$, ρ_1 are replaced by three new variables $\mathbf{w} = (w_0, w_1, w_2)$, one of which, w_0, is effectively the potential vorticity $\tilde{q}$. The second variable, w_1, is the divergent velocity potential $\chi = (-\Delta_h)^{-1}\partial U_3/\partial x_3$. The third component, w_2, is related in a simple way to the vertical motion or omega equation (e.g. [40], [34], [35]). Up to a normalization, w_2 is precisely the geostrophic departure/thermal wind imbalance $-\nabla_h^2(\text{buoyancy}) + f\frac{\partial}{\partial x_3}(\text{vertical vorticity})$. It characterises *imbalance* in the vertical motion or omega equation. This change of variables requires constant f and N. However, the projection can be generalised. In certain cases it is possible to use this projection with f and N allowed to vary. This gives the so-called 'implicit normal mode' introduced by Temperton ([25]). We refer to [25] for additional discussion of the Craya basis.

This article is organised as follows. In Section 2 we recast the primitive equations in the Craya cyclic basis and present the limit resonant equations. In Section 3, we investigate in depth resonances and quasi-resonances for both 3-waves and 2-waves. We also describe the uniform, in η, a_1, a_2, a_3, convergence results. These results require much less differentiability than those in BMN [7]. In fact, the approach of BMN [7] can be applied to the case $\eta \geq 0$; here in the case $1/\eta_0 \leq \eta \leq \eta_0$ with fixed $\eta_0 > 1$, we obtain better and simpler estimates of small divisors which result in milder smoothness restrictions: now only six derivatives on initial data are required for the uniform (in η, a_1, a_2, a_3) convergence results to the 3D QG component, with arbitrarily large AG initial data.

In Section 4, we give regularity properties of the 3DQG equations with nonlocal limit operators $\nu_{\mathrm{QG}} > 0$ (these are resonant limits whenever both ν_1, $\nu_2 > 0$); we also consider the partially inviscid case $\nu_2 = 0$, $\nu_{\mathrm{QG}} \geq 0$ where now the limit operator ν_{QG} is only non-negative. In Section 5, we study the 'catalytic' AG equations and establish their regularity in H_s, $s \geq 0$, in the inviscid case. For $s > 0$, this is not trivial, as only the energy is conserved (L^2 norm). Although the inviscid AG equations (1.24) with $\nu_{\mathrm{AG}} = 0$, are linear, their coefficients involve the time-dependent $\mathbf{w}_{\mathrm{QG}}(t)$ and the solutions need not be bounded globally for all times in H_s norms, $s \geq 1$; their properties are very different from the pure rotating Euler case, contrary to the assertion of [32].

In Section 6 we demonstrate the global existence of strong solutions of the limit Navier–Stokes equations ($\nu_1 > 0$, $\nu_2 > 0$) for all domain aspect ratios and all small Froude and Rossby numbers, including the case of 3-wave resonances which yield nonlinear '$2\frac{1}{2}$-dimensional' limit equations. In Section 7, we give new regularity and existence theorems for all times of the viscous primitive equations of geophysics with 'unprepared' initial data in H_α, $\alpha \geq 3/4$, including all 3-wave resonances. We establish the regularization of Leray's classical weak solutions, for N and f finite, albeit large enough. In Section 8, we establish arbitrarily long-time existence results for the inviscid Euler–Boussinesq equations with initial conditions in H_α, $\alpha > 5/2$, again with arbitrarily large 'unprepared' AG initial data, but excluding the case of 3-wave resonances.

In Section 9 we present our most important physical results on the nonlinear dynamics of strongly stratified weakly rotating flows; we describe classes of nonlinear anisotropic AG baroclinic waves which are generated by the strong nonlinear interactions between the QG modes and inertio-gravity waves. In the asymptotic regime of strong stratification and weak rotation we show how switching on weak rotation triggers frontogenesis. The mechanism of front formation is contraction in the horizontal dimension balanced by vertical shearing through coupling of large horizontal and small vertical scales by weak rotation. Vertical slanting of these fronts is proportional to $\sqrt{\eta}$ where η is the ratio of Coriolis and Brunt-Väisälä parameters. These fronts select slow baroclinic waves through the nonlinear adjustment of the horizontal to vertical scale by weak rotation, and are the envelope of inertio-gravity waves. Mathematically, this is generated by asymptotic hyperbolic systems describing the strong nonlinear interactions between waves and potential vorticity dynamics. This frontogenesis yields vertical 'glueing' of pancake dynamics, in contrast to the independent dynamics of horizontal layers in strongly-stratified turbulence without rotation.

2 The limit resonant equations in the Craya–Fourier basis

In this section we present the limit asymptotic resonant equations in the Craya basis. The Craya basis was originally introduced in [24].

We use Fourier series expansions for fields

$$\mathbf{U}^{\dagger}(x) = (U_1(x), U_2(x), U_3(x), \rho(x)), \quad x = (x_1, x_2, x_3):$$

$$\mathbf{U}^{\dagger}(x) = \sum_n \exp(i(n_1 x_1 + n_2 x_2/a_2 + n_3 x_3/a_3))\mathbf{U}_n^{\dagger} = \sum_n \exp(i\check{n}\cdot x)\mathbf{U}_n^{\dagger} \quad (2.1)$$

where $\mathbf{U}_n^{\dagger}$ are the (4-component) Fourier coefficients, $[n_1, n_2, n_3] \in \mathbf{Z}^3$, $\check{n} = [n_1, n_2/a_2, n_3/a_3]$ are wavenumbers ($a_1 = 1$). We introduce the space of functions H_s with the norm defined on Fourier coefficients $\mathbf{U}_n^{\dagger}$ as follows (where $|\check{n}| = (n_1^2 + n_2^2/a_2^2 + n_3^2/a_3^2)^{1/2}$):

$$||\mathbf{U}^{\dagger}||_{H_s}^2 = \sum_n |\check{n}|^{2s} |\mathbf{U}_n^{\dagger}|^2. \quad (2.2)$$

We assume that all functions have zero average over the periodic parallepiped. Stress-free boundary conditions at $x_3 = 0, 2\pi a_3$ correspond to U_1, U_2 even in x_3 and U_3, ρ odd in x_3. Sobolev spaces are restricted to such functions. Here, $\mathbf{R}_n$, $\mathbf{S}_n$ will denote the action of the operators $\mathbf{R}$ and $\mathbf{S}$ defined in (1.21) on nth Fourier component, $\eta\mathbf{R}_n + \mathbf{S}_n = \mathbf{M}_n$.

We take into account the divergence-free condition by applying the Helmholtz–Leray projection $\mathbf{P}^d$ onto divergence-free vector fields. The matrix $(\mathbf{P}^d\mathbf{M}\mathbf{P}^d)_n$ is a real skew-symmetric matrix; the corresponding operator restricted to the 3D subspace of divergence-free vectors $\mathbf{U}_n^\dagger$ has one zero eigenvalue and two complex conjugate eigenvalues $\pm i\omega_n \neq 0$. We introduce the divergence-free vectors (2.4) which form a real cyclic basis for it:

$$\mathbf{P}_n^d\mathbf{M}q_{0n} = 0,\ \mathbf{P}_n^d\mathbf{M}q_{1n} = -\omega_n q_{2n},\ \mathbf{P}_n^d\mathbf{M}q_{2n} = \omega_n q_{1n}, \tag{2.3}$$

where $\mathbf{P}_n^d q_{jn} = q_{jn}$,

$$q_{0n} = \frac{1}{\omega_n}(\phi_n p_{0n} + \eta\xi_n p_{2n}), \quad q_{1n} = p_{1n}, \quad q_{2n} = \frac{1}{\omega_n}(\phi_n p_{2n} - \eta\xi_n p_{0n}). \tag{2.4}$$

Here p_{0n}, p_{1n}, p_{2n} form an orthonormal basis of the divergence-free subspace for nth Fourier mode; the p_{jn} are the Craya basis for the purely stratified problem, already used by Riley *et al.* [66]:

$$p_{0n} = \left[-\frac{\check{n}_2}{|\check{n}'|}, \frac{n_1}{|\check{n}'|}, 0, 0\right], \quad p_{1n} = \left[\frac{n_1\check{n}_3}{|\check{n}|\,|\check{n}'|}, \frac{\check{n}_2\check{n}_3}{|\check{n}|\,|\check{n}'|}, \frac{-n_1^2 - \check{n}_2^2}{|\check{n}|\,|\check{n}'|}, 0\right], \tag{2.5}$$
$$p_{2n} = e_4 = [\,0,\ 0,\ 0,\ 1\,].$$

The eigenvalues $\pm i\omega_n$ are given by

$$\omega_n = \sqrt{\phi_n^2 + \eta^2\xi_n^2}, \qquad \xi_n = \frac{\check{n}_3}{|\check{n}|}, \qquad \phi_n = \frac{|\check{n}'|}{|\check{n}|}, \qquad \eta = f/N, \tag{2.6}$$

where $|\check{n}|^2 = n_1^2 + n_2^2/a_2^2 + n_3^2/a_3^2$, $|\check{n}'|^2 = n_1^2 + n_2^2/a_2^2$. We consider the case when the ratio $\eta = f/N$ is bounded by a finite $\eta_0 > 1$:

$$1/\eta_0 \leq \eta = f/N \leq \eta_0, \tag{2.7}$$

$$1/\eta_0 \leq \min(1,\eta) \leq \omega_n \leq \max(1,\eta) \leq \eta_0. \tag{2.8}$$

In the case $n_1 = n_2 = 0$ (corresponding to taking horizontal averages) we choose the basis which is obtained from (2.4) by setting $n_1 = n_2 \neq 0$ and taking $n_1 \to 0$. In particular, when $n_1 = n_2 = 0$, we obtain $\omega_n = \eta$ and the eigenvectors are

$$q_{0\overline{n}} = (0,0,0,1), \quad q_{1\overline{n}} = \left(\frac{1}{\sqrt{2}}, \frac{1}{\sqrt{2}}, 0, 0\right), \quad q_{2\overline{n}} = \left(\frac{1}{\sqrt{2}}, -\frac{1}{\sqrt{2}}, 0, 0\right), \tag{2.9}$$

where $\overline{n} = (0,0,n_3)$ denotes wavenumbers for which $n_1 = n_2 = 0$.

Any arbitrary divergence-free vector field $\mathbf{U}_n^\dagger$ can be written as

$$\mathbf{U}_n^\dagger = V_n^0 q_{0n} + V_n^1 q_{1n} + V_n^2 q_{2n}. \tag{2.10}$$

We shall use the variables V to denote the vector of coefficients corresponding to $\mathbf{U}_n^\dagger$: $\mathbf{V}_n = [V_n^0, V_n^1, V_n^2] = [V_n^0, \mathbf{V}'_n]$, $\mathbf{V}'_n = [V_n^1, V_n^2]$. Note that the relation between the $\mathbf{U}^\dagger$ and $\mathbf{V}$ variables is given by

$$V_n^0 = \mathbf{U}_n^\dagger \cdot q_{0n}, \qquad V_n^1 = \mathbf{U}_n^\dagger \cdot q_{1n}, \qquad V_n^2 = \mathbf{U}_n^\dagger \cdot q_{2n}. \tag{2.11}$$

Clearly, $V_n^{0*} = -V_{-n}^0$ and $V_n^{i*} = V_{-n}^i$, $i = 1, 2$, for real $\mathbf{U}(x)$ and $\rho(x)$. We denote by Π_n^{QG} the projection onto q_{0n} and call it as usual the QG mode:

$$\Pi^{\mathrm{QG}}\mathbf{U}^\dagger(x) = \sum_n V_n^0 q_{0n} e^{i\check{n}\cdot x}, \qquad \Pi_n^{\mathrm{QG}}\mathbf{U}_n^\dagger = V_n^0 q_{0n}.$$

The projection onto the 2D subspace corresponding to $\pm i\omega_n$ is denoted by Π^{AG} and defines the AG component:

$$\Pi_n^{\mathrm{AG}}\mathbf{U}_n^\dagger = V_n^1 q_{1n} + V_n^2 q_{2n}.$$

The case when $\eta \to 0$ or $\eta \to \infty$ was discussed in BMN [7]; detailed mathematical consideration of this can be done along the lines of that paper and BMN [9], but requires additional non-trivial considerations; in particular the structure of resonant sets and smoothness conditions are different from those imposed here.

Eqs. (1.20) in Fourier representation in the V variables can be written in the cyclic basis (2.4) as

$$\partial_t V_n^{i_3} = -i \sum_{k+m=n, i_1, i_2} Q_{kmn}^{i_1 i_2 i_3} V_k^{i_1} V_m^{i_2} - N\omega_n (M'_n V_n)^{i_3} - (\hat{\boldsymbol{\nu}}|\check{n}|^2 \mathbf{V}_n)^{i_3} + F_n^{i_3}, \tag{2.12}$$

where $i_1, i_2, i_3 = 0, 1, 2$, M' is the matrix M in the V variables given by (2.13); $\hat{\boldsymbol{\nu}}$ is the viscosity matrix $\overline{\boldsymbol{\nu}}$ in the V-basis. Here $\mathbf{J}$, $\mathbf{M}'_n$ are given by

$$\mathbf{M}'_n = \begin{pmatrix} 0 & 0 & 0 \\ 0 & 0 & -1 \\ 0 & 1 & 0 \end{pmatrix}, \qquad \mathbf{J} = \begin{pmatrix} 0 & -1 \\ 1 & 0 \end{pmatrix}, \tag{2.13}$$

$$\hat{\boldsymbol{\nu}} = \frac{1}{\omega_n^2} \begin{pmatrix} \nu_1\phi_n^2 + \eta^2\xi_n^2\nu_2 & 0 & (\nu_1 - \nu_2)\eta\xi_n\phi_n \\ 0 & \nu_1\omega_n^2 & 0 \\ (\nu_1 - \nu_2)\eta\xi_n\phi_n & 0 & \nu_2\phi_n^2 + \eta^2\xi_n^2\nu_1 \end{pmatrix}. \tag{2.14}$$

The coefficients $Q_{kmn}^{i_1 i_2 i_3}$ are determined from the equations using (2.4), see BMN[12]:

$$Q_{kmn}^{i_1 i_2 i_3} = (q_{i_1 k} \cdot m)(q_{i_2 m} \cdot q_{i_3 n}). \tag{2.15}$$

We use the following notation for the skew-symmetric product: $n' \wedge m' \equiv n_1 m_2 - n_2 m_1$. To save space, we only give formulas for 0-wave interactions; see BMN [12] for general coefficients:

$$Q^{000}_{kmn} = \frac{\omega_m |\check{m}|\ \check{n}' \wedge \check{m}'}{\omega_k \omega_n |\check{k}||\check{n}|}, \tag{2.16}$$

when $n = k + m$; clearly $|Q^{i_1 i_2 i_3}_{kmn}| \leq |\check{m}|$. Into Eq. (2.12),

$$\partial_t \mathbf{V}_n + N\omega_n \mathbf{M}'_n \mathbf{V}_n = (\mathbf{B}(\mathbf{V},\mathbf{V}))_n - A_n \mathbf{V}_n + \mathbf{F}_n,\ A_n = \hat{\boldsymbol{\nu}}|\check{n}|^2, \tag{2.17}$$

we introduce the change of variables

$$\mathbf{V} = \mathbf{E}(-Nt)\mathbf{v}, \qquad \mathbf{V}_n = \exp(-N\omega_n \mathbf{M}'_n t)\mathbf{v}_n, \tag{2.18}$$

where $\mathbf{v} = [v^0, v^1, v^2]$ and $\mathbf{M}'$ is defined by (2.13). The action of the linear propagator on the Fourier components $\mathbf{E}(Nt)$ can be written in the V variables in the Craya cyclic basis as

$$\mathbf{E}(Nt)[V^0, \mathbf{V}']_n = \exp(N\omega_n t \mathbf{M}'_n)[V^0, \mathbf{V}']_n = [V^0, \exp(N\omega_n t \mathbf{J})\mathbf{V}']. \tag{2.19}$$

Obviously, $\mathbf{E}(Nt)$ represents vector rotation in the V^1–V^2-plane; the orthogonal V^0 component (called QG) along the axis of rotation is not affected. To save space, we always write $\mathbf{V}_n = [V^0, \mathbf{V}']_n$ as a row, understanding that it is a column in the matrix multiplication. Eq. (2.17), written in the $\mathbf{v}$ variables, has the form

$$\begin{aligned} \partial_t \mathbf{v} &= \mathbf{B}_p(\mathbf{N}t, \mathbf{v}, \mathbf{v}) - \mathbf{E}(Nt)A\mathbf{E}(-Nt)\mathbf{v} + \mathbf{F}_{\mathrm{QG}} + \mathbf{E}(Nt)\mathbf{F}_{\mathrm{AG}}, && (2.20)\\ \mathbf{B}_p(Nt, \mathbf{v}, \mathbf{v}) &= \mathbf{E}(Nt)\mathbf{B}(\mathbf{E}(-Nt)\mathbf{v}, \mathbf{E}(-Nt)\mathbf{v}), && (2.21) \end{aligned}$$

where $\mathbf{F}^\dagger = \mathbf{F}_{\mathrm{QG}} + \mathbf{F}_{\mathrm{AG}}$ in the Craya basis and A is the non-local operator such that $A_n = \hat{\boldsymbol{\nu}}|\check{n}|^2$. Eq. (2.20) is explicitly time-dependent with rapidly varying coefficients. The corresponding equations for Fourier coefficients have the form:

$$\begin{aligned} \partial_t v^{i_3}_n = & \sum_{n=k+m, i_1, i_2} \tilde{Q}^{i_1,i_2,i_3}_{kmn} v^{i_1}_k v^{i_2}_m \\ & + \sum_{n=k+m, i_1, i_2} \hat{Q}^{i_1,i_2,i_3}_{kmn}(Nt) v^{i_1}_k v^{i_2}_m - \tilde{A}_n v^{i_3}_n - \hat{A}_n(Nt) v^{i_3}_n \\ & + \mathbf{F}_{\mathrm{QG},n} + \mathbf{E}_n(Nt)\mathbf{F}_{\mathrm{AG},n} \end{aligned} \tag{2.22}$$

where the first sum consists of resonant terms. In the second sum every matrix element $\hat{Q}^{i_1,i_2,i_3}_{kmn}(Nt)$ of the non-resonant part, as well as $\hat{A}_n(Nt)$, equals a sum of terms of the form $C\exp(\pm i D_\ell N t)$ with $D_\ell \neq 0$. Generally, $D_\ell = \pm\omega'_n \pm \omega'_m \pm \omega'_k$, $\ell = 1, \ldots, 8$, where either $\omega'_n = \omega_n$ or $\omega'_n = 0$. When $D_\ell = 0$ we call these interactions resonant, and when $D_\ell \neq 0$, the interactions are non-resonant; see BMN [9] for more details. When all three $\omega'_n, \omega'_m, \omega'_k$ are non-zero we have strict 3-wave resonances; when exactly two of $\omega'_n, \omega'_m, \omega'_k$ are non-zero we have 2-wave resonances; when exactly one of $\omega'_n, \omega'_m, \omega'_k$ is non-zero we have 1-wave resonances. We have shown in BMN [7], [12] that for all but a countable non-dense set of η, all 3-wave interactions are non-resonant and thus do not contribute to the limit equations; see Section 3 for details.

The resonant contribution $\tilde{A}_n$ from the viscous term does not coincide with the original operator $\hat{\boldsymbol{\nu}}\Delta$ since $\hat{\boldsymbol{\nu}}$ does not commute with M. Simple computation gives the resonant terms. Let ν_1 and ν_2 be the kinematic viscosity and the heat conductivity, respectively. We have, in the V-basis

$$\begin{aligned}\exp(P^d M' P^d Nt)\hat{\boldsymbol{\nu}}\exp(-P^d M' P^d Nt) &= \operatorname{diag}(\nu_{\mathrm{QG}}(n), \nu_{\mathrm{AG}}(n), \nu_{\mathrm{AG}}(n)) \\ &\quad + \Re(2N\omega_n t),\end{aligned} \tag{2.23}$$

where all elements of the non-resonant matrix $\Re$ include factors $\exp(\pm i2N\omega_n t)$; $\hat{\boldsymbol{\nu}}$ is given by (2.14). Thus we obtain the diagonal non-resonant matrix $\tilde{\boldsymbol{\nu}}(n) = \operatorname{diag}(\nu_{\mathrm{QG}}(n), \nu_{\mathrm{AG}}(n), \nu_{\mathrm{AG}}(n))$ in terms of the QG and AG viscosities ν_{QG} and ν_{AG} given by (see BMN [12])

$$\begin{aligned}\nu_{\mathrm{QG}}(n) &= \nu_2 + (\nu_1 - \nu_2)\frac{|\check{n}'|^2}{|\check{n}'|^2 + \eta^2 \check{n}_3^2}, \\ \nu_{\mathrm{AG}}(n) &= \nu_1 + (\nu_2 - \nu_1)\frac{|\check{n}'|^2}{|\check{n}'|^2 + \eta^2 \check{n}_3^2}\end{aligned} \tag{2.24}$$

where $\eta = f/N$, $|\check{n}'|^2 = n_1^2 + n_2^2/a_2^2, \check{n}_3 = n_3/a_3$. Clearly

$$\nu_{\mathrm{QG}}(n) = \nu_1 \frac{|\check{n}'|^2}{|\check{n}'|^2 + \eta^2 \check{n}_3^2} + \nu_2 \frac{\eta^2 \check{n}_3^2}{|\check{n}'|^2 + \eta^2 \check{n}_3^2} = \nu_1 \frac{\phi_n^2}{\omega_n^2} + \nu_2 \frac{\eta^2 \xi_n^2}{\omega_n^2}, \tag{2.25}$$

$$\nu = \min(\nu_1, \nu_2) \leq \nu_{\mathrm{QG}}(n) \leq \max(\nu_1, \nu_2) \tag{2.26}$$

and the same inequality holds for $\nu_{\mathrm{AG}}(n)$.

The following equations describe the reduced dynamics which are obtained by annihilating all terms in (2.22) that contain fast oscillating factors:

$$\partial_t \mathbf{w} = \tilde{\mathbf{B}}(\mathbf{w}, \mathbf{w}) - \tilde{\mathbf{A}}\mathbf{w} + \tilde{\mathbf{F}}, \tag{2.27}$$

where $\tilde{\mathbf{A}} = -\tilde{\boldsymbol{\nu}}\Delta$ and where $\tilde{\boldsymbol{\nu}}$ is the non-local linear matrix operator with symbol $\tilde{\boldsymbol{\nu}}(n)$ (when $\mathbf{F}$ depends on Nt, the limit equations may include $\tilde{F}_1$- and $\tilde{F}_2$-resonant components). Clearly, when represented in Fourier modes, the operator $\tilde{\mathbf{B}}$ on the right-hand side of (2.27) has coefficients $\tilde{Q}_{kmn}^{i_1,i_2,i_3}$ in (2.22) and the resonant reduced equations are

$$\partial_t w_n^{i_3} = \sum_{n=k+m, i_1, i_2} \tilde{Q}_{kmn}^{i_1,i_2,i_3} w_k^{i_1} w_m^{i_2} - \tilde{A}_n w_n^{i_3} + \tilde{F}_n^{i_3}, \tag{2.28}$$

where $\tilde{A}_n = \tilde{\boldsymbol{\nu}}(n)\check{n}^2$ and the summation is over resonant terms. In Eq. (2.28) $\tilde{F}_n^{i_3} = \tilde{F}_{\mathrm{QG},n}$ for $i_3 = 0$ and $\tilde{F}_n^{i_3}$ in the appropriate component of $\tilde{\mathbf{F}}_{\mathrm{AG},n}$ for $i_3 = 1, 2$. Projecting Eq. (2.28) on the QG mode (with $i_3 = 0$) and projecting to the AG subspace we obtain separately the equations for the QG and AG components. In this article we usually consider the case $\mathbf{F}_{\mathrm{AG}}$ non-resonant.

We note that projection of (2.27) onto the QG mode (which corresponds to zero eigenvalue of the linear problem) leads to the additional constraint $\omega'_n = 0$. Then the conditions $\pm\omega'_k \pm \omega'_m \pm \omega'_n = 0$ and $\omega'_n = 0$ reduce to 2-wave interactions $\omega_k = \omega_m$. For $\eta \neq 1$ the condition $\omega_k = \omega_m$ is equivalent to $|\check{k}'|/|\check{k}| = |\check{m}'|/|\check{m}|$ (equivalently, $\phi_k = \phi_m$ – see (2.6)). Clearly, reduced equations (2.27) projected onto the QG mode involve only the coefficients $Q_{kmn}^{i_1 i_2 i_3}$ with $i_3 = 0$ ($n = k+m$). One trivial solution of $\omega'_k = \omega'_m$ is $\omega'_k = \omega'_m = 0$ which corresponds to the QG coefficient Q_{kmn}^{000}. An important observation is that other terms involving the coefficients $Q_{kmn}^{i_1 i_2 0}$ ($i_1 \neq 0$ or $i_2 \neq 0$) in (2.27) are annihilated for all n, m when the resonance condition $\phi_k = \phi_m$ is used (see BMN [7], [12]). Therefore the *quasigeostrophic component of the resonant equations* (2.27) *completely decouples.* This fact was proved in BMN [7], [12] by direct computation and also by Embid & Majda [28] using Ertel's theorem.

The QG equation (1.22) is given by

$$\begin{aligned} \partial_t w_n^0 &= \mathbf{B}_0(w^0, w^0)_n - \tilde{A}_n^{\mathrm{QG}} w_n^0 + \tilde{F}_{\mathrm{QG},n}, \\ \mathbf{B}_0(w^0, w^0)_n &= -i \sum_{k+m=n} Q_{kmn}^{000} w_k^0 w_m^0. \end{aligned} \tag{2.29}$$

We introduce variables $\tilde{q}$, $\tilde{\mathbf{U}}_{\mathrm{QG}}$, $\tilde{\Psi}^0$ (QG potential, velocity and stream function):

$$\tilde{q}_m = \omega_m |\check{m}| w_m^0, \quad \tilde{\mathbf{U}}_k = [-k_2/a_2, k_1, 0, 0]\tilde{\Psi}_k^0, \quad \tilde{\Psi}_k^0 - \frac{\tilde{q}_k}{(\omega_k^2 |\check{k}|^2)}. \tag{2.30}$$

Recalling that $\omega_k^2 |\check{k}|^2 = |\check{k}'|^2 + \eta^2 \check{k}_3^2$, $\eta = f/N$, we have the familiar formula which relates $\tilde{\Psi}^0$ and $\tilde{q}$ in physical space:

$$-(\nabla_h^2 + \eta^2 \partial_3^2)\tilde{\Psi}^0 = \tilde{q}. \tag{2.31}$$

Using (2.30), Eq. (2.29) can be written in the form

$$\partial_t \tilde{q}_n = -i \sum_{k+m=n} (\tilde{\mathbf{U}}_k \cdot m)\tilde{q}_m - \tilde{A}_n^{\mathrm{QG}} \tilde{q}_n + \omega_n |\check{n}| F_n^0 \tag{2.32}$$

where $F_n^0 = F_{\mathrm{QG},n}$ in the Craya basis. The field $\tilde{q}(t, x)$ obeys in physical space the 3D QG equations (see [18] for the inviscid case) where the viscous dissipation operator $\tilde{A}^{\mathrm{QG}}$ is a linear pseudo-differential operator which in Fourier representation is multiplication by $\tilde{A}_n^{\mathrm{QG}} = \nu_{\mathrm{QG}}(n)|n|^2$, with $\nu_{\mathrm{QG}}(n)$ given by (2.24).

Thus, in the asymptotic limit equations (2.27) with $\mathbf{w} = (w^0, w^1, w^2)$ splits into the limit QG field $w^0(t) = \mathbf{w}_{\text{QG}}$ satisfying (1.22), which is uncoupled from $\mathbf{w}_{\text{AG}}$ to the lowest order, and into the AG component $\mathbf{w}_{\text{AG}} = (w^1, w^2)$, which satisfies in general equations of the type

$$\partial_t \mathbf{w}_{\text{AG}} = \mathbf{B}_2(\mathbf{w}_{\text{QG}}(t), \mathbf{w}_{\text{AG}}) + \mathbf{B}_3(\mathbf{w}_{\text{AG}}, \mathbf{w}_{\text{AG}}) - \tilde{A}^{\text{AG}} \mathbf{w}_{\text{AG}} + \tilde{\mathbf{F}}_{\text{AG}}. \quad (2.33)$$

In Fourier representation this is

$$\partial_t \mathbf{w}^{i_3}_{\text{AG},n} = \sum_{\substack{k+m=n, i_1, i_2; \\ \pm\omega'_k \pm \omega'_m \pm \omega'_n = 0}} \tilde{Q}^{i_1 i_2 i_3}_{kmn} w^{i_1}_k w^{i_2}_m - \nu_{\text{AG}}(n) |\check{n}|^2 w^{i_3}_n + \tilde{F}^{i_3}_{\text{AG},n}, \quad (2.34)$$

where $i_3 \neq 0$; $i_1, i_2 = 0, 1, 2$. These are the '$2\frac{1}{2}$-dimensional' resonant equations in the Craya basis for the AG component $\mathbf{w}_{\text{AG}} = (w^1, w^2)$. In (2.34), the catalytic operator $\mathbf{B}_2$ corresponds to either $i_1 = 0$ or $i_2 = 0$, and $\omega_m = \omega_n$ or $\omega_k = \omega_n$.

Eqs. (2.34) for $\mathbf{w}_{\text{AG}}$ without 3-wave interactions (that is, where either $i_1 = 0$ or $i_2 = 0$) always include two invariant subsystems: the first consists of modes with $n_3 = 0$ (it corresponds to invariance with respect to vertical averaging). This follows from the condition for 2-wave resonances $\omega_m = \omega_n$ which is equivalent to $|m_3|/|\check{m}| = |n_3|/|\check{n}|$; therefore $n_3 = 0$ implies $m_3 = 0$. The second subsystem corresponds to $n_1 = n_2 = 0$ since the condition for 2-wave resonances $\omega_m = \omega_n$ is equivalent to $|\check{m}'|/|\check{m}| = |\check{n}'|/|\check{n}|$ and $n' = 0$ implies $m' = 0$ (horizontal averaging). In [12] it was shown that horizontally-averaged velocity and density are adiabatic invariants of 3D primitive equations in the strongly-stratified limit in the absence of rotation (they are exactly conserved by the asymptotic limit equations). Note that horizontally-averaged density is still an adiabatic invariant for the case with rotation, $\eta \neq 0$ ([12]), as long as $\text{Bu} \neq 0$.

3 Small divisors and uniform convergence results

In this section we detail the algebraic structure of resonant and quasi-resonant sets and present new regularity and strong uniform (in η, a_2, a_3) convergence results for the limit resonant equations. The uniform convergence results are substantial improvements over the corresponding ones in BMN [7] in that much less regularity is required for the initial data, with the Sobolev space $H_{17/2}$ being the worst and H_4 being the best case. Since 3-wave interactions in (2.33) discontinously depend on η, any strong convergence of AG-components on a fat set of η, uniform in the parameter η, can only be to equations (1.24) (see the discussion of uniform convergence in BMN [9]). Uniform error estimates cannot circumvent control of small divisors and sharp estimates of both near 3-wave and near 2-wave resonances. Here we investigate the density and probability of both 3-wave and 2-wave resonances as functions of η and two geometric parameters a_2, a_3 (with $a_1 = 1$).

We recall some facts on the geometry of resonances (see [13]). We denote

$$\eta = f/N, \qquad \theta_1 = 1/a_1^2, \qquad \theta_2 = 1/a_2^2, \qquad \theta_3 = 1/a_3^2. \tag{3.1}$$

Let K denote the set of resonant wavenumbers k, m, n for given η, a_1, a_2, a_3:

$$K = \{(k, m, n) : \pm\omega'_k \pm \omega'_m \pm \omega'_n = 0\}, \tag{3.2}$$

where $\omega'_n = 0$ for QG modes and $\omega'_n = \omega_n$ for AG modes and similarly for k and m.

For non-generic values of these parameters, the limit inviscid equations for the AG field $\mathbf{w}_{\mathrm{AG}}$ are nonlinear; for $i_3 \neq 0$,

$$\partial_t \mathbf{w}_{\mathrm{AG}} = \mathbf{B}_2(\mathbf{w}_{\mathrm{QG}}(t), \mathbf{w}_{\mathrm{AG}}) + \mathbf{B}_3(\mathbf{w}_{\mathrm{AG}}, \mathbf{w}_{\mathrm{AG}}); \tag{3.3}$$

$$\mathbf{B}_3(\mathbf{w}_{\mathrm{AG}}, \mathbf{w}_{\mathrm{AG}})_n^{i_3} = \sum_{\substack{(k,m) \in K^* \\ k+m+n=0 \\ i_1, i_2 = 1 \text{ or } 2}} \tilde{Q}_{kmn}^{i_1,i_2,i_3} w_k^{i_1} w_m^{i_2};$$

we may easily computer $\mathbf{B}_3$ in the Craya cyclic basis; we give no detials here for the sakes of conciseness. For $\mathbf{B}_3$, the domain of summation K^* (3-wave resonances) $(k, m) \in K^*$ is defined by the condition $\pm\omega_n \pm \omega_m \pm \omega_k = 0$ where each $\omega_n, \omega_m, \omega_k \neq 0$ depends on $(\eta, 1/a_2^2, 1/a_3^2) = (\eta, \theta_2, \theta_3)$; that is $K^* = K^*(\eta, \theta_2, \theta_3)$. For every fixed $\theta_2 = 1/a_2^2$ and $\theta_3 = 1/a_3^2$, the summation set $K^*(\eta, \theta_2, \theta_3)$ is not empty when $\eta \in \Theta^*(\theta_2, \theta_3)$; the singular set $\Theta^*(\theta_2, \theta_3)$ is very thin, indeed it is countable. We call it a strict 3-wave resonant set. When $\eta \in \Theta^*(\theta_2, \theta_3)$ is strictly resonant, $\mathbf{B}_3$ is non-zero and depends *strongly* on η; the sets $K^*(\eta, \theta_2, \theta_3)$ with different, but nearby, η do not intersect (a non-trivial result from the study of the small divisor problem, cf. [13]). This implies that the operator $\mathbf{B}_3$ *depends on resonant* η *discontinuously* at every point $\eta \in \Theta^*(\theta_2, \theta_3)$ that is a point of discontinuity of the operator $\mathbf{B}_3$. Since $\mathbf{B}_3$ is not zero, *solutions of the limit system with general initial data discontinuously depend on* η *as well.* The solutions of the original rotating Euler–Boussinesq equations depend on η continuously (on a small time interval $[0, T_1]$), uniform in η, a_2, a_3. When $(k, m, n) \notin K^*$ (i.e. $\eta \notin \Theta^*$), only the catalytic operator $\mathbf{B}_2$ is present. We refer to [11], [12] and [13] for an extensive study of the analytic form and properties of $\mathbf{B}_2$; see also Section 5.

The uniform estimates of convergence following BMN[9] are reduced to the small divisor estimates. The case $\eta = 1$ is very special. It is much simpler, but requires separate considerations since $\omega_n = 1$, for all n in this case; when $1/9 < \eta < 9$ 3-wave resonances are absent and small divisors for 3-wave resonances are very well estimated, but estimates for 2-wave resonances become worse as $\eta \to 1$; therefore we assume $|1 - \eta| \geq c_0 > 0$.

For $\eta \notin \Theta^*$, Eqs. (3.3) reduces to the linear equations for catalytic interactions. Eqs. (3.3) strongly depend on $\theta_2 = 1/a_2^2$, since the 2-wave resonance condition $\omega_m = \omega_n$ is equivalent to $|\check{m}'|/|m_3| = |\check{n}'|/|n_3|$ which is in turn equivalent to the equation

$$(m_3^2 n_2^2 - n_3^2 m_2^2)\theta_2 + (m_3^2 n_1^2 - n_3^2 m_1^2) = 0 \tag{3.4}$$

which for irrational $\theta_2 = 1/a_2^2$ implies $m_3^2 n_1^2 - n_3^2 m_1^2 = 0$, $m_3^2 n_2^2 - n_3^2 m_2^2 = 0$ (clearly these two relations imply $\omega_m - \omega_n = 0$). Therefore, the vectors $(|m_1|, |m_2|, |m_3|)$ and $(|n_1|, |n_2|, |n_3|)$ are colinear, hence for every irrational θ_2, we find $\mathbf{B}_2$ splits in Fourier space into uncoupled, restricted interaction operators on 4-ray families as in (1.25). For every resonant rational point $\theta_2 \in \Theta_2^*$, the term $\mathbf{B}_2$ includes more interactions (much larger, but finite number of Fourier rays), so indeed *rational* θ_2 *are also points of discontinuity* for $\mathbf{B}_2$ and further contribute to the non-uniformity of convergence. An important observation is that 2-wave resonances are controlled by θ_2 only, *not* by η *or* a_3. This follows from the fact that $\omega_m = \omega_n$ implies $|m_3|/|\check{m}'| = |n_3|/|\check{n}'|$. In contrast, *3-wave resonances are controlled by* η *uniformly in* θ_2 *and* θ_3; although strict resonant values $\eta = \Theta^*(\theta_2, \theta_3)$ depend on θ_2 and θ_3, we prove below that the estimate of the measure of almost resonant η does not depend on θ_2 and is uniform in θ_3.

We first describe the set $\Theta^*(\theta_2, \theta_3)$ of strict 3-wave resonances. Recall that D_l is given by (1.17)–(1.18) where $\omega'_k, \omega'_m, \omega'_n \neq 0$ for strict 3-wave resonances.

Lemma 3.1 *For every* a_2, a_3, k, m *there exists at most two positive roots* $\eta^{\pm}(\theta_2, \theta_3, k, m)$ *of the equation* $\pm\omega_n \pm \omega_m \pm \omega_k = 0$, *every root analytically depends on* θ_2, θ_3.

Definition 3.1 We put $\Theta^*(\theta_2, \theta_3, k, m) = \eta^+(\theta_2, \theta_3, k, m) \bigcup \eta^-(\theta_2, \theta_3, k, m)$ and $\Theta^*(\theta_2, \theta_3) = \bigcup_{k,m} \Theta^*(\theta_2, \theta_3, k, m)$.

Proof of Lemma 3.1. Consider the case $D_\ell(k, m, n) = (\omega_k + \omega_n - \omega_m)$. Other cases are treated in a similar way. We have the identity

$$\frac{1}{(\omega_k + \omega_n - \omega_m)} = \frac{(\omega_k + \omega_m + \omega_n)(-\omega_k + \omega_m + \omega_n)(-\omega_n + \omega_m + \omega_k)}{(\omega_k + \omega_m + \omega_n)(\omega_k - \omega_m + \omega_n)(-\omega_k + \omega_m + \omega_n)(\omega_k + \omega_m - \omega_n)}.$$

The denominator is $-P = ((\omega_k)^2 + (\omega_m)^2 - (\omega_n)^2)^2 - 4(\omega_k)^2(\omega_m)^2$. It is a polynomial of degree 2 of $\lambda = \eta^2$. We write ω_k in the form (see Eq. (2.6))

$$\omega_k = \sqrt{\chi_k + \lambda(1 - \chi_k)}, \quad \chi_k = \frac{|\check{k}'|^2}{|\check{k}|^2} = \phi_k^2, \quad 1 - \chi_k = \xi_k^2.$$

With $\lambda = \eta^2$, the polynomial takes the form $P = P_2\lambda^2 + P_1\lambda + P_0$, where

$$\begin{aligned}
P_2 &= \chi_k^2 + \chi_m^2 + \chi_n^2 - 2\chi_k\chi_n - 2\chi_k\chi_m - 2\chi_m\chi_n - 3 + 2(\chi_k + \chi_m + \chi_n) \\
P_1 &= -2(\chi_k^2 + \chi_m^2 + \chi_n^2 - 2\chi_k\chi_n - 2\chi_k\chi_m - 2\chi_m\chi_n + \chi_k + \chi_m + \chi_n) \\
&= -[(\chi_k - \chi_n)^2 + (\chi_m - \chi_n)^2 + (\chi_k - \chi_m)^2] - \chi_k(1 - \chi_k) \\
&\qquad -\chi_m(1 - \chi_m) - \chi_n(1 - \chi_n) \\
&< 0; \\
P_0 &= \chi_k^2 + \chi_m^2 + \chi_n^2 - 2\chi_k\chi_n - 2\chi_k\chi_m - 2\chi_m\chi_n.
\end{aligned}$$

The discriminant Δ_P of the quadratic polynomial satisfies the symmetric formula

$$\Delta_P = P_1^2 - 4P_2P_0 = 8[(\chi_k - \chi_n)^2 + (\chi_m - \chi_n)^2 + (\chi_k - \chi_m)^2] \geq 0.$$

We are interested in resonant values of λ that are solutions of the equation $P(\lambda) = 0$. There are no more than two solutions $\lambda_\pm(k, m)$. Clearly, only positive solutions $\lambda_\pm$ satisfy the constraint $\lambda = \eta^2$; with $\eta^* = +\sqrt{\lambda_\pm}$, whenever $\lambda_\pm > 0$.

We also investigate quasi-3-wave resonances. For this, take a small neighborhood of $\lambda_\pm(k, m)$ (quasiresonant η), defined by

$$P(\lambda) = \delta,$$

with a small δ. Using the quadratic formula we easily obtain the derivative at $\delta = 0$:

$$\left|\frac{d\lambda}{d\delta}\right| = \frac{1}{\sqrt{\Delta_P}} = \frac{1}{\sqrt{8[(\chi_k - \chi_n)^2 + (\chi_m - \chi_n)^2 + (\chi_k - \chi_m)^2]}}.$$

Since $\omega_k^2 - \omega_m^2 = (1 - \lambda)(\chi_k - \chi_m)$, we can rewrite this formula as

$$\left|\frac{d\lambda}{d\delta}\right| = \frac{|1 - \lambda|}{\sqrt{8[(\omega_k^2 - \omega_n^2)^2 + (\omega_m^2 - \omega_n^2)^2 + (\omega_k^2 - \omega_m^2)^2]}}.$$

Now we are interested in estimating $(\omega_k + \omega_n - \omega_m)^{-1} = D_\ell(k, m, n)^{-1}$. We consider for given η two cases:

Case I. Let $|\omega_n - \omega_m| \leq |\omega_k|/2$; then

$$\frac{1}{|\omega_k + \omega_n - \omega_m|} \leq \frac{2}{|\omega_k|} \leq 2\eta_0$$

and in this case the divisor is not small.

Case II. Let $|\omega_n - \omega_m| \geq |\omega_k|/2$; then

$$|\omega_n^2 - \omega_m^2| \geq |\omega_k|\frac{(|\omega_n| + |\omega_m|)}{2} \geq \frac{|\omega_k||\omega_m|}{2},$$

and we have the estimate

$$\left|\frac{d\lambda}{d\delta}\right| = \frac{1}{\sqrt{\Delta_P}} \le \frac{|1-\lambda|}{\sqrt{2}|\omega_k||\omega_m|} \le \frac{(1+\eta_0^2)\eta_0^2}{\sqrt{2}}, \tag{3.5}$$

and this estimate implies that every root $\lambda_\pm\ (\theta_2, \theta_3, k, m)$ belongs to an *analytic surface*. This also implies that both solutions $\eta^*(\theta_2, \theta_3)$ are on a smooth analytic surface $\Theta^*_{k,m}$ which consists of either one or two separate sheets (in the latter case, we still denote them by $\Theta^*_{k,m}$).

Definition 3.2 For a given measure μ_3, and for a given summable sequence $\zeta^\kappa \ge 0$ (here κ is upper index), with $\sum_\kappa \zeta^\kappa \le 1$, $\kappa = (k, m)$, define the 3-wave quasi-resonant set $\Theta_3^{\mu_3}(\theta_2, \theta_3)$ as

$$\Theta_3^{\mu_3} = \bigcup_\kappa \{\eta :\ 2|\eta - \eta^*(\kappa, \theta_2, \theta_3)| < \mu_3 \zeta^\kappa\}. \tag{3.6}$$

Obviously, the Lebesgue measure

$$\text{meas}(\Theta_3^{\mu_3}(\theta_2, \theta_3)) \le \mu_3, \quad \text{for all } \theta_2,\ \theta_3.$$

We now use (3.5) and find an appropriate choice of ζ^κ to estimate the small divisor $D_\ell(k, m, n)$ for θ_2, θ_3 in the complement of the quasi-resonant set $\Theta_3^{\mu_3}$ of given measure μ_3. From Definition 3.2:

$$\frac{1}{|\eta - \eta^*|} \le \frac{2}{\mu_3 \zeta^\kappa} \quad \text{if } (\theta_2, \theta_3) \notin \Theta_3^{\mu_3}. \tag{3.7}$$

Following the notation of Lemma 3.1,

$$D_\ell^{-1}(k, m, n) = \prod_{q \ne \ell} \frac{D_q(k, m, n)}{P(\lambda)}.$$

For a given k, m, θ_2, θ_3, the quasi-resonant set is implicitly defined by the small neighbourhood of $\lambda^*(k, m, \theta_2, \theta_3)$ defined by:

$$P(\lambda) \le \delta, \quad \delta \text{ small}.$$

Outside such a quasi-resonant neighborhood we have

$$D_\ell^{-1} \le \frac{27\eta_0^3}{\delta}.$$

With the help of (3.5), we have for $\delta \ll 1$:

$$\delta \sim \left|\frac{d\lambda}{d\delta}\right|^{-1} |\lambda - \lambda^*| \sim 2\left|\frac{d\lambda}{d\delta}\right|^{-1} \eta^* |\eta - \eta^*|$$

and

$$D_\ell^{-1} \le \frac{27\eta_0^5(1+\eta_0^2)}{\sqrt{8}|\eta-\eta^*||\omega_k||\omega_m|}$$

and similar expressions for other 3-wave resonances. With (3.7) either $D_\ell^{-1} \le 2\eta_0$ or

$$D_\ell^{-1} \le \frac{27\eta_0^5(1+\eta_0^2)}{\sqrt{2}\mu_3\zeta^\kappa|\omega_k||\omega_m|} \quad \text{if } \theta_2, \theta_3 \notin \Theta_3^{\mu_3}. \tag{3.8}$$

We fix the sequence ζ^κ as follows:

$$\begin{aligned} \zeta^\kappa &= \frac{\zeta_3^\kappa}{\zeta_3^*}, \quad \zeta_3^\kappa = |k_3|^{-1-\epsilon_0}|m_3|^{-1-\epsilon_0}|\check{m}'|^{-2-\epsilon_0}, \quad |\check{k}'|^{-2-\epsilon_0}, \\ \zeta_3^* &= \sum_\kappa \zeta_3^\kappa, \epsilon_0 > 0. \end{aligned} \tag{3.9}$$

In this definition, 0 to any negative power is set equal to 1.

Theorem 3.1 *Let $\epsilon_0 > 0$, and the sequence ζ^κ defined by (3.9); then for every η, θ_2, θ_3, $\eta \notin \Theta_3^{\mu_3}(\theta_2, \theta_3)$, we have $D_\ell(k,m,n) \neq 0$ for all ℓ, k, m, n, $k+m=n$ and*

$$|D_\ell(k,m,n)|^{-1} \le \max\left(\frac{C_3}{\mu_3} a_3^{2+2\epsilon_0}(|\check{k}||\check{m}|)^{3+2\epsilon_0}, 2\eta_0\right) \tag{3.10}$$

where $C_3 = C_3(\eta_0, \epsilon_0)$.

Proof. Simply compound estimate (3.8) with

$$\frac{1}{\omega_k\omega_m} \le \eta_0^2 a_3^2 \frac{|\check{k}|}{|k_3|}\frac{|\check{m}|}{|m_3|}.$$

We now focus on 2-wave resonances. Consider the case $\omega_k = \omega_m$, equivalently $\frac{|\check{k}'|}{|k_3|} = \frac{|\check{m}'|}{|m_3|}$. Since $n = k+m$, (3.4) is equivalent to

$$(m_3^2k_2^2 - k_3^2m_2^2)\theta_2 + m_3^2k_1^2 - k_3^2m_1^2 = 0. \tag{3.11}$$

Strict 2-wave resonant θ_2 are possible only when θ_2 is rational; we denote the solutions of the equation (3.11) (that is strict 2-wave resonant θ_2) by $\theta_2^*(k,m)$. For given summable sequences ζ_2^κ with $\sum_\kappa \zeta_2^\kappa \le 1$, and κ from the union of three sets $\kappa = (k,m)$ or $\kappa = (k,n)$ or $\kappa = (m,n)$, we introduce strict 2-wave resonant sets Θ_2^* and almost 2-wave resonant sets $\Theta_2^{\mu_2}$ by

$$\Theta_2^* = \bigcup_\kappa \theta_2^*(\kappa), \quad \Theta_2^{\mu_2} = \bigcup_\kappa \{\theta_2 : 2|\theta_2 - \theta_2^*(\kappa)| < \mu_2\zeta_2^\kappa\}, \quad \Theta_2^0 = \bigcap_{\mu_2} \Theta_2^{\mu_2}. \tag{3.12}$$

Clearly, meas$(\Theta_2^{\mu_2}) \le \mu_2$. We choose here with $\epsilon_1 = \epsilon_2 = \epsilon_0 > 0$ for $\kappa = (k,m) \notin \Theta_2^*$

$$\zeta_2^\kappa = \frac{|m_3^2 k_2^2 - k_3^2 m_2^2|^{-1-\epsilon_2} |k_1 m_1|^{-1-\epsilon_1}}{(3C_4')}, \tag{3.13}$$

$$C_4' = \sum_\kappa |m_3^2 k_2^2 - k_3^2 m_2^2|^{-1-\epsilon_2} |k_1 m_1|^{-1-\epsilon_1} \tag{3.14}$$

for κ such that $k_3 m_3 n_3 \neq 0$; we sum over such k, m that $|m_3^2 k_2^2 - k_3^2 m_2^2| \neq 0$, and similarly with the same C_4' for $\kappa = (k,n), \kappa = (m,n)$.

Theorem 3.2 *Let $\epsilon_0 > 0$, $\mu_2 > 0$, $D_\ell\ (k,m,n) \neq 0$, and $\Theta_2^{\mu_2}$ (almost resonant parameter set) with the Lebesgue measure $\le \mu_2$ be defined in (3.12). Then for every $\theta_2 \notin \Theta_2^{\mu_2}$*

$$\frac{1}{|\omega_k \pm \omega_m|} \le \frac{C_4 (a_3)^{2+2\epsilon_0}}{\mu_2} |\check{k}|^{3+3\epsilon_0} |\check{m}|^{3+3\epsilon_0}, \tag{3.15}$$

when $\omega_n = 0$, $\omega_k \pm \omega_m \neq 0$. Moreover

$$\frac{1}{|\omega_n \pm \omega_m|} \le \frac{C_4 (a_3)^{2+2\epsilon_0}}{\mu_2} |\check{k}|^{1+\epsilon_0} (|\check{k}|^{2+2\epsilon_0} + |\check{m}|^{2+2\epsilon_0}) |\check{m}|^{3+3\epsilon_0}, \tag{3.16}$$

when $\omega_k = 0$, $\omega_n \pm \omega_m \neq 0$; the same estimate holds for $|\omega_k \pm \omega_n|$, when $\omega_m = 0$, $\omega_k \pm \omega_n \neq 0$. In the above, $k_3 m_3 n_3 \neq 0$; C_4 depends only on ϵ_0 and η_0.

Remark 3.1 Better estimates hold if $k_3 m_3 n_3 = 0$.

Proof of Theorem 3.2. For $\omega_n = 0$ (the two other cases are similar) we obtain, using $\omega_k^2 = (1-\eta^2)\phi_k^2 + \eta^2$ in the case where $\pm$ is $-$ (the $+$ case is trivial):

$$\begin{aligned} \frac{1}{|D_l(k,m,n)|} &= \frac{1}{|\omega_k - \omega_m|} = \frac{|\omega_k + \omega_m|}{|\omega_k^2 - \omega_m^2|} = \frac{|\omega_k + \omega_m|}{|(1-\eta^2)(\phi_k^2 - \phi_m^2)|} \\ &= \frac{|\check{m}|^2 |\check{n}|^2 |\omega_k + \omega_m|}{|(1-\eta^2)\theta_3((m_3^2 k_2^2 - k_3^2 m_2^2)\theta_2 + m_3^2 k_1^2 - k_3^2 m_1^2)|} \\ &\le \frac{2\eta_0 |\check{m}|^2 |\check{n}|^2}{\left|(1-\eta^2)\theta_3 (m_3^2 k_2^2 - k_3^2 m_2^2) \left(\theta_2 + \frac{(m_3^2 k_1^2 - k_3^2 m_1^2)}{(m_3^2 k_2^2 - k_3^2 m_2^2)}\right)\right|}, \end{aligned} \tag{3.17}$$

in the case when $(m_3^2 k_2^2 - k_3^2 m_2^2) \neq 0$; when $|m_3^2 k_2^2 - k_3^2 m_2^2| = 0$, $m_3^2 k_1^2 - k_3^2 m_1^2 \neq 0$ since $\omega_m - \omega_k \neq 0$, so $|m_3^2 k_1^2 - k_3^2 m_1^2| \ge 1$ and the estimate of $1/|D_l(k,m,n)| \le C|m|^2|n|^2$ is trivial. By the definition of $\Theta_2^{\mu_2}$

$$\left| \frac{\theta_2 + (m_3^2 k_1^2 - k_3^2 m_1^2)}{(m_3^2 k_2^2 - k_3^2 m_2^2)} \right| \ge \frac{\mu_2}{2} \frac{|m_3^2 k_2^2 - k_3^2 m_2^2|^{-1-\epsilon_2} |k_1 m_1|^{-1-\epsilon_1}}{3C_4'},$$

for θ_2 outside the set $\Theta_2^{\mu_2}$. Therefore

$$
\begin{aligned}
|D_\ell|^{-1} &\leq \frac{2\eta_0 a_3^2}{|1-\eta^2|} \frac{6C_4'}{\mu_2} |\check{m}|^2 |\check{n}|^2 |k_1 m_1|^{1+\epsilon_1} |m_3^2 k_2^2 - k_3^2 m_2^2|^{\epsilon_2} \\
&\leq \frac{C_4 a_3^{2+2\epsilon}}{} \mu_2 |\check{k}|^{3+3\epsilon_0} |\check{m}|^{3+3\epsilon_0}. \qquad (3.18)
\end{aligned}
$$

A similar proof may be carried over for $\omega_k = \omega_n$ and $\omega_m = \omega_n$; estimates are worse, as one needs $|n| \leq |k| + |m|$.

The main uniform convergence result of this section (Theorem 3.3) shows that the convergence is uniform and the error is of order $(1/\mu_2 + 1/\mu_3)/N$ when $\theta_2 \notin \Theta_2^{\mu_2}$, $\eta \notin \Theta_3^{\mu_3}(\theta_2, \theta_3)$, with the Lebesgue measure meas$(\Theta_3^{\mu_3}) \leq \mu_3$, meas$(\Theta_2^{\mu_2}) \leq \mu_2$ with μ_2, μ_3 arbitrarily small. Here $\Theta_3^{\mu_3}$ and $\Theta_2^{\mu_2}$ are the sets of *near resonant* 3-waves and 2-waves, and $||\cdot||_\alpha$ designates the norm in the Sobolev space H_α.

Theorem 3.3 *Let* $0 \leq \nu_1$, $\nu_2 \leq 1$ *(including* $\nu_1 = \nu_2 = 0$*),* $\eta \notin \Theta_3^{\mu_3}(\theta_2)$, $\theta_2 \notin \Theta_2^{\mu_2}$. *Let* $\alpha > 3/2$, $\sigma - \alpha > 7$, $M_{0\sigma} > 0$, $\mu_2, \mu_3 \leq 1$. *Let* $||\mathbf{U}^\dagger(0)||_\sigma \leq M_{0\sigma}$. *Let* $\mathbf{U}^\dagger(t)$ *be an exact solution of the 3D Euler–Boussinesq equations. Let* $\mathbf{W}_{\mathrm{QG}}(t)$ *be the solution to the QG equations* (1.22), (2.29)–(2.32) *with initial data* $\Pi^{\mathrm{QG}}\mathbf{U}^\dagger(0)$, *and* $\mathbf{w}_{\mathrm{AG}}(t)$ *the solution to the limit AG equations on the 4-rays* (1.25) *with initial data* $\Pi^{\mathrm{AG}}\mathbf{U}^\dagger(0)$. *Let* $\mathbf{E}(Nt)$ *be the inertio-gravity waves linear propagator. Then for* $0 \leq t \leq T_1$

$$
||\mathbf{U}^\dagger(t) - \mathbf{W}_{\mathrm{QG}}(t) - \mathbf{E}(-Nt)\mathbf{w}_{\mathrm{AG}}(t)||_\alpha \leq \frac{C a_3^2 \left(\frac{1}{\mu_2} + \frac{1}{\mu_3}\right)}{N}, \qquad (3.19)
$$

where T_1 *depends on only on* $M_{0\sigma}$; C *depends only on* $M_{0\sigma}$, α, η_0.

Theorem 3.4 *For* $\Pi^{\mathrm{QG}}\mathbf{U}^\dagger(t) - \mathbf{W}_{\mathrm{QG}}(t)$, *under the same conditions as in Theorem* 3.3, *but with the weaker smoothness* $\sigma - \alpha > 5$ *we have the estimate, for both inviscid and viscous cases:*

$$
||\Pi^{\mathrm{QG}}\mathbf{U}^\dagger(t) - \mathbf{W}_{\mathrm{QG}}(t)||_\alpha \leq \frac{C a_3^2}{N\mu_2}. \qquad (3.20)
$$

The same estimates hold for

$$
||\overline{\mathbf{U}}^\dagger(t) - \overline{\mathbf{W}}_{\mathrm{QG}}(t) - \overline{\mathbf{E}(-Nt)\mathbf{w}_{\mathrm{AG}}(t)}||_\alpha,
$$

but with $\sigma - \alpha > 4$; *here* $\overline{\mathbf{U}}^\dagger$ *designates vertical averaging.*

Proofs of Theorems 3.3 and 3.4 are similar to those given in [9], together with estimates for $\mathbf{w}_{\mathrm{QG}}$ in the next Section 4 and $\mathbf{w}_{\mathrm{AG}}$ in Section 5.

Remark 3.2 For the full error, the above requires smoothness of initial data in $H_{17/2}$. For the convergence of the QG component, only $H_{13/2}$. This is a substantial improvement over the previous estimates in BMN [6] and [7]. The uniform convergence under the $H_{11/2}$ smoothness for the vertically averaged fields is rather remarkable, as it involves both the QG and the AG components. It clearly shows that the dynamical Taylor–Proudman theorem established in BMN [6], [9], has a modified version for the $\mathrm{Bu} = O(1)$ case, coupling the QG and AG components. In Theorems 3.3 and 3.4, the measures μ_2 and μ_3 are equal to the measures of the excluded sets of $\theta_2 = 1/a_2^2$ and η.

4 3D quasigeostrophic equations

In this section we study the quasigeostrophic Eqs. (1.22). We recall the structure of the 3D QG equations. We introduce variables $\tilde{q}$, $\tilde{\mathbf{U}}$ (also called $\mathbf{U}_{\mathrm{QG}}$), $\tilde{\Psi}^0$ (QG potential, velocity and stream function)

$$\tilde{q}_m = \omega_m|\check{m}|w_m^0, \qquad \tilde{\mathbf{U}}_k = \left[\frac{-k_2}{a_2}, k_1, 0, 0\right]\tilde{\Psi}_k^0, \qquad \tilde{\Psi}_k^0 = \frac{\tilde{q}_k}{\omega_k^2|\check{k}|^2}. \tag{4.1}$$

Recalling that $\omega_k^2|\check{k}|^2 = |\check{k}'|^2 + \eta^2\check{k}_3^2$, $\eta = f/N$, we find a familiar formula which relates $\tilde{\Psi}^0$ and $\tilde{q}$ in physical space

$$-(\nabla_h^2 + \eta^2\partial_3^2)\tilde{\Psi}^0 = \tilde{q}. \tag{4.2}$$

Using (4.1), Eq. (1.22) is written in the form

$$\partial_t\tilde{q}_n = -i\sum_{k+m=n}(\tilde{\mathbf{U}}_k\cdot m)\tilde{q}_m - \tilde{A}_{\mathrm{QG},n}\tilde{q}_n + \tilde{F}_{0n}. \tag{4.3}$$

In physical space $\tilde{q}(t,x)$ obeys the 3D QG equations (see Bourgeois & Beale, [18], for the inviscid case)

$$\partial_t\tilde{q} = \tilde{\mathbf{B}}_0(\tilde{q},\tilde{q}) - \tilde{A}_{\mathrm{QG}}\tilde{q} + \tilde{F}_0, \qquad \tilde{\mathbf{B}}_0(\tilde{q},\tilde{q}) = -\tilde{\mathbf{U}}\cdot\nabla_h\tilde{q}, \tag{4.4}$$

where $\tilde{A}_{\mathrm{QG}}$ is a linear pseudo-differential operator which in Fourier representation is multiplication by $A_{\mathrm{QG},n} = \nu_{\mathrm{QG}}(n)|n|^2$, with $\nu_{\mathrm{QG}}(n)$ given by (2.24). For Fourier coefficients (4.4) becomes

$$\partial_t\tilde{q}_n = -i\sum_{k+m=n}\tilde{q}_k\tilde{q}_m\frac{\check{k}'\wedge\check{m}'}{\omega_k^2|\check{k}|^2} - \nu_{\mathrm{QG}}(n)|\check{n}|^2\tilde{q}_n + \tilde{F}_{0n}, \quad \tilde{F}_{0n} = F_n^0|\check{n}|\omega_n \tag{4.5}$$

where $F_n^0 = F_{\mathrm{QG},n}$ in the Craya basis. We note that the coefficients in Eq. (4.5) are skew-symmetric in m, n when replacing n by $-n$ and $k+m+n=0$. Local existence of regular solutions of (4.4) can be proved in a standard way as for 3D Navier–Stokes equations. When the kinematic viscosity ν_1 is positive, we obtain the global existence of regular solutions of (4.4):

Theorem 4.1 *Let $\nu_1 > 0, \nu_2 \geq 0$ ($\tilde{F}_0 = 0$ when $\nu_2 = 0$). Let $T^* > 0$ be arbitrarily large, so that (1.19) holds for $T < T^* - 1$, and let $s = \alpha - 1 \geq 0$. Then there exists a solution $\tilde{q}(t)$ of the quasigeostrophic equations which belongs to H_s for $0 \leq t < T^*$; this solution is unique, $||\tilde{q}(t)||_s \leq M_s, 0 \leq t \leq T^*$. If $\nu_1 > 0, \nu_2 > 0, \alpha \geq 3/4$, the statement holds with $T^* = +\infty$ and*

$$||\tilde{q}(t)||_s \leq M_{1F} \qquad \text{for all } t \geq 0 \tag{4.6}$$

where M_{1F} depends on $\nu, M_{0F}, ||\tilde{q}(0)||_s$. The equation has the following smoothing property: if $\tilde{q}(0) \in H_0$ then $\tilde{q}(t) \in H_\alpha$ for $t > 0$.

Proof. The proof is similar to the proof of the regularity of a 2D Navier–Stokes system in stream-function representation. For simplicity, we give a formal proof assuming all functions are smooth. All estimates can be justified in a standard way using Galerkin approximations. First, multiplying (4.4) by $\tilde{q}$ and integrating in x we obtain

$$\partial_t ||\tilde{q}(t)||_0^2 + 2(\tilde{A}_{\mathrm{QG}}\tilde{q}, \tilde{q}) = 2(\tilde{F}_0, \tilde{q}). \tag{4.7}$$

Note that $\tilde{A}_{\mathrm{QG}}$ is a positive second-order pseudo-differential operator which commutes with Δ; one easily obtains in Fourier representation, for every s,

$$|(\tilde{A}_{\mathrm{QG}}u, (\ \ \Delta)^s u)| \geq c(\nu_1 ||\nabla_h u||_s^2 + \nu_2 ||\partial_3 u||_s^2) \tag{4.8}$$

with $c = 1/\eta_0^2$ and $\nabla_h u = (\partial_1 u, \partial_2 u)$. Therefore

$$\partial_t ||\tilde{q}(t)||^2 + c\nu_1 ||\nabla_h \tilde{q}||_0^2 + c\nu_2 ||\partial_3 \tilde{q}||_0^2 \leq \nu_1^{-1} ||\tilde{F}_0||_0^2. \tag{4.9}$$

This implies the estimates

$$||\tilde{q}(t)||_0^2 \quad \leq \quad (c\nu_1)^{-2} ||\tilde{F}_0||_0^2 + ||\tilde{q}(0)||_0^2 e^{-c\nu_1 t}, \tag{4.10}$$

$$\int_0^T \nu_1 ||\nabla_h \tilde{q}(t)||_0^2 dt \quad \leq \quad C_0 T \nu_1^{-1} ||\tilde{F}_0||_0^2 + ||\tilde{q}(0)||^2. \tag{4.11}$$

Now we prove that the solution remains in H_s for all $t \geq 0$.

We give the proof for $0 < s < 2$; the general case is quite similar, see BMN[9] for an analagous situation. We have to obtain an estimate of $||\tilde{q}(t)||_s$. Multiplying (4.5) by $|n|^{2s}\tilde{q}_n^*(t) = |n|^{2s}\tilde{q}_{-n}(t)$ and summing in n we obtain

$$\partial_t ||\tilde{q}(t)||_s^2 + \nu_1 ||\nabla_h \tilde{q}||_s^2 \leq |(\tilde{\mathbf{B}}_0(\tilde{\mathbf{U}}, \tilde{q}(t)), (-\Delta)^s \tilde{q}(t))| + \nu^{-1} ||\tilde{F}_0||_0^2. \tag{4.12}$$

We have

$$|(\tilde{\mathbf{B}}_0(\tilde{q}(t), \tilde{q}(t)), (-\Delta)^s \tilde{q}(t))| = \left| \sum_{k+m+n=0} \tilde{Q}_{kmn}^{000} \tilde{q}_k(t) \tilde{q}_m \tilde{q}_n |\check{n}|^{2s} \right|.$$

Note that when $s \geq 0$

$$| \, |\check{m} + \check{k}|^s - |\check{m}|^s| \leq C(|\check{k}|^s + |\check{k}||\check{m}|^{s-1}) \tag{4.13}$$

with C depending only on s. By skew-symmetry of $\tilde{Q}^{000}_{kmn} = (n' \wedge m')/(\omega_k^2 |\check{k}|^2)$ in n, m, we have

$$\sum_{k+m+n=0} \tilde{Q}^{000}_{kmn} \tilde{q}_k(t) \tilde{q}_m \tilde{q}_n |\check{n}|^s |\check{m}|^s = 0,$$

and we obtain

$$|(\tilde{\mathbf{B}}_0(\tilde{q}(t), \tilde{q}(t)), (-\Delta)^s \tilde{q}(t))| \leq C \sum_{k+m+n=0} |\check{k}'| \, |\check{k}|^{-2+s} \, |\check{m}'| |\tilde{q}_k| \, |\tilde{q}_m| \, |\tilde{q}_n| \, |\check{n}|^s.$$

Since the sum above represents a scalar product in L^2 of a convolution of functions Z_k, Z'_m with coefficients $Z_k = |\check{k}'| \, |\check{k}|^{-2+s} |\tilde{q}_k|$ and $Z'_m = |\check{m}'| \, |\tilde{q}_m|$, with a function $Z', Z'_n = |\tilde{q}_n| \, |\check{n}|^s$, and which by Parseval's equality equals an integral of the product $ZZ'Z'(x)$, we obtain

$$|(\tilde{\mathbf{B}}_0(\tilde{q}, \tilde{q}), (-\Delta)^s \tilde{q})| \leq C_1 ||Z'||_{L_{p_1}} \, ||Z||_{L_{p_2}} \, ||Z'||_{L_{p_3}} \tag{4.14}$$

when $1/p_1 + 1/p_2 + 1/p_3 = 1$. Using Sobolev's embedding theorem we obtain

$$|(\tilde{\mathbf{B}}_0(\tilde{\mathbf{U}}, \tilde{q}), (-\Delta) \tilde{q})| \leq C'_1 ||Z||_{s_1} \, ||Z'||_{s_2} \, ||Z'||_{s_3} \tag{4.15}$$

with $-3/p_1 \leq s_1 - 3/2$, $-3/p_2 \leq s_2 - 3/2$, $-3/p_3 \leq s_3 - 3/2$. We use the definition of Z, Z', Z' to obtain estimates in terms of Sobolev norms of $\tilde{q}$. We take $p_1 = \infty$ when $s < 1/2$, $p_2 = p_3 = 2$,

$$\begin{aligned} |(\tilde{\mathbf{B}}_0(\tilde{\mathbf{U}}, \tilde{q}), (-\Delta) \tilde{q})| &\leq C_2 ||\nabla_h \tilde{q}||_0^2 \, ||\tilde{q}||_s, & (4.16) \\ \partial_t ||\tilde{q}(t)||_s^2 + c\nu_1 ||\nabla_h \tilde{q}||_s^2 &\leq C_2 \nu_1^{-1} ||\nabla_h \tilde{q}||_0^2 \, ||\tilde{q}||_s^2 + \nu^{-1} ||\tilde{F}_0||^2. & (4.17) \end{aligned}$$

When $s \geq 1/2$ we take $s_1 + s - 2 > 0$

$$|(\tilde{\mathbf{B}}_0(\tilde{q}(t), \tilde{q}), (-\Delta) \tilde{q})| \leq C_2 ||\nabla_h \tilde{q}||_{s_1+s-2} \, ||\nabla_h \tilde{q}||_{s_2} \, ||\tilde{q}||_{s_3+s}. \tag{4.18}$$

We take $s_2 = s, s_3 = 0$; we have $1/p_2 = 1/2 - s/3$, $p_3 = 2$. So $1/p_3 = s/3$. These conditions make sense if $s < 3/2$. Since $s_1 + s_2 + s_3 \geq 3/2$, $s_1 = 3/2 - s$ and $s_1 + s - 2 = -1/2$, we have

$$\partial_t ||\tilde{q}(t)||_s^2 + c\nu_1 ||\nabla_h \tilde{q}||_s^2 + c\nu_1 ||\partial_3 \tilde{q}||_s^2 \leq C_2 \nu_1^{-1} ||\nabla_h \tilde{q}||_0^2 ||\tilde{q}||_s^2 + \nu^{-1} ||\tilde{F}_0||^2, \tag{4.19}$$

by (4.11), and, using Gronwall's inequality, we deduce the boundedness of $||\tilde{q}(t)||_s$ on $[0, T]$ for any T.

To obtain smoothing, one has to multiply by $t(-\Delta)^s q$ and make estimates as before; see the similar situation in BMN [9]. From (4.10) we deduce boundedness for all $t \geq 0$ in H_s, using smoothing as in [9].

When $3/4 \le \alpha < 1$ we introduce the auxiliary function $g(t)$ which is a solution of a linear equation

$$\partial_t g = -\tilde{A}_{\mathrm{QG}} g + \tilde{F}_0,\ g(0) = \tilde{q}(0). \tag{4.20}$$

This solution satisfies the estimate

$$||g(t)||^2_{\alpha-1} \le C'_1,\ \int_0^T ||g(t)||^2_\alpha dt \le C'_1(1+T). \tag{4.21}$$

We put $\tilde{q}(t) = g + \hat{q}(t)$. This function satisfies

$$\partial_t \hat{q} = \tilde{\mathbf{B}}_0(g,\hat{q}) + \tilde{\mathbf{B}}_0(\hat{q},g) + \tilde{\mathbf{B}}_0(\hat{q},\hat{q}) - A_{\mathrm{QG}}\hat{q} + \hat{F}_0,\ \hat{F}_0 = \tilde{\mathbf{B}}_0(g,g). \tag{4.22}$$

One can easily check that $\hat{F}_0$ satisfies (1.19) with $\alpha = 0$. Making estimates with $\alpha = 0$ similar to the above, we deduce

$$\partial_t ||\hat{q}(t)||^2_0 \le C\nu^{-1}||\hat{F}_0||^2_{-1} + C_1||g(t)||^2_{3/4}||\hat{q}(t)||^2_0. \tag{4.23}$$

Therefore $||\hat{q}(t)||^2_0$ is bounded for bounded t, and so is $||\tilde{q}(t)||^2_{\alpha-1}$, $0 \le t \le T_0$. To prove uniform boundedness for all $t \ge 0$ we note first that in (2.29) the coefficient Q^{000}_{kmn} is skew-symmetric in k,n (see (2.16)). This implies, similarly to (4.9),

$$\begin{aligned} ||w^0(t)||^2_0 &\le (c\nu_1)^{-2}||F_0||^2_{-1} + ||w^0(0)||^2_0 e^{-c\nu t}, \\ \int_T^{T+1} \nu||w^0(t)||^2_1 dt &\le C_0\nu^{-1}||F_0||^2_{-1} + ||w^0(0)||^2. \end{aligned} \tag{4.24}$$

Thanks to (1.19) this implies that on every interval $[T, T+1]$ there exists a point t_0 at which $||\tilde{q}(t_0)||^2_{\alpha-1} < ||\tilde{q}(t_0)||^2_0 \le ||w^0(t_0)||^2_1 \le \hat{C}$, where $\hat{C}$ depends only on M_{0F} and $||\tilde{q}||_s$. Now we change t for $t - t_0$ and use the boundedness of $||\tilde{q}(t)||^2_{\alpha-1}$ on a finite interval $[0,2]$; since $[1,2]$ includes another point t'_0 such that $||w^0(t'_0)||^2_1 \le \hat{C}$, this implies uniform boundedness of $||\tilde{q}(t)||^2_{\alpha-1}$. This implies the statement of the theorem.

Note that, since $\tilde{q}_n = w^0_n|n|\omega_n$, Theorem 4.1 gives existence for $w^0 \in H_\alpha$, $\alpha \ge 1$. In the case $\nu_1 = \nu_2 = 0$, the global regularity in H_s with any $s > 5/2$ follows from Bourgeois & Beale [18]. In fact condition $s \ge 3$ is imposed in [18], but one obtains the case $s > 5/2$ from the proof similar to that in [18] by continuity.

5 Ageostrophic catalytic equations

In this section we investigate regularity of the 'catalytic' AG equations (1.24). We assume for simplicity that the horizontally-averaged buoyancy is zero in the reduced equations. The fact that this quantity is an adiabatic invariant (see [12]) allows us to do that. In general, we will have phase corrections associated with the buoyancy time scale as described in [12]. The 'catalytic' system for $\mathbf{w}_n = (w_n^0, w_n')$, $\mathbf{w}_n' = (w_n^1, w_n^2)$ consists of the equations for w_n^0 and the following equations for $\mathbf{w}_n'$

$$\partial_t \mathbf{w}_n' = - \sum_{\substack{\phi_m = \phi_n \\ k+m=n}} \tilde{\Psi}_k^0(t)(D_{mn}(\eta)\mathbf{I} - G_{mn}(\eta)\mathbf{J})\mathbf{w}_m' - A_{\mathrm{AG},n}\mathbf{w}_n' \tag{5.1}$$

where

$$\mathbf{I} = \begin{pmatrix} 1 & 0 \\ 0 & 1 \end{pmatrix}, \qquad \mathbf{J} = \begin{pmatrix} 0 & -1 \\ 1 & 0 \end{pmatrix}, \tag{5.2}$$

$$\begin{aligned} D_{mn}(\eta) &= 2(\check{n}' \wedge \check{m}')\left(\frac{\check{n}_3\check{m}_3\check{n}' \cdot \check{m}'}{|\check{m}'|\,|\check{n}'|\,|\check{m}|\,|\check{n}|}\left(1 + \frac{\eta^2}{\omega_n^2}\right) + \frac{|\check{n}'|^4 - \eta^2\check{n}_3^4}{|\check{n}|^4\omega_n^2}\right), \\ G_{mn}(\eta) &= \eta\frac{3\check{n}_3\check{m}_3(\check{n}' \wedge \check{m}')^2 + ((\check{n} \times \check{m}) \cdot \check{m}^\perp)((\check{n} \times \check{m}) \cdot \check{n}^\perp)}{|\check{m}'|\,|\check{m}|\,|\check{n}'|\,|\check{n}|\,\omega_n}. \end{aligned} \tag{5.3}$$

Here $\check{m}^\perp = (-\check{m}_2, \check{m}_1, 0)$, $\check{n}^\perp = (-\check{n}_2, \check{n}_1, 0)$, $\check{n}' \wedge \check{m}' = \check{n}_1\check{m}_2 - \check{n}_2\check{m}_1$, $\check{n} \times \check{m} = (\check{n}_2\check{m}_3 - \check{n}_3\check{m}_2, \check{n}_3\check{m}_1 - \check{n}_1\check{m}_3, \check{n}_1\check{m}_2 - \check{n}_2\check{m}_1)$ and $\tilde{\Psi}^0(t)$ is the QG streamfunction. Eqs. (5.3) are obtained from formulas presented in [12] with the help of resonant conditions $\phi_m = \phi_n$.

Now we state a theorem on existence and regularity of solutions of the AG system (5.1)–(5.3). The proof is needed since energy is only conserved (when $\nu = 0$) for this equation with general $\tilde{\Psi}^0(t)$; higher Sobolev norms $||\cdot||_s$ *generally are not preserved.* One has to use the special structure of the AG system to obtain the following theorem on well-posedness.

Theorem 5.1 *Let* $\nu_1 \geq 0, \nu_2 \geq 0, s \geq 1, 0 < \delta \ll 1$. *Let* $\int_0^T [||\tilde{q}(t)||_{3/2+\delta} + ||\tilde{q}(t)||_{s-3/2}]dt \leq C, 0 \leq t \leq T$; $\mathbf{w}'(0) \in H_s$. *Then there exists a unique regular solution* $\mathbf{w}'(t), 0 \leq t < T$ *of the AG equation* (5.1) *which belongs to* H_s *for* $0 \leq t \leq T$.

Proof. First we consider the case $\nu_1 \geq 0, \nu_2 \geq 0$ and obtain results uniformly in $1 \geq \nu_1 \geq 0, 1 \geq \nu_2 \geq 0$; to save space we set $\nu_1 = 0, \nu_2 = 0$. Approximate solutions can be obtained by truncating the Fourier series, so to prove existence of the system we have to obtain *a priori* estimates. We obtain them formally for the full system; the truncated case is similar. Multiplying by $|\check{n}|^{2s}(\mathbf{w}_n')^*$ and using that, for real fields $(\mathbf{w}_n')^* = \mathbf{w}_{-n}'$, we obtain

$$1/2\partial_t|\mathbf{w}_n'|^2|\check{n}|^{2s} = \sum_{\substack{\phi_m = \phi_n \\ k+m+n=0}} \tilde{\Psi}_k^0(t)(D_{mn}\mathbf{I} - G_{mn}\mathbf{J})\mathbf{w}_m' \cdot \mathbf{w}_n'|\check{n}|^{2s}. \tag{5.4}$$

We have $|\check{n}|^{2s} = |\check{n}|^s|\check{m}|^s + R_{kmn}$, with $R_{kmn} = |\check{n}|^s(|\check{n}|^s - |\check{m}|^s)$ and $|R_{kmn}| \leq C(s)(|\check{m}|^{s-1}|\check{k}| + |\check{k}|^s)|\check{n}|^s)$. Note that $G_{mn} = G_{nm}$, $D_{mn} = -D_{nm}$, $\mathbf{J} = -\mathbf{J}^*$. Therefore the following sum is skew-symmetric with respect to interchange of n, m and so equals zero:

$$\sum_{\substack{\phi_m = \phi_n \\ k+m+n=0}} \tilde{\Psi}_k^0(t)(D_{mn}\mathbf{I} - G_{mn}\mathbf{J})\mathbf{w}'_m \cdot \mathbf{w}'_n |\check{m}|^s|\check{n}|^s = 0. \tag{5.5}$$

Since $k + n + m = 0$ we have $\check{n}' \wedge \check{m}' = -\check{k}' \wedge \check{m}', \check{n} \wedge \check{m} = -\check{k} \wedge \check{m}$. Therefore by Eqs. (5.1)–(5.3) we have

$$|D_{mn}| \leq 8|\check{k}|\,|\check{m}|, \qquad |G_{mn}| \leq 4\eta|\check{k}|\,|\check{m}|,$$

$$\begin{aligned}
&\left|\sum_{\substack{\phi_m = \phi_n \\ k+m+n=0}} \tilde{\Psi}_k^0(t)(D_{mn}\mathbf{I} - G_{mn}\mathbf{J})\mathbf{w}'_m \cdot \mathbf{w}'_n(R_{mnk})\right| \\
&\leq C \sum_{\substack{\phi_m = \phi_n \\ k+m+n=0}} |\tilde{\Psi}_k^0(t)||\check{k}|\,|\check{m}|\,|\check{n}|^s|\mathbf{w}'_m||\mathbf{w}'_n||R_{mnk}| \\
&\leq C'\left(\sum_{k+m+n=0} |\tilde{\Psi}_k^0(t)||\check{k}|^2\,|\check{m}|^s\,|\check{n}|^s|\mathbf{w}'_m|\,|\mathbf{w}'_n|\right. \\
&\qquad \left. + \sum_{k+m+n=0} |\tilde{\Psi}_k^0(t)|\,|\check{k}|^s\,|\check{m}|\,|\check{n}|^s|\mathbf{w}'_m|\,|\mathbf{w}'_n|\right).
\end{aligned} \tag{5.6}$$

In the above sums we do not include $k = 0$, $m = 0$, $n = 0$ since corresponding coefficients are zero according to zero average condition. Therefore, estimating the convolutions in a standard way we obtain

$$\begin{aligned}
\sum_n \sum_{\substack{\phi_m = \phi_n \\ k+m+n=0}} |\tilde{\Psi}_k^0(t)||\mathbf{w}'_m||\mathbf{w}'_n|\,|\check{k}|^2\,|\check{m}|^s\,|\check{n}|^s &\leq \sum_n |\mathbf{w}'_n|^2|\check{n}|^{2s}\sum_k |\tilde{\Psi}_k^0(t)|\,|\check{k}|^2 \\
&\leq C||\mathbf{w}'||_s^2\,||\tilde{q}||_{3/2+\delta};
\end{aligned} \tag{5.7}$$

We can estimate the second sum in (5.6) as in Theorem 4.1:

$$\sum_n \sum_{\substack{\phi_m = \phi_n \\ k+m+n=0}} |\tilde{\Psi}_k^0(t)|\,|\mathbf{w}'_m|\,|\mathbf{w}'_n|\,|\check{m}|\,|\check{k}|^s\,|\check{n}|^s \leq C||\mathbf{w}'||_s^2\,||\tilde{q}||_{s-3/2+\delta}, \tag{5.8}$$

$$\frac{1}{2}\partial_t||\mathbf{w}'||_s^2 \;\leq\; C_1[\,||\tilde{q}^0(t)||_{3/2+\delta} + ||\tilde{q}||_{s-3/2+\delta}]\,||\mathbf{w}'||_s^2. \tag{5.9}$$

Using Gronwall's inequality we obtain the estimate

$$||\mathbf{w}'(T)||_s^2 \leq C_2 \exp\left(\int_{0\leq t\leq T}[\,||\tilde{q}(t)||_{3/2+\delta} + ||\tilde{q}||_{s-3/2+\delta}]\,dt\right), \tag{5.10}$$

and using this one easily obtains the existence of a solution $\mathbf{w}'(t)$ of (5.1).

Theorem 5.2 *Let* $\nu_1 > 0$, $\nu_2 > 0$, $s \geq 0$, *let* $\int_T^{T+1}[||\tilde{q}(t)||_0^4 + ||\tilde{q}||_{s-3/2+\delta}]dt \leq M_{2F}$, *for all* $T \geq 0$, $\mathbf{w}'(0) \in H_s$. *Then there exists a unique regular solution* $\mathbf{w}'(t)$, $0 \leq t < T$ *of the AG equation* (5.1) *which belongs to* H_s *and*

$$||\mathbf{w}'(t)||_s \leq M_s, \qquad \text{for all } t \geq 0 \tag{5.11}$$

where M_s *depends on* $\nu, M_{2F}, ||\mathbf{w}'(0)||_s$.

Proof. The case $\nu_1 > 0$, $\nu_2 > 0$ is similar to Theorem 5.1; now instead of (5.9) we derive the estimate

$$\frac{1}{2}\partial_t||\mathbf{w}'||_s^2 + \nu||\mathbf{w}'||_{s+1}^2 \leq C_2[\,||\tilde{q}||_0^4 + ||\tilde{q}||_{s-3/2+\delta}]\,||\mathbf{w}'||_s^2 + \frac{\nu||\mathbf{w}'||_{s+1}^2}{2}. \tag{5.12}$$

This inequality implies

$$||\mathbf{w}'(T)||_s^2 \leq C_2'||\mathbf{w}'(0)||_s^2 \exp\left(\int_{0\leq t\leq T} 2C_2\,[\,||\tilde{q}||_0^4 + ||\tilde{q}||_{s-3/2+\delta}]\,dt\right). \tag{5.13}$$

To obtain (5.12) we use instead of (5.7) the following estimates

$$\sum_n \sum_{\substack{\phi_m = \phi_n \\ k+m+n=0}} |\tilde{\Psi}_k^0(t)|\,|\mathbf{w}'_m|\,|\mathbf{w}'_n|\,|k|^2\,|m|^s\,|n|^s \leq C||\mathbf{w}'||_{s+3/4}\,||\mathbf{w}'||_{s+3/4}\,||\tilde{q}||_0; \tag{5.14}$$

and for $1/2 \leq s \leq 3/2$ (the case $s \geq 3/2$ is simpler) instead of (5.8) we have

$$\sum_n \sum_{\substack{\phi_m = \phi_n \\ k+m+n=0}} |\tilde{\Psi}_k^0(t)|\,|\mathbf{w}'_m|\,|\mathbf{w}'_n|\,|m|\,|k|^s\,|n|^s \leq C||\mathbf{w}'||_{s+1}\,||\mathbf{w}'||_s\,||\tilde{q}||_{s-3/2}. \tag{5.15}$$

We have also

$$||\mathbf{w}'(T)||_0^2 - ||\mathbf{w}'(0)||_0^2 + \nu\int_0^T ||\mathbf{w}'(T)||_1^2 dt \leq 0.$$

This implies uniform boundedness of $||\mathbf{w}'(t)||_0$. To prove boundedness in H_1 for $t \geq 0$, we use boundedness of the solution in H_0 and, for $t \geq 1$, the smoothing argument based on multiplication by $t(-\Delta)\mathbf{w}'$ similar to that given in BMN[9], and similarly for boundedness in H_s. Theorem 5.2 is proved.

6 Global regularity of the limit $2\frac{1}{2}D$ resonant equations

In this section and in Section 7 we remove all restrictions to non-resonant domains (any parameters θ_2, θ_3) as well as restrictions to non-resonant $\eta = f/N$. We treat the case of quadratic resonant operators for the asymptotic limit equations ($\mathbf{B}_3 \neq 0$ in Eq. (1.23)). From this, in Section 7, we give the most general theorems on existence of strong solutions on infinite time intervals for the 3D Navier–Stokes primitive equations of geophysics in regimes of small Rossby and/or Froude numbers.

The limit resonant operator $\tilde{\mathbf{B}}$ defined in (1.16) inherits properties of the operator $\mathbf{B}$. This statement follows from the following

Lemma 6.1 *Let* $(\mathbf{u}, \mathbf{v}, \mathbf{z}) \in H_{3/4} \times H_{3/4} \times H_1$ *given in the Craya basis. Then*

$$\left(\tilde{\mathbf{B}}(\mathbf{u}, \mathbf{v}), \mathbf{z}\right) = \lim_{N\to\infty} \frac{1}{T} \int_0^T (\mathbf{B}_p(Ns, \mathbf{u}, \mathbf{v}), \mathbf{z}) ds. \tag{6.1}$$

Here $\mathbf{u}$, $\mathbf{v}$ and $\mathbf{z}$ denote generic *time-independent* vectors in the Craya basis, [24]; $\mathbf{B}_p$ is the non-autonomous oscillating operator defined in (1.14). From now on, we shall omit the index p in $\mathbf{B}(Nt, \mathbf{u}, \mathbf{v})$.

Proof. We introduce projections π_R on the finite-dimensional subspace of Fourier modes with $|n| \leq R$. We fix $(\mathbf{u}, \mathbf{v}, \mathbf{z}) \in H_{3/4} \times H_{3/4} \times H_1$. We put $\mathbf{u}_R = \pi_R \mathbf{u}$ and similarly for $\mathbf{v}$ and $\mathbf{z}$. Clearly,

$$\begin{aligned}(\mathbf{B}(Nt, \mathbf{u}, \mathbf{v}), \mathbf{z}) - (\tilde{\mathbf{B}}(\mathbf{u}, \mathbf{v}), \mathbf{z}) = {} & [(\mathbf{B}(Nt, \mathbf{u}, \mathbf{v}), \mathbf{z}) - (\mathbf{B}(Nt, \mathbf{u}_R, \mathbf{v}_R), \mathbf{z}_R)] \\ & + [(\mathbf{B}(Nt, \mathbf{u}_R, \mathbf{v}_R), \mathbf{z}_R) - (\tilde{\mathbf{B}}(\mathbf{u}_R, \mathbf{v}_R), \mathbf{z}_R)] \\ & + [(\tilde{\mathbf{B}}(\mathbf{u}_R, \mathbf{v}_R), \mathbf{z}_R) - (\tilde{\mathbf{B}}(\mathbf{u}, \mathbf{v}), \mathbf{z}].\end{aligned}$$

The operators $\mathbf{B}$ and $\tilde{\mathbf{B}}$ are continuous on $H_{3/4} \times H_{3/4} \times H_1$. Moreover, since the unitary Poincaré propagator $\mathbf{E}(Nt)$ preserves all Sobolev norms, the operator $\mathbf{B}(Nt, \mathbf{u}, \mathbf{v})$ is continuous uniformly in Nt. Therefore, the first and third brackets on the right-hand side tend to zero as $R \to \infty$. Let $\epsilon > 0$; we find an R such that the absolute values of the first and third brackets are less than ϵ.

After that we consider the second bracket

$$\begin{aligned}[(\mathbf{B}(Nt, \mathbf{u}_R, \mathbf{v}_R), \mathbf{w}_R) - (\tilde{\mathbf{B}}(\mathbf{u}_R, \mathbf{v}_R), \mathbf{w}_R)] &= (\mathbf{B}(Nt, \mathbf{u}_R, \mathbf{v}_R) - \tilde{\mathbf{B}}(\mathbf{u}_R, \mathbf{v}_R), \mathbf{w}_R) \\ &= (\mathbf{B}_{\text{osc}}(Nt, \mathbf{u}_R, \mathbf{v}_R), \mathbf{w}_R).\end{aligned}$$

Since $\mathbf{B}_{\text{osc}}$ contains only non-resonant terms, we obtain, after integrating by parts as in [9], that

$$\frac{1}{T} \int_0^T (\mathbf{B}_{\text{osc}}(Ns, \mathbf{u}_R, \mathbf{v}_R), \mathbf{w}_R) ds = O(1/N) \to 0 \quad \text{as} \quad N \to \infty.$$

Therefore, the integrals of all three brackets are less than ϵ when N is large, and the lemma is proven.

Corollary 6.1 *Let $\sigma \geq 1$ and $\mathbf{w}$ be the Craya vector variable, $\mathbf{w} = (w^0, w^1, w^2)$ (w^0 corresponds to QG modes and w^1, w^2 correspond to AG modes). Then*

$$\left(\tilde{\mathbf{B}}(\mathbf{w}, (-\Delta)^{\sigma/2}\mathbf{w}), (-\Delta)^{\sigma/2}\mathbf{w}\right) = 0. \tag{6.2}$$

Proof. From Lemma 6.1, it suffices to prove the similar identity for the general operator $\mathbf{B}(Nt, \mathbf{v}, \mathbf{v})$ in Eqs. (2.20)–(2.22) for the non-averaged Eqs. (2.22) written in Fourier space in the Craya basis. Using Eqs. (2.15)

$$\begin{aligned}
&(\mathbf{B}(Nt, \mathbf{v}, (-\Delta)^{\sigma/2}\mathbf{v}), (-\Delta)^{\sigma/2}\mathbf{v}) \\
&= \sum_{\substack{k+m+n=0;\\ i_1,i_2,i_3;\\ l=1,\dots,8}} c_l e^{iD_l(k,m,n)Nt}\, Q^{i_1 i_2 i_3}_{kmn}\, v^{i_1}_k\, v^{i_2}_m\, |\check{m}|^\sigma\, |\check{n}|^\sigma\, v^{i_3}_n \\
&= \sum_{\substack{k+m+n=0;\\ i_1,i_2,i_3;\\ l=1,\dots,8}} c_l e^{iD_l(k,m,n)Nt}\, (q_{i_1 k}\cdot \check{m})(q_{i_2 m}\cdot q_{i_3 n})\, |\check{m}|^\sigma\, |\check{n}|^\sigma\, v^{i_1}_k v^{i_2}_m\, v^{i_3}_n \\
&= \sum_{\substack{k+m+n=0;\\ i_1,i_2,i_3;\\ l=1,\dots,8}} c_l e^{iD_l(k,m,n)Nt}\, (q_{i_1 k}\cdot \check{n})\,(q_{i_3 n}\cdot q_{i_2 m})\, |\check{m}|^\sigma\, |\check{n}|^\sigma\, v^{i_1}_k\, v^{i_2}_m\, v^{i_3}_n \\
&\quad - \sum_{\substack{k+m+n=0;\\ i_1,i_2,i_3;\\ l=1,\dots,8}} c_l e^{iD_l(k,m,n)Nt}\, (q_{i_1 k}\cdot (\check{k}+\check{m}))\,(q_{i_3 n}\cdot q_{i_2 m})\, |\check{m}|^\sigma\, |\check{n}|^\sigma\, v^{i_1}_k\, v^{i_2}_m\, v^{i_3}_n \\
&\quad - \sum_{\substack{k+m+n=0;\\ i_1,i_2,i_3;\\ l=1,\dots,8}} c_l e^{iD_l(k,m,n)Nt}\, (q_{i_1 k}\cdot \check{m})\,(q_{i_3 n}\cdot q_{i_2 m})\, |\check{m}|^\sigma\, |\check{n}|^\sigma\, v^{i_1}_k\, v^{i_2}_m\, v^{i_3}_n \\
&= -(\mathbf{B}(Nt, \mathbf{v}, (-\Delta)^{\sigma/2}\mathbf{v}), (-\Delta)^{\sigma/2}\mathbf{v}),
\end{aligned}$$

where in the above sum we interchanged indices m and n, i_2 and i_3 and used the divergence-free condition $\check{k}\cdot q_{i_1 k} = 0$. Here c_l are absolute constants indexed by i_1, i_2, i_3 with values $\pm 1/8$ (cf. Eqs. (2.19), (2.20)). We use $(\mathbf{E}(Nt))^* = \mathbf{E}(-Nt)$ to ensure symmetry of the terms $(\mathbf{E}(-Nt)\mathbf{v})_m$ and $(\mathbf{E}(-Nt)\mathbf{v})_n$. Then from (6.3) we have

$$(\mathbf{B}(Nt, \mathbf{v}, (-\Delta)^{\sigma/2}\mathbf{v}), (-\Delta)^{\sigma/2}\mathbf{v}) = 0 \tag{6.3}$$

Eqs. (6.2) follow from (6.3) and Lemma 6.1.

We follow with the estimate for the resonant operator $\tilde{\mathbf{B}}(\mathbf{w}, \mathbf{w})$ in the Craya basis:

Corollary 6.2 *Let $\sigma \geq 1$ and $\mathbf{w}$ be the Craya vector variable. Then*

$$|(\tilde{\mathbf{B}}(\mathbf{w},\mathbf{w}),(-\Delta)\mathbf{w})| \leq C_{\tilde{B}} \sum_{\substack{k+m+n=0,\\ \pm\omega'_k \pm \omega'_m \pm \omega'_n = 0}} |\check{k}|\,|\mathbf{w}_k|\,|\check{m}|\,|\mathbf{w}_m|\,|\check{n}|\,|\mathbf{w}_n|. \quad (6.4)$$

Proof. Recall that the Craya basis vectors q are normalised with norm 1. We first prove such an estimate for the non-resonant general operator $\mathbf{B}(Nt,\mathbf{v},\mathbf{v})$ for every fixed t. We have

$$\begin{aligned}
&(\mathbf{B}(Nt,\mathbf{v},\mathbf{v}),(-\Delta)\mathbf{v}) \\
&= \sum_{\substack{k+m+n=0;\\ i_1,i_2,i_3;\\ l=1,\dots,8}} c_l e^{iD_l(k,m,n)Nt}\,(q_{i_1k}\cdot\check{m})\,(q_{i_2m}\cdot q_{i_3n})\,|\check{n}|^2\, v_k^{i_1}\, v_m^{i_2}\, v_n^{i_3} \quad (6.5)\\
&= \sum_{\substack{k+m+n=0;\\ i_1,i_2,i_3;\\ l=1,\dots,8}} c_l e^{iD_l(k,m,n)Nt}\,(q_{i_1k}\cdot\check{m})\,(q_{i_2m}\cdot q_{i_3n})\,|\check{n}|\,(|\check{n}|-|\check{m}|)v_k^{i_1}\, v_m^{i_2}\, v_n^{i_3}. \quad (6.6)
\end{aligned}$$

Note that $|\,|\check{n}|-|\check{m}|\,| = |\,|\check{k}+\check{m}|-|\check{m}|\,| \leq 7|\check{k}|$, yielding

$$|(\mathbf{B}(Nt,\mathbf{v},\mathbf{v}),(-\Delta)\mathbf{v})| \leq C_B \sum_{k+m+n=0} |\check{k}|\,|\mathbf{v}_k|\,|\check{m}|\,|\mathbf{v}_m|\,|\check{n}|\,|\mathbf{v}_n|. \quad (6.7)$$

The same estimate follows for $\tilde{\mathbf{B}}(\mathbf{w},\mathbf{w})$, from the skew-symmetry Corollary 6.1 and from averaging Eqs. (6.5); this only further restricts the k,m,n interactions to the set

$$\pm\omega'_k \pm \omega'_m \pm \omega'_n = 0, \quad (6.8)$$

where $\omega'_n = 0$ for QG modes and $\omega'_n = \omega_n$ for AG modes, and similarly for k,m. In Eqs. (2.22), the resonant operators $\tilde{Q}^{i_1i_2i_3}_{kmn}$ are first-order Fourier integral operators. For $\tilde{\mathbf{B}}(\mathbf{w},\mathbf{w})$ we obtain

$$|(\tilde{\mathbf{B}}(\mathbf{w},\mathbf{w}),(-\Delta)\mathbf{w})| \leq C_{\tilde{B}} \sum_{\substack{k+m+n=0,\\ \pm\omega'_k \pm \omega'_m \pm \omega'_n = 0}} |\check{k}|\,|\mathbf{w}_k|\,|\check{m}|\,|\mathbf{w}_m|\,|\check{n}|\,|\mathbf{w}_n|. \quad (6.9)$$

Remark 6.1 Since $\mathbf{w}_{\mathrm{AG}}$ is orthogonal to $\mathbf{w}_{\mathrm{QG}}$ (orthogonality of the q's) we also have

$$\left(\mathbf{B}_3(\mathbf{w}_{\mathrm{AG}},(-\Delta)^{\sigma/2}\mathbf{w}_{\mathrm{AG}}),(-\Delta)^{\sigma/2}\mathbf{w}_{\mathrm{AG}}\right) = 0 \quad (6.10)$$

and the estimate (6.9) holds for $|(\mathbf{B}_2(\mathbf{w}_{\mathrm{QG}},\mathbf{w}_{\mathrm{AG}})+\mathbf{B}_3(\mathbf{w}_{\mathrm{AG}},\mathbf{w}_{\mathrm{AG}}),(-\Delta)\mathbf{w}_{\mathrm{AG}})|$.

Remark 6.2 The above estimate will be used together with the Lemma 6.2 on restricted convolutions to obtain global regularity in H_1 for Eqs. (2.27)–(2.28).

In this section we present new estimates for the nonlinear '$2\frac{1}{2}$-dimensional' operator $\mathbf{B}_3$ which ensure global existence of strong solutions of the limit AG viscous equations (2.33)–(2.34) and, consequently, Eqs. (2.27)–(2.28) for *all* domain parameters. The following theorem which will be proved below provides the main estimate for the resonant operator $\mathbf{B}_3$ for the 'worst' case of all interactions on the '$2\frac{1}{2}$-dimensional' interaction manifold K^* defined in Section 3.

Theorem 6.1 *Let $\mathbf{w}_{\mathrm{AG}}(x_1, x_2, x_3) \in H_2$ (the Sobolev space of periodic vector fields with zero mean). Then the following estimate holds*

$$|(\mathbf{B}_3(\mathbf{w}_{\mathrm{AG}}, \mathbf{w}_{\mathrm{AG}}), (-\Delta)\mathbf{w}_{\mathrm{AG}})| \leq C_{\mathrm{III}} ||\mathbf{w}_{\mathrm{AG}}||_2 \, ||\mathbf{w}_{\mathrm{AG}}||_1^2, \tag{6.11}$$

where C_{III} is some constant.

Remark 6.3 Estimate (6.11) is of the same type as the classical estimate of Ladyzhenskaya [46] in the 2D case with Dirichlet boundary conditions. For the periodic boundary conditions in 2D it is well-known that the analogue of the left-hand side of (6.11) is identically zero ([23]). Of course, in (6.11) the divergence-free vector field $\mathbf{w}_{\mathrm{AG}}(x_1, x_2, x_3)$ and the Sobolev spaces H_α are 3D with space variables x_1, x_2 and x_3.

From the estimate (6.11) we immediately obtain the following theorem in a standard way (cf. [4], [23], [73]) (note that if the force $\mathbf{F}^\dagger(t, x)$ in the original equation does not depend on N and f, then $\tilde{\mathbf{F}}_{\mathrm{AG}} = 0$).

Theorem 6.2 *Let $\nu_1, \nu_2 > 0$, $\nu = \min(\nu_1, \nu_2)$, $||\mathbf{w}_{\mathrm{AG}}(0)||_\alpha \leq M_\alpha$, $1 \geq \alpha > 3/4$; $\tilde{\mathbf{F}}_{\mathrm{AG}}$ satisfies with $\alpha = 1$:*

$$\sup_T \int_T^{T+1} ||\tilde{\mathbf{F}}_{\mathrm{AG}}||_{\alpha-1}^2 dt \leq M_{\alpha F}^2. \tag{6.12}$$

Then there exists a unique regular solution $\mathbf{w}_{\mathrm{AG}}(t)$ of the '$2\frac{1}{2}$-dimensional' primitive Navier–Stokes Eqs. (2.33)–(2.34),

$$||\mathbf{w}_{\mathrm{AG}}(t)||_1 \leq M_1'(\nu, M_{1F}, M_\alpha, a_1, a_2, a_3)$$

for all $t \geq 0$.

Proof. A local regular solution to the '$2\frac{1}{2}$-dimensional' equations (2.33)–(2.34) exists on a small interval of time $0 \leq t \leq t_1$ (see BMN [9]) and belongs to H_1 for $0 < t < t_1$ thanks to the smoothing property which follows from

$$\frac{\nu_{\mathrm{AG}}}{2} \int_0^t ||\mathbf{w}_{\mathrm{AG}}||_{1+\gamma}^2 d\tau + ||\mathbf{w}_{\mathrm{AG}}(t)||_\gamma^2 \leq C(t), \qquad 0 \leq t < t_1. \tag{6.13}$$

Therefore, it is sufficient to consider $\gamma = 1$.

Multiplying Eqs. (2.33) for $\mathbf{w}_{\mathrm{AG}}$ by $(-\Delta)\mathbf{w}_{\mathrm{AG}}$ we obtain

$$\begin{aligned}\partial_t\|\mathbf{w}_{\mathrm{AG}}\|_1^2 = & -2\nu_{\mathrm{AG}}\|\mathbf{w}_{\mathrm{AG}}\|_2^2 + 2\left(\mathbf{B}_3(\mathbf{w}_{\mathrm{AG}}, \mathbf{w}_{\mathrm{AG}}), (-\Delta)\mathbf{w}_{\mathrm{AG}}\right) \\ & +2\left(\mathbf{B}_2(\mathbf{w}_{\mathrm{QG}}, \mathbf{w}_{\mathrm{AG}}), (-\Delta)\mathbf{w}_{\mathrm{AG}}\right) + 2\left(\tilde{\mathbf{F}}^{\mathrm{AG}}, (-\Delta)\mathbf{w}_{\mathrm{AG}}\right). \quad (6.14)\end{aligned}$$

For the 3-wave resonant operator $\mathbf{B}_3(\mathbf{w}_{\mathrm{AG}}, \mathbf{w}_{\mathrm{AG}})$ we have from Theorem 6.1

$$|(\mathbf{B}_3(\mathbf{w}_{\mathrm{AG}}, \mathbf{w}_{\mathrm{AG}}), (-\Delta)\mathbf{w}_{\mathrm{AG}})| \leq C_{\mathrm{III}}\|\mathbf{w}_{\mathrm{AG}}\|_2\, \|\mathbf{w}_{\mathrm{AG}}\|_1^2. \quad (6.15)$$

We have according to [13]

$$|(\mathbf{B}_2(\mathbf{w}_{\mathrm{QG}}, \mathbf{w}_{\mathrm{AG}}), (-\Delta)\mathbf{w}_{\mathrm{AG}})| \leq C_{\mathrm{II}}\|\tilde{\mathbf{U}}_{\mathrm{QG}}\|_1\, \|\mathbf{w}_{\mathrm{AG}}\|_1^2 \quad (6.16)$$

where the QG velocity $\tilde{\mathbf{U}}_{\mathrm{QG}}$ was defined in Eqs. (4.1), and where C_{II} is some constant. Estimates for the viscous QG equation are derived in [13]. Using the above estimate, a standard Gronwall inequality yields the estimate in H_1 for all $t \geq 0$, and uniqueness of the solutions $\mathbf{w}_{\mathrm{AG}}$ follows in a standard way (cf. [23], [73]). Theorem 6.2 is proved.

Remark 6.4 Using Theorems 6.1 and 6.2 one can develop regularity theory for solutions of '$2\frac{1}{2}$-dimensional' Navier–Stokes equations in $H_\gamma, \gamma > 1$ spaces. This is done in a way similar to the well-known higher regularity theory for sufficiently regular solutions of the 3D Navier–Stokes equations (see Temam [73]).

Now we prove Theorem 6.1; the proof is based on the following lemma on restricted convolutions. Here without loss of generality we assume $\theta_1 = \theta_2 = \theta_3 = 1$.

Lemma 6.2 (Lemma on Restricted Convolutions) *Let $\chi(k, m, n)$ be the characteristic function of some set K^* in $(\mathbf{Z}^3)^3$ such that*

$$\chi(k, m, n) = \chi(m, k, n) = \chi(k, n, m)$$

is symmetric. Let $\alpha \geq 0$, β, be fixed and

$$\sup_n \sum_{k:k+m+n=0,\ k\in\Sigma_i} \chi(k, m, n)|k|^{-\alpha} \leq C_0 2^{i\beta} \quad (6.17)$$

for every $i = 0, 1, 2, \ldots$, where

$$\Sigma_i = \left\{k = (k_1, k_2, k_3) \mid 2^i \leq |k| < 2^{i+1},\ |k| = \sqrt{k_1^2 + k_2^2 + k_3^2}\right\}. \quad (6.18)$$

Then for any sequence u_n with $u_{(0,0,0)} = 0$

$$\sum_{k+m+n=0} |u_k|\,|u_m|\,|u_n|\,\chi(k,m,n)$$
$$\leq C\left(\sum_n |n|^\beta |u_n|^2\right)^{1/2}\left(\sum_k |k|^\alpha |u_k|^2\right)^{1/2}\left(\sum_m |u_m|^2\right)^{1/2} \quad (6.19)$$

where $C = 6\sqrt{2C_0}$.

Proof. Let $\alpha \geq 0$. We first give the proof for $\beta \geq 0$. Since the left sum in Eq. (6.19) is symmetric with respect to k, m, n, we have

$$\sum_{k+m+n=0} |u_k u_m u_n|\,\chi(k,m,n) \leq 6 \sum_{\substack{k+m+n=0,\\ |n|\geq|k|\geq|m|}} |u_k u_m u_n|\,\chi(k,m,n), \quad (6.20)$$

and it is sufficient to take k, m, n such that $|n| \geq |k| \geq |m|$. After that, we apply the Littlewood–Paley technique of dyadic decomposition (Stein [71]). We estimate

$$\begin{aligned} S &= \sum_{\substack{k+m+n=0,\\ |n|\geq|k|\geq|m|}} |u_k u_m u_n|\,\chi(k,m,n) \\ &= \sum_i \sum_n |u_n| \sum_{k\in\Sigma_i, |n|\geq|k|\geq|m|} |u_k u_{-k-n}|\,\chi(k,-k-n,n). \end{aligned}$$

Since $|n| \geq |k| \geq |m|$ and $k+m+n=0$, we have $2|k| \geq |n| \geq |k|$. Therefore,

$$\begin{aligned} S &\leq \sum_i \sum_{n\in\Sigma_i\cup\Sigma_{i+1}} |u_n| \sum_{k\in\Sigma_i} |u_k u_{-k-n}|\,\chi(k,-k-n,n) \\ &\leq \sum_i \sum_{n\in\Sigma_i\cup\Sigma_{i+1}} |u_n| \left(\sum_{k\in\Sigma_i} |u_k u_{-k-n}|^2 |k|^\alpha\right)^{\frac{1}{2}} \left(\sum_{k\in\Sigma_i} |k|^{-\alpha}\chi(k,-k-n,n)\right)^{\frac{1}{2}} \\ &\leq \sum_i \left(\sum_{n\in\Sigma_i\cup\Sigma_{i+1}} |u_n|^2\right)^{\frac{1}{2}} \left(\sum_{n\in\Sigma_i\cup\Sigma_{i+1}} \sum_{k\in\Sigma_i} |u_k u_{-k-n}|^2\,|k|^\alpha\right)^{\frac{1}{2}} \\ &\quad \times \sup_n \left(\sum_{k\in\Sigma_i} |k|^{-\alpha}\,\chi(k,-k-n,n)\right)^{\frac{1}{2}} \\ &\leq C_0^{\frac{1}{2}} \sum_i \left(\sum_{n\in\Sigma_i\cup\Sigma_{i+1}} |u_n|^2\right)^{\frac{1}{2}} \left(\sum_{n\in\Sigma_i\cup\Sigma_{i+1}} \sum_{k\in\Sigma_i} |u_k u_{-k-n}|^2 |k|^\alpha\right)^{\frac{1}{2}} 2^{\beta i/2} \end{aligned}$$

$$\leq C_0^{1/2}\sum_i\left(\sum_{n\in\Sigma_i\cup\Sigma_{i+1}}|u_n|^2 2^{\beta i}\right)^{\frac{1}{2}}\left(\sum_m\sum_{k\in\Sigma_i}|u_k u_m|^2|k|^\alpha\right)^{\frac{1}{2}}$$

$$\leq C_0^{1/2}\sum_i\left(\sum_{n\in\Sigma_i\cup\Sigma_{i+1}}|u_n|^2 2^{\beta i}\right)^{\frac{1}{2}}\left(\sum_{k\in\Sigma_i}|u_k|^2|k|^\alpha\right)^{\frac{1}{2}}\left(\sum_m|u_m|^2\right)^{\frac{1}{2}}.$$

Therefore,

$$\begin{aligned}
S &\leq C_0^{1/2}\sum_i\left(\sum_{n\in\Sigma_i\cup\Sigma_{i+1}}|n|^\beta|u_n|^2\right)^{\frac{1}{2}}\left(\sum_{k\in\Sigma_i}|u_k|^2|k|^\alpha\right)^{\frac{1}{2}}\left(\sum_m|u_m|^2\right)^{1/2}\\
&\leq C_0^{1/2}\left(\sum_i\sum_{n\in\Sigma_i\cup\Sigma_{i+1}}|n|^\beta|u_n|^2\right)^{\frac{1}{2}}\left(\sum_i\sum_{k\in\Sigma_i}|u_k|^2|k|^\alpha\right)^{\frac{1}{2}}\left(\sum_m|u_m|^2\right)^{1/2}\\
&\leq (2C_0)^{1/2}\left(\sum_n|n|^\beta|u_n|^2\right)^{\frac{1}{2}}\left(\sum_k|k|^\alpha|u_k|^2\right)^{\frac{1}{2}}\left(\sum_m|u_m|^2\right)^{\frac{1}{2}}.
\end{aligned}$$

Considering in the same manner other permutations of $|k|$, $|m|$ and $|n|$, we obtain (6.19) with $C = 6\sqrt{2C_0}$. The proof extends to $\beta < 0$ with a different constant C.

We note that one obtains similar results for general θ_1, θ_2, θ_3 bounded away from 0 and $+\infty$. In that case the constants depend on θ_1, θ_2, θ_3.

Proof of Theorem 6.1. From Corollary 6.2 we obtain the following inequality

$$\begin{aligned}
&|(\mathbf{B}_3(\mathbf{w}_{\mathrm{AG}},\mathbf{w}_{\mathrm{AG}}),(-\Delta)\mathbf{w}_{\mathrm{AG}})|\\
&\quad\leq c'\sum_{k+m+n=0}|\check{k}|\,|\mathbf{w}_{\mathrm{AG},k}|\,|\check{m}|\,|\mathbf{w}_{\mathrm{AG},m}|\,|\check{n}|\,|\mathbf{w}_{\mathrm{AG},n}|\,\chi(k,m,n).
\end{aligned}\tag{6.21}$$

Here $\chi(k,m,-n)$ is the characteristic function of the resonant set K^* of strict 3-wave resonances:

$$\pm\omega_k\pm\omega_m\pm\omega_n=0,\qquad \omega_k\omega_m\omega_n\neq 0.\tag{6.22}$$

This set lies in the manifold of solutions of a polynomial equation $P(k,m,n) = 0$. Indeed, we have the identity

$$\begin{aligned}
&\frac{1}{\omega_k+\omega_n-\omega_m}\\
&\quad=\frac{(\omega_k+\omega_m+\omega_n)(-\omega_k+\omega_m+\omega_n)(-\omega_n+\omega_m+\omega_k)}{(\omega_k+\omega_m+\omega_n)(\omega_k-\omega_m+\omega_n)(-\omega_k+\omega_m+\omega_n)(\omega_k+\omega_m-\omega_n)}.
\end{aligned}$$

The denominator is $-P = ((\omega_k)^2+(\omega_m)^2-(\omega_n)^2)^2-4(\omega_k)^2(\omega_m)^2$. Thus P is a polynomial of degree 2 of $\lambda=\eta^2$. We write ω_k in the form (see Eq. (2.6))

$$\omega_k=\sqrt{\chi_k+\lambda(1-\chi_k)},\quad \chi_k=\frac{|\check{k}'|^2}{|\check{k}|^2}=\phi_k^2,\quad 1-\chi_k=\xi_k^2.$$

The polynomial takes the form (with $\lambda = \eta^2$) $P = P_2\lambda^2 + P_1\lambda + P_0$, where

$$\begin{aligned} P_2 &= \chi_k^2 + \chi_m^2 + \chi_n^2 - 2\chi_k\chi_n - 2\chi_k\chi_m - 2\chi_m\chi_n - 3 + 2(\chi_k + \chi_m + \chi_n); \\ P_1 &= -2(\chi_k^2 + \chi_m^2 + \chi_n^2 - 2\chi_k\chi_n - 2\chi_k\chi_m - 2\chi_m\chi_n + \chi_k + \chi_m + \chi_n) \\ P_0 &= \chi_k^2 + \chi_m^2 + \chi_n^2 - 2\chi_k\chi_n - 2\chi_k\chi_m - 2\chi_m\chi_n. \end{aligned}$$

Instead of considering P as a polynomial in η^2, we renormalise it as

$$\Pi(k,m,n) = |\check{k}|^4\,|\check{m}|^4\,|\check{n}|^4\,P(k,m,n), \tag{6.23}$$

where Π is a homogeneous polynomial of degree 12 in the variables k, m, n, and η is considered as a parameter. For a given η, θ_1, θ_2, θ_3, $\Pi(k,m,n) = 0$ is equivalent to $(k,m,n) \in K^*$ (*vice versa*, fixing k, m, n as parameters, and solving for η as a function of $\theta_1 = 1$, θ_2, θ_3 defines the singular values of $\eta \in \Theta_3^*(\theta_2,\theta_3)$). It follows that for fixed η, θ_2, θ_3, $\Pi(k,-k-n,-n)$ is a polynomial of degree at most 8 in k_3. The leading power in k_3^8 is:

$$-k_3^8\left(|n'|^2 + \eta^2 n_3^2\right)\left(3\eta^2 n_3^2 + (4\eta^2-1)|n'|^2\right), \tag{6.24}$$

where m was eliminated via $m = -k-n$. If this leading term is not zero, there are at most 8 k_3 satisfying $\Pi(k,-m-n,-n) = 0$ for given k_1, k_2, n; this holds provided that

$$3\eta^2 n_3^2 + (4\eta^2-1)|n'|^2 \neq 0. \tag{6.25}$$

Note that if $n' = 0$, the condition (6.25) is trivially satisfied as $n \neq 0$. Also, the condition is satisfied whenever $4\eta^2 - 1 \geq 0$. If the condition (6.25) is not satisfied, i.e. if n belongs to the manifold

$$3\eta^2 n_3^2 + (4\eta^2-1)|n'|^2 = 0, \qquad n' \neq 0, \tag{6.26}$$

then we must verify that the polynomial Π is not identically null. This is not trivial, as one verifies that the coefficient of k_3^7 is null under the condition (6.26). Under the latter condition the coefficient of k_3^6 reduces to

$$k_3^6\frac{|n'|^4}{\eta^2}\left(-4\eta^4\frac{1-4\eta^2}{\eta^2}|n'|^2 + \frac{8}{9}\frac{(\eta^2-1)^3}{\eta^2}(|m'|^2+|k'|^2)\right), \tag{6.27}$$

which is strictly negative whenever $0 < \eta^2 \leq 1/4$ and $n' \neq 0$. Therefore, the polynomial $\Pi(k_3)$ does not vanish for any value of admissible parameters, for fixed k_1, k_2, n. Then there are at most 8 k_3 satisfying $\chi(k,-k-n,-n) = 0$.

Now we estimate the sum in (6.17) with $\alpha = 1$ as follows

$$\begin{aligned} \sum_{2^i \leq |k| < 2^{i+1}} \left(k_1^2 + k_2^2 + k_3^2\right)^{-1/2} \chi(k,-k-n,n) &\leq 8 + 8 \sum_{0<|k'|<2^{i+1}} \left(k_1^2 + k_2^2\right)^{-1/2} \\ &\leq C_0 2^i, \end{aligned}$$

where C_0 is an absolute constant; $i = 0, 1, 2, \ldots$. The first 8 on the right-hand side of the above inequality accounts for $k' = 0$. Therefore, the inequality (6.17) holds with $\alpha = \beta = 1$. Let $v_k = |k||\mathbf{w}_{\mathrm{AG},k}|$ and similarly for m and n. Since $||v||_{1/2} = ||\mathbf{w}_{\mathrm{AG}}||_{3/2}$, $||v||_0 = ||\mathbf{w}_{\mathrm{AG}}||_1$, Eqs. (6.19)–(6.21) imply

$$
\begin{aligned}
|(\mathbf{B}_3(\mathbf{w}_{\mathrm{AG}}, \mathbf{w}_{\mathrm{AG}}), (-\Delta)\mathbf{w}_{\mathrm{AG}})| &\leq c' \sum_{k+m+n=0} |v_k|\,|v_m|\,|v_n|\,\chi(k,m,n) \\
&\leq c'C||v||^2_{1/2}\,||v||_0 \\
&= c'C||\mathbf{w}_{\mathrm{AG}}||_1\,||\mathbf{w}_{\mathrm{AG}}||^2_{3/2}. \qquad (6.28)
\end{aligned}
$$

Applying the interpolation inequality $||\mathbf{w}_{\mathrm{AG}}||^2_{3/2} \leq \mathrm{const}||\mathbf{w}_{\mathrm{AG}}||_1\,||\mathbf{w}_{\mathrm{AG}}||_2$ we obtain from (6.28) the estimate (6.11) (where the constant depends on a_1, a_2, a_3 in general case). This concludes the proof of Theorem 6.1.

We note that the operator $\mathbf{B}_3$ is a bilinear convolution-type operator with the domain of summation K^* given by (6.22). The estimate (6.11) for $\mathbf{B}_3$ is for the 'worst case' of all interactions on the '$2\frac{1}{2}$-dimensional' interaction manifold K^*.

7 Infinite time regularity of the 3D Navier–Stokes primitive equations of geophysics for finite large N

In this section we establish the global existence and regularity of solutions of Eqs. (1.1)–(1.3) (equivalently, Eqs. (1.20)) for N large enough, including the case of *all* 3-wave resonances, where $\mathbf{B}_3(\mathbf{w}_{\mathrm{AG}}, \mathbf{w}_{\mathrm{AG}})$ is present in the limit equations (1.23). The proof of global regularity of the 3D 'primitive' Navier–Stokes equations (1.1)–(1.3) for resonant domains presented in this section relies on the global regularity of the '$2\frac{1}{2}$-dimensional'-limit nonlinear Navier–Stokes equations (1.23), (2.33), (2.34) and techniques for convergence theorems as $N \to \infty$ developed in [9], [3], [13]. We impose in our regularity theorems only an integral regularity condition on the forcing term $\mathbf{F}^\dagger$ of the type

$$
\sup_T \int_T^{T+1} ||\mathbf{F}^\dagger||^2_{\alpha-1} dt \leq M^2_{\alpha F} \qquad (7.1)
$$

where $\alpha > 3/4$.

In BMN [9], [6], [13] we proved the regularity of solutions for smooth enough initial data $\mathbf{U}^\dagger(0)$ and forcing term $\mathbf{F}^\dagger$ for almost all aspect ratios (no strict 3-wave resonances were allowed in regularity theorems). Now, following Avrin & BMN [3] we relax the smoothness conditions on $\mathbf{U}^\dagger(0,x)$ and $\mathbf{F}^\dagger(t,x)$ using a simple argument based on our previous results on equations with smooth data and approximating the data by smooth functions. In fact, we show that we can extend our previous results with very smooth initial data and forcing terms to the non-smooth case by continuity. First, we replace $\mathbf{U}^\dagger(0)$ and $\mathbf{F}^\dagger$ respectively by smooth initial data $\mathbf{U}_s^\dagger(0) \in H_\sigma$ and force $\mathbf{F}_s^\dagger(t)$ with $\mathbf{F}_s^\dagger(t) \in H_\sigma$, $\partial_t \mathbf{F}_s^\dagger(t) \in H_\sigma, \sigma > \alpha + 2$, which are close to $\mathbf{U}^\dagger(0)$ and $\mathbf{F}^\dagger$. Our approximation of initial data is thus

$$\|\mathbf{U}^\dagger(0) - \mathbf{U}_s^\dagger(0)\|_\alpha \le \epsilon. \tag{7.2}$$

Further, we assume that $\mathbf{F}^\dagger$ is approximated by $\mathbf{F}_s^\dagger$. We denote $\mathbf{F}' = \mathbf{F}^\dagger - \mathbf{F}_s^\dagger$ and assume

$$\sup_T \int_T^{T+1} \|\mathbf{F}'\|_{\alpha-1}^2 dt \le \epsilon^2 \tag{7.3}$$

with $\alpha > 3/4$.

Of course norms in H_σ, $\sigma > \alpha + 2$, of approximations $\mathbf{U}_s^\dagger(0)$, $\mathbf{F}_s^\dagger$ and $\partial_t\ \mathbf{F}_s^\dagger$ tend to infinity as $\epsilon \to 0$, but they are bounded for every non-zero ϵ. Using results of [9] we will find a solution $\mathbf{U}_s^\dagger(t)$ of (1.1)–(1.2) with mollified data which satisfies ϵ-independent estimates in H_α for large N (see Theorem 7.2 below). The solutions $\mathbf{U}_s^\dagger$, $\mathbf{U}^\dagger$ satisfy equations of the form

$$\begin{aligned} \partial_t \mathbf{U}^\dagger(t) &= \mathbf{B}(\mathbf{U}^\dagger, \mathbf{U}^\dagger) + \overline{\boldsymbol{\nu}}\Delta \mathbf{U} - N\mathbf{P}^\dagger \mathbf{M} \mathbf{P}^\dagger \mathbf{U}^\dagger + \mathbf{F}^\dagger \\ \partial_t \mathbf{U}_s^\dagger(t) &= \mathbf{B}(\mathbf{U}_s^\dagger, \mathbf{U}_s^\dagger) + \overline{\boldsymbol{\nu}}\Delta \mathbf{U}_s^\dagger - N\mathbf{P}^\dagger \mathbf{M} \mathbf{P}^\dagger \mathbf{U}_s^\dagger + \mathbf{F}_s^\dagger \end{aligned} \tag{7.4}$$

with the same bilinear operator $\mathbf{B}$ and different (but close) initial data. In the above equations $\overline{\boldsymbol{\nu}} = \mathrm{diag}(\nu_1, \nu_1, \nu_1, \nu_2)$ is the viscosity matrix, $\mathbf{P}^\dagger = (\mathbf{P}, \mathrm{Id})$ where $\mathbf{P}$ is the Leray projection operator. The difference $\Xi(t) = \mathbf{U}^\dagger(t) - \mathbf{U}_s^\dagger(t)$ satisfies the equation

$$\begin{aligned} \partial_t \Xi(t) &= \mathbf{B}(\Xi, \mathbf{U}_s^\dagger) + \mathbf{B}(\mathbf{U}_s^\dagger, \Xi) + \mathbf{B}(\Xi, \Xi) + \overline{\boldsymbol{\nu}}\Delta\Xi - N\mathbf{P}^\dagger \mathbf{M} \mathbf{P}^\dagger \Xi + \mathbf{F}', \\ \Xi(0) &= \mathbf{U}^\dagger(0) - \mathbf{U}_s^\dagger(0) \end{aligned} \tag{7.5}$$

with a small forcing term $\mathbf{F}' = \mathbf{F}^\dagger - \mathbf{F}_s^\dagger$, and small initial data $\Xi(0)$.

Theorem 7.1 *Let $\alpha > 3/4$, $\nu_1, \nu_2 > 0$, $T_0 > 0$. Let (7.2)–(7.3) hold and*

$$\|\mathbf{U}_s^\dagger(t)\|_\alpha \le M_{s,\alpha}, \quad 0 \le t \le T_0, \qquad \nu \int_0^{T_0} \|\mathbf{U}_s^\dagger(t)\|_{\alpha+1}^2 \le M_{s,\alpha}^2; \tag{7.6}$$

let $\epsilon \le \epsilon_0$, where ϵ_0 depends on $M_{s,\alpha}$,α, ν, T_0. Then a regular solution Ξ of Eqs. (7.5) exists and

$$\|\Xi(t)\|_\alpha \le C_0\epsilon, \quad 0 \le t \le T_0; \qquad \nu \int_0^{T_0} \|\Xi(t)\|_{\alpha+1}^2 \le C_0^2\epsilon^2. \tag{7.7}$$

where C_0 depends on $M_{s,\alpha}, \alpha, \nu$; $\nu = \min(\nu_1, \nu_2)$.

Proof. Multiplying (7.5) by $2(-\Delta)^\alpha\Xi$ we obtain, using, from Lemma 7.1 below, Eqs. (7.12)–(7.14),

$$\partial_t\|\Xi(t)\|^2_\alpha + 2\nu\|\Xi(t)\|^2_{1+\alpha}$$
$$\leq C\|\Xi(t)\|^{2-\delta}_{1+\alpha}\,\|\Xi(t)\|^{1+\delta}_\alpha + C\|\mathbf{U}^\dagger_0(t)\,\|_{1+\alpha}\|\,\Xi(t)\|_{1+\alpha}\,\|\Xi(t)\|_\alpha$$
$$+2\|\mathbf{F}'\|_{\alpha-1}\|\,\Xi(t)\|_{1+\alpha}$$

where $\delta > 0$ when $\alpha > 1/2$. This equation implies

$$\begin{aligned}\partial_t\|\Xi(t)\|^2_\alpha + \nu\|\Xi(t)\|^2_{\alpha+1} \;\leq\; & C_1\|\Xi(t)\|^{2/\delta+2}_\alpha \\ & +C_2\|\Xi(t)\|^2_\alpha\,\|\mathbf{U}^\dagger_0(t)\|^2_{\alpha+1} + C_3\|\mathbf{F}'\|^2_{\alpha-1},\end{aligned} \tag{7.8}$$

where C_1, C_2, C_3 depend on ν, $M_{0\alpha}$ and $\mathbf{F}' = \mathbf{F}^\dagger(t) - \mathbf{F}^\dagger_0(t)$.

We have $\|\Xi(t)\|^2_\alpha \leq y(t)$, where $y(t)$ is a solution of

$$y(t) = C_1\int_0^t y^{1/\delta+1}dt + C_2\int_0^t h(t)ydt + \epsilon^2(C_3+1), \tag{7.9}$$

where $h(t) = \|\mathbf{U}^\dagger_0(t)\|^2_{\alpha+1}$. The solution of the corresponding differential equation $\partial_t y = C_1 y^{1/\delta+1} + C_2h(t)y$, $y(0) = \epsilon^2(C_3+1)$, has a large interval of existence when $\epsilon \to 0$, this solution being close to zero on a fixed interval $[0, T_0]$. Using the generalised Gronwall lemma we obtain the statement of the theorem.

We rely on techniques from BMN[9], a substantial difference being that one has to use the following inequalities which replace Lemma 4.3 and Lemma 4.3′ of BMN[9].

Lemma 7.1 *Let* $\mathbf{X} \subset (\mathbf{Z}^9\backslash\{0\})\times\{1,\ldots,8\} = ((\mathbf{Z}^3_k\times\mathbf{Z}^3_m\times\mathbf{Z}^3_n)\backslash\{0\})\times\{1,\ldots,8\}$ *be a set,* $\mathbf{Q}_{kmn}$ *be a bilinear function from* $\mathbf{C}^3\times\mathbf{C}^3$ *to* $\mathbf{C}^3$ *which depends on* $(k,m,n)\in\mathbf{Z}^9$ *and on* $l = 1,\ldots,8$, *and satisfies, for* $n = k+m$,

$$|\mathbf{Q}_{kmn,l}(\mathbf{u}_k,\mathbf{v}_m)| \leq C_Q|\check{m}|\,|\mathbf{u}_k|\,|\mathbf{v}_m|. \tag{7.10}$$

Let

$$(\mathbf{B}_Q(\mathbf{u},\mathbf{v}))_n = \sum_{\substack{k+m=n,\\(k,m,n,l)\in\mathbf{X}}} \mathbf{Q}_{kmn,l}(\mathbf{u}_k,\mathbf{v}_m). \tag{7.11}$$

Let $\alpha \geq 0$. *Then with* $C_8 = C_8(\alpha)$ *we have*

$$\begin{aligned}|(\mathbf{B}_Q(\mathbf{u},\mathbf{v})),(-\Delta)^\alpha\mathbf{w})| \;\leq\; & C_QC_8\|\mathbf{w}\|_{\alpha+1}\,[\|\mathbf{u}\|_{1/2}\,\|\mathbf{v}\|_{\alpha+1}\|\mathbf{u}\|_{\alpha+1/2}\,\|\mathbf{v}\|_1] \\ \leq\; & \nu\frac{\|\mathbf{w}\|^2_{\alpha+1}}{8} + C[\|\mathbf{u}\|^2_\sigma\,\|\mathbf{v}\|^2_\sigma]/\nu,\end{aligned} \tag{7.12}$$

with arbitrary $\nu > 0$ *when* $\sigma \geq \max(\alpha+1, 1)$. *Moreover, if* $0 \leq \alpha < 3/2$,

$$\begin{aligned} |(\mathbf{B}_Q(\mathbf{u},\mathbf{v})), (-\Delta)^\alpha \mathbf{w})| &\leq C_Q C_8 ||\mathbf{w}||_{\alpha+1} ||\mathbf{u}||_{3/4+\alpha/2} \, ||\mathbf{v}||_{3/4+\alpha/2} \\ &\leq \nu \frac{||\mathbf{w}||^2_{\alpha+1}}{8} + C\left(\frac{1}{\nu}\right) ||\mathbf{u}||^2_{3/4+\alpha/2} \, ||\mathbf{v}||^2_{3/4+\alpha/2}. \end{aligned} \tag{7.13}$$

If $0 \leq \alpha < 3/2$, $\epsilon \ll 1$, *we have*

$$\begin{aligned} |(\mathbf{B}_Q(\mathbf{v},\mathbf{w})), (-\Delta)^\alpha \mathbf{w})| &\leq C_Q C_7 ||\mathbf{w}||_{\alpha+1} \, ||\mathbf{w}||_{1+\alpha-\epsilon} \, ||\mathbf{v}||_{1/2+\epsilon} \\ &\leq \nu \frac{||\mathbf{w}||^2_{\alpha+1}}{8} + C\nu^{1-2/\epsilon} ||\mathbf{u}||^{2/\epsilon}_{1/2+\epsilon} \, ||\mathbf{w}||^2_{\alpha}. \end{aligned} \tag{7.14}$$

Proof. The proof uses Sobolev's embedding theorem and Holder's inequality as in the proof of Theorem 4.1. We also use $|m| \leq |k| + |n|$ in the proof of (7.14).

Remark 7.1 In BMN [9] we used Lemma 4.3′ instead of (7.14), which gives inequality

$$|(\mathbf{B}(\mathbf{v},\mathbf{u}), \mathbf{A}^\sigma \mathbf{u})| \leq C_B(\nu) ||\mathbf{v}||_\sigma \, ||\mathbf{u}||^2_\sigma \tag{7.15}$$

for $\sigma > 5/2$ in the 3D case. In the 2D case in [9], the reference to Lemma 4.3′ in particular to inequality (7.15) with $\sigma > 3/2$ was incorrect, when $2 \geq \sigma > 3/2$; in the proofs in [9], (7.15) was not in fact used, but rather the following inequality, with arbitrary small $\delta > 0$:

$$|(\mathbf{B}(\mathbf{v},\mathbf{u}), \mathbf{A}^\sigma \mathbf{u})| \leq C_B(\nu) ||\mathbf{v}||_{2+\delta} \, ||\mathbf{u}||^2_\sigma. \tag{7.16}$$

We now give a proof of the existence for all times for the primitive equations (1.20) in the viscous case $\nu_1, \nu_2 > 0$, with smooth initial data and forces. We sketch the existence theorem for Eqs. (1.1)–(1.3) with smooth in H_σ initial data extending the results of [13] to cover the 3-wave resonance operator.

Theorem 7.2 *Let* $\eta = f/N$ *and the domain parameters* a_1, a_2 *and* a_3 *be arbitrary and fixed; let* $\alpha > 3/4$, $\nu_1, \nu_2 > 0$, $\sigma > \alpha + 2$, $T_0 > 0$ *and let* $\mathbf{U}_s(t)$ *be a solution of* (1.20) *with smooth initial data and forcing term such that*

$$\|\mathbf{U}_s^\dagger(0)\|_\alpha \;\leq\; M_\alpha, \qquad \sup_T \int_T^{T+1} \|\mathbf{F}_s^\dagger\|^2_{\alpha-1} dt \;\leq\; (M_{\alpha F})^2; \tag{7.17}$$

$$\begin{aligned} \|\mathbf{U}_s^\dagger(0)\|_\sigma \;\leq\; M_{\sigma F}, \quad &\sup_T \int_T^{T+1} \|\mathbf{F}^\dagger\|^2_{\sigma-1} \, dt \;\leq\; (M_{\sigma F})^2, \\ &\sup_T \int_T^{T+1} \|\partial_t \mathbf{F}_s^\dagger\|^2_\sigma \, dt \;\leq\; (M_{\sigma F})^2. \end{aligned} \tag{7.18}$$

Then for every $N \geq N_0(M_\alpha, M_{\alpha F}, M_{\sigma F}, \nu, a_1, a_2, a_3)$, *there exists a unique solution* $\mathbf{U}_s^\dagger(t)$ *to Eqs.* (1.20) *for* $0 \leq t \leq T_0$ *such that*

$$\|\mathbf{U}_s^\dagger(t)\|_\alpha \leq M'_\alpha, \qquad 0 \leq t \leq T_0, \tag{7.19}$$

$$\nu \int_0^{T_0} \|\mathbf{U}_s^\dagger(t)\|_{\alpha+1}^2 dt \leq (M'_\alpha)^2, \tag{7.20}$$

where M'_α *depends only on* M_α, $M_{\alpha F}$, ν, a_1, a_2, a_3 *and* T_0; *and* $\nu = \min(\nu_1, \nu_2)$.

Proof. The proof is similar to that of Theorem 8.2 given in BMN [9] for the 3D rotating Euler and Navier–Stokes system with $\nu \geq 0$ and the proof of Theorem 5.2 in [14]. Since the proofs are based on energy estimates, one can almost literally repeat the proof, with obvious modifications, for the primitive equations. Note that solutions $\overline{\mathbf{w}}$ of the 2D3C (two-dimensional, three-component) Euler equations are replaced now by $\mathbf{U}_{\text{QG}}$ and solutions $\mathbf{w}^\perp$ of the $2\frac{1}{2}$D equations in [14] are replaced by $\mathbf{U}_{\text{AG}}$, the large parameter Ω being replaced by a large parameter N. We also use our new Theorem 6.2 which yields global regularity of $\mathbf{w}_{\text{AG}}$ for all values of parameters, including 3-wave resonances.

We have now, according to Theorems 4.1, 5.2 and 6.2, solutions of the limit QG and AG equations bounded for $T_0 \geq t \geq 0$ in H_α by M_α for $\alpha \geq 3/4$.

The only essential difference in proofs is that now thanks to the condition $\nu > 0$ we can replace the condition $\alpha > 3/2$ imposed in BMN [9] by $\alpha > 1/2$ (we can do it everywhere, but for global existence of solutions of QG equations we need $\alpha \geq 3/4$, according to Theorem 4.1). Note that (7.19) follows from the estimate (6.11) of BMN [9] of the difference between the solution $\mathbf{v}(t)$ of (2.20) and the solution $\mathbf{w}(t)$ of the limit equations (2.27)–(2.28); this estimate now takes the form

$$||\mathbf{v}(t) - \mathbf{w}(t)||_\alpha \leq \delta(N) \qquad \text{for all } t \in [0, T_\sigma], \tag{7.21}$$

where $\delta(N) \to 0$ as $N \to \infty$. Here T_σ is the classical local time existence of Eqs. (1.20) with smooth initial data in H_σ. The estimate (7.21) follows from (8.14) of BMN [9] which here takes the form

$$\begin{aligned} ||\mathbf{U}^\dagger(t)||_\alpha &\leq ||\mathbf{U}^\dagger - \mathbf{W}_{\text{QG}} - \mathbf{E}(-Nt)\mathbf{w}_{\text{AG}}||_\alpha + ||\mathbf{W}_{\text{QG}} + \mathbf{E}(-Nt)\mathbf{w}_{\text{AG}}||_\alpha \\ &\leq 2M_\alpha^0 \\ &\leq M'_\alpha/4, \qquad 0 \leq t \leq T. \end{aligned} \tag{7.22}$$

Here $\alpha > 1/2$; this estimate holds as long as $||\mathbf{w}_{\text{QG}}||_\alpha$ and $||\mathbf{w}_{\text{AG}}||_\alpha$ are bounded; they are so, from Theorems 4.1, 5.1, 5.2 and 6.2, if $\alpha \geq 3/4$. Here $\mathbf{w}_{\text{QG}}, \mathbf{w}_{\text{AG}}$ are solutions of the reduced QG–AG equations, with initial data $\Pi^{\text{QG}}\mathbf{U}^\dagger(0)$, $\Pi^{\text{AG}}\mathbf{U}^\dagger(0)$ and $\mathbf{W}_{\text{AG}} = \mathbf{E}(-Nt)\mathbf{w}_{\text{AG}}$. Inequality (7.22) follows from estimation of the error term $\hat{\mathbf{y}}$ in (6.14) of BMN [9]. We sketch the error estimate below.

We recall the following notation adopted from [9]: $\hat{\mathbf{v}} = \pi_R \mathbf{v}$, $\hat{\mathbf{w}} = \pi_R \mathbf{w}$ where $\mathbf{v}$ now stands for the Craya representation of the exact solution in Poincaré's slow variable, Eq. (2.18), whereas $\mathbf{w}$ is the corresponding one for the limit equations. Here $\pi_R \mathbf{v}$ is the projection of $\mathbf{v}_n$ onto the Fourier modes with $|\check{n}| \leq R$ (similarly $\pi_R \mathbf{w}$). The truncated fields $\hat{\mathbf{v}}$ and $\hat{\mathbf{w}}$ satisfy the equations of the same form as $\mathbf{v}$ and $\mathbf{w}$ but with the extra forcing term

$$\begin{aligned} g_{\mathrm{tr}} &= \pi_R \left[\mathbf{B}(Nt, \mathbf{v}, \mathbf{v} - \pi_R \mathbf{v}) + \mathbf{B}(Nt, \mathbf{v} - \pi_R \mathbf{v}, \hat{\mathbf{v}}) \right], \\ \tilde{g}_{\mathrm{tr}} &= \pi_R \left[\tilde{\mathbf{B}}(\mathbf{v}, \mathbf{v} - \pi_R \mathbf{v}) + \tilde{\mathbf{B}}(\mathbf{v} - \pi_R \mathbf{v}, \hat{\mathbf{v}}) \right]. \end{aligned} \tag{7.23}$$

Now we define $\hat{\mathbf{r}} = \hat{\mathbf{v}} - \hat{\mathbf{w}}$; we rewrite the non-resonant terms in Eqs. (2.22) in the form $(\partial_t \hat{\mathbf{r}}_{1n} + \hat{\mathbf{g}}_n^1)/N$ where

$$\begin{aligned} \hat{\mathbf{r}}_{1n} &= -i \sum_{\mathrm{n.r.}, l, k+m=n} e^{iND_l(k,m,n)t} \frac{Q_{kmn,l}(\hat{\mathbf{v}}_k, \hat{\mathbf{v}}_m)}{D_l(k,m,n)} \\ &\qquad - \frac{1}{\omega_n} \mathbf{J} \exp(N\omega_n \mathbf{J} t) \mathbf{F}_n^{\mathrm{AG}} - \frac{\Re_n(2N\omega_n t)\hat{\mathbf{v}}_n}{2\omega_n}, \end{aligned} \tag{7.24}$$

$$\begin{aligned} \hat{\mathbf{g}}_n^1 &= i \sum_{\mathrm{n.r.}, l, k+m=n} e^{iND_l(k,m,n)t} \frac{(Q_{kmn,l}(\partial_t \hat{\mathbf{v}}_k, \hat{\mathbf{v}}_m) + + Q_{kmn,l}(\hat{\mathbf{v}}, \partial_t \hat{\mathbf{v}}_m))}{D_l(k,m,n)} \\ &\qquad + \frac{1}{\omega_n} \mathbf{J} \exp(N\omega_n \mathbf{J} t) \partial_t \mathbf{F}_n^{\mathrm{AG}} - \frac{\Re_n(2N\omega_n t)\partial_t \hat{\mathbf{v}}_n}{2\omega_n} \end{aligned} \tag{7.25}$$

where n.r. stands for non-resonant terms and $\Re(2N\omega_n t)$ is defined in equation (2.23). Recall the definition of $\nu = \min(\nu_1, \nu_2)$ and $\tilde{\mathbf{A}}$ given by (2.27). We can rewrite the equation for the truncated $\hat{\mathbf{r}}$:

$$\partial_t \hat{\mathbf{r}} + \tilde{\mathbf{A}} \hat{\mathbf{r}} + \mathbf{L}(\hat{\mathbf{v}}, \hat{\mathbf{w}}) \hat{\mathbf{r}} = \frac{\hat{\mathbf{g}}^1}{N} + \frac{\partial_t \hat{\mathbf{r}}_1}{N} + g_{\mathrm{tr}} - \tilde{g}_{\mathrm{tr}}, \qquad \hat{\mathbf{r}}(0) = 0, \tag{7.26}$$

with $\mathbf{L}(\hat{\mathbf{v}}, \hat{\mathbf{w}})\hat{\mathbf{r}}$ defined as

$$\mathbf{L}(\hat{\mathbf{v}}, \hat{\mathbf{w}})\hat{\mathbf{r}} = -(\tilde{\mathbf{B}}(\hat{\mathbf{v}}, \hat{\mathbf{r}}) + \tilde{\mathbf{B}}(\hat{\mathbf{r}}, \hat{\mathbf{w}})). \tag{7.27}$$

We denote $\hat{\mathbf{y}} = \hat{\mathbf{r}} - \hat{\mathbf{r}}_1/N$ and obtain

$$\partial_t \hat{\mathbf{y}} + \mathbf{L}(\mathbf{v}, \mathbf{w})\hat{\mathbf{y}} + \tilde{\mathbf{A}}\hat{\mathbf{y}} = \frac{\hat{\mathbf{G}}}{N} + g_{\mathrm{tr}} - \tilde{g}_{\mathrm{tr}}, \quad \hat{\mathbf{y}}(0) = -\frac{\hat{\mathbf{r}}_1(0)}{N}, \tag{7.28}$$

$$\hat{\mathbf{G}} = \hat{\mathbf{g}}^1 - \mathbf{L}(\mathbf{v}, \mathbf{w})\hat{\mathbf{r}}_1 - \tilde{\mathbf{A}}\hat{\mathbf{r}}_1. \tag{7.29}$$

To estimate $1/|D_l(k,m,n)|$, $k, m, n \notin K$ in Eq. (7.24), we use the elementary estimate $1/|D_l(k,m,n)| \leq C_0(R)$ for $|\check{k}|, |\check{m}|, |\check{n}| \leq R$. Since $\hat{\mathbf{v}} = \hat{\mathbf{w}} + \hat{\mathbf{y}} + \hat{\mathbf{r}}_1/N$ it is sufficient to establish that $||\hat{\mathbf{y}}||_\alpha \leq \delta(N)$.

The estimate (7.22) follows in a similar way to (6.13) and (6.14) in [9] from the differential inequality

$$\begin{aligned}&\partial_t||\hat{\mathbf{y}}||^2_\alpha + \nu||\hat{\mathbf{y}}||^2_{\alpha+1}\\ &\quad\le C(M_\sigma,\nu)||\hat{\mathbf{y}}||^2_\alpha + \frac{\nu||\hat{\mathbf{y}}||^2_{\alpha+1}}{2} + + \frac{||\hat{\mathbf{G}}||^2_\alpha}{N^2} + 2||g_{\text{tr}}||^2_\alpha + 2||\tilde{g}_{\text{tr}}||^2_\alpha, \qquad (7.30)\\ &||\hat{\mathbf{y}}(0)||_\alpha = \frac{||\hat{\mathbf{r}}_1(0)||_\alpha}{N}.\end{aligned}$$

Here the truncation error terms satisfy the estimate

$$||g_{\text{tr}}||^2_\alpha + ||\tilde{g}_{\text{tr}}||^2_\alpha \le C^2 M^4_\sigma C_0(R) R^{2(\alpha+1-\sigma)}, \qquad 0 \le t \le T_\sigma, \qquad (7.31)$$

with arbitrary truncation parameter R; this estimate is true for any value of α including $\alpha \ge 1/2$. We also have the estimate

$$||\hat{\mathbf{G}}||_\alpha + ||\hat{\mathbf{r}}_1||_\alpha \le \frac{CR^{\alpha+2}C_0(R)C(M_\sigma)}{N}, \qquad 0 \le t \le T_\sigma.$$

Here one can take $\alpha > 1/2$ since one can use estimates of solutions in a smooth space H_σ and we take here $\sigma > \alpha + 2$ (compared with $\sigma > \alpha + 1$ as was done in BMN [9]). The term $C(M_\sigma,\nu)||\hat{\mathbf{y}}||^2_\alpha + \nu||\hat{\mathbf{y}}||^2_{\alpha+1}/2$ is obtained from Lemma 7.1. To obtain estimate (7.30) from (7.28) we multiply the latter by $(-\Delta)^\alpha \mathbf{y}$ and apply Lemma 7.1 instead of Lemma 4.3 and Lemma 4.3$'$ of BMN [9]. Taking into account the remarks given, one can follow the proof of Theorem 8.2 in BMN [9] to obtain the statement of Theorem 7.2 by bootstraping from $[0,T_\sigma]$ to $[0,T_0]$. The theorem is proved.

We now conclude with the existence and regularity theorem for less smooth $\mathbf{U}^\dagger(0)$ and $\mathbf{F}^\dagger$, by bootstrapping local existence with the help of Theorems 7.1 and 7.2.

Theorem 7.3 *Let* $\eta = f/N$ *and the domain parameters* a_1, a_2, a_3 *be fixed but arbitrary. Let* ν_1, $\nu_2 > 0$, $\nu = \min(\nu_1,\nu_2)$, $\alpha > 3/4$, *and let the condition* (7.1) *on the force* $\mathbf{F}(t,x)$ *hold. Let*

$$||\mathbf{U}^\dagger(0)||_\alpha \le \tilde{M}_\alpha,$$

and N *be large:* $N \ge N_1(\tilde{M}_\alpha, M_{\alpha F}, \nu, a_1, a_2, a_3)$. *Then solutions of the 3D primitive Navier–Stokes system* (1.1)–(1.3) *are regular for all* $t \ge 0$, *and*

$$||\mathbf{U}^\dagger(t)||_\alpha \le \tilde{M}'_\alpha \quad \text{for all} \quad t \ge 0.$$

Proof. We have a regular solution $\mathbf{U}^\dagger(t) \in H_\alpha$ with $||\mathbf{U}^\dagger(t)||_\alpha \le M_\alpha(\tilde{M}_\alpha, M_{\alpha F}, \nu)$ on a small time interval $[0,T_\alpha]$, $T_\alpha = T_\alpha(\tilde{M}_\alpha, M_{\alpha F}, \nu)$. We consider the case $3/4 < \alpha \le 1$. We have the energy estimate for regular solutions:

$$||\mathbf{U}^\dagger(t)||_0 \le M_0(\tilde{M}_\alpha, M_{\alpha F}, \nu) \quad \text{for all} \quad t \ge 0 \qquad (7.32)$$

$$\nu \int_T^{T+\tau} ||\mathbf{U}^\dagger(t)||_1^2 \leq M_0^2 \quad \text{for all} \quad T \geq 0; \quad 0 \leq \tau \leq 1, \tag{7.33}$$

where $\nu = \min(\nu_1, \nu_2)$.

Remark 7.2 Uniform boundedness of the energy for the condition (7.1) on $\mathbf{F}(t,x)$ follows from the usual Gronwall inequality estimate:

$$||\mathbf{U}^\dagger(t)||_0^2 \leq ||\mathbf{U}^\dagger(0)||_0^2 \, e^{-\nu\lambda_1 t} + \frac{1}{\nu\lambda_1^\alpha} \int_0^t e^{-\nu\lambda_1(t-s)} ||\mathbf{F}^\dagger(s)||_{\alpha-1}^2 ds, \tag{7.34}$$

and

$$||\mathbf{U}^\dagger(t)||_0^2 \leq ||\mathbf{U}^\dagger(0)||_0^2 \, e^{-\nu\lambda_1 t} + \frac{M_{\alpha F}^2}{\nu\lambda_1^\alpha} \frac{1}{1 - e^{-\nu\lambda_1}}, \tag{7.35}$$

whence

$$M_0^2 = C_\alpha^2 \tilde{M}_\alpha^2 + M_{\alpha F}^2 \left(1 + \frac{1}{\nu\lambda_1^\alpha (1 - e^{-\nu\lambda_1})}\right). \tag{7.36}$$

Here C_α is an embedding constant from H_0 to H_α and λ_1 is the first eigenvalue of the Stokes operator.

For every $t \geq \tau$, Eq. (7.33) implies that on every interval $[t-\tau, t]$ including $t = \tau$, we have a point t^* for which:

$$||\mathbf{U}^\dagger(t^*)||_1 \leq \frac{M_0}{\sqrt{\nu\tau}}.$$

From now on, we choose $\tau = T_\alpha$, the local existence time defined above. For every $t \geq T_\alpha$, we take $\mathbf{U}^\dagger(t^*)$ as new initial data, with $t - T_\alpha \leq t^* \leq t$. To prove that the solution is uniformly bounded in H_α for all $t \geq T_\alpha$, it suffices to derive a uniform bound for $t \in [t^*, t^* + T_\alpha]$, with the help of Theorems 7.1 and 7.2: in both theorems we set $T_0 = T_\alpha$. At $t = t^*$, the initial condition (7.17) of Theorem 5.2 becomes

$$M_\alpha = C_{\alpha,1} \frac{M_0}{\sqrt{\nu T_\alpha}}, \tag{7.37}$$

with $C_{\alpha,1}$ an embedding constant from H_α to H_1.

Approximating the force $\mathbf{F}^\dagger(t)$, $\partial_t \mathbf{F}^\dagger(t)$ and the initial data $\mathbf{U}^\dagger(t^*)$ by smooth functions $\mathbf{F}_s^\dagger(t)$, $\partial_t \mathbf{F}_s^\dagger(t)$, $\mathbf{U}_s^\dagger(t^*)$ in H_α we obtain

$$||\Xi(t^*)||_\alpha = ||\mathbf{U}^\dagger(t^*) - \mathbf{U}_s^\dagger(t^*)||_\alpha \leq \epsilon, \qquad ||\Xi(t^*)||_\sigma \leq M_\epsilon.$$

Moreover, the inequality (7.3) holds and, with $\mathbf{F}' = \mathbf{F}^\dagger - \mathbf{F}_s^\dagger$, we get

$$||\mathbf{F}'(t)||_\sigma \leq M_\epsilon, \quad ||\partial_t \mathbf{F}'(t)||_\sigma \leq M_\epsilon, \qquad t^* \leq t \leq t^* + T_\alpha, \tag{7.38}$$

where M_ϵ depends on M_α, $M_{\alpha F}$ and ϵ only; of course, M_α in (7.37) depends on the original $\tilde{M}_\alpha$, and $M_{\alpha F}$, ν, a_1, a_2, a_3. We choose ϵ so small that we have, by Theorem 7.1 (where $M_{\sigma F}$ is replaced by $M_{\alpha F}$), a regular solution $\Xi(t)$ on $[t^*, t^* + T_\alpha]$ which is bounded in H_α by 1 when the initial data are in H_α. After that we consider the Navier–Stokes equations with smooth initial data $\mathbf{U}_s^\dagger(t^*)$ and force $\mathbf{F}_s$ which satisfy (7.18) for $t^* \leq t \leq t^* + T_\alpha$.

The H_σ-norms of these smooth functions are bounded by (a possibly large) constant M_ϵ depending on this fixed ϵ and M_α (hence $\tilde{M}_\alpha$) and $M_{\alpha F}$. After that we choose $N \geq N_1(\epsilon, M_\alpha, M_{\alpha F})$ so large that we have (7.19) and (7.20) for solutions $\mathbf{U}_s^\dagger(t)$ of equations with smooth data. By (7.19) and (7.7) with $C_0\epsilon \leq 1$ we have $||\mathbf{U}^\dagger(t)||_\alpha \leq ||\mathbf{U}_s^\dagger(t)||_\alpha + ||\mathbf{U}^\dagger(t) - \mathbf{U}_s^\dagger(t)||_\alpha \leq M'_\alpha + 1$ with $M'_\alpha = M'_\alpha(\tilde{M}_\alpha, M_{\alpha F}, \nu, a_1, a_2, a_3)$. Setting $\tilde{M}'_\alpha = \max(M_\alpha, M'_\alpha + 1)$ completes the proof of boundedness of $\mathbf{U}^\dagger(t)$ in H_α for all $t \geq 0$. We also have

$$\nu \int_T^{T+1} \|\mathbf{U}^\dagger(t)\|_{\alpha+1}^2 dt \leq (M''_\alpha)^2 \tag{7.39}$$

for every $T \geq 0$ and $3/4 < \alpha \leq 1$. To extend the above to the case $\alpha > 1$ we use uniform-in-t boundedness in H_1 already proven and then apply the smoothing property for solutions of Navier–Stokes equations (see Theorem 8.2 in [9]) and obtain that the solutions are bounded for $t \geq t_* > 0$ in H_α $\alpha > 1$; we get the statement of Theorem 7.3 in this case as well. Theorem 7.3 is proved.

Finally, as in [9], we obtain regularity for all large enough times for weak solutions of the 3D 'primitive' Navier–Stokes equations (1.1)–(1.3) with a force $\mathbf{F}^\dagger(t)$. This theorem describes the situation when N is fixed, and large enough (depending only on the magnitude of $\mathbf{F}^\dagger(t)$ and independent of the initial data). The situation is that of non-smooth and arbitrary large initial data in H_0. Then weak Leray solutions $\mathbf{U}^\dagger(t)$ always exist (with a possible blow-up in H_1 at some values of $t < t^*$, see [19]); here we show that blow-up cannot happen if t is large.

Theorem 7.4 *Let $\eta = f/N$ and the domain parameters a_1, a_2, a_3 be fixed but arbitrary. Let $\nu_1, \nu_2 > 0$, $\nu = \min(\nu_1, \nu_2)$, $\alpha > 3/4$ and the condition* (1.19) *on the force be satisfied. Let $||\mathbf{U}^\dagger(0)||_0 \leq M_0$, $\hat{T} = \hat{T}(M_0, M_{\alpha F}, \nu)$. Then for every $N \geq N'(a_1, a_2, a_3, \nu, M_{\alpha F})$, N' independent of M_0, and for every weak solution $\mathbf{U}^\dagger(t, x_1, x_2, x_3)$ of the 3D 'primitive' Navier–Stokes equations* (1.20) *defined on $[0, \hat{T}]$ which satisfies the classical energy estimates on $[0, \hat{T}]$, the following holds: $\mathbf{U}^\dagger(t, x_1, x_2, x_3)$ can be extended to $0 < t < +\infty$ and is regular for every $t : \hat{T} \leq t < +\infty$; it belongs to H_α and $||\mathbf{U}^\dagger(t, x_1, x_2, x_3)||_\alpha \leq C_1(a_1, a_2, a_3, M_{\alpha F}, \nu)$ for every $t \geq \hat{T}$ where $M_{\alpha F}$ is the H_α-norm of $\mathbf{F}^\dagger$. If $\mathbf{F}^\dagger$ is independent of t then there exists a global attractor for the 3D primitive Navier–Stokes equations of geophysics* (1.1)–(1.3) *bounded in H_α; such an attractor has a finite fractal dimension and attracts every weak Leray solution as $t \to +\infty$.*

8 Regularity results for inviscid case

Now we give inviscid versions of the above theorems from Section 7, valid only when there are no 3-wave resonances. We again consider (7.5), now $\nu_1 = \nu_2 = 0$, $\mathbf{F}^\dagger = \mathbf{F}_0^\dagger = 0$.

Theorem 8.1 *Let* $\alpha > 5/2, \nu = 0$, $\mathbf{F}^\dagger = 0$, $\eta \notin \Theta^*(\theta_2, \theta_3)$, $\Xi(t) = \mathbf{U}^\dagger(t) - \mathbf{U}_0^\dagger(t)$, *and*

$$\|\mathbf{U}_0^\dagger(t)\|_{\alpha+1} \leq M_0, \quad 0 \leq t \leq T; \qquad ||\Xi(0)||_\alpha \leq \epsilon. \tag{8.1}$$

Let $\epsilon \leq \epsilon_0$. *Then a regular* $\Xi(t)$ *exists and* $\|\Xi(t)\|_\alpha \leq 3\epsilon$, $0 \leq t \leq T_g$, *where* $T_g = 2/(C_1 M_0)$.

Proof. The proof is similar to that of Theorem 7.1. Multiplying the inviscid form of (7.5) by $2(-\Delta)^\alpha \Xi$ we now obtain by using Lemma 4.3′ of BMN [9] in the 3D case:

$$\partial_t \|\Xi(t)\|_\alpha^2 \leq C_1 \|\mathbf{U}_0^\dagger(t)\|_{\alpha+1} \|\Xi(t)\|_\alpha^2 \; + \; C_2 \|\Xi(t)\|_\alpha^3. \tag{8.2}$$

This implies

$$\partial_t \|\Xi(t)\|_\alpha^2 \leq C_2 \|\Xi(t)\|_\alpha^3 + C_1 M_0 \|\Xi(t)\|_\alpha^2, \quad \|\Xi(0)\|_\alpha \leq \epsilon \tag{8.3}$$

where C_1 and C_2 depend on α. We easily obtain an inequality similar to (7.8) and (7.9): specifically, setting $z(t) = ||\Xi(t)||_\alpha$, we have

$$\frac{z(t)}{z(t) + \tilde{C} M_0} \leq \frac{z(0)}{z(0) + \tilde{C} M_0} \exp\left(\frac{C_1 M_0 t}{2}\right) \tag{8.4}$$

with $\tilde{C} = C_1/C_2$. Then for $0 \leq t \leq T_g$, where $T_g = 2/(C_1 M_0)$, we obtain $z(t) \leq 3\epsilon$ for small enough ϵ. Note that we cannot bound $\Xi(t)$ on an arbitrary $[0, T]$, as the above estimate for $z(t)$ blows-up at

$$T_b = \frac{2}{C_1 M_0} \log\left(1 + \frac{\tilde{C} M_0}{z(0)}\right).$$

Theorem 8.2 *Let* $\alpha > 5/2$, $\nu_1 = \nu_2 = 0$, $\sigma > \alpha + 3$, $\eta \notin \Theta^*(\theta_2, \theta_3)$ *and*

$$\|\mathbf{U}_0^\dagger(0)\|_\sigma \leq M_\sigma, \qquad \| \mathbf{U}_0^\dagger(0)\|_\alpha \leq M_\alpha. \tag{8.5}$$

Let the solution $\mathbf{w}_{\mathrm{QG}}(t)$ *of the 3D QG system be bounded in* H_α *for* $0 \leq t \leq T$; *let* $N \geq N_0(M_\alpha, M_\sigma, T, a_1, a_2, a_3)$. *Then*

$$\|\mathbf{U}_0^\dagger(t)\|_\alpha \leq M'_\alpha, \tag{8.6}$$

for $0 \leq t \leq T$ *where* M'_α *depends only on* $M_\alpha, T, a_1, a_2, a_3$.

Proof. In the proof of Theorem 8.1 in BMN [9] it is shown that the estimate (8.6) follows from (8.6) of BMN [9]. One has to replace global regularity of the 2D Euler equation by global regularity of inviscid 3D QG equations (regularity for the QG equation is proved in Bourgeois & Beale, [18], for $\alpha \geq 3$, but the same proof works for $\alpha > 5/2$); one also has to replace global regularity of the extended rotating Euler equations by global regularity of the 'catalytic' AG equations (Theorem 5.1). Now one also can use a theorem similar to Theorem 8.1.

Theorem 8.3 *Let* $\alpha > 5/2$, $\nu_1 = \nu_2 = 0$, $\eta \notin \Theta^*(\theta_2, \theta_3)$, $\mathbf{F}^\dagger = 0$,

$$\|\mathbf{U}^\dagger(0)\|_\alpha \leq M_\alpha; \tag{8.7}$$

let $T > 0$, $N \geq N_*(M_\alpha, T, a_1, a_2, a_3)$. *Then solution of the inviscid primitive system is regular for* $t \leq T$*:*

$$\|\mathbf{U}^\dagger(t)\|_\alpha \leq M'_\alpha + 1, \qquad 0 \leq t \leq T. \tag{8.8}$$

Proof. Solutions of the limit equations are bounded for all $0 \leq t \leq T$ in H_α by M_α. The bound for 3D QG equation is proved in Bourgeois & Beale, [18], for $\alpha \geq 3$, but the same proof works for $\alpha > 5/2$. After that we proceed as in the proof of Theorem 7.2, but now from the very beginning we restrict all considerations to a fixed interval $[0, T]$ and do not use any smoothing arguments involving $\nu > 0$. The major difference is the lack of existence of a time t^* where the H_1-norm is bounded. One must find ϵ small enough so that $3^n \epsilon \leq 1$ where $n = T/T_g$, and where T_g is defined in Theorem 8.1. This is clearly satisfied if

$$T(\log 3) C_1 ||\mathbf{U}_0^\dagger||_{\alpha+1} \leq 2 \log(1/\epsilon), \tag{8.9}$$

which restricts ϵ, given T and $||\mathbf{U}_0^\dagger||_{\alpha+1}$. Technically, one needs Theorem 8.2 with both α and $\alpha+1$ estimates so that $||\mathbf{U}_0^\dagger||_{\alpha+1} \leq M'_{\alpha+1}$ on $[0, T]$, T arbitrary large. However, $M'_{\alpha+1}$ appears only in the inequality $3^n \epsilon \leq 1$ and does not appear in the final statement of Theorem 8.3.

Now we give a uniform theorem on regularity; such theorem requires more smoothness of initial data. It improves the result of BMN [7] using more precise small divisors estimates given in Section 3, where μ_2 (respectively, μ_3) denote the measures of the sets of quasi-2-waves (respectively, quasi-3-waves) resonances.

Theorem 8.4 *Let* $\eta \notin \Theta_3^{\mu_3}(\theta_2, \theta_3)$, $\theta_2 \notin \Theta_2^{\mu_2}$. *Let* $\nu_1 = \nu_2 = 0$. *Let* $\sigma > 19/2$, *and* $M_\sigma > 0$, *let* $T^* > 0$ *be arbitrarily large. Then there exists* $N^* = N^*(M_\sigma, T^*, \mu_3, \mu_2)$ *such that for* $||\mathbf{U}^\dagger(0)||_\sigma \leq M_\sigma$ *and* $N \geq N^*$, *there exists a unique regular solution* $\mathbf{U}^\dagger(t)$ *of the 3D Euler primitive equations which belongs* *to H_σ for $0 \le t \le T^*$. For M_σ fixed, $T^* \to +\infty$ as $N^* \to +\infty$ with explicit uniform dependence of T^* on M_σ, μ_3, μ_2, N^*. Simultaneously, we can take an arbitrarily large (but bounded) set of initial data: $M_\sigma \to +\infty$ if $N^* \to +\infty$, for fixed T^*.*

The proof is similar to the proof of Theorem 6.2 in BMN [9], [7], and relies on Theorems 3.3, 3.4 above.

9 Baroclinic wave dynamics and AG wave fronts

In this section we further analyse the 3D Euler–Boussinesq equations for Bu $= O(1)$ flows and in the asymptotic regime of strong stratification and weak rotation. In Section 9.1 we describe classes of nonlinear anisotropic AG baroclinic waves which are generated by the strong nonlinear interactions between the QG modes and inertio-gravity waves. The problem that we study in Section 9.2 is the initial stage of scale adjustment of strongly stratified turbulence to geostrophic turbulence.

For intermediate scales of motion, rotation is present but not dominant, so that the Rossby number is neither very large nor very small. Such systems *have not* been investigated by turbulence closure models (Lilly, [49]). In the asymptotic regime of strong stratification and weak rotation (no hydrostatic assumption) we show how switching on weak rotation triggers AG fronts. Vertical slanting of these fronts is proportional to $\sqrt{\eta}$, where η is the ratio of the Coriolis and Brunt-Väisälä parameters. These slowly moving fronts select the slowest baroclinic waves through adjustment of the horizontal to vertical scale through rotation, and are the envelope of inertio-gravity waves. The fronts effectively balance the frequencies of baroclinic waves uniformly to $O(\sqrt{\eta})$. This frontogenesis yields the vertical 'glueing' of pancake dynamics by weak rotation. The mechanism of its formation is contraction in horizontal dimension balanced by vertical stretching. This agrees with the conclusions of Hoskins & Bretherton ([41]) and Hoskins ([42]), that the vertical deformation field is crucial in the dynamics of frontal systems, as it balances strong horizontal density gradients. In their study of atmospheric frontogenesis models, smaller scale AG motions embedded in the baroclinic flow lead to the rapid formation of a front. It has been recognised that the evolution of baroclinic waves provide the dynamical environment for upper-level frontogenesis at the tropopause and lower stratosphere (Keyser & Shapiro, [45]).

The importance of even weak rotation on mesoscale flows and its fundamental role in the scale adjustment process is emphasised in Newley, Pearson & Hunt (henceforth NPH) ([62]) and Rotunno ([67]). This is a singular perturbation problem where weak Coriolis accelerations may have large effects on long horizontal scales. The latter are coupled to *small vertical scales* by rotation. Such singular perturbation effects can be treated using our asymptotic analysis which is based on careful studies of resonances and quasi-resonances.

9.1 Baroclinic wave dynamics

The limit catalytic resonant equations (5.1) (for $N \to +\infty$, $\eta = f/N$ fixed) for the AG field $\mathbf{w}'_n = (w_n^1, w_n^2)$ include w_m^1, w_m^2 and the already found QG component $w_k^0(t)$. In this section we analyse inviscid Eqs. (5.1), $A_{\mathrm{AG}} = 0$.

We recall that

$$\omega_m^2 = \frac{|\check{m}'|^2}{|\check{m}|^2} + \eta^2 \frac{\check{m}_3^2}{|\check{m}|^2} = \eta^2 + (1-\eta^2)\frac{|\check{m}'|^2}{|\check{m}|^2}$$

(with a similar expression for ω_n^2) and, therefore, the resonance condition $\omega_m = \omega_n$ is equivalent to the condition $\frac{|\check{m}'|}{|\check{m}|} = \frac{|\check{n}'|}{|\check{n}|}$ (or $\phi_m = \phi_n$) and $\frac{|\check{m}_3|}{|\check{m}|} = \frac{|\check{n}_3|}{|\check{n}|}$ or $|\xi_m| = |\xi_n|$, see (2.6). These conditions can be written as $\frac{|\check{m}'|}{|\check{m}_3|} = \frac{|\check{n}'|}{|\check{n}_3|}$ and they define resonant rays on cones in Fourier space:

$$\frac{1}{m_3^2}\left(\frac{m_1^2}{a_1^2} + \frac{m_2^2}{a_2^2}\right) = \frac{1}{n_3^2}\left(\frac{n_1^2}{a_1^2} + \frac{n_2^2}{a_2^2}\right). \tag{9.1}$$

We can further simplify the resonant rays on the cone (9.1) (although, this is not strictly necessary for the following considerations and will be lifted in the last part of Section 9.2): for all domain aspect ratios a_1, a_3 and for all $a_2 \neq a_2(j)$, $j = 1, 2, \ldots$, $\mathbf{B}_2(\mathbf{w}_{\mathrm{QG}}, \mathbf{w}_{\mathrm{AG}})$ in (5.1) (note $\mathbf{w}_{\mathrm{QG}} \equiv w^0$ and $\mathbf{w}_{\mathrm{AG}} \equiv \mathbf{w}'$) splits in Fourier space into uncoupled, restricted interaction operators on 4-ray families given by Eqs. (1.25). This is obtained by further reducing the resonances $\omega_m = \omega_n$ to $a_2 \neq \{a_2(j)\}$ and Eq. (9.1) reduces to $m_1/n_1 = \pm\lambda$, $m_2/n_2 = \pm\lambda$, $m_3/n_3 = \pm\lambda$, with λ rational. Then m and n are related by (1.25).

Four fundamental rays describing interactions in reduced equations (5.1) are defined by choosing the minimal length integer vector $l = (l_1, l_2, l_3)$ along a resonant ray. Then

$$\begin{aligned} R_{++} &= \{\gamma l_1, \gamma l_2, \gamma l_3\}, & R_{--} &= \{-\gamma l_1, -\gamma l_2, \gamma l_3\}, \\ R_{-+} &= \{-\gamma l_1, \gamma l_2, \gamma l_3\}, & R_{+-} &= \{\gamma l_1, -\gamma l_2, \gamma l_3\}. \end{aligned} \tag{9.2}$$

Here R_{-+} and R_{+-} are the 'polarised' pair relative to R_{++}. It follows from (5.3) that the coefficients $D_{mn}(\eta)$ and $G_{mn}(\eta)$ are identically zero if both n and m are on the same ray since $\check{n} \times \check{m} = 0$ and $\check{n}' \wedge \check{m}' = 0$ in this case. We note that $D_{mn}(\eta) = 0$ if m lies on the rays R_{--} and R_{++} relative to n (since $n' \wedge m' = 0$ in this case). Thus $D_{mn}(\eta)$ impacts on the AG dynamics only through the 'polarised rays' R_{-+} and R_{+-}.

Suppose that on R_{++} we have $n = (\gamma_1 l_1, \gamma_1 l_2, \gamma_1 l_3)$ fixed, where γ_1 is any integer (positive or negative), and consider four different cases ($m \in R_{++}$, $m \in R_{--}$, $m \in R_{+-}$, $m \in R_{-+}$). Then we have, for any positive or negative integer γ_2:

$$\begin{aligned} &m = m_{++} = \gamma_2(l_1, l_2, l_3) \text{ on } R_{++}, && m = m_{--} = \gamma_2(-l_1, -l_2, l_3) \text{ on } R_{--} \\ &m = m_{-+} = \gamma_2(-l_1, l_2, l_3) \text{ on } R_{-+}, && m = m_{+-} = \gamma_2(l_1, -l_2, l_3) \text{ on } R_{+-}, \end{aligned} \tag{9.3}$$

and with $\lambda = \gamma_2/\gamma_1$ the vectors m and n are related by Eqs. (1.25). The operators D and G given by (5.3) reduce to the following expressions on the rays:

$$\begin{aligned}
&D_{m_{++}n}(\eta) = D_{m_{--}n}(\eta) = G_{m_{++}n}(\eta) = 0,\\
&G_{m_{--}n}(\eta) = -4\lambda\eta\frac{|\check{n}'|}{|\check{n}|}\frac{|\check{n}'|\check{n}_3^2}{\sqrt{|\check{n}'|^2+\eta^2\check{n}_3^2}},\\
&G_{m_{-+}n}(\eta) = 4\lambda\eta\frac{\check{n}_1^2\check{n}_3^2(3\check{n}_2^2-\check{n}_1^2)}{|\check{n}'|^2|\check{n}|\sqrt{|\check{n}'|^2+\eta^2\check{n}_3^2}},\\
&G_{m_{+-}n}(\eta) = 4\lambda\eta\frac{\check{n}_2^2\check{n}_3^2(3\check{n}_1^2-\check{n}_2^2)}{|\check{n}'|^2|\check{n}|\sqrt{|\check{n}'|^2+\eta^2\check{n}_3^2}},\\
&D_{m_{-+}n}(\eta) =\\
&4\lambda\check{n}_1\check{n}_2\left(\frac{|\check{n}'|^2}{|\check{n}|^2}-\frac{\eta^2\check{n}_3^2}{|\check{n}'|^2+\eta^2\check{n}_3^2}+\frac{\check{n}_3^2(-\check{n}_1^2+\check{n}_2^2)}{|\check{n}'|^2}\left(\frac{1}{|\check{n}|^2}+\frac{\eta^2}{|\check{n}'|^2+\eta^2\check{n}_3^2}\right)\right),\\
&D_{m_{+-}n}(\eta) =\\
&-4\lambda\check{n}_1\check{n}_2\left(\frac{|\check{n}'|^2}{|\check{n}|^2}-\frac{\eta^2\check{n}_3^2}{|\check{n}'|^2+\eta^2\check{n}_3^2}+\frac{\check{n}_3^2(\check{n}_1^2-\check{n}_2^2)}{|\check{n}'|^2}\left(\frac{1}{|\check{n}|^2}+\frac{\eta^2}{|\check{n}'|^2+\eta^2\check{n}_3^2}\right)\right).
\end{aligned} \tag{9.4}$$

Our notation in (9.4) is understood as follows. For example, $G_{m_{--}n}(\eta)$ simply means that we evaluate G_{mn} for fixed $n = (n_1, n_2, n_3) = (\gamma_1 l_1, \gamma_1 l_2, \gamma_1 l_3)$ and $m \equiv m_{--} \in R_{--}$. Since $m \in R_{--}$, we have $m = (-\gamma_2 l_1, -\gamma_2 l_2, \gamma_2 l_3)$ for some γ_2. Then $\frac{|m|}{|n|} = \left|\frac{\gamma_2}{\gamma_1}\right| = |\lambda|$, $k = n - m = ((1+\lambda)n_1, (1+\lambda)n_2, (1-\lambda)n_3)$.

Eqs. (5.1) describing interactions on four polarised rays (1.25) have the following form, since self-interactions are zero:

$$\begin{aligned}
\partial_t \mathbf{w}'_n \;=\; &-\sum_{\gamma_2}\tilde{\Psi}_k^0(t)(D_{m_{--}n}(\eta)\mathbf{I}-G_{m_{--}n}(\eta)\mathbf{J})\mathbf{w}'_{m_{--}}\\
&-\sum_{\gamma_2}\tilde{\Psi}_k^0(t)(D_{m_{+-}n}(\eta)\mathbf{I}-G_{m_{+-}n}(\eta)\mathbf{J})\mathbf{w}'_{m_{+-}}\\
&-\sum_{\gamma_2}\tilde{\Psi}_k^0(t)(D_{m_{-+}n}(\eta)\mathbf{I}-G_{m_{-+}n}(\eta)\mathbf{J})\mathbf{w}'_{m_{-+}}
\end{aligned}$$

where $n = (\gamma_1 l_1, \gamma_1 l_2, \gamma_1 l_3)$, $\lambda = \gamma_2/\gamma_1$, $k = n - m$. Eqs. (9.5) are obtained after substitution of (5.2) in (5.1). Here $m_{--} \in R_{--}$, $m_{+-} \in R_{+-}$, $m_{-+} \in R_{-+}$. Thus $m_{--} = (-\gamma_2 l_1, -\gamma_2 l_2, \gamma_2 l_3) = (-\lambda n_1, -\lambda n_2, \lambda n_3)$ with similar expressions for m_{+-} and m_{-+}. Furthermore, $D_{m_{--}n}(\eta) = 0$. Only the polarised rays see interactions between D and G operators.

There exist special time-independent solutions of the QG equations which have only one non-zero Fourier mode $\tilde{\Psi}^0_{\pm k_0}$ ($\tilde{\Psi}_k = 0$ for $k \neq \pm k_0$). We call such solutions *monochromatic*. Now we consider special cases for monochromatic steady $\tilde{\Psi}^0$ for which Eqs. (9.5) for the AG field $\mathbf{w}'$ can be solved explicitly:

(1) $\tilde{\Psi}^0$ coupling rays R_{++} and R_{--};

(2) $\tilde{\Psi}^0$ coupling rays R_{++} and R_{-+};

(3) $\tilde{\Psi}^0$ coupling rays R_{++} and R_{+-}.

First, suppose that $\tilde{\Psi}^0_k$ is independent of t and monochromatic so that $\tilde{\Psi}^0_k = \tilde{\Psi}^0_{k_0}$, $k_0 = ((1+\lambda_0)n_1, (1+\lambda_0)n_2, (1-\lambda_0)n_3)$, for some λ_0, and its complex conjugate are the only two non-zero modes. This special choice of $\tilde{\Psi}^0_k$ reduces interactions in (9.5) to a single term on one ray R_{--}. Recalling that $D_{m_{--}n}(\eta) = 0$, we have from Eqs. (9.5)

$$\partial_t \mathbf{w}'_n = \tilde{\Psi}^0_{k_0} G_{m_{--}n}(\eta) \mathbf{J} \mathbf{w}'_{m_{--}} \tag{9.5}$$

where $m_{--} = (-\lambda_0 n_1, -\lambda_0 n_2, \lambda_0 n_3)$, $k_0 = ((1+\lambda_0)n_1, (1+\lambda_0)n_2, (1-\lambda_0)n_3)$. An equation for $\mathbf{w}'_{m_{--}}$ is obtained similarly. Only one term remains in the summation in (9.5) and it reduces to

$$\partial_t \mathbf{w}'_{m_{--}} = \tilde{\Psi}^0_{-k_0} G_{m_{--}n}(\eta) \mathbf{J} \mathbf{w}'_n. \tag{9.6}$$

Then Eqs. (9.5) reduce to a system of coupled 1-rays R_{++} and R_{--} equations (9.5)–(9.6). Using the properties $\tilde{\Psi}^0_{-k_0} = (\tilde{\Psi}^0_{k_0})^*$ and $\mathbf{J}^2 = -\mathbf{I}$, we have from (9.5)–(9.6)

$$\frac{\partial^2}{\partial t^2} \mathbf{w}'_n = -|G_{m_{--}n}|^2 |\tilde{\Psi}^0_{k_0}|^2 \mathbf{w}'_n. \tag{9.7}$$

Eqs. (9.7) describe baroclinic wave propagation coupled with a QG state given by the monochromatic streamfunction $\tilde{\Psi}^0_{k_0}$. Since

$$|G_{m_{--}n}| = 4|\lambda_0|\eta \frac{|\check{n}'|}{|\check{n}|} \frac{|\check{n}'|\check{n}_3^2}{\sqrt{|\check{n}'|^2 + \eta^2 \check{n}_3^2}}$$

by (9.4) their dispersion relation is found from (9.7) as

$$\theta^{--}_{n,G} = |G_{m_{--}n}||\tilde{\Psi}^0_{k_0}| = 4|\lambda_0|\,\eta\,\frac{|\check{n}'|}{|\check{n}|} \frac{|\check{n}'|\check{n}_3^2}{\sqrt{|\check{n}'|^2 + \eta^2 \check{n}_3^2}}\,|\tilde{\Psi}^0_{k_0}| \tag{9.8}$$

where $k_0 = ((1+\lambda_0)n_1,\ (1+\lambda_0)n_2,\ (1-\lambda_0)n_3)$. Because of the special choice of the monochromatic QG potential, no baroclinic waves are excited on the polarised rays R_{-+} and R_{+-}, and the $\theta^{--}_{n,G}$ waves correspond to pure rotation effects via the shearing operators G. They switch on as $\eta \neq 0$.

Now suppose that $\tilde{\Psi}^0$ is independent of t and monochromatic, so that $\tilde{\Psi}^0_{k_0}$ with $k_0 = ((1+\lambda_0)n_1, (1-\lambda_0)n_2, (1-\lambda_0)n_3)$, and its complex conjugate are the only two non-zero modes. Then from Eqs. (9.5) we obtain a system of coupled equations for modes $\mathbf{w}'_n$ and $\mathbf{w}'_{m_{-+}}$ belonging to the ray R_{++} and the polarised ray R_{-+}, respectively:

$$\begin{aligned}
\partial_t \mathbf{w}'_n &= -\tilde{\Psi}^0_{k_0}(D_{m_{-+}n}(\eta)\mathbf{I} - G_{m_{-+}n}(\eta)\mathbf{J})\mathbf{w}'_{m_{-+}}, \\
\partial_t \mathbf{w}'_{m_{-+}} &= \tilde{\Psi}^0_{-k_0}(D_{m_{-+}n}(\eta)\mathbf{I} + G_{m_{-+}n}(\eta)\mathbf{J})\mathbf{w}'_n
\end{aligned} \tag{9.9}$$

where $m_{-+} = (-\lambda_0 n_1, \lambda_0 n_2, \lambda_0 n_3)$ and the coefficients D and G are given by (9.4). We have $(\alpha\mathbf{I} - \beta\mathbf{J})(\alpha\mathbf{I} + \beta\mathbf{J}) = (\alpha^2 + \beta^2)\mathbf{I}$ since $\mathbf{J}^2 = -\mathbf{I}$. Using again $\tilde{\Psi}^0_{-k_0} = (\tilde{\Psi}^0_{k_0})^*$, Eqs. (9.9) imply

$$\begin{gathered}
\frac{\partial^2}{\partial t^2}\mathbf{w}'_n = -((\theta^{-+}_{n,D})^2 + (\theta^{-+}_{n,G})^2)\mathbf{w}'_n, \\
(\theta^{-+}_{n,D})^2 = D^2_{m_{-+}n}|\tilde{\Psi}^0_{k_0}|^2, \quad (\theta^{-+}_{n,G})^2 = G^2_{m_{-+}n}|\tilde{\Psi}^0_{k_0}|^2,
\end{gathered} \tag{9.10}$$

where $k_0 = ((1+\lambda_0)n_1, (1-\lambda_0)n_2, (1-\lambda_0)n_3)$, $m_{-+} = (-\lambda_0 n_1, \lambda_0 n_2, \lambda_0 n_3)$. Eqs. (9.10) describe baroclinic wave propagation on R_{++}, R_{-+}, coupled with a QG state given by the monochromatic streamfunction $\tilde{\Psi}^0_{k_0}$. Because of the special choice of the QG potential, no baroclinic waves are excited on R_{--} or R_{+-}. A dispersion relation is obtained from (9.10) and is given by

$$\theta^{-+}_{n,DG} = \sqrt{(\theta^{-+}_{n,D})^2 + (\theta^{-+}_{n,G})^2}. \tag{9.11}$$

Finally, choosing a monochromatic $\tilde{\Psi}^0_{k_0}$ with

$$k_0 = ((1-\lambda_0)n_1, (1+\lambda_0)n_2, (1-\lambda_0)n_3)$$

in (9.5) will only excite AG dynamics on R_{++} and the polarised ray R_{+-}; there

$$\theta^{+-}_{n,DG} = \sqrt{D^2_{m_{+-}n} + G^2_{m_{+-}n}}|\tilde{\Psi}^0_{k_0}| = \sqrt{(\theta^{+-}_{n,D})^2 + (\theta^{+-}_{n,G})^2}, \tag{9.12}$$

where $\theta^{+-}_{n,D} = |\tilde{\Psi}^0_{k_0}|\,|D_{m_{+-}n}|$ and

$$\theta^{+-}_{n,G} = 4|\lambda_0|\,|\eta|\,|\tilde{\Psi}^0_{k_0}|\,\frac{\check{n}_2^2}{|\check{n}'|^2}\,\frac{\check{n}_3^2}{|\check{n}|}\,\frac{|3\check{n}_1^2 - \check{n}_2^2|}{\sqrt{|\check{n}'|^2 + \eta^2\check{n}_3^2}}. \tag{9.13}$$

Notice that for $\eta = 0$, both $\theta^{+-}_{n,G}$ and $\theta^{-+}_{n,G}$ are null. The baroclinic waves (9.11)-(9.12) then reduce to 'horizontal' waves of frequencies

$$\begin{aligned}
\theta^{-+}_{n,D}|_{\eta=0} &= 4|\lambda_0|\,|\tilde{\Psi}^0_{k_0}|\,|\check{n}_1\check{n}_2|\left|\frac{|\check{n}'|^2}{|\check{n}|^2} + \frac{\check{n}_3^2(-\check{n}_1^2 + \check{n}_2^2)}{|\check{n}|^2|\check{n}'|^2}\right|, \\
\theta^{+-}_{n,D}|_{\eta=0} &= 4|\lambda_0|\,|\tilde{\Psi}^0_{k_0}|\,|\check{n}_1\check{n}_2|\left|\frac{|\check{n}'|^2}{|\check{n}|^2} + \frac{\check{n}_3^2(\check{n}_1^2 - \check{n}_2^2)}{|\check{n}|^2|\check{n}'|^2}\right|.
\end{aligned} \tag{9.14}$$

In some sense the above baroclinic waves $\theta^{+-}_{n,DG}$ and $\theta^{-+}_{n,DG}$ are the continuation of the AG horizontally propagating gravity waves of the pure stratified case;

switching on $\eta > 0$ couples the wave dynamics of both w^1 and w^2 through interactions of both operators $D(\eta)$ and $G(\eta)$. One can repeat the above analysis with time independent stream functions of the form $\tilde{\Psi}^0 = \tilde{\Psi}^0(s_1x_1 + s_2x_2, x_3)$, where $\tilde{\Psi}^0$ is an arbitrary function (they are exact steady state solutions of 3DQG equations for arbitrary s_1 and s_2). In this case the dynamics for each Fourier component of $\tilde{\Psi}^0$ decouples into wave problems described above which can be solved independently.

In the general case (no restrictions whatsoever on $\tilde{\Psi}^0$ and arbitrary $\eta = f/N$), the dynamics of baroclinic field along any given 'beam' of rays will fully couple four systems of the type (9.5) on R_{++}, R_{--}, R_{+-}, R_{-+}. Technically, these are coupled non-local hyperbolic systems in the variables t and λ $(= \gamma_2/\gamma_1)$. What makes them unusual is the extreme variability of the coefficients $\tilde{\Psi}^0_k$. Wave frequencies such as $\theta^{--}_{n,G}$, $\theta^{+-}_{n,G}$, $\theta^{-+}_{n,G}$, $\theta^{+-}_{n,DG}$ and $\theta^{-+}_{n,DG}$ become incommensurably mixed. The dynamics of the total baroclinic field that is obtained from a full spectrum of $\tilde{\Psi}^0_k$ can be very complex.

9.2 Genesis of fronts in the regime of strong stratification and weak rotation

In this section we analyse the intermediate asymptotic regime of strong stratification and weak rotation. It will be shown that the effect of weak rotation is to couple large horizontal and small vertical scales leading to vertically slanted AG wave fronts.

For $\eta = 0$ (no rotation) the splitting (1.22)–(1.23), (5.1) is also valid (see [12]). In this pure stratified case Eqs. (1.22) coincide with the familiar quasi-2D Euler systems which can be seen by introducing variables $\tilde{q}$ and $\mathbf{U}_{\mathrm{Q2D}}$ (quasi-2D potential and velocity):

$$\tilde{q}_m = i|\check{m}'|w^0_m, \; m \neq \overline{m} \quad \text{and} \quad \mathbf{U}_{Q2D,k} = -i[-\check{k}_2, \check{k}_1, 0, 0]\frac{\tilde{q}_k}{|\check{k}'|^2}, \; k \neq \overline{k}. \tag{9.15}$$

In this notation (1.22) is written in the form of the 2D Euler equations which depend on x_3 as a parameter:

$$\partial_t \tilde{q}_n = -i \sum_{k+m=n} \left((\overline{\mathbf{U}}_{Q2D,\overline{k}} + \mathbf{U}_{Q2D,k}) \cdot \check{m} \right) \tilde{q}_m, \tag{9.16}$$

with

$$\partial_t \overline{\mathbf{U}}_{Q2D,\overline{n}} = 0. \tag{9.17}$$

To derive (9.17), we have used $Q^{000}_{km\overline{n}}(0) = 0$, $Q^{000}_{k\overline{m}n}(0) = 0$ and the appropriate limit for $Q^{000}_{\overline{k}mn}(0)$. Then in physical space the velocity $\mathbf{U}_{\mathrm{Q2D}}(t, x_1, x_2, x_3)$ satisfies the quasi-2D Euler systems

$$\partial_t \mathbf{U}_{\mathrm{Q2D}} + (\overline{\mathbf{U}}_{\mathrm{Q2D}} + \mathbf{U}_{\mathrm{Q2D}}) \cdot \nabla_h \mathbf{U}_{\mathrm{Q2D}} = -\nabla_h \tilde{p}, \qquad \nabla_h \cdot \mathbf{U}_{\mathrm{Q2D}} = 0 \tag{9.18}$$

which depends on x_3 as a parameter, $\nabla_h = [\partial_1, \partial_2]$. In Eqs. (9.18) $\overline{\mathbf{U}}_{\text{Q2D}}$ denotes horizontally-averaged velocity which is an adiabatic invariant of the 3D Boussinesq equations in the strongly stratified limit in the absence of rotation as shown in [12]. Note that $\overline{\rho}(x_3)$ is still an adiabatic invariant for the case $\eta = 0$.

It follows from (5.3) that for $\eta = 0$ (no rotation) $G_{mn}(0) = 0$ and $D_{mn}(0)$ reduces to

$$D_{mn}(0) = (\check{n}' \wedge \check{m}') \frac{2(\check{n}_3 \check{m}_3 \check{n}' \cdot \check{m}' + |\check{m}'|^2 |\check{n}'|^2)}{|\check{m}'||\check{n}'||\check{m}||\check{n}|}. \tag{9.19}$$

In the non-rotating case ($\eta = 0$) the AG dynamics is described by $D_{mn}(0)\mathbf{I}$ in (5.1), since $G_{mn}(0) = 0$ by (5.3).

In the purely stratified regime ($\eta = 0$, Bu $= +\infty$), the QG flow degenerates into quasi-2D flows parametrised in x_3: Eqs. (9.18). Without viscosity, there is no bound on vertical shearing associated with the dynamics of 3D2C (3-dimensional, 2-component) decoupled pancakes (parametrised in x_3) with different pressures at every level; this leads to unbalanced dynamics at the lowest order. There is no saturation of the exponential build-up of vertical enstrophy (in small vertical scales) for the AG dynamics as the latter is coupled to the quasi–2D-field through $\partial U^1_{\text{Q2D}}/\partial x_3$, $\partial U^2_{\text{Q2D}}/\partial x_3$. The major problem is lack of boundedness of vertical shearing in quasi-2D equations (see [49]). Of course, control of vertical shearing can be achieved trivially by introducing vertical viscosity; however, this corresponds to a non-physical laboratory set-up rather than the real atmosphere, or a poorly-resolved in the vertical x_3-scale numerical model.

In such a purely stratified context ($\eta = 0$), the 4-ray AG dynamics Eqs. (9.5) reduce to

$$\partial_t \mathbf{w}'_n = -\sum_{\gamma_2} \tilde{\Psi}^0_k(t) D_{m_{+-}n}(0)\mathbf{I}\mathbf{w}'_{m_{+-}} - \sum_{\gamma_2} \tilde{\Psi}^0_k(t) D_{m_{-+}n}(0)\mathbf{I}\mathbf{w}'_{m_{-+}}, \tag{9.20}$$

with uncoupling of w^1 and w^2; there are only interactions between the polarised rays R_{-+}, R_{+-} on the right-hand side of Eqs. (9.20) (recall that $D_{m_{--}n}(\eta) = 0$). For $\eta = 0$ the operators D and G given by Eqs. (9.4) become

$$\begin{aligned}
G_{m_{++}n}(0) = G_{m_{--}n}(0) &= G_{m_{-+}n}(0) = G_{m_{+-}n}(0) = 0, \\
D_{m_{-+}n}(0) &= 4\lambda \check{n}_1 \check{n}_2 \left(\frac{|\check{n}'|^2}{|\check{n}|^2} + \frac{\check{n}_3^2(-\check{n}_1^2 + \check{n}_2^2)}{|\check{n}|^2|\check{n}'|^2} \right), \\
D_{m_{+-}n}(0) &= -4\lambda \check{n}_1 \check{n}_2 \left(\frac{|\check{n}'|^2}{|\check{n}|^2} + \frac{\check{n}_3^2(\check{n}_1^2 - \check{n}_2^2)}{|\check{n}|^2|\check{n}'|^2} \right).
\end{aligned} \tag{9.21}$$

A key role in readjusting these purely stratified pancake dynamics is played by weak rotation, $\eta = f/N \ll 1$ but fixed (say $O(10^{-2})$). This can be immediately inferred from the frequencies of inertio-gravity waves. Strong vertical shearing on small vertical scales can excite inertial Poincaré waves which can balance gravity waves: this is exactly the balance between a spectral Froude number $(N|\check{n}'|/|\check{n}|)^{-1}$ and a spectral anisotropic Rossby number $(f|\check{n}_3|/|\check{n}|)^{-1}$. An approach such as in [28] and [55], which treats rotation scales as *uniformly slow* ('only large scale horizontal rotation'), misses this readjustment mechanism. Rotation does impact on small vertical scales. Of course, weak rotation does control (bound) the vertical shearing of the quasi-2D3C flow through conservation of the QG potential $\tilde{q}$: $\partial_t \sum_n |\tilde{q}_n(t)|^2 = 0$, where $\tilde{q}_n = (|\check{n}'|^2 + \eta^2 \check{n}_3^2)\tilde{\Psi}_n^0$. As shown in [11] and here, weak rotation regularises vertical shearing and lets us control (bound) AG vertical scales for all times; this is especially important for regimes where the local Richardson number is small enough to generate Kelvin–Helmholtz instabilities. There is no need to resort to vertical viscosity as the principal stabilising mechanism (Reynolds number Re $\sim 10^{10} - 10^{12}$ in atmospheric flows).

The effect of weak rotation is to trigger interactions between the polarised pairs of rays R_{-+}, R_{+-} and the ray R_{--} through the $O(\eta)$ operators $G(\eta)$ given by Eqs. (9.4). It also follows from (9.4) that for $\eta \neq 0$

$$
\begin{aligned}
&D_{m_{-+}n}(\eta) = \\
&4\lambda \check{n}_1 \check{n}_2 \left(\frac{|\check{n}'|^2}{|\check{n}|^2} - \frac{\eta^2 \check{n}_3^2}{|\check{n}'|^2 + \eta^2 \check{n}_3^2} + \frac{\check{n}_3^2(-\check{n}_1^2 + \check{n}_2^2)}{|\check{n}'|^2} \left(\frac{1}{|\check{n}|^2} + \frac{\eta^2}{|\check{n}'|^2 + \eta^2 \check{n}_3^2} \right) \right), \\
&D_{m_{+-}n}(\eta) = \\
&-4\lambda \check{n}_1 \check{n}_2 \left(\frac{|\check{n}'|^2}{|\check{n}|^2} - \frac{\eta^2 \check{n}_3^2}{|\check{n}'|^2 + \eta^2 \check{n}_3^2} + \frac{\check{n}_3^2(\check{n}_1^2 - \check{n}_2^2)}{|\check{n}'|^2} \left(\frac{1}{|\check{n}|^2} + \frac{\eta^2}{|\check{n}'|^2 + \eta^2 \check{n}_3^2} \right) \right).
\end{aligned}
\tag{9.22}
$$

In our search for hyperbolic fronts in Eqs. (9.5), we look for quasi-standing, slow-moving waves along special 4-ray bundles: such fronts are not discontinuities but rather correspond to strong gradients (shearing) of the field. The above discussion of monochromatic waves shows that in general the wave frequencies do couple the operators $G(\eta)$ and $D(\eta)$, Eqs. (9.11)–(9.12). Only in the special case where $D(\eta)$ does not contribute to the wave dispersion law can we expect to get slow-moving waves of frequency $O(\eta)$, since only $G(\eta)$ is $O(\eta)$; see Eqs. (9.4) and (9.8). For $\eta = 0$ the coefficients $D_{m_{-+}n}(0)$ and $D_{m_{+-}n}(0)$ are given by (9.21) with the corresponding wave frequencies given by (9.14). These are stratified AG waves propagating horizontally. Their frequencies depend on the vertical variability of $\tilde{\Psi}^0$. Even weak rotation triggers interactions between the polarised pairs of rays R_{-+}, R_{+-} and the ray R_{--} through the $O(\eta)$ operators $G(\eta)$ in Eqs. (9.4). From Eqs. (9.8) and (9.13), the natural frequencies $\theta^{--}_{n,G}$, $\theta^{+-}_{n,G}$, $\theta^{-+}_{n,G}$ are $O(\eta)$; whereas $\theta^{-+}_{n,D}$ (resp. $\theta^{-+}_{n,DG}$) and $\theta^{+-}_{n,D}$ (resp. $\theta^{+-}_{n,DG}$) are all $O(1)$ on polarised rays R_{-+}, R_{+-}. Apparently, 'horizontal' waves associated to $D_{m_{-+}n}(0)$ through $\theta^{-+}_{n,DG}$ and $\theta^{+-}_{n,DG}$ still seem to dominate wave dynamics. However, there is one remarkable set of wave vectors given by $\check{n}_1^2 = \check{n}_2^2$ and the vertical slanting condition (9.24) which do generate *much slower frequencies* on the polarised rays; specifically those which annihilate $D_{m_{-+}n}(\eta)$ and $D_{m_{+-}n}(\eta)$, given by (9.4), and reduce $\theta^{+-}_{n,DG}$, $\theta^{-+}_{n,DG}$ to $\theta^{+-}_{n,DG} = \theta^{+-}_{n,G}$, $\theta^{-+}_{n,DG} = \theta^{-+}_{n,G}$. Up to terms of order $\sqrt{\eta}$ these are now standing waves.

Following these remarks, we search for more general solutions of (9.5), without any restriction on the turbulent streamfunction $\tilde{\Psi}^0(t)$. We look for special 4-ray bundles such that the symbols $D_{m_{+-}n}(\eta)$ and $D_{m_{-+}n}(\eta)$ of the hyperbolic system (9.5) are both null (recall that $D_{m_{--}n}(\eta) = 0$ and $G(\eta)$ is $O(\eta)$). Careful inspection of (9.22) show that this can be achieved with the characteristic conditions $\check{n}_1^2 = \check{n}_2^2$ and

$$\frac{|\check{n}'|^2}{|\check{n}|^2} = \frac{\eta^2 \check{n}_3^2}{|\check{n}'|^2 + \eta^2 \check{n}_3^2}; \tag{9.23}$$

these conditions annihilate $D_{m_{-+}n}(\eta)$ and $D_{m_{+-}n}(\eta)$ in (9.22). Eq. (9.23) is equivalent to $|\check{n}'|^4 + \eta^2 \check{n}_3^2 |\check{n}'|^2 = \eta^2(\check{n}_3^4 + \check{n}_3^2 |\check{n}'|^2)$ and to $|\check{n}'|^4 = \eta^2 \check{n}_3^4$. This defines a family of cones with a special vertical direction for the family of resonant rays:

$$\frac{\check{n}_3^2}{|\check{n}'|^2} = \frac{1}{\eta} \Leftrightarrow \frac{|\check{n}'|}{|\check{n}_3|} = \sqrt{\eta}. \tag{9.24}$$

These are indeed families of wave fronts generated by weak rotation. For $\eta = 0$, the wave-front vector is strictly vertical and this is the limit of purely-horizontal AG-stratified wave propagation (but with frequency $O(1)$). For $\eta > 0$, $\eta \ll 1$, the wave-front vector is strongly vertically slanted and the front is nearly horizontal. As η increases the front becomes oblique. Up to terms of order $\sqrt{\eta}$ (rather than η as shown below; see Eqs. (9.26)–(9.27)) this leads to equations describing *zero-frequency standing waves in the envelope equations* for the AG field. Note that the horizontal characteristic condition $\check{n}_1^2 - \check{n}_2^2 = 0$ (which annihilates the second term in (9.22)) is strictly equivalent to $\check{n}' \cdot \check{m}'_{+-} = \check{n}' \cdot \check{m}'_{-+} = 0$ for the polarised rays (which annihilates $\check{n}_3 \check{m}_3 \check{n}' \cdot \check{m}'$ in (5.3)). These conditions are intrinsic in terms of the horizontal rotation of the computational box. In practice, given a horizontal front propagation vector, the computational box must be adjusted to fit the horizontal orientation of the front vector.

The rigorous justification of (9.24) involves four coupled hyperbolic equations of the type (9.5) on R_{++}, R_{--}, R_{+-}, R_{-+} with *general time-dependent* $\tilde{\Psi}^0_k(t)$ (recall that $D_{m_{--}n}(\eta) \equiv 0$). Conditions (9.24) ensure that the right-hand side of (9.5) is uniformly of order $O(\sqrt{\eta})$ for small η (weak rotation). This leads to equations describing *slowly-moving AG wave fronts.* A sketch of the rigorous proof in the general $\tilde{\Psi}^0_k(t)$ context goes as follows. The front condition (9.24) is equivalent to setting

$$D_{m_{-+}n}(\eta) = D_{m_{+-}n}(\eta) = 0 \tag{9.25}$$

in (9.4) (characteristic equations for the hyperbolic operators restricted to $D(\eta)$ on the polarised rays). However, the wave interactions generated by the shearing operators $G(\eta)$ perturb these exact standing wave fronts and must be estimated together with the front condition (9.24). Using $\eta \check{n}_3^2 = |\check{n}'|^2$, we find that:

$$G_{m_{--}n}(\eta) = -\frac{4\lambda\eta^{3/2}}{1+\eta}\check{n}_3^2 = -\frac{4\lambda\eta^{1/2}}{1+\eta}|\check{n}'|^2,$$
$$G_{m_{-+}n}(\eta) = G_{m_{+-}n}(\eta) = \frac{8\lambda\eta^{1/2}}{1+\eta}\frac{\check{n}_1^2\check{n}_2^2}{|\check{n}'|^2}, \tag{9.26}$$

and the $D(\eta)$ operators disappear in the rays equations (9.5) on the fronts provided that $\check{n}_1^2 - \check{n}_2^2 = 0$ and (9.24) are exactly satisfied. In practice, these are approximately satisfied and we must estimate the error in $D(\eta)$. On the front, the G-waves contribute frequencies of order $\sqrt{\eta}$. This must be compatible with Eqs. (9.25) and the contribution of $D(\eta)$ near the characteristics (9.24), where (9.25) is satisfied up to an arbitrary small error. We find

$$D_{m_{-+}n}(\eta) = \frac{4\lambda\check{n}_1\check{n}_2(|\check{n}'|^4 - \eta^2\check{n}_3^4)}{|\check{n}|^2(|\check{n}'|^2 + \eta^2\check{n}_3^2)} = \frac{16\lambda\sqrt{\eta}}{(1+\eta)^2}\check{n}_1\check{n}_2\left\{\frac{|\check{n}'|}{|\check{n}_3|} - \sqrt{\eta}\right\}. \tag{9.27}$$

Setting $\epsilon = \frac{|\check{n}'|}{|\check{n}_3|} - \sqrt{\eta}$ and noting that $\lambda|\check{n}'|^2 = |\check{m}'||\check{n}'|$ and $\lambda\check{n}_1\check{n}_2 = \check{m}_1\check{n}_2 = \check{m}_2\check{n}_1$ on the rays, we conclude that the condition $\epsilon = O(\eta)$ (or $\epsilon = o(\eta)$) ensures that the front is nearly standing, propagating with a slow frequency $O(\sqrt{\eta})$ *on large horizontal scales. The front balances all wave frequencies to* $O(\sqrt{\eta})$. Similarly, the contribution of $D_{m_{-+}n}(\eta)$ where $\check{n}_2^2 - \check{n}_1^2 = \varrho$ is

$$D_{m_{-+}n}(\eta) = \frac{4\lambda\check{n}_1\check{n}_2}{|\check{n}'|^2}\varrho \tag{9.28}$$

so that it suffices to take $\varrho = O(\eta^{3/2})$.

We now present equations for wave fronts in pseudo-physical space. For cones satisfying the condition (9.24) we have $\eta^2/\omega_n^2 = \eta$ and the coefficients $D(\eta)$ and $G(\eta)$ given by (5.3) become

$$D_{mn}(\eta) = \frac{2(1+\eta)(\check{n}' \wedge \check{m}')(\check{n}' \cdot \check{m}')(\check{n}_3\check{m}_3)}{|\check{m}'||\check{n}'||\check{m}||\check{n}|},$$
$$G_{mn}(\eta) = \sqrt{\eta}\frac{3\check{n}_3\check{m}_3(\check{n}' \wedge \check{m}')^2 + ((\check{n} \times \check{m}) \cdot \check{m}^{\perp})((\check{n} \times \check{m}) \cdot \check{n}^{\perp})}{|\check{m}'||\check{n}'||\check{m}||\check{n}|}. \tag{9.29}$$

For the wave front dynamics described above, the coefficient $D(\eta)$ is annihilated by the condition $\check{n}' \cdot \check{m}'_{+-} = \check{n}' \cdot \check{m}'_{-+} = 0$ for the polarised rays. For the coefficients $G_{mn}(\eta)$, we obtain on the polarised rays ($n = m_{++}$)

$$G_{m_{--}n}(\eta) = -\frac{4\gamma_1\gamma_2\sqrt{\eta}}{1+\eta}|\check{l}'|^2, \qquad G_{m_{-+}n}(\eta) = \frac{2\gamma_1\gamma_2\sqrt{\eta}}{1+\eta}|\check{l}'|^2,$$
$$G_{m_{+-}n}(\eta) = \frac{2\gamma_1\gamma_2\sqrt{\eta}}{1+\eta}|\check{l}'|^2. \tag{9.30}$$

We define direction vectors ($\check{l} \equiv \check{l}_{++}$):

$$\begin{aligned} \check{l}_{++} &= (\check{l}_1, \check{l}_2, \check{l}_3), & \check{l}_{+-} &= (\check{l}_1, -\check{l}_2, \check{l}_3), \\ \check{l}_{-+} &= (-\check{l}_1, \check{l}_2, \check{l}_3), & \check{l}_{--} &= (-\check{l}_1, -\check{l}_2, \check{l}_3). \end{aligned} \tag{9.31}$$

After substitution of the coefficients (9.30) in Eqs. (9.5) we obtain

$$\begin{gathered} \partial_t \mathbf{w}'_{\gamma_1 l_{++}} = \frac{2\sqrt{\eta}}{1+\eta} |\check{l}'|^2 \gamma_1 \sum_{\gamma_2} \tilde{\Psi}^0_k(t) \gamma_2 \mathbf{J} (-2\mathbf{w}'_{\gamma_2 l_{--}} + \mathbf{w}'_{\gamma_2 l_{-+}} + \mathbf{w}'_{\gamma_2 l_{+-}}), \\ \mathbf{J} = \begin{pmatrix} 0 & -1 \\ 1 & 0 \end{pmatrix}. \end{gathered} \tag{9.32}$$

In Eqs. (9.32) it is understood that $k = n - m$ where $n = \gamma_1 l_{++}$ and $m = \gamma_2 l_{--}$, $m = \gamma_2 l_{+-}$ or $m = \gamma_2 l_{-+}$.

We introduce averaging for the AG field $\mathbf{w}' = (w_1, w_2)$ in the phase planes orthogonal to the direction vectors $\check{l}_{++}$, $\check{l}_{+-}$, $\check{l}_{-+}$ and $\check{l}_{--}$ (averaging across the front). For example, averaging across $\check{l}_{++}$ is defined as

$$\mathbf{w}'_{++} = \int_{x:\check{l}_{++}\cdot x = 0} \mathbf{w}'(t,x)\, dx, \tag{9.33}$$

where $\check{l}_{++} \cdot x = \check{l} \cdot x = \check{l}_1 x_1 + \check{l}_2 x_2 + \check{l}_3 x_3 = 0$ is the plane (in physical space) orthogonal to the front vector $\check{l} = \check{l}_{++} = (\check{l}_1, \check{l}_2, \check{l}_3)$. It is easily seen that the Fourier transform of $\mathbf{w}'_{++}$ (in physical space) is $\mathbf{w}'_n$ restricted to the ray R_{++} in Fourier space. Similarly, we define $\mathbf{w}'_{+-}$, $\mathbf{w}'_{-+}$ and $\mathbf{w}'_{--}$. For instance, $\mathbf{w}'_{+-}$ is the physical field $\mathbf{w}'$ averaged over the plane orthogonal to $\check{l}_{+-}$ and its Fourier transform is $\mathbf{w}'_{n_{+-}}$ restricted to the Fourier ray R_{+-}. Here it is understood that Fourier series defined by the lattice vector $\check{l}_{++}$ are used in the definition of $\mathbf{w}'_{++}$ (similarly, for $\mathbf{w}'_{+-}$, $\mathbf{w}'_{-+}$ and $\mathbf{w}'_{--}$):

$$\begin{gathered} \mathbf{w}'_{++} = \sum_{\gamma_1} e^{i\gamma_1(\check{l}_{++}\cdot x)} \mathbf{w}'_{\gamma_1 l_{++}}, \qquad s^{++} = \check{l}_{++} \cdot x, \\ \frac{\partial \mathbf{w}'_{++}}{\partial s^{++}} = \sum_{\gamma_1} i\gamma_1 e^{i\gamma_1(\check{l}_{++}\cdot x)} \mathbf{w}'_{\gamma_1 l_{++}}. \end{gathered} \tag{9.34}$$

Eqs. (9.32) are in pseudo-convolution form. Multiplying these equations by $e^{i\gamma_1(l_{++}\cdot x)}$ and summing over γ_1, we obtain equations in pseudo-physical space. Then equations for the AG wave front coupling four fields $\mathbf{w}'_{++}$, $\mathbf{w}'_{+-}$, $\mathbf{w}'_{-+}$ and $\mathbf{w}'_{--}$ have the following form in physical space

$$\begin{aligned} \frac{\partial \mathbf{w}'_{++}}{\partial \tau} &= |\check{l}'|^2\, \mathbf{J} \frac{\partial}{\partial s^{++}} \left(\tilde{\Psi}^0(t,x) \left(2\frac{\partial \mathbf{w}'_{--}}{\partial s^{--}} - \frac{\partial \mathbf{w}'_{-+}}{\partial s^{-+}} - \frac{\partial \mathbf{w}'_{+-}}{\partial s^{+-}} \right) \right), \\ \frac{\partial \mathbf{w}'_{--}}{\partial \tau} &= |\check{l}'|^2\, \mathbf{J} \frac{\partial}{\partial s^{--}} \left(\tilde{\Psi}^0(t,x) \left(2\frac{\partial \mathbf{w}'_{++}}{\partial s^{++}} - \frac{\partial \mathbf{w}'_{-+}}{\partial s^{-+}} - \frac{\partial \mathbf{w}'_{+-}}{\partial s^{+-}} \right) \right), \\ \frac{\partial \mathbf{w}'_{+-}}{\partial \tau} &= |\check{l}'|^2\, \mathbf{J} \frac{\partial}{\partial s^{+-}} \left(\tilde{\Psi}^0(t,x) \left(2\frac{\partial \mathbf{w}'_{-+}}{\partial s^{-+}} - \frac{\partial \mathbf{w}'_{++}}{\partial s^{++}} - \frac{\partial \mathbf{w}'_{--}}{\partial s^{--}} \right) \right), \\ \frac{\partial \mathbf{w}'_{-+}}{\partial \tau} &= |\check{l}'|^2\, \mathbf{J} \frac{\partial}{\partial s^{-+}} \left(\tilde{\Psi}^0(t,x) \left(2\frac{\partial \mathbf{w}'_{+-}}{\partial s^{+-}} - \frac{\partial \mathbf{w}'_{++}}{\partial s^{++}} - \frac{\partial \mathbf{w}'_{--}}{\partial s^{--}} \right) \right), \end{aligned} \tag{9.35}$$

where $\tau = \frac{2\sqrt{\eta}}{1+\eta}t$ is a slow time (recall that the parameter η is small in the regime considered in this section).

In Eqs. (9.35) $\frac{\partial}{\partial s^{++}}$ is the directional derivative along the ray l_{++} after averaging in the phase plane $l_{++} \cdot x = 0$ (the $\frac{\partial}{\partial s^{++}}$ operator is defined by multiplication by $i\gamma_1$ in Fourier space). Similarly $\frac{\partial}{\partial s^{+-}}$, $\frac{\partial}{\partial s^{-+}}$ and $\frac{\partial}{\partial s^{--}}$ are directional derivatives along the rays l_{+-}, l_{-+} and l_{--} after averaging in the phase planes $l_{+-} \cdot x = 0$, $l_{-+} \cdot x = 0$ and $l_{--} \cdot x = 0$, respectively. Eqs. (9.35) form a stochastic hyperbolic system of coupled wave equations. These AG equations are for ageostrophic fields w_1, which is the divergent velocity potential, and w_2, which is the geostrophic departure (thermal wind imbalance) variable. This is a hyperbolic system of conservation laws, as the sum of the energy of the $\mathbf{w}'$ is conserved in (9.35).

We now consider an important special class of solutions of (5.1) which corresponds to similar hyperbolic wave fronts on a cone with the front condition (9.24) for $\eta \neq 0$, but $\eta \ll 1$. The only difference is that the condition $\check{n}_1^2 = \check{n}_2^2$ is not required in this case and our considerations are valid for *all* rays, not just in the context of Eqs. (1.25). This branch of solution originates from a single horizontal layer for $\eta = 0$ leading to the formation of a vertically-slanted AG wave front for $\eta \neq 0$. We recall that for $\eta = 0$ the dynamics of quasi-2D equations (9.18) is independent in every horizontal layer. An important property of the inviscid Eqs. (1.22), (2.32), is that when $\eta > 0$ they admit exact solutions in the potential vorticity $\tilde{q}$ which are delta-functions $\delta(x_3)$ in x_3. These solutions correspond to a single horizontal layer located at $x_3 = 0$ (potential vorticity anomaly). That means that the Fourier transform of $\tilde{q}$ does not depend on n_3: $\tilde{q}(n_1, n_2) = \tilde{q}(n_1, n_2, n_3)$. For this special single-layer class of initial data, the inviscid Eqs. (1.22) can be written in the form:

$$\begin{aligned}\partial_t \tilde{q}(n_1, n_2) &= -\sum_{k'} \left(\sum_{k_3} \frac{\check{k}' \wedge \check{m}'}{|\check{k}'|^2 + \eta^2 \check{k}_3^2} \right) \tilde{q}(k_1, k_2)\tilde{q}(n_1 - k_1, n_2 - k_2) \\ &= -\sum_{k'} (\check{k}' \wedge \check{m}') S\left(|k'|^2, \frac{\eta^2}{a_3^2}\right) \tilde{q}(k_1, k_2)\tilde{q}(n_1 - k_1, n_2 - k_2), \quad (9.36)\end{aligned}$$

where the relation $\tilde{q}_k = \omega_k |\check{k}| w_k^0$ was used and the function $S(|k'|^2, \eta^2/a_3^2)$ is defined by

$$S\left(|k'|^2, \frac{\eta^2}{a_3^2}\right) = \sum_{k_3} \frac{1}{|\check{k}'|^2 + \eta^2 k_3^2/a_3^2}. \quad (9.37)$$

As above, $\tilde{\Psi}^0$ is related to $\tilde{q}$ by $\tilde{\Psi}^0(k_1, k_2, k_3) = \tilde{q}(k_1, k_2)/(|\check{k}'|^2 + \eta^2 \check{k}_3^2)$ (in physical space, $-\left(\nabla_h^2 + \eta^2 \frac{\partial^2}{\partial x_3^2}\right) \tilde{\Psi}^0 = \tilde{q}$). Then we have

$$\tilde{\Psi}^0(k) = \frac{\tilde{q}(k_1, k_2)}{|\check{k}'|^2 + \eta^2 \check{k}_3^2} = \frac{\tilde{q}(k_1, k_2)}{|\check{k}'|^2} \left(1 - \frac{\eta^2 \check{k}_3^2}{|\check{k}'|^2 + \eta^2 \check{k}_3^2}\right). \quad (9.38)$$

The second term in the last parenthesis in (9.38) is small for scales satisfying $\eta\check{k}_3^2/|\check{k}'|^2 = O(1)$. In particular, for scales exactly satisfying the front condition (9.24) one has

$$\frac{\eta^2\check{k}_3^2}{|\check{k}'|^2+\eta^2\check{k}_3^2} = \frac{\eta}{1+\eta}.$$

It follows from (9.38) that for such scales $\tilde{\Psi}^0(k)$ is independent of k_3 up to terms of order η: $\tilde{\Psi}^0(k) = \tilde{\Psi}^0(k_1,k_2) + O(\eta)$, $\tilde{\Psi}^0(k_1,k_2) = \tilde{q}(k_1,k_2)/|k'|^2$.

In Eqs. (5.1) we split Fourier coefficients $\mathbf{w}'_n$ of a general AG field into the sum of fields $\mathbf{w}'^{even}_{n_1,n_2,n_3}$ and $\mathbf{w}'^{odd}_{n_1,n_2,n_3}$, even and odd in $n_3$respectively. We have $\mathbf{w}'^{even}_{n_1,n_2,n_3} = \mathbf{w}'^{even}_{n_1,n_2,-n_3}$. Note that such a splitting for $\mathbf{w}'$ is generally true for (5.1) provided that $\tilde{\Psi}^0(n_1,n_2,-n_3) = \tilde{\Psi}^0(n_1,n_2,n_3)$. Since all coefficients in (5.1) are invariant under simultaneous change of sign of n_3, k_3, m_3 even and odd components separately satisfy this linear equation up to terms $O(\eta)$. Below we show that the even AG field contains a slow-varying component which corresponds to vertically slanted fronts. We consider the case of small η and recall that the coefficients $G_{mn}(\eta)$ are of order $O(\eta)$ ($G_{mn}(0) = 0$) in (5.1). Collecting the terms with m_3 and $-m_3$ in inviscid Eqs.(5.1), we obtain

$$\begin{aligned}\partial_t \mathbf{w}_n^{even'} = &\\ - \sum_{\substack{\phi_m=\phi_n\\ k+m=n}} &\tilde{\Psi}^0_k(t)(D^{\text{even}}_{mn}(\eta)\mathbf{I} - G^{\text{even}}_{mn}(\eta)\mathbf{J})\mathbf{w}'^{\text{even}}_m \\ + \sum_{\substack{\phi_m=\phi_n\\ k+m=n}} &(\tilde{\Psi}^0_{n_1-m_1,n_2-m_2,n_3-m_3}(t) - \tilde{\Psi}^0_{n_1-m_1,n_2-m_2,n_3+m_3}(t)) \\ &\times(D^{\text{odd}}_{mn}(\eta)\mathbf{I} - G^{\text{odd}}_{mn}(\eta)\mathbf{J})\mathbf{w}'^{\text{even}}_m. \end{aligned} \quad (9.39)$$

It follows from (9.38) that the second term in (9.39) is small for small η and it is identically zero if $\tilde{\Psi}^0_k$ is independent of k_3. Thus, cancelling odd terms in m_3, we derive from (5.3)

$$D^{\text{even}}_{mn}(\eta) = D_{m',m_3,n}(\eta) + D_{m',-m_3,n}(\eta) = \frac{4(\check{n}'\wedge\check{m}')(|\check{n}'|^4 - \eta^2\check{n}_3^4)}{|\check{n}|^2(|\check{n}'|^2+\eta^2\check{n}_3^2)}. \quad (9.40)$$

Similarly, using resonant relations, we find that

$$G^{\text{even}}_{mn}(\eta) = G_{m',m_3,n}(\eta) + G_{m',-m_3,n}(\eta) = \frac{4\eta(\check{m}'\cdot\check{n}')\check{n}_3^2}{|\check{n}|\sqrt{|\check{n}'|^2+\eta^2\check{n}_3^2}}. \quad (9.41)$$

Clearly, the operator $D^{\text{even}}_{mn}(\eta) = 0$, if $\eta = |\check{n}'|^2/|\check{n}_3|^2$, which gives again the condition for the slow-moving (velocity $O(\eta)$) AG wave fronts. Thus, we again obtain the wave front relation (9.24). If $\eta = |\check{n}'|^2/|\check{n}_3|^2$ then the coefficient G^{even}_{mn} becomes

$$G^{\text{even}}_{mn}(\eta) = \frac{4\sqrt{\eta}}{\eta+1}(\check{n}'\cdot\check{m}'). \quad (9.42)$$

Cancelling the odd terms in (9.39) gives an error of order $\eta/(1+\eta)$ on the wave front.

This analysis shows that a single horizontal layer bifurcates to vertically-slanted fronts (with the magnitude of the angle depending only on η) for the AG component, without the constraint $\check{n}_1^2 = \check{n}_2^2$, and for *all rays*.

It follows from (9.42) that the wave front cone principal term in Eqs. (9.39) has the form

$$\partial_t \mathbf{w}_n'^{\text{even}} = \frac{4\sqrt{\eta}}{1+\eta} \sum_{k'+m'=n'} (\check{n}' \cdot \check{m}') \tilde{\Psi}_k^0(t) \mathbf{J}) \mathbf{w}_m'^{\text{even}} \tag{9.43}$$

where $\mathbf{w}_n'^{\text{even}} = \mathbf{w}'^{\text{even}}(\check{n}_1, \check{n}_2, \pm|\check{n}'|/\sqrt{\eta})$ and where $\tilde{\Psi}_k^0(t)$ is a streamfunction for a singular potential vorticity sheet located at $x_3 = 0$. It is obtained from $\tilde{q}$ using Eqs. (9.36)–(9.38). Eqs. (9.43) can be written in physical space for $\mathbf{w}' = \mathbf{w}'^{\text{even}}$ as follows:

$$\frac{\partial \mathbf{w}'}{\partial \tau} = -\text{div}_h \left(\tilde{\Psi}^0(t,x) \mathbf{J} \nabla_h \mathbf{w}' \right) \tag{9.44}$$

where $\tau = \frac{4\sqrt{\eta}}{1+\eta} t$ is again slow time. In Eqs. (9.44) div_h and ∇_h are differential operators in x_1 and x_2 applied to every component (e.g. $\nabla_h \mathbf{w}' = (\nabla_h w_1, \nabla_h w_2)$). Solutions of Eqs. (9.44) in fact give us $\mathbf{w}'(\check{n}_1, \check{n}_2, \pm|\check{n}'|/\sqrt{\eta})$ where, for fixed horizontal wavenumbers $\check{n}_1$ and $\check{n}_2$, the vertical wavenumber $\check{n}_3$ is found from the cone equation (9.24) for fixed η (here we assume that $\mathbf{w}'(\check{n}_1, \check{n}_2, \check{n}_3) = \mathbf{w}'(\check{n}_1, \check{n}_2, -\check{n}_3)$).

The relation $\frac{1}{\sqrt{\eta}} = \frac{|\check{n}_3|}{|\check{n}'|}$ given by (9.24) implies that vertical slanting of the wave-front vector is proportional to $1/\sqrt{\eta}$ for small η. This relation links vertical shear, horizontal gradients and rotation (via the parameter $\eta = f/N = f_0/N_0$ which is approximately 10^{-2} at mid-latitudes). The front orientation in the horizontal plane depends on the initial conditions of the AG field. Other effects breaking horizontal isotropy include β term effects (see [10] in the context of rotating shallow-water equations where β terms induce linear fast-fast resonances impacting on to the AG field).

Across the front, there are sharp gradients of the buoyancy variable ρ and the velocity component U_3. From the perspective of Lilly's pancake dynamics ([49]), the fronts appear as vertically-curved sharp pancake edges that 'slow-down' the unrestrained horizontal propagation: this is effectively vertical glueing of the pancakes by rotation. The vertical and horizontal scales readjust, leading to an effective decrease of the Burger number from $\sqrt{\text{Bu}} \gg 1$ to $\sqrt{\text{Bu}} = H/L\eta$. An approach such as in Embid & Majda ([28]) and Majda & Grote ([55]) which treats rotation scales as *uniformly slow* ('only large scale horizontal rotation') misses this readjustment mechanism. Rotation impacts on small vertical scales. We predict a definite change in the nature of AG dynamics under the impact of weak rotation in this problem. Mathematically, our AG reduced-asymptotic equations change from parabolic to hyperbolic type when switching on $\eta \neq 0$. The wave fronts that we construct saturate local Kelvin–Helmholtz instabilities (for the local fluctuating AG field) with small Richardson numbers (strong shearing) through adjustment of vertical and horizontal scales on the slanted fronts.

We conclude with a discussion of viscosity on this frontogenesis. It has been argued (Majda & Grote, 1997) that vertical viscosity is the only mechanism of relevance for saturation of vertical shear in strongly stratified pancake dynamics with uniformly slow rotation. Below we establish for large realistic Reynolds numbers that viscosity has an impact next to nil on *front dynamics on large horizontal scales.* Only for a Reynolds number (to be defined below) smaller than 10^3 does viscosity wash out frontogenesis. Moreover, our wave fronts handle the context of locally unstable small Richardson numbers, for the fluctuating AG velocities.

As the G-waves are primarily responsible for the slow $O(\sqrt{\eta})$ motion of the front, we look for the impact of viscosity ν on $\theta^{--}_{n,G}$ (see Eq. (9.8) for example). Other cases can be similarly treated. One finds that the wave is damped and its frequency shifted as

$$\varpi^{--}_{n,G} = i\frac{\nu}{2}\left((\lambda^0)^2+1\right)|\check{n}|^2 \;\pm\; \frac{1}{2}\left(4|\theta^{--}_{n,G}|^2 - \nu^2(1-(\lambda^0)^2)^2|\check{n}|^4\right)^{1/2}. \quad (9.45)$$

Note that $(\lambda^0)^2|n|^2 = |m_{--}|^2$, and $k_0 = n - m_{--}$ in $\tilde{\Psi}^0_{k_0}$. Hence $\lambda^0 = O(1)$ corresponds to large horizontal scales $k_0' = (1+\lambda_0)n'$ for the QG coefficient $\tilde{\Psi}^0_{k_0}$, as long as $|n'| = O(1)$. In this large scale context, the damping $\frac{\nu}{2}(|m_{--}|^2 + |n|^2)$ is evanescent in the atmosphere. Now the viscosity will affect the original frequencies $\theta^{--}_{n,G}$ only at critical damping:

$$2\theta^{--}_{n,G} \sim \nu\left|1-(\lambda^0)^2\right|\,|\check{n}|^2. \quad (9.46)$$

Using the relation $\eta\check{n}_3^2 = |\check{n}'|^2$ on the front, this is shown to be equivalent to

$$\frac{8\lambda_0}{|(\lambda^0)^2-1|} \sim \frac{\nu}{|\tilde{\Psi}^0_{k_0}|}\frac{(1+\eta)^2}{\eta^{3/2}}. \quad (9.47)$$

This formula can be considered as an instantaneous critical damping condition for an adiabatically frozen $|\tilde{\Psi}^0_{k_0}|$. For a QG regime dominated by large scale structures and strong inverse cascades at typical large scales $\sim k_0$, $|\tilde{\Psi}^0_{k_0}| \sim L^{\mathrm{QG}}U^{\mathrm{QG}}$ and the ratio $\langle|\tilde{\Psi}^0_{k_0}|\rangle/\nu$ can define an effective Reynolds number for the large scale dominated QG dynamics:

$$Re^{\mathrm{QG}} = \frac{\langle|\tilde{\Psi}^0_{k_0}|\rangle}{\nu}. \quad (9.48)$$

The condition (9.47) for critical damping of the G-wave by viscosity becomes

$$\frac{8\lambda^0}{|(\lambda^0)^2-1|} \sim \frac{\eta^{-3/2}}{\mathrm{Re}^{\mathrm{QG}}}. \quad (9.49)$$

For $1 < \lambda^0 < 10$, $|\check{n}'| = O(1)$, $|\check{m}'| = |\lambda^0 \check{n}'| = O(1)$, $\eta^{-3/2} \sim 10^3$ this requires $\mathrm{Re}^{\mathrm{QG}} \leq 10^3$ for critical damping of the wave front on large horizontal scales. This is not satisfied in the atmosphere. The frontogenesis is not impacted by viscosity on large horizontal scales, at large atmospheric QG Reynolds numbers. Of course formula (9.49) confirms the obvious fact that at small horizontal scales and large wave frequencies, viscous damping prevails. The exact nature of baroclinic wave turbulence must involve the derivation of an effective viscosity for the stochastic hyperbolic systems on the resonant rays, Eqs. (9.5), with time- and space- dependent coefficients $\tilde{\Psi}^0$. The in-depth study of these stochastic wave systems on resonant rays and their correlation with QG turbulence are the focus of our continuing investigations.

Our asymptotic theory presented in this section quantitatively describe anisotropy in AG wave turbulence with strong AG energy cascades along the front direction: this can be checked against experimental measurements provided that the latter distinguish between wave turbulence and the ambient potential vorticity turbulence. That the impact of rotation triggers mechanisms which allow an internal adjustment of horizontal scales has been already demonstrated in Rotunno ([67]) and NPH ([62]), in a linear theory. Internal radii of deformation determine the horizontal extent of motion and circulation; they are not resolved accurately by the usual numerical models with coarse gridding in the vertical direction. Current numerical models smear out sharp vertical gradients especially with *ad hoc* vertical eddy viscosities. Paradoxically, they should be benchmarked against our exact asymptotic dynamics to gauge for resolution of vertical stiffness. Any balanced/unbalanced model and/or any DNS must resolve the hierarchy of time scales: $1/N \ll t \ll \tau = \frac{2\sqrt{\eta}}{1+\eta} t$, $\eta \ll 1$ which rule the formation of slow AG energy cascades, as demonsrated in this section. With realistic potential vorticity configurations, the direct numerical simulations of our *non-stiff* nonlinear asymptotic limit equations for the AG wave fronts should be compared with actual experiments on frontal dynamics provided that the latter are unconstrained by boundaries and beyond small Reynolds numbers.

Acknowledgements

The authors wish to thank for their support the AFOSR (grant F49620–96–0–0165) and the ASU Center for Environmental Fluid Dynamics. The hospitality of the Newton Institute (Cambridge) under the special programme *Mathematics of Atmosphere and Ocean Dynamics* is gratefully acknowledged as well as the hospitality of the Ecoles Normales of Paris and Cachan. We would like to thank Professors V.I. Arnold, C.W. Bardos, P. Bartello, Y. Brenier, M.J.P. Cullen, M. Farge, C. Foias, F. Golse, J.C.R. Hunt and H.K. Moffatt for very useful discussions.

References

[1] V.I. Arnold (1965), Small denominators. I. Mappings of the circumference onto itself, *Amer. Math. Soc. Transl. Ser. 2*, **46**, 213–284.

[2] V.I. Arnold and B.A. Khesin (1997), *Topological Methods in Hydrodynamics, Applied Mathematical Sciences*, **125**, Springer.

[3] J. Avrin, A. Babin, A. Mahalov and B. Nicolaenko (1999), On regularity of solutions of 3D Navier–Stokes equations, *Applicable Analysis*, **71**, 197–214.

[4] A.V. Babin and M.I. Vishik (1992), *Attractors of Evolution Equations*, North-Holland.

[5] A. Babin, A. Mahalov, and B. Nicolaenko (1995), Long-time averaged Euler and Navier–Stokes equations for rotating fluids, In *Structure and Dynamics of Nonlinear Waves in Fluids*, 1994 IUTAM Conference, K. Kirchgässner and A. Mielke (eds), World Scientific, 145–157.

[6] A. Babin, A. Mahalov, and B. Nicolaenko (1996), Global splitting, integrability and regularity of 3D Euler and Navier–Stokes equations for uniformly rotating fluids, *Europ. J. of Mech., B/Fluids*, **15** (3), 291–300.

[7] A. Babin, A. Mahalov, and B. Nicolaenko (1996), Resonances and regularity for Boussinesq equations, *Russian J. Math. Phys.*, **4** (4), 417–428.

[8] A. Babin, A. Mahalov, and B. Nicolaenko (1997), Regularity and integrability of rotating shallow-water equations, *Proc. Acad. Sci. Paris*, **324**, Ser. 1, 593–598.

[9] A. Babin, A. Mahalov, and B. Nicolaenko (1997), Global regularity and integrability of 3D Euler and Navier–Stokes equations for uniformly rotating fluids, *Asymptotic Analysis*, **15**, 103–150.

[10] A. Babin, A. Mahalov, and B. Nicolaenko (1997), Global splitting and regularity of rotating shallow-water equations, *Eur. J. Mech., B/Fluids*, **16** (1), 725–754.

[11] A. Babin, A. Mahalov, and B. Nicolaenko (1998), On the nonlinear baroclinic waves and adjustment of pancake dynamics, *Theor. and Comp. Fluid Dynamics*, **11**, 215–235.

[12] A. Babin, A. Mahalov, B. Nicolaenko and Y. Zhou (1997), On the asymptotic regimes and the strongly stratified limit of rotating Boussinesq equations, *Theor. and Comp. Fluid Dyn.*, **9**, 223–251.

[13] A. Babin, A. Mahalov and B. Nicolaenko (1999), On the regularity of three-dimensional rotating Euler–Boussinesq equations, *Mathematical Models and Methods in Applied Sciences*, **9** (7), 1089–1121.

[14] A. Babin, A. Mahalov and B. Nicolaenko (1999), Global regularity of 3D rotating Navier–Stokes equations for resonant domains, *Indiana University Mathematics Journal*, **48**, 1133–1176.

[15] A. Babin, A. Mahalov and B. Nicolaenko (2000), Fast singular oscillating limits and global regularity for the 3D primitive equations of geophysics, *Mathematical Modelling and Numerical Analysis*, **34** (2), 201–222.

[16] Bartello (1995), Geostrophic adjustment and inverse cascades in rotating stratified turbulence, *J. Atm. Sci.*, **52** (24), 4410–4428.

[17] J.-M. Bony (1981), Calcul symbolique et propagation des singularités pour les équations aux dérivées partielles non-linéaires, *Ann. Sci. Ecole Norm. Sup.*, **14**, 209–246.

[18] A.J. Bourgeois and J.T. Beale (1994), Validity of the quasigeostrophic model for large-scale flow in the atmosphere and the ocean, *SIAM J. Math. Anal.*, **25** (4), 1023–1068.

[19] L. Caffarelli, R. Kohn and L. Nirenberg (1982), Partial regularity of suitable weak solutions of the Navier–Stokes equations, *Comm. Pure Appl. Math.*, **35**, 771–831.

[20] J.G. Charney (1948), On the scale of atmospheric motions, *Geofys. Publ. Oslo*, **17** (2), 1–17.

[21] J.-Y. Chemin (1995), A propos d'un probleme de pénalisation de type antisymétrique, *Proc. Paris Acad. Sci.*, **321**, 861–864.

[22] Constantin (1997), The Littlewood-Paley spectrum in two-dimensional turbulence, *Theor. and Comp. Fluid Dyn.*, **9** (3/4), 183–191.

[23] Constantin and C. Foias (1988), *Navier–Stokes Equations*, The University of Chicago Press.

[24] A. Craya (1958), Contribution à l'analyse de la turbulence associée à des vitesses moyennes, *P.S.T. Ministere de l'Air* (Paris), **345**.

[25] M.J.P. Cullen (2000), New mathematical developments in atmosphere and ocean dynamics and their application to computer simulations; this volume.

[26] P.G. Drazin and W.H. Reid (1981), *Hydrodynamic Stability*, Cambridge University Press.

[27] D.G. Dritschel and de la Torre Juarez (1996), The instability and breakdown of tall columnar vortices in a quasi-geostrophic fluid, *J. Fluid Mech.*, **328**, 129–160.

[28] P.F. Embid and A.J. Majda (1996), Averaging over fast gravity waves for geophysical flows with arbitrary potential vorticity, *Comm. Partial Diff. Eqs.*, **21**, 619–658.

[29] M. Farge and R. Sadourny (1989), Wave-vortex dynamics in rotating shallow layer, *J. Fluid Mech.*, **206**, 433–462.

[30] H.J.S. Fernando and J.C.R. Hunt (1996), Some aspects of turbulence and mixing in stably-stratified layers, *Dyn. of Atm. and Oceans*, **23**, 35.

[31] I. Gallagher (1997), Un résultat de stabilité pour les équations des fluides tournants, *C.R. Acad. Sci. Paris*, Série I, 183–186.

[32] I. Gallagher (1998), Asymptotics of the solutions of hyperbolic equations with a skew-symmetric perturbation, *J. Diff. Eq.*, **150**, 363–384.

[33] I. Gallagher (1998), Applications of Schochet's methods to parabolic equations, *J. Math. Pures Appl.* **77**, 989–1054.

[34] P.R. Gent and J.C. McWilliams (1983), Consistent balanced models in bounded and periodic domains, *Dyn. Atm. and Oceans*, **7**, 67–93.

[35] P.R. Gent and J.C. McWilliams (1983), Regimes of validity for balanced models, *Dyn. Atm. and Oceans*, **7**, 167–183.

[36] A. E. Gill (1982), *Atmosphere-Ocean Dynamics*, Academic Press.

[37] F.S. Godeferd and C. Cambon (1994), Detailed investigation of energy transfer in homogeneous stratified turbulence, *Phys. of Fluids*, **6** (6), 2084–2100.

[38] E. Grenier, Rotating fluids and inertial waves, *Proc. Acad Sci. Paris*, t. 321, ser. 1, (1995), 711–714.

[39] J.R. Herring and O. Métais (1989), Numerical experiments in forced stably-stratified turbulence, *J. Fluid Mech.*, **202**, 97.

[40] J.R. Holton (1992), *An Introduction to Dynamic Meteorology*, Academic Press.

[41] B.J. Hoskins and F.P. Bretherton (1972), Atmospheric frontogenesis models: Mathematical formulation and solution, *J. Atm. Sci.*, **29**, 11–37.

[42] B.J. Hoskins (1982), The mathematical theory of frontogenesis, *Annu. Rev. of Fluid Mech.*, **14**, 131–151.

[43] J.L. Joly, G. Métivier and J. Rauch (1993), Generic rigorous asymptotic expansions for weakly nonlinear multidimensional oscillatory waves, *Duke Math. J.* **70**, 373–404. See more references in bibliography of BMN [9].

[44] D.A. Jones, A. Mahalov and B. Nicolaenko (1998), A numerical study of an operator splitting method for rotating flows with large ageostrophic initial data, *Theor. and Comp. Fluid Dyn.*, **13** (2), 143–159.

[45] D. Keyser and M.A. Shapiro (1986), A review of the structure and dynamics of upper-level frontal zones, *Monthly Weather Review*, **114**, 452–499.

[46] O.A. Ladyzhenskaya (1969), *The Mathematical Theory of Viscous Incompressible Flow*, 2nd ed., Gordon and Breach.

[47] M.-P. Lelong and J. Riley (1991), Internal wave-vortical mode interactions in strongly stratified flows, *J. Fluid Mech.*, **232**, 1–19.

[48] M. Lesieur (1987), *Turbulence in Fluids*, Martinus Nijhoff Publishers.

[49] D.K. Lilly (1983), Stratified turbulence and the mesoscale variability of the atmosphere, *J. Atm. Sc.*, **40**, 749.

[50] J.L. Lions, R. Temam and S. Wang, Geostrophic asymptotics of the primitive equations of the atmosphere, *Topological Methods in Nonlinear Analysis*, **4**, (1994), 253–287, special issue dedicated to J. Leray.

[51] L.R.M. Maas and J.J.M. van Haren (1987), Observations on the vertical structure of tidal and inertial currents in central North Sea, *J. Mar. Res.*, **45**, 293–318.

[52] A. Mahalov, S. Leibovich and E.S. Titi (1990), Invariant helical subspaces for the Navier–Stokes Equations, *Arch. for Rational Mech. and Anal.*, **112** (3), 193–222.

[53] A. Mahalov and P.S. Marcus (1995), Long-time averaged rotating shallow-water equations, *Proc. of the First Asian Computational Fluid Dynamics Conference*, eds. W.H. Hui, Y.-K. Kwok and J.R. Chasnov, vol. 3, 1227–1230, Hong Kong University of Science and Technology.

[54] A. Mahalov and Y. Zhou (1996), Analytical and phenomenological studies of rotating turbulence, *Phys. of Fluids*, **8** (8), 2138–2152.

[55] A.J. Majda and M.J. Grote (1997) Model dynamics and vertical collapse in decaying strongly stratified flows, *Phys. Fluids*, **9** (10), 2932–2940.

[56] J.C. McWilliams (1985), A note on a uniformly valid model spanning the regimes of geostrophic and isotropic stratified turbulence: balanced turbulence, *J. Atm. Sci.*, **42**, 1773–1774.

[57] J.C. McWilliams and P.R. Gent (1980), Intermediate models of planetary circulations in the atmosphere and ocean, *J. Atm. Sci.*, **37**, 1657–1678.

[58] J.C. McWilliams, J.B. Weiss and I. Yavneh (1994), Anisotropy and coherent vortex structures in planetary turbulence, *Science*, **264**, 410–413.

[59] O. Métais and J.R. Herring (1989), Numerical experiments of freely evolving turbulence in stably-stratified fluids, *J. Fluid Mech.*, **202**, 117.

[60] O. Métais, Bartello, E. Garnier, J.J. Riley and M. Lesieur (1996), Inverse cascades in stably-stratified rotating turbulence, *Dyn. of Atm. and Oceans*, **23**, 193–203.

[61] Müller, G. Holloway, F. Henyey and N. Pomphrey (1986), Nonlinear interactions among internal gravity waves, *Reviews of Geophysics*, **24** (3), 493–536.

[62] T.M.J. Newley, H.J. Pearson and J.C.R. Hunt (1991), stably-stratified rotating flow through a group of obstacles, *Geophys. Astrophys. Fluid Dyn.*, **58**, 147–171.

[63] J. Pedlosky, *Geophysical Fluid Dynamics*, 2nd edition, Springer-Verlag, (1987).

[64] H. Poincaré (1910), Sur la précession des corps déformables, *Bull. Astronomique*, **27**, 321.

[65] G. Raugel and G. Sell (1993), Navier–Stokes equations on thin 3D domains. I. Global attractors and global regularity of solutions, *J. Amer. Math. Soc.*, **6** (3), 503–568.

[66] J.J. Riley, R.W. Metcalfe and M.A. Weisman (1981), Direct numerical simulations of homogeneous turbulence in density-stratified fluids, *Proc. AIP Conf. Nonl. Prop. of Internal Waves*, B.J. West (ed.), 79.

[67] R. Rotunno (1983), On the linear theory of the land and sea breeze, *J. of the Atmospheric Sciences*, **40**, 1999–2009.

[68] R. Sadourny (1975), The dynamics of finite-difference models of the shallow-water equations, *J. Atm. Sci.*, **3**, 680–689.

[69] S. Schochet (1994), Fast singular limits of hyperbolic PDEs, *J. Diff. Eq.*, **114**, 476–512.

[70] J. Sidi and J. Barat (1986), Observational evidence of an inertial wind structure in the stratosphere, *J. Geophys. Res.*, **91**, 1209–1217.

[71] E.M. Stein (1970), *Singular Integrals and Differentiability Properties of Functions*, Princeton University Press.

[72] S. L. Sobolev (1954), Ob odnoi novoi zadache matematicheskoi fiziki, *Izvestiia Akademii Nauk SSSR, Ser. Matematicheskaia*, **18** (1), 3–50.

[73] R. Temam (1983), *Navier–Stokes Equations and Nonlinear Functional Analysis*, SIAM, Philadelphia.

[74] T. Warn (1986), Statistical mechanical equilibria of the shallow-water equations, *Tellus*, **38A**, 1–11.

References added in Proof

[75] A. Babin, A. Mahalov and B. Nicolaenko (2001), 3D Navier–Stokes and Euler equations with initial data characterized by uniformly large vorticity, *Indiana University Mathematics Journal*, **50**, 1–35.

[76] A. Babin, A. Mahalov and B. Nicolaenko (2001), Strongly stratified limit of 3D primitive equations in an infinite layer, in *Advances in Wave Interaction and Turbulence, Contemporary Mathematics*, **283**, 1–11.

New Mathematical Developments in Atmosphere and Ocean Dynamics, and their Application to Computer Simulations

M.J.P. Cullen

1 Introduction

This chapter reviews new developments in the mathematical theory of the partial differential equations which govern the large scale behaviour of the atmosphere and ocean. It then discusses how these can be applied to the models which are used to predict weather and climate, and to define the initial state of the atmosphere or ocean from limited observations. Particular topics include: (i) the use of Lagrangian conservation properties in establishing the existence of solutions to, and properties of, simpler equations which describe the atmosphere and ocean in the limits of rapid rotation and/or strong stratification; (ii) demonstrations that the complete equations can be proved to have solutions close to those of the simpler equations in these asymptotic limits; and (iii) how the statistical effect of small scale motions interacts with the large scale solutions. It is demonstrated how these techniques can be exploited in the design of computer models suitable for operational applications, including numerical methods for representing the resolved dynamics and the coupling of sub-grid models to the resolved dynamics. The latter is particularly important where moisture has a strong influence on the dynamics. These techniques can also be used to exploit observed data more fully by improving the design of data assimilation systems. In particular, they can also be used to refine estimates of error growth. These are important in data assimilation, where model errors have to be allowed for and corrected, and in predictability studies.

Weather forecasting and climate modelling have advanced tremendously since the pioneering studies of Richardson (1922) and Charney, Fjortoft and von Neumann (1950) first proposed that it was possible to simulate and predict the behaviour of the ocean and the atmosphere by direct computer solution of the classical equations of dynamics and thermodynamics. Since then, computers have been developed which can run operational forecast programmes at 10^{11} floating point operations per second, and much faster in research mode, enabling real time forecasts of the global atmosphere to be made in a few minutes per simulated day on a grid of points about 40 km apart in the horizontal and 500 m apart in the vertical. While in the early days of atmosphere and ocean modelling, lack of computer power was clearly the main limiting factor in the quality of the simulations, it is now becoming clear that greater scientific knowledge of the phenomena being simulated should also be exploited in the design of the computer models. Similarly, it is necessary to exploit to the maximum the information contained in routine operational observations of the atmosphere and ocean, since the cost of such observations is increasing and the maintenance of even the current level of observations is difficult.

It was therefore timely to hold a six month study programme at the Isaac Newton Institute for Mathematical Sciences in Cambridge to review and discuss the knowledge that mathematicians could bring to bear on meteorological and oceanographic problems, and to advertise to mathematicians the wide scope for quite rigorous studies of atmosphere and ocean dynamics. This chapter summarises some of the material presented, discussed, and developed both during and following the programme.

2 Flow regimes in the atmosphere and ocean

We start from the basic equations for the dynamics and thermodynamics of an ideal fluid. We write them in the form used by Shutts and Cullen (1987). These are fully compressible, and allow the axis of rotation not to be parallel to the gravitational force. We use Cartesian coordinates to simplify the presentation, but the inclusion of the general direction of the axis of rotation allows immediate application of the results in spherical geometry. The equations are as follows:

$$\begin{aligned} \frac{D\mathbf{v}}{Dt} + 2\Omega \times \mathbf{v} + \nabla\Phi + \alpha\nabla p &= \nu\nabla^2\mathbf{v} \\ \frac{D\alpha}{Dt} &= \alpha\nabla.\mathbf{v} \\ C_v\frac{DT}{Dt} + p\frac{D\alpha}{Dt} &= \kappa\nabla^2 T \\ p\alpha &= RT \end{aligned} \tag{2.1}$$

where $\mathbf{v}$ is the wind velocity vector, α the specific volume (or inverse density), p the pressure, C_v the specific heat at constant volume and T the temperature; $\mathbf{\Omega}$ is the rotation vector, and $\nabla\Phi$ a specified geopotential force, representing gravitational and centrifugal forces; ν, κ represent molecular viscosity and thermal conductivity. The temperature equation can be replaced by the equation

$$\frac{D\theta}{Dt} = \kappa\nabla^2 T \tag{2.2}$$

where θ is the 'potential temperature', which is a function of the entropy and proportional to $p\alpha^\gamma$, where γ is the ratio of specific heats. The effect of moisture would be included as a source term in equation (2.2), together with a small modification to the equation of state and extra equations for the conservation of water substance and the conversion of water between vapour and liquid or solid states.

In the atmospheric context, it is most useful to consider periodic boundary conditions in x, y and a lower boundary in z, with no normal flow across the lower boundary. At the upper boundary we have $p \to 0$ as $z \to \infty$.

The equivalent equations for the ocean are similar, but it is usually sufficient to use incompressible equations. The effect of salinity on the buoyancy has to be included.

$$\begin{aligned} \frac{D\mathbf{v}}{Dt} + 2\Omega \times \mathbf{v} + \nabla\Phi + \alpha\nabla p &= \nu\nabla^2\mathbf{v} \\ \nabla.\mathbf{v} &= 0 \\ \frac{DT}{Dt} &= \kappa\nabla^2 T \\ \frac{DS}{Dt} &= \sigma\nabla^2 S \\ \alpha &= \alpha(S,T). \end{aligned} \tag{2.3}$$

Here, S represents the salinity and σ its diffusivity. The appropriate boundary conditions are to have a basin Γ with depth $h(x, y)$ and rigid boundaries. There is no normal flow on all boundaries except the upper boundary which is a free surface $z = \eta(x, y) \simeq 0$ on which the pressure is a constant small value (representing the atmospheric pressure).

Energy conservation for the inviscid ($\nu = \kappa = 0$) form of (2.1) is expressed as

$$E = \int_\Gamma \left\{ \frac{1}{2}|\mathbf{v}|^2 + \Phi + C_v T \right\} \alpha^{-1} d\tau = \text{a constant} \tag{2.4}$$

where Γ indicates the volume of integration. For the inviscid form of (2.3) we have

$$E = \int_\Gamma \left\{ \frac{1}{2}|\mathbf{v}|^2 + \Phi \right\} \alpha^{-1} d\tau = \text{a constant}. \tag{2.5}$$

These equations describe the complete motion of the atmosphere and ocean. When used for weather and climate forecasting, they have to be averaged in space and time to make computer solution practical. The averaging scale needs to at least twice the smallest affordable grid size in order to prevent excessive numerical errors. In the atmosphere, the length scale equivalent to unit Reynolds number for typical flow speeds is about 10^{-6} m, and in the ocean 10^{-5} m. Accurate direct solutions require this scale to be resolved. High resolution global atmospheric models use an averaging scale of at least 50 km in the horizontal and about 1 km in the vertical, less near the Earth's surface. Models used for limited area forecasting have much shorter averaging scales, down to around 1 km. In the ocean, the highest resolution global models currently

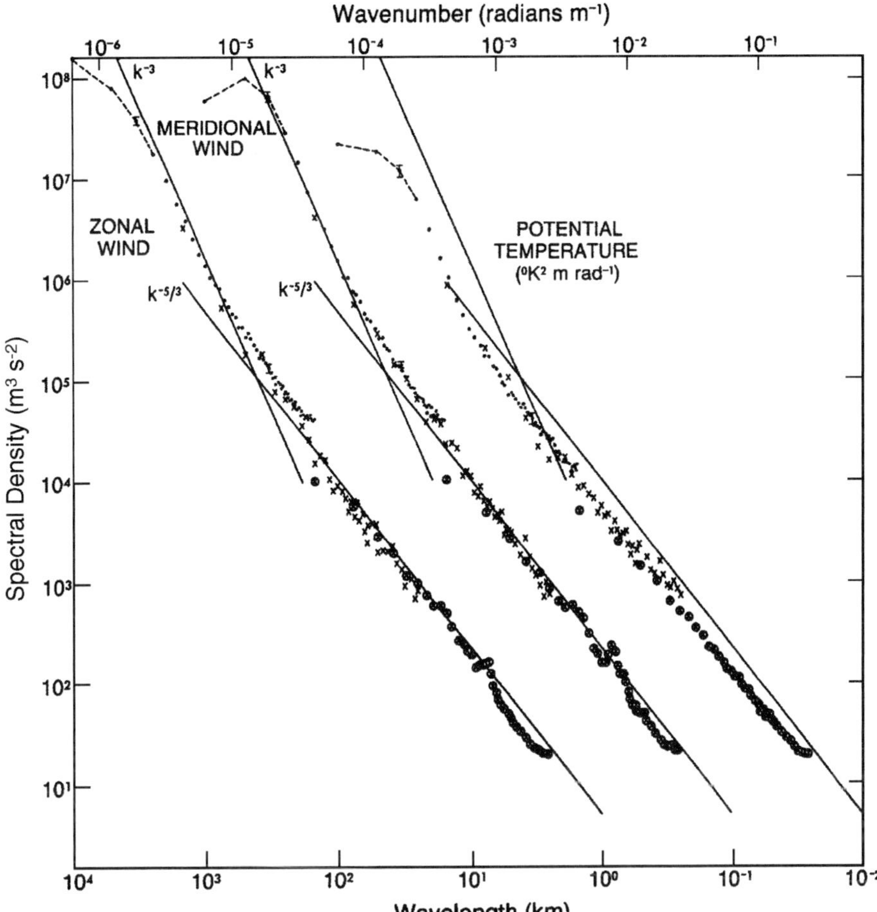

Figure 1: Summary components of wavenumber spectra of potential temperature and zonal and meridional velocity, composited from three groups of flight segments of different lengths. The 3 types of symbol show results from each group. The straight lines indicate slopes of -3 and $-5/3$. The meridional wind spectra are shifted one decade to the right and the potential temperature spectra are shifted two decades to the right (after Gage and Nastrom, 1985).

used have an averaging scale of about 25 km in the horizontal. The resolutions used in practice are thus short by a factor of 10^9–10^{11} of those that would be required for accurate solutions. Success in practice depends on being able to define the effect of unresolved motions in terms of the resolved motions, i.e. correct sub-grid modelling (see Pielke (1984) for a review applied to weather forecasting models). This would be much easier if there was a 'spectral gap', i.e. a range of scales where the energy in atmospheric motions was much less than in larger or smaller scales. Observational evidence (Figure 1) shows a

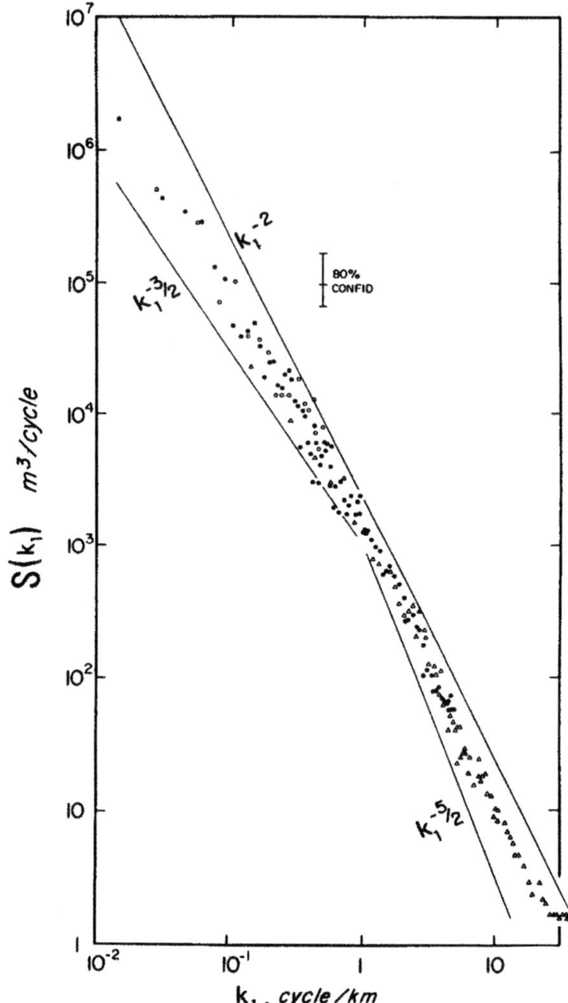

Figure 2: Vertical displacement spectrum $S_\xi(k_1)$ about the 12° isotherm in the MODE area. Power law slopes of the from k_1^{-q} are also shown for comparison (after Katz (1975), © American Geophysical Union).

gradual change in spectral slope from k^{-3} at large scales to $k^{-5/3}$ at small scales, but no gaps. There will however be a minimum in the enstrophy (mean squared vorticity) spectrum associated with this change in slope. In the ocean there is similarly no sign of a spectral gap between wavelengths of 0.1 and 100 km (Figure 2).

The effect of this averaging for the atmosphere is illustrated in Figure 3. This shows the typical space (l) and time (T) scales associated with different atmospheric phenomena. The line given by $T = l^{2/3}$ associated with the energy spectrum $k^{-5/3}$ is plotted along the diagonal. The phenomena occurring

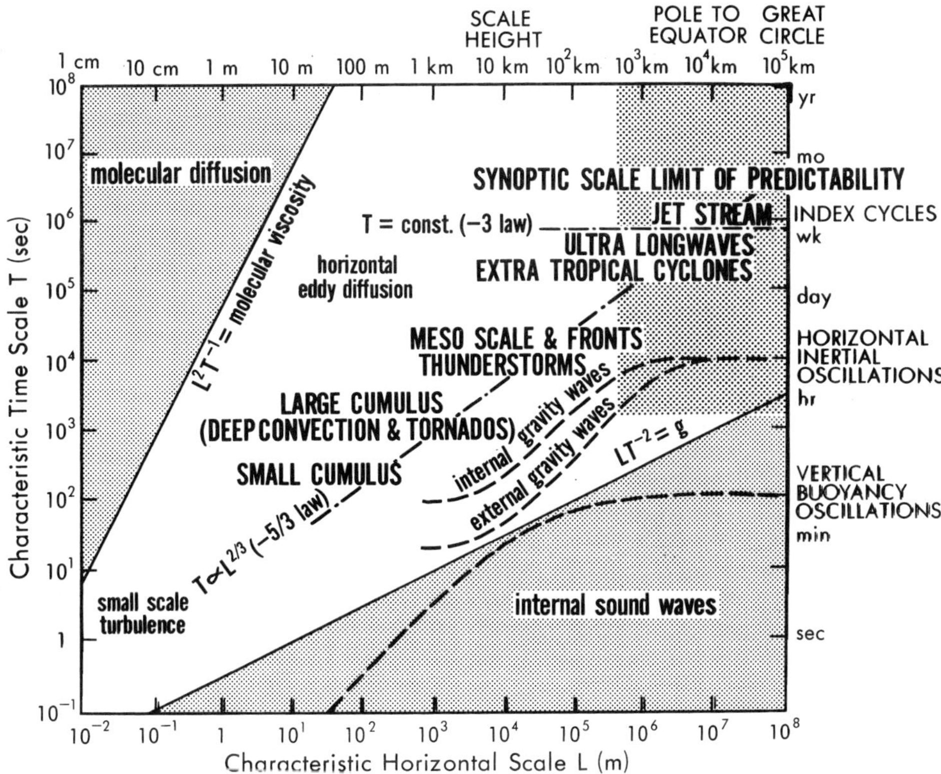

Figure 3: Typical space and time scales of atmospheric phenomena (following Smagorinsky 1974).

along this diagonal are those traditionally associated with 'weather'. Other motions such as internal gravity waves are elsewhere on the diagram. Figure 3 illustrates that there is a difficulty associated with simply averaging the equations in space and time, in that any choice of averaging scale will cut across the peak scales of some phenomena, resulting in them being partly resolved and thus inaccurately represented. If the averaging scale is reduced, then these phenomena may be well predicted, but a new set will become partly resolved. We can also see this from atmospheric observations. Figure 5 shows two satellite pictures for a similar time. The first covers the North Atlantic and Western Europe. It shows large scale organised phenomena, such as the waving cloud bands extending south-westwards into the Atlantic from Ireland, and regions of highly structured small scale flows, such as south of Iceland. If we zoom in on the United Kingdom, new phenomena become visible, such as the small scale wave train running south-west to north-east across the centre of the British Isles.

The time-scales of typical ocean surface phenomena are illustrated in Figure 4. There is an equally complex mix of phenomena affecting the internal

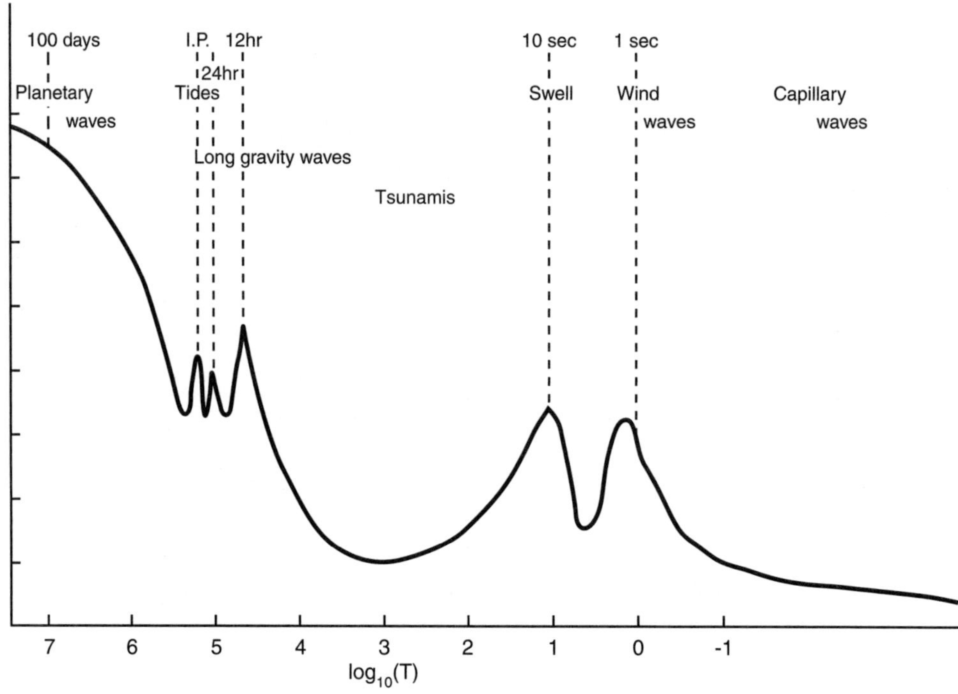

Figure 4: Schematic energy spectrum of ocean surface variability, showing the different types of wave. I.P. denotes the inertial period, defined as $\pi/(2\Omega|\sin\phi|)$, where Ω is the magnitude of the Earth's rotation vector and ϕ is the latitude. The diagram illustrates latitude $\pm 20°$ where I.P. $=$ 35 hours. After LeBlond and Mysak (1978).

ocean structure. Though there is a spectral gap for periods of around 1 hour in surface phenomena, Figure 2 shows that there is no evidence for any such gap in the time-scales of the ocean interior.

There is thus a need to choose averaging scales which make sense in terms of phenomena to be simulated. Having done this, it is necessary to ensure that the equations, with the sub-grid model, have solutions which respect this scale. This will usually mean that they stay smooth on the averaging scale in both space and time. Mathematical results are therefore required to prove that this is so for a given sub-grid formulation. Practical success in this approach requires a clear separation in terms of space and time scale between phenomena that will be simulated and those that will be excluded. The numerical solution should use a resolution significantly finer than the averaging scale, to ensure that the model is integrated accurately.

An alternative approach, which avoids the problem of scale separation to some extent, is to choose a 'sub grid model' which restricts the solutions to specific phenomena, though not requiring smoothness on specific scales. This allows treatment of phenomena such as fronts, which have a small space scale

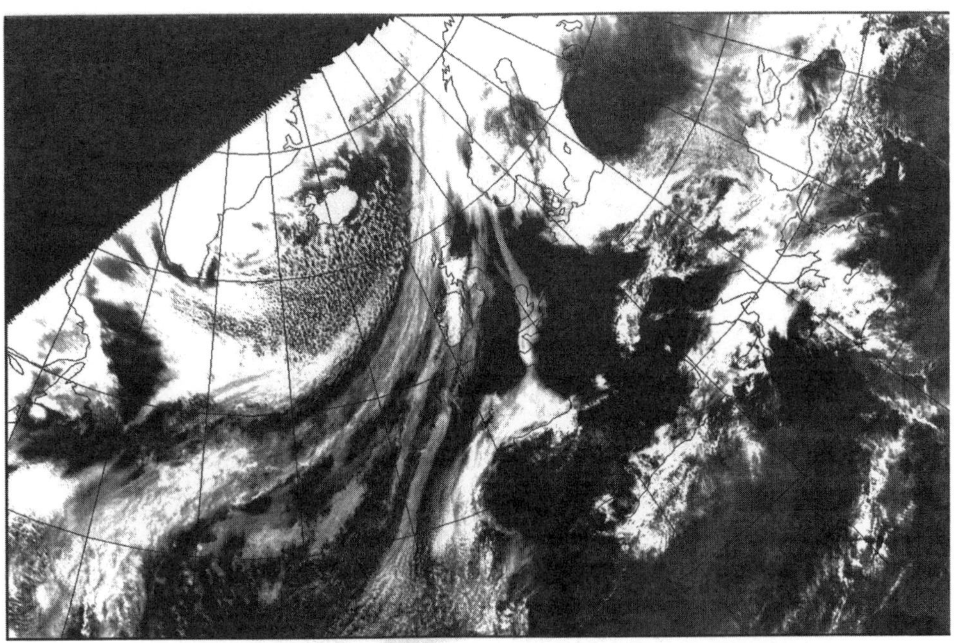

Figure 5: (a) METEOSAT visible picture of the North Atlantic and Europe for 1500 UTC on 13 March 1995. (b) High resolution visible picture of Western Europe for 1315 UTC on 13 March 1995.

in one direction, but have a large scale in other directions. An illustration of this is given in Figure 6, which shows a time series of wind speed. There is a lot of small time-scale variability, but a large jump in mean values at 1500EDST which is almost as rapid as the oscillations, but represents a large scale 'coherent' change. Modelling the rapid, but coherent, changes while excluding the oscillations can be achieved by 'reduced' systems of equations, which have a simpler set of solutions than the full equations, only describing the phenomena desired. Success in this type of modelling depends on showing that the interaction between resolved and unresolved phenomena is weak. This can be achieved by the geometry of the flow, as well as by scale separation.

One effect of the latter approach is that it is necessary to exclude atmospheric or oceanic states which in reality would be unstable to motions not permitted by the reduced equations. These are likely to include, for instance, states which are statically unstable. In the horizontal, inertial stability is often required. Knox (1997) reviews this issue for a wide set of 'reduced' equations. We can therefore assess the usefulness of the reduced equations by studying

Figure 5: Continued.

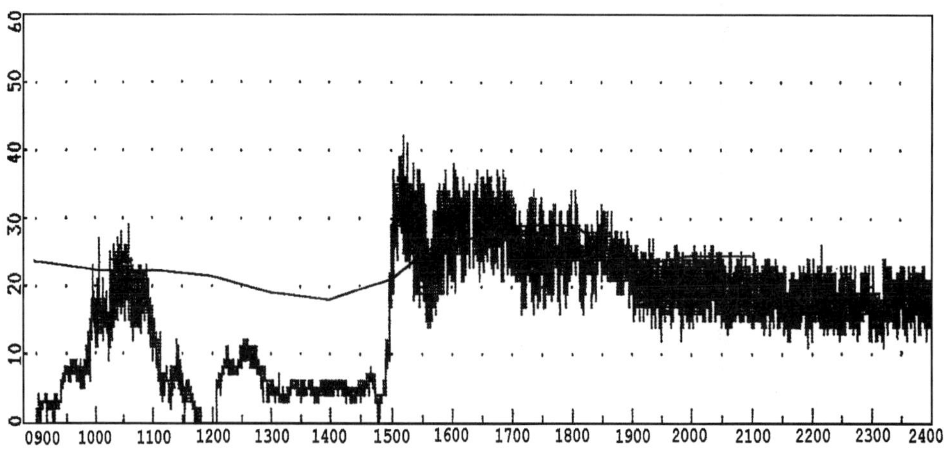

Figure 6: Anemograph for Bellambi point (Australia) on 26 December 1996 (wind speed in knots). (The solid smooth line is a model forecast.) After Batt and Leslie (1998).

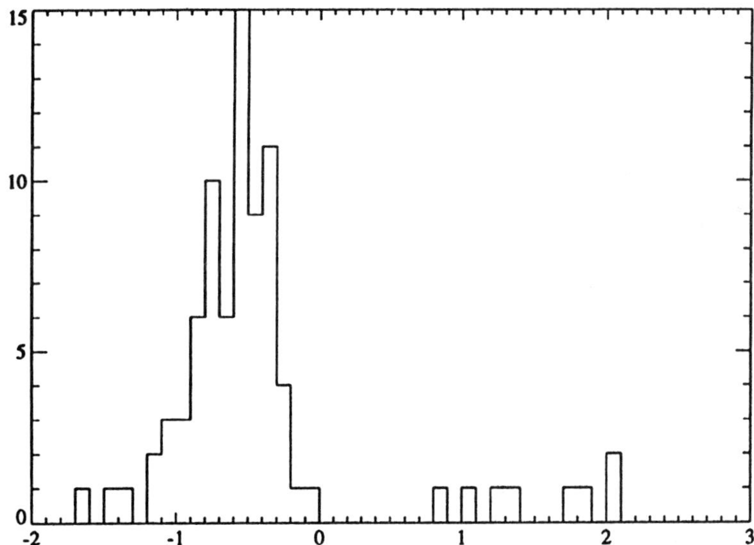

Figure 7: Frequency histogram of inertial stability deduced from automated aircraft reports between 30N and 30S using dates between June 1992 and June 1993. The parameter β plotted is defined as $1 + f^{-1}\zeta$, where ζ is the relative vorticity. Inertial stability requires $\beta > -1$. Only reliable estimates are plotted; sometimes there is more than one for the same flight.

how widespread such unstable states are in atmospheric or oceanic observations. An example from such a study is shown in Figure 7, due to Veitch and Mawson (1993), showing automated aircraft wind data from cases between 30N and 30S where the aircraft track crossed regions of straight flow at close to a right angle. The observations are typically 80 km apart. The inertial stability condition is most likely to be violated in low latitudes, because the time-scale of inertial instability is $(2\Omega \sin \phi)^{-1}$, where ϕ is the latitude. This is of the order of 1 day at 5 degrees from the equator. It was found that in only 5 out of 121 cases was the inertial stability condition violated, suggesting that those reduced systems of equations which require inertial stability may be useful at this horizontal scale even in low latitudes.

2.1 Definition of asymptotic regimes

We define the most important asymptotic regimes in the atmosphere and ocean that we will be considering in terms of the following parameters.

$$\begin{aligned} Ro &= U/fL \text{ Rossby number} \\ Fr &= U/NH \text{ Froude number} \\ B &= Ro/Fr \text{ Burger number} \end{aligned} \tag{2.6}$$

$$\begin{aligned} Ma &= U/c \text{ Mach number} \\ A &= H/L \text{ aspect ratio} \end{aligned}$$

where L, H are horizontal and vertical length scales, U is a horizontal velocity scale, f is the Coriolis parameter, N the Brunt-Väisälla frequency, and c the speed of sound. For weather systems in the atmosphere, typical values are $L = 10$–10^4 km, $H = 1$–10 km, $U = 1$–$100\,\mathrm{m\,s^{-1}}$, $f = 0$–$10^{-4}\,\mathrm{s^{-1}}$, $g = 10\,\mathrm{m\,s^{-2}}$, $N \geq 10^{-1}\,\mathrm{s^{-1}}$, and $c = 300\,\mathrm{m\,s^{-1}}$. In the ocean $U = 0.1$–$1\,\mathrm{m\,s^{-1}}$. We will use T for the time scale. A systematic scale analysis of the governing equations in terms of these parameters is set out, for instance, in Holton (1992). The approximations made in the various regimes were discussed in more detail in the chapter by White earlier in this volume. We summarise only the main points here.

The asymptotic regimes relevant to weather systems and large scale ocean circulations are characterised by long time scales, and thus balances between the forces occurring in (2.1) and (2.3). As discussed by Bartello and Thomas (1996), for instance, there is an issue as to whether the Eulerian or Lagrangian time scale is greater. In the absence of geographically fixed forcing, such as mountains, their arguments suggest that the Lagrangian time scale is greater, especially in regimes where the k^{-3} spectrum holds. If this is so, we have $T = L/U$, where L is treated as a length scale along trajectories. Other length scales may be shorter.

The horizontal extent of the atmosphere and ocean are much greater than the vertical extent, allowing the existence of flow regimes with very small aspect ratio. To allow these to be identified, we write (2.1) in terms of horizontal and vertical components of momentum, where the direction $\nabla\Phi$ is defined as vertical. Let W be a vertical velocity scale, then $T = H/W$. In the atmospheric case, the vertical variations of the thermodynamic quantities p, θ, α are large. Assume these quantities have background scales p_0, θ_0, α_0, related by the equation of state, and scales of horizontal spatial variations p', θ', α'. In the vertical momentum equation, we expect the vertical particle acceleration Dw/Dt to be much less than g. It is therefore appropriate to subtract a state of rest, depending only on the vertical, from the equations, as discussed in the chapter by White earlier in this volume. This is given by values $\overline{p}, \overline{\alpha}, \overline{\theta}$ satisfying

$$\begin{aligned} \overline{p\alpha^\gamma} &= R_0 \overline{\theta}^\gamma \\ \frac{\partial \Phi}{\partial z} + \overline{\alpha}\frac{\partial \overline{p}}{\partial z} &= 0 \end{aligned} \tag{2.7}$$

where the first equation is the equation of state rewritten in terms of potential temperature with R_0 a modified gas constant.

Therefore, the vertical pressure gradient term has to be of a size comparable to g, so $\frac{\partial p}{\partial z}$ is of order $g\alpha_0^{-1}$. The equation of state then allows us to estimate

$\frac{\partial \alpha}{\partial z}$ and $\frac{\partial \theta}{\partial z}$. Observed large scale atmospheric flows have large variability in $\frac{\partial \theta}{\partial z}$, including regions where it is zero. It is thus appropriate to estimate $\frac{\partial \alpha}{\partial z}$ to balance $\frac{\partial p}{\partial z}$; with $\frac{\partial \theta}{\partial z}$ assumed smaller in magnitude than either. This gives $\frac{\partial \alpha}{\partial z} \simeq g\alpha_0^{\gamma-1}/(\gamma p_0) = g\alpha_0^{\gamma}/c^2$. Similarly, linearising the first equation of (2.7) with θ held fixed gives $p' = c^2\alpha'$.

Subtracting this state from equations (2.1) and (2.2), using primes to denote perturbations from the rest state gives:

$$\begin{array}{rl}
\dfrac{D\mathbf{v}_h}{Dt} & +2(\Omega \times (\mathbf{v}_h, w))_h + \alpha \nabla_h p' = \nu \nabla^2 \mathbf{v}_h \\
UT^{-1} & (fU, fW) \quad c^2\alpha_0\alpha' L^{-1} \\
\dfrac{Dw}{Dt} & +2(\Omega \times (\mathbf{u}, w))_z + \dfrac{\alpha'}{\bar{\alpha}}\dfrac{\partial \Phi}{\partial z} + \alpha \dfrac{\partial p'}{\partial z} = \nu \nabla^2 w \\
WT^{-1} & (fU, fW) \quad g\dfrac{\alpha'}{\alpha} \quad c^2\alpha_0\alpha' H^{-1} \\
\dfrac{\partial \alpha}{\partial t} & +(\mathbf{v}.\nabla\alpha)_h + w\dfrac{\partial \alpha}{\partial z} = \alpha \nabla.\mathbf{v} \\
\alpha' T^{-1} & \alpha_0 U L^{-1} \quad wg\alpha_0^{-\gamma}/c^2 \\
\dfrac{\partial \theta}{\partial t} & +(\mathbf{v}.\nabla\theta)_h + w\dfrac{\partial \theta}{\partial z} = \kappa \nabla^2 T \\
\theta' T^{-1} & \alpha_0 U L^{-1} \quad wN^2\theta_0/g \\
p\alpha^\gamma - \bar{p}\bar{\alpha}^\gamma & = R_0\theta'.
\end{array} \tag{2.8}$$

Subscript h indicates horizontal components of vectors, subscript z the vertical component.

We first identify a scaling associated with low Mach number. If we assume that $U = L/T$ and that the horizontal acceleration is of similar size to the horizontal pressure gradient, taking the divergence of the momentum equation gives $\nabla.\mathbf{v} \sim c^2\alpha_0\alpha' L^{-2}T$, so that the ratio of $\frac{\partial \alpha}{\partial t}$ to $\alpha\nabla.\mathbf{v}$ is $L^2T^{-2}c^{-2} = U^2/c^2 = Ma^2$. The continuity equation thus requires the local divergence to be small. In the atmosphere, the appropriate approximate continuity equation and boundary conditions are given by the 'anelastic' approximation

$$\begin{aligned}
\nabla.\alpha^{-1}\mathbf{v} &= 0 \\
\alpha^{-1} w &= 0 \text{ at } z = 0, \infty.
\end{aligned} \tag{2.9}$$

Similar equations apply in the ocean.

Define the Brunt-Väisälla frequency N by

$$N^2 = \frac{g}{\theta}\frac{\partial \theta}{\partial z}. \tag{2.10}$$

We can use (2.7) and a number of manipulations to give

$$\frac{\alpha'}{\bar{\alpha}}\frac{\partial \Phi}{\partial z} + \alpha\frac{\partial p'}{\partial z} = \frac{\theta'}{\bar{\theta}}\frac{\partial \Phi}{\partial z} + \alpha\frac{\partial p'}{\partial z}. \tag{2.11}$$

Use the thermodynamic equation to estimate

$$w \sim \frac{g\theta' T^{-1}}{N^2\theta_0}.$$

If the time-scale $T \gg N^2$, and $f^{-1} \gg N^2$, then the dominant terms in the vertical momentum equation are

$$\frac{\alpha'}{\alpha}\frac{\partial \Phi}{\partial z} + \alpha\frac{\partial p'}{\partial z} = 0, \qquad (2.12)$$

representing hydrostatic balance. Using (2.9), we can set $U/L = W/H$, so that $W = UA$.

We now consider regimes where, additionally, either the Rossby number or the Froude number is small. Small Rossby numbers are associated with rapid rotation, and small Froude numbers with strong stratification. We classify these regimes according to the Burger number, (2.6), as illustrated in Figure 8. We treat the analysis of the flow in each regime separately. It is also useful to consider the horizontal scale as a function of Burger number with H, f and N^2 fixed, so that

$$L = (NH)/(fB). \qquad (2.13)$$

If there is rotation, but no stratification, then $B = 0$. In the atmospheric case, this corresponds to uniform potential temperature θ_0. The assumption of small Ro gives geostrophic balance, which using the hydrostatic relation (2.12) as well can be written:

$$2\Omega \times \mathbf{v} + \nabla\Phi + \alpha\nabla p = 0. \qquad (2.14)$$

The condition $\theta = \theta_0$ means that α is a function of p, so that $\alpha\nabla p$ can be written as a gradient. Using (2.9) and (2.14), we derive the generalized Taylor–Proudman theorem that $\mathbf{v}$ does not vary in the direction of Ω. Thus the flow becomes 2-dimensional, with no component along the axis of rotation. In the atmosphere, this regime only occurs in limited regions, though it can occur in neutrally stratified layers (for instance observations of Karman vortex wakes behind islands or, probably, the vortex downstream of Greenland in Fig 5(a)). A similar regime occurs in tropical cyclones, though here it is due to rapid system rotation.

If the rotation is much stronger than the stratification, then $B \ll 1$ and the horizontal length scale is large compared with the Rossby radius NH/f. This is the appropriate regime for the large scale planetary waves in the extra-tropical atmosphere. However, the geometry of the atmosphere as a thin layer on a spherical surface requires that such motions are horizontal, and not normal to the axis of rotation. In this situation it is appropriate to modify the problem (2.8) by making the 'shallow atmosphere' approximation (see chapter by White earlier in this volume). Write the vertical component of the rotation vector

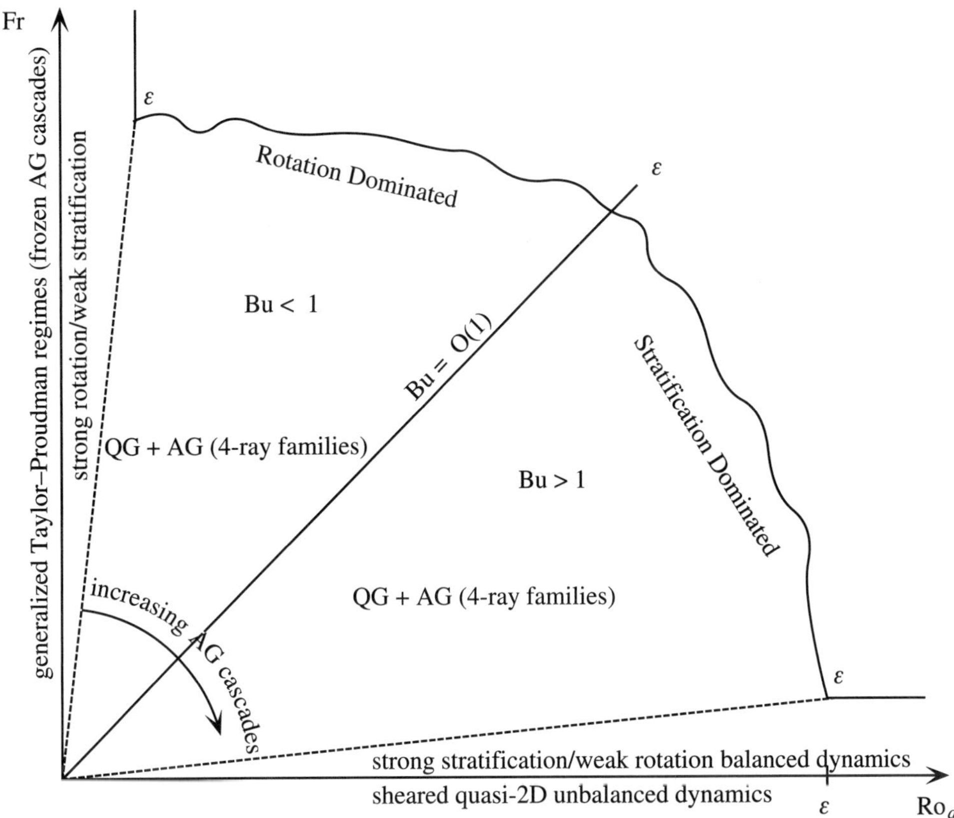

Figure 8: Definition of main asymptotic regimes in the atmosphere and ocean presented as plot against Froude number Fr and 'anisotropic' Rossby number $Ro_a = Ro/A$, where A is the aspect ratio H/L. QG refers to balanced motion, and AG to inertio-gravity waves.

as $f\hat{\mathbf{k}}$, where $f = 2\Omega \sin\phi$ and ϕ is the latitude, and write $\nabla\Phi = (0, 0, g)$. The other components of the rotation vector are neglected. If, in addition, we neglect the small molecular viscosity and thermal conductivity, the equations then become

$$\begin{aligned} \frac{D\mathbf{v}_h}{Dt} + (-fv, fu) + \alpha\nabla_h p &= 0 \\ g + \alpha\frac{\partial p}{\partial z} &= 0 \\ \nabla.\alpha^{-1}\mathbf{v} &= 0 \\ \frac{\partial\theta}{\partial t} + (\mathbf{u}.\nabla\theta)_h + w\frac{\partial\theta}{\partial z} &= 0. \end{aligned} \tag{2.15}$$

The geostrophic relation (2.14) becomes

$$(-fv, fu) + \alpha\nabla_h p = 0 \tag{2.16}$$

Combining the geostrophic relation, (2.16), with the continuity equation (2.9) shows that the horizontal velocity is independent of z if θ is uniform. The vertical scale of such flows, H, will thus tend to be the scale height of the atmosphere H_0 (around 10 km). This regime is thus only robust for horizontal scales large compared with NH_0/f. A typical value of this in the middle latitude troposphere is 1000 km.

The condition $B = O(1)$, so that the horizontal length scale is comparable with the Rossby radius, and $Ro \simeq Fr \ll 1$ is the natural scaling for 'synoptic' motions in the atmosphere and ocean. For instance, as shown in Gill (1982 chapter 13), it is the natural scale for the baroclinic waves which form the essential dynamics of developing weather systems.

Non-rotating, stratified flow corresponds to $B = \infty$. The first equation of (2.15) becomes

$$\frac{D\mathbf{v}_h}{Dt} + \alpha\nabla_h p = 0 \tag{2.17}$$

and the two terms must balance. Standard manipulations can then be used to show that $W \simeq FrUH/L$, so that, if $Fr \ll 1$, the flow is approximately horizontally non-divergent. The large scale balance condition is simply $\nabla_h p = 0$, so that there is no flow. The natural solutions will have horizontal density surfaces with gravity waves or gravity currents superposed on them.

If $B \gg 1$ we are considering sub-synoptic horizontal scales, or most low latitude circulations. As in the case $B = \infty$, the flow will be approximately horizontally non-divergent, and the vertical velocity will be small. There will therefore be a tendency for variables to become uncoupled in the vertical and the vertical scale to reduce, destroying the condition $B \gg 1$. There are thus doubts about the sustainability of such a regime.

The best example of such a regime is given by the shallow water equations with a large equivalent depth h. These equations can be obtained by averaging (2.15) over the depth of the atmosphere. The Froude number is then $U/\sqrt{(gh)}$, and the condition $B \gg 1$ is satisfied for mid-latitude values of f and parameters appropriate for the depth averaged atmosphere if $L \ll 3 \times 10^6$ m. This is smaller than typical deep atmospheric circulations. In low latitudes, this becomes $L \ll 3 \times 10^7$ m, which is almost always satisfied. Since there is a single layer of fluid, the vertical scale cannot be reduced and the regime will be sustained. The shallow water equations are widely regarded as a useful model for the depth averaged behaviour of the atmosphere and ocean. A similar argument holds for a fluid with a small number of deep homogeneous layers with significantly different density, which can be described by a set of shallow water equations. Such models have been widely used for understanding the development of weather systems, but their accuracy in describing the more continuously stratified flows that occur in the real atmosphere is not known.

There are many other important asymptotic regimes in meteorology and oceanography. We refer to the books by Holton and Gill for a more comprehensive introduction.

3 Lagrangian analysis of the governing equations

There is an extensive literature analysing solutions of (2.1) relevant to meteorology and of (2.3) relevant to oceanography, e.g. Gill (1982), based primarily on linearisation of the governing equations. In this chapter, we emphasise nonlinear analysis. The main nonlinearities in equations (2.1) and (2.3) are associated with the advection operator D/Dt. The effect of the advection operator is to perform a 'rearrangement' of the quantity being advected. The mathematical definitions and properties of rearrangements are discussed in the chapter by Douglas earlier in this volume. Here we recall the basic definitions and show how they can be used to discuss the solutions of (2.1) and (2.3). Note that the concepts are simpler if the rearrangements are volume-preserving, rather than mass-preserving. This will be the case in the ocean, and is a good approximation for large scale flows in the atmosphere if the volume is defined in terms of a stretched vertical coordinate which allows for a basic state vertical density variation.

3.1 Basic definitions

We first recall the basic definition of a rearrangement of a scalar-valued function given in the chapter by Douglas:

Let $f, g : \Omega \to \mathbb{R}$, where Ω is a bounded subset of $\mathbb{R}^n$, be two non-negative integrable functions, that is $\int_\Omega f(\mathbf{x})d\mathbf{x} < \infty$, $\int_\Omega g(\mathbf{x})d\mathbf{x} < \infty$. We say f is a *rearrangement* of g if

$$\int_\Omega (f(\mathbf{x}) - \alpha)_+ d\mathbf{x} = \int_\Omega (g(\mathbf{x}) - \alpha)_+ d\mathbf{x} \tag{3.1}$$

for each $\alpha > 0$, where h_+ denotes the non-negative part of the function h. Douglas gives the natural extension of this definition to vector-valued functions. Essentially, any two functions are rearrangements if they take any given set of values on sets of the same size. Unless f is constant, there is more than one rearrangement of f. We write $\mathcal{R}(f)$ for the set of all functions which are rearrangements of f.

When using rearrangement theory to study the properties of an evolving flow we consider the trajectory mapping that transforms initial positions $\mathbf{x}(\mathbf{a}, 0)$ of fluid particles $\mathbf{a}$ into positions $\mathbf{x}(\mathbf{a}, t)$ at a later time. We write this mapping as $\nu(0, t)$. Any conserved particle property q is transported by the flow, so that $q(\mathbf{x}(\mathbf{a}, t), t) = q(\mathbf{x}(\mathbf{a}, 0), 0)$. We write this more concisely as $q(t) \circ \nu(0, t) = q(0)$. If the fluid is incompressible it follows that, for any t, $q(\mathbf{x}, t)$ is a rearrangement of the initial data $q(\mathbf{x}, 0)$. This is because for each particle in a given subset U of the fluid at time t, we can find the original position of the particles at time zero and the set of these points has the same size as U. It follows that the trajectory mapping is a measure preserving mapping, the definition of which we recall from the chapter by Douglas:

Definition A *measure preserving mapping* from Ω to itself is a mapping $\nu : \Omega \to \Omega$ such that for each (measurable) set $U \subset \Omega$, $\mu(\{x : \nu(x) \in U\}) = \mu(U)$, where μ measures the size of the set. We must restrict our definition to sets where we can measure the size of the set — this excludes some pathological choices of U. Halmos (1950) shows that this definition is equivalent to requiring that for every integrable function f,

$$\int_\Omega f \circ \nu d\mu = \int_\Omega f d\mu. \tag{3.2}$$

3.2 Interpretation of partial differential equations using rearrangements

The ability to generate rearrangements by measure-preserving mappings suggests that they can be used to give a more general interpretation of partial differential equations governing incompressible flow. Consider the inviscid form of equations (2.3). These contain no spatial derivatives except the gradients of Φ and p, the divergence operator applied to $\mathbf{v}$, and the total derivative operator $\frac{D}{Dt}$. The value of $\nabla\Phi$ will be given explicitly, as in (2.15). There is no explicit evolution equation for p, the incompressibility condition on $\mathbf{v}$ makes up the number of equations.

A natural way to define a generalised solution of (2.3) is through the method of characteristics. Assume the equations are being solved in a region Γ with rigid boundary $\partial\Gamma$. (Note that, as written, (2.3) has a free surface upper boundary, which would require the procedure to be generalised.) The condition $\nabla.\mathbf{v} = 0$ and boundary conditions $\mathbf{v}.\mathbf{n} = 0$ on the boundary $\partial\Gamma$ are interpreted as stating that the time evolution of fluid particle positions is a rearrangement within Γ. Then the trajectory mapping $\nu(0, t)$ defined in section 3.1 is a measure-preserving mapping for any t. Solution of the equations depends on showing that there is a unique trajectory map of the fluid over any time interval $(0, t)$, giving the correct rates of change following fluid particles. We will see two examples of this in the following subsections.

Important points about this procedure are:

(i) There is no requirement for the position map to be a smooth function of the initial map. It may be possible to prove that it is smooth in a particular case. It is thus most useful if there are no spatial derivatives other than the $\frac{D}{Dt}$ operator. The term ∇p has to be determined implicitly, and we will give two interpretations of it applicable to non-smooth position maps in the subsequent sections.

(ii) Similarly, there is no need for particles initially in contact with the boundary to stay there. The rigid boundary condition simply requires all particles to stay inside Γ.

(iii) The rearrangement property is imposed, we seek a solution where $\mathbf{x}(a,t)$ is in $\mathcal{R}(\mathbf{x}(a,0))$ for all t, and do not consider 'solutions' outside this space.

(iv) It should be possible to generalise this procedure to compressible equations or free surface boundary conditions, where we consider the flow map as performing a 'mass' rearrangement.

3.3 Solution of the incompressible equations

We first illustrate the procedure described above with a simplified model problem which can be derived from (2.1) or (2.3). We will use this problem extensively in the rest of this chapter. Assume the axis of rotation and the gravitational force are in the same direction, setting $\nabla\Phi = (0,0,g)$ with g constant. We write $2\Omega = (0,0,f)$ and take f as constant. We discuss some aspects of the case with f variable, as in (2.15), in section 3.5. We make a form of Boussinesq approximation (Hoskins (1975)), which only incorporates the effect of buoyancy through variations in the gravitational term. In the atmospheric context, this allows a transformation of the vertical coordinate, allowing the equations to be written in incompressible form. The derivation is much more straightforward in the oceanic case from (2.3), see Gill (1982). We also neglect viscosity and thermal conductivity. Then set

$$z = \left(1 - \left(\frac{p}{p_0}\right)^{\frac{\gamma-1}{\gamma}}\right) H_s, \qquad (3.3)$$

where $H_s = \frac{\gamma\alpha_0 p_0}{g(\gamma-1)}$ with α_0, p_0 defined as in section 2.1. The continuity equation then becomes

$$\nabla.r(z)\mathbf{v} = 0, \qquad (3.4)$$

where r is a fixed function. Make the Boussinesq approximation to set $\nabla.\mathbf{v} = 0$. Use Cartesian coordinates (x,y,z), and solve in a finite bounded region Γ, with boundary $\partial\Gamma$. Write the volume element as $d\tau$. The incompressible equations then become

$$\begin{aligned} \frac{D\mathbf{v}_h}{Dt} + (-fv, fu) + \nabla_h p' &= 0 \\ \frac{Dw}{Dt} - g\theta/\theta_0 + \frac{\partial p'}{\partial z} &= 0 \\ \nabla.\mathbf{v} &= 0 \\ \frac{D\theta}{Dt} &= 0 \\ \mathbf{v.n} &= 0 \text{ on } \partial\Gamma. \end{aligned} \qquad (3.5)$$

Equation (3.5) includes explicit equations for the evolution of the velocity components and θ. The perturbation pressure is determined implicitly by the incompressibility constraint and the boundary conditions. As a first step to solving (3.5), ignore the pressure gradient term, then (3.5) can be solved explicitly. Given initial particle positions $(x(0), y(0), z(0))$ and velocities $(u(0), v(0), w(0))$, calculate a trajectory centre (x_c, y_c) for each particle as $(x(0) + v(0)/f, y(0) - u(0)/f)$. The particle trajectories consist of circles around the trajectory centres, together with a uniformly accelerating motion in the z direction. The solution at time t is then

$$\begin{aligned}
\theta(t) &= \theta(0) \\
x(t) &= x_c + (x(0) - x_c)\cos(ft) + (y(0) - y_c)\sin(ft) \\
y(t) &= y_c + (y(0) - y_c)\cos(ft) - (x(0) - x_c)\sin(ft) \\
z(t) &= z(0) + w(0)t + \frac{1}{2}g\theta'(0)t^2/\theta_0 \\
u(t) &= u(0)\cos(ft) + v(0)\sin(ft) \\
v(t) &= v(0)\cos(ft) - u(0)\sin(ft) \\
w(t) &= w(0) + g\theta'(0)t/\theta_0.
\end{aligned} \tag{3.6}$$

This solution does not satisfy the continuity equation. Equivalently, the associated mapping $\nu(0, t)$ of initial particle positions $\mathbf{x}(0)$ to later positions $\mathbf{x}(t)$ is not measure-preserving. To obtain a solution of (3.5), ν must be projected onto the set of measure-preserving mappings. If we use the 'polar factorisation' theorem of Brenier (1991), we can write

$$\nu = \nabla\chi \circ s \tag{3.7}$$

where χ is a convex function and s a measure-preserving mapping from Γ to itself. Note that this decomposition only exists under certain restrictions. Burton and Douglas (1998) have reduced these restrictions compared with those required by Brenier (and have conjectured existence for every integrable function). They show that s is as close to ν in L^2 as any other measure-preserving mapping $\Gamma \to \Gamma$; moreover Douglas and McCann (unpublished note) have demonstrated that all closest measure-preserving mappings arise from a polar factorisation. Thus, when the factorisation is unique, we can talk about the L^2-projection onto the set of measure-preserving mappings. If ν is non-degenerate, that is it maps no set of positive measure to a set of zero size, the polar factorisation is unique, and we can write $s = \nabla\chi^* \circ \nu$, where χ^* is the Legendre-Fenchel conjugate convex function of χ. In order to obtain a solution of (3.5), s must be identified with a continuous trajectory linking initial and final particle positions. Given initial particle positions $\mathbf{x}_0$, we can seek solutions by splitting the time interval $(0, t)$ into n equal parts $0 = t_0 < t_1 < \cdots < t_n = t$. Now find the particle mapping $\nu(0, t_1)$ using (3.6), and project onto the set of measure-preserving mappings as described above to obtain $s(0, t_1)$. Starting at positions given by $s(0, t_1)\mathbf{x}_0$, we find the particle mapping to time t_2 and project this mapping onto the set of measure-preserving mappings to obtain $s(t_1, t_2)$. Continue in this way, building a 'discrete trajectory mapping' s_n. If it can be proved that the sequence (s_n) converges (in a suitable sense) as $n \to \infty$, then we call the result a generalised solution of (3.5).

A proof of this type has not yet been achieved, though such a procedure underlies many standard and successful numerical methods. However, some remarks can be made about what results may be possible. The two-dimensional version of equations (3.5) is known to have solutions which stay as regular as the initial data, Kato and Ponce (1986). Consider a single step of the iteration procedure set out above:

(i) Given initial particle positions $\mathbf{x}_0$, calculate estimates of particle positions at time t_1 using (3.6). Write the solutions as $\nu(0, t_1)\mathbf{x}_0$. Since $\mathbf{v}_0$ satisfies the continuity equation and boundary condition, we have

$$\det D_{\mathbf{x}}(\nu(0, t_1)\mathbf{x}_0) = 1 + O((t/n)^2) \equiv \rho(\nu(0, t_1)), \tag{3.8}$$

where $D_{\mathbf{x}}(\nu(0, t_1)\mathbf{x}_0)$ denotes the Jacobian matrix.

(ii) Project this onto a measure preserving mapping $s(0, t_1)$ using the polar factorisation.

Equation (3.8) states that $|\rho - 1| \leq C((t/n)^2)$ where C is a bound on the velocity gradients. If $\mathbf{v} \in W_{1,\infty}$, then ρ is bounded away from 0 and ∞ for sufficiently large n. Caffarelli (1996) shows (in two dimensions with smooth boundaries) that this would imply that s and $\nabla\chi^*$ are one derivative smoother than ρ, while (3.8) shows that ρ is one derivative less smooth than $\nu(0, t_1)$. (3.6) shows that $\nu(0, t_1)$ is as smooth as $\mathbf{v}$. More recent work suggests that these results are true in three dimensions if the boundary conditions are periodic. Hence the solution will be as smooth as the initial data if the bound on the velocity gradients can be maintained. In two dimensions, this follows from vorticity conservation, Gerard (1992).

In the 3-dimensional problem, there is no known way of maintaining a bound on the velocity gradients. However, the solution procedure can still be followed through in a formal sense if $\mathbf{v}$ is bounded. The projection onto a measure preserving mapping will not be unique, but it would be possible to choose the member of the family of possible projections which minimised a suitable norm of $s(0, t_1)$. The difficulty now is to maintain the bound on $\mathbf{v}$ as the number of time intervals n tends to infinity. This requires $\| \nabla\chi^*(t_i, t_{i+1}) \| \leq C(t/n)^2$ in a suitable norm, with C independent of n.

If the above results can be achieved, s will be identified with a continuous trajectory and particle positions will vary smoothly in time. This is a minimum requirement for the solution to make physical sense. However, even if this can be achieved, the discussion above suggests that the resulting velocity field may not vary smoothly in space. This gives some plausibility to the idea that typical length scales along trajectories may be greater than length scales in other directions and is consistent with the results of Bartello and Thomas quoted above which were based on Fourier expansions. A velocity field of bounded variation can be consistent with the $-5/3$ power spectrum if discontinuities are present, but intermittent.

Brenier (1991) discusses the relation between this procedure and a solution procedure relying on the Helmholz decomposition of a general velocity field into an incompressible velocity field and the gradient of a scalar. This decomposition is the linearisation of the polar decomposition about the identity map. Though the polar decomposition uses a convex potential, there is no convexity implied of the scalar in the Helmholz decomposition, and thus no implied restriction on the form of the pressure in the solution of (3.5).

3.4 Extremisation with respect to rearrangements

The scale analysis of section 2 highlighted the importance of the geostrophic and hydrostatic relations (2.14). We now show how (2.14) can be derived by a variational argument based on rearrangements, following Shutts and Cullen (1987) and Cullen *et al.* (1991). The energy associated with solutions of (2.1) is given by (2.4). We seek to minimise this with respect to fluid displacements $\Xi = (\xi, \eta, \zeta)$ which satisfy the boundary conditions, with changes to the flow variables being given by

$$\begin{aligned} \delta \mathbf{v} &= -2\Omega \times \Xi \\ C_v \delta T + p\delta\alpha &= 0 \\ \delta\alpha &= \alpha \nabla . \Xi \\ \delta\Phi &= \Xi . \nabla \Phi . \end{aligned} \tag{3.9}$$

These changes are consistent with the evolution equations (2.1) apart from the 'freezing' of the pressure gradient ∇p. Shutts and Cullen show that the geostrophic and hydrostatic relations (2.14) are the condition for the energy to be stationary with respect to (3.9). The condition for a minimum is that

$$\delta^2 E = \Xi . \Lambda . \Xi \geq 0 \tag{3.10}$$

where

$$\Lambda_{ij} = 2|\Omega| \frac{\partial M_i}{\partial x_j} - \alpha \frac{\partial \ln\theta}{\partial x_i} \frac{\partial p}{\partial x_j} \tag{3.11}$$

and $M = 2|\Omega|(\mathbf{x} - (\hat{\Omega}.\mathbf{x})\hat{\Omega}) - \hat{\Omega} \times \mathbf{v}$. Shutts and Cullen show that this corresponds to the condition for a fluid parcel to be stable against displacements in a 'frozen' pressure field. (3.10) implies the condition

$$\det \Lambda \geq 0. \tag{3.12}$$

The physical justification for this is the natural model of a stable steady state being an energy minimising state. Increased energy corresponds to oscillations about the steady state. If the basic state is time-dependent, but the time-scale is much greater than the period of the oscillations about the stable state, then the argument should remain accurate. In this example, the oscillations all have a frequency greater than 2Ω, so that the argument requires the geostrophic and hydrostatic basic state to evolve on a time scale much slower than this. If the time scales become comparable, the procedure can still be carried out, but will not give useful results.

States which are stationary points of the energy, but not minima, are unstable to motions with a growth rate greater than $(2\Omega)^{-1}$. The evolution will then not be usefully described by oscillations about such a state. However, it is possible that the instability will be released quickly, so that there is a transition on a time-scale less than $(2\Omega)^{-1}$ to a state close to a minimum energy state. For this to happen, the motions which achieve this transition must either allow energy loss, or else not satisfy (3.9). In practice, both are likely.

To illustrate the methods further, we use the simplified model problem (3.5) introduced in the previous subsection. The conserved energy for these equations, where we write θ for θ' henceforth, is

$$E = \int_\Gamma \left\{ \frac{1}{2}(u^2 + v^2 + w^2) - g\theta z/\theta_0 \right\} d\tau. \tag{3.13}$$

Shutts and Cullen show that if we minimise the energy subject to displacements satisfying

$$\begin{aligned} \delta \mathbf{v} &= -f\mathbf{k} \times \xi \\ \delta\theta &= 0 \\ \nabla.\Xi &= 0 \\ \Xi.\mathbf{n} &= 0 \text{ on } \partial\Gamma \end{aligned} \tag{3.14}$$

then the condition for the energy to be stationary is geostrophic and hydrostatic balance in the form

$$(fv, -fu, g\theta/\theta_0) = \nabla s \tag{3.15}$$

for some scalar s. (3.15) still holds if f is not constant, but subsequent derivations do not. The necessary reformulations are discussed in the next subsection. The condition for a minimum is that the matrix $\mathbf{Q}$ whose components are

$$Q_{i,j} = \frac{\partial^2 S}{\partial x_i \partial x_j} \tag{3.16}$$

$$S = f^{-2}s + \frac{1}{2}(x^2 + y^2) \tag{3.17}$$

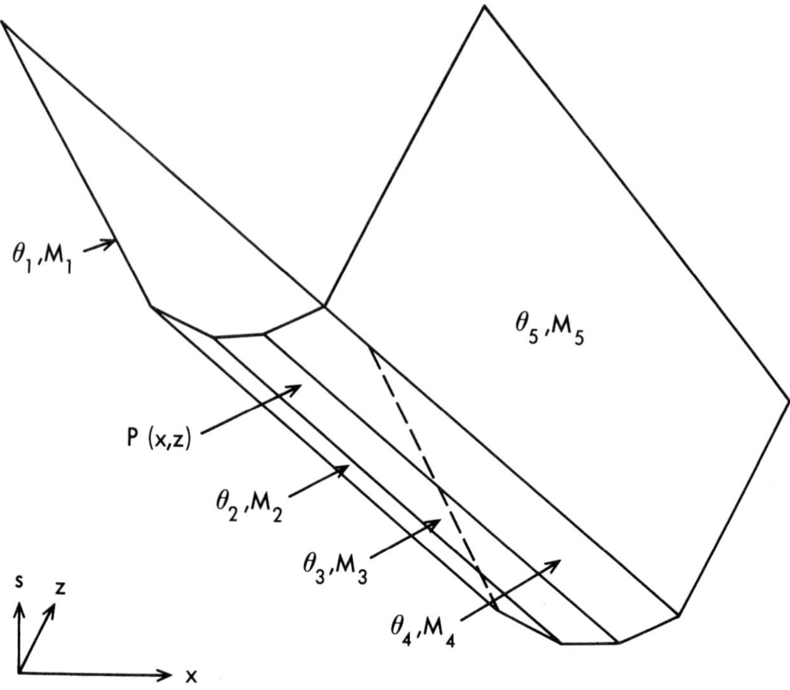

Figure 9: Construction of a convex polyhedral surface $P(x, z)$ from faces with given gradients $(X_i, Z_i) = (M_i, \theta_i)$ and areas.

is positive definite. Define

$$\begin{aligned} X &= x + f^{-1}v \\ Y &= y - f^{-1}u \\ Z &= g\theta/f^2\theta_0 \\ \mathbf{X} &= (X, Y, Z). \end{aligned} \tag{3.18}$$

(3.15) shows that

$$(X, Y, Z) = \nabla S. \tag{3.19}$$

Equation (3.14) thus implies $\delta X = \delta Y = \delta Z = 0$. The solution of the minimisation problem can be interpreted geometrically as finding a convex surface S whose gradients $\mathbf{X}$ take prescribed ranges of values on sets of specified measure. The condition that (3.16) is positive definite is equivalent to convexity of S. A simple example is the problem of finding a polyhedral surface with faces with specified gradients and areas (Figure 9), corresponding to choosing piecewise constant values of (X, Y, Z) in (3.18). A proof that this construction is possible for any finite set of values is given in Cullen and Purser (1984). A more general set of such results on constructing polyhedral surfaces is given by Pogorelov (1973).

In general, we can specify the problem as to find $S(x, y, z)$ over a specified region Γ in (x, y, z) given that $\mathbf{X} = \nabla S$ satisfies (implicitly), for given ρ,

$$\begin{aligned} \rho &= \det \frac{\partial(x, y, z)}{\partial(X, Y, Z)} \\ \int_{\mathbb{R}^3} \rho d\mathbf{X} &= \mu(\Gamma) \end{aligned} \tag{3.20}$$

where $\mu(\Gamma)$ is the volume of Γ. The proof that this problem can be solved is a result of theorem 1.1 of Brenier (1991), subsequently extended by McCann (1995), which shows that a general mapping f from a bounded subset of $\mathbb{R}^n$ to $\mathbb{R}^n$ has a unique rearrangement equal to the gradient of a convex function. In particular it should be noted that S can have discontinuous gradients, even when ρ is a smooth function. Cullen and Purser proposed this as a simple model of atmospheric fronts. Extension of this identification of the problem to periodic domains and shallow water models is given in Cullen and Purser (1989). A fuller discussion of these analytical results is given in the chapter by Douglas earlier in this volume.

3.5 Relation to the Monge–Ampère equation and Monge mass transport problem

Cullen and Purser (1989) showed that (3.19) has a dual form for $R(\mathbf{X})$ dual to $S(\mathbf{x})$

$$\begin{aligned} (x, y, z) &= \nabla_{\mathbf{X}} R \\ R &= \mathbf{x}.\mathbf{X} - S. \end{aligned} \tag{3.21}$$

This results from the identification of the transformation from $\mathbf{x}$ to $\mathbf{X}$ as a Legendre transformation (see the chapter by Sewell). Using this, (3.20) becomes

$$\rho = \det \frac{\partial^2 R}{\partial(X, Y, Z)^2}. \tag{3.22}$$

This is a Monge–Ampère equation. Brenier (1991) shows that his 'polar factorization theorem' is equivalent to an existence proof for a generalised solution of (3.22), subject to the compatibility and boundary conditions

$$\begin{aligned} \int_{\mathbb{R}^3} \rho d\mathbf{X} &= \mu(\Gamma) \\ \nabla R &\in \Gamma. \end{aligned} \tag{3.23}$$

(See the chapter by Douglas earlier in this volume for a more complete derivation of the duality structure and its consequences.) The theory of this problem, sometimes called the 'assignment problem' when it arises in other contexts, has been taken much further, see Caffarelli (1996).

Our previous work has assumed the existence of a (unique) minimum energy state (under variations (3.14)). If the vertical contribution to the kinetic energy is ignored (it is not varied under (3.14)), the energy (3.13) can be written as

$$E = f^2 \int_\Gamma \left\{ \frac{1}{2}((x-X)^2 + (y-Y)^2) - zZ \right\} d\tau. \tag{3.24}$$

We noted earlier that variations (3.14) are those which conserve $\mathbf{X}$ on particles, and satisfy conservation of mass. Such variations are described by $\mathcal{R}(\mathbf{X}_0)$, for some prescribed $\mathbf{X}_0$. We demonstrate that (3.24) has a unique minimiser relative to this set. Noting that $\int_\Gamma (z^2 + Z^2) d\tau$ is conserved under rearrangements of $\mathbf{X}$, we can study the equivalent problem of minimising

$$E = \frac{1}{2} f^2 \int_\Gamma \{ |\mathbf{X}(\mathbf{x}) - \mathbf{x}|^2 \} d\tau \tag{3.25}$$

for all $\mathbf{X} \in \mathcal{R}(\mathbf{X}_0)$. This may be rewritten as a Monge mass transport problem as follows. Define $\nu(B) = \mu(\mathbf{X}_0^{-1}(B))$ for all subsets B of $\mathbb{R}^3$ which are sufficiently well behaved. Then the set of measure preserving mappings between (Γ, μ) and $(\mathbb{R}^3, \nu)$ is exactly $\mathcal{R}(\mathbf{X}_0)$. The minimiser of (3.25) can be written as

$$\inf_{s \in \mathcal{R}(\mathbf{X}_0)} \int_\Gamma c(\mathbf{x}, s(\mathbf{x})) d\mu \tag{3.26}$$

where the cost function $c : \mathbb{R}^3 \times \mathbb{R}^3 \to \mathbb{R}$ is defined by $c(\mathbf{x}, \mathbf{y}) = |\mathbf{x} - \mathbf{y}|^2$. Gangbo and McCann (1996) showed that the infimum in (3.26) is uniquely attained by a measure-preserving mapping equal to the gradient of a convex function. They also discuss conditions under which these problems can be solved for a more general cost function. This justifies our earlier claim.

Cullen and Douglas (1998) show that the energy minimisation argument described above can be used to define the geostrophic transformation (by identifying the convex function appropriately), and then demonstrate that a more general form of this problem can be used to extend the theory to the surface of a sphere S^2. Assume that (x, y) and (X, Y) represent coordinates on the spherical surface, and z and Z represent coordinates normal to the surface. The Coriolis parameter f is now equal to $2\Omega \sin \phi$, where ϕ is the latitude. Let g_{ij} denote the metric for S^2. Define $\hat{g}^{ij} = f^2 g_{ij}$, a conformal rescaling of the metric for $f > 0$. Let S^+ denote the upper hemisphere excluding the equator, and write $\hat{S}^+$ for the conformal rescaling of S^+. The pair $(\hat{S}^+, \hat{g}^{ij})$ is a Riemannian manifold. Let $\hat{d}$ be the Riemannian distance induced on $\hat{S}^+$ by $\hat{g}^{ij}$. Let $\hat{\mu}$ be surface area on a closed set $\hat{N} \subset \hat{S}^+$. Then, given a prescribed $\mathbf{X}_0 : \hat{S}^+ \to \hat{N}$, define $\hat{\nu}(B) = \hat{\mu}(\mathbf{X}_0^{-1}(\hat{B}))$ for well behaved sets $B \subset \hat{S}^+$. By analogy with the constant rotation case, we minimise the energy over $\mathcal{R}(\mathbf{X}_0)$, (extending the definition of rearrangement in the obvious way). Cullen and

Douglas (1998) show that this energy can be written as

$$E = \frac{1}{2}\int_{\hat{S}+} \hat{d}(\mathbf{x}, \mathbf{X}(\mathbf{x}))^2 d\hat{\mu}(\mathbf{x}). \tag{3.27}$$

We can rewrite the energy minimisation problem in mass transport form by noting that the set of measure preserving mappings from $(\hat{S}^+, \hat{\mu})$ to $(\hat{S}^+, \hat{\nu})$ is exactly $\mathcal{R}(\mathbf{X}_0)$. Then the minimiser of (3.27) becomes

$$\inf_{s\in\mathcal{R}(\mathbf{X}_0)} \frac{1}{2}\int_{\hat{S}+} \hat{d}(\mathbf{x}, s(\mathbf{x}))^2 d\hat{\mu}(\mathbf{x}). \tag{3.28}$$

McCann (2001) has proved that this problem admits a unique minimiser if all pairs of points in $\hat{N}$ are linked by a minimal geodesic. We define a transformation $\hat{\mathbf{X}}$ to be this unique minimiser. Note that this generalisation is different from the theories discussed in the chapter by Purser.

3.6 Rearrangements and mixing

The energy minimisation problem of section 3.5 is a case where the problem of minimising a functional with respect to rearrangements of functions can be solved uniquely. However, such results are difficult to prove because the set of rearrangements of a given function is not compact, except in the trivial case of a constant function. There can be infinite sequences of rearrangements of a given function which do not converge to a limit which is a rearrangement. An example of this is a sequence of arbitrarily fine-grained rearrangements. Let $f_0 : [0, 1] \to \mathbb{R}$ be defined by

$$f_0(x) = \begin{cases} 0 & \text{if } x \in [0, 1/2], \\ 1 & \text{if } x \in [1/2, 1], \end{cases} \tag{3.29}$$

Define, for $n \in \mathbb{Z}$,

$$f_n(x) = \begin{cases} 0 & \text{if } x = 0, \\ 0 & \text{if } x \in (m/n, (2m+1)/2n], \\ 1 & \text{if } x \in ((2m+1)/2n, (m+1)/n], \end{cases} \tag{3.30}$$

where $m = 0, 1, \ldots, n-1$. The functions f_3 and f_8 are illustrated in Figure 10. For each $n \in \mathbb{Z}$, f_n is equal to zero on a set of length 1/2, and equal to 1 on a set of length 1/2, therefore f_n is a rearrangement of f_0 for each $n \in \mathbb{Z}$. However, given any $g \in L^2(0, 1)$, it may be shown that $\int_0^1 f_n g dx \to \frac{1}{2}\int_0^1 g dx$ as $n \to \infty$, that is, f_n converges weakly to the constant function with value 1/2, which is not a rearrangement of f_0.

In applications where we need to take limits and thus ensure that we work with a compact set, we can use the weak closure of the set of rearrangements of a given function. This is the smallest weakly compact set that contains

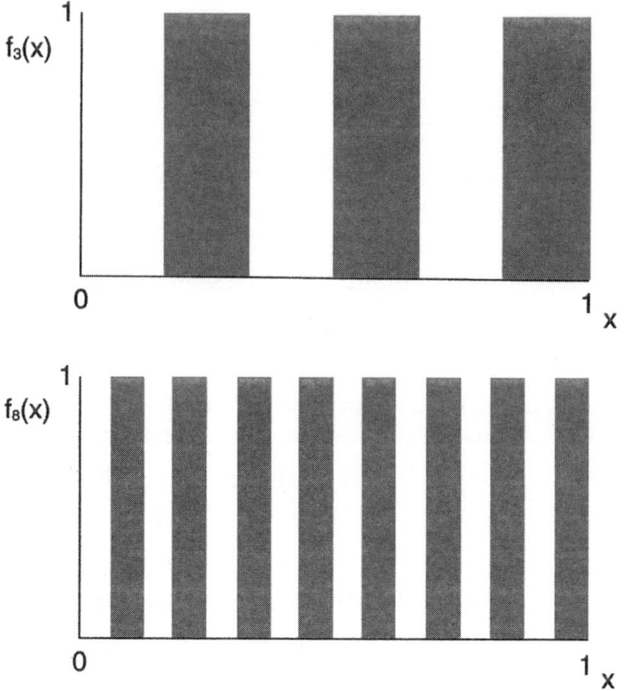

Figure 10: The rearrangements $f_3 = f_3(x)$ and $f_8 = f_8(x)$ defined by equation (3.30) of the function $f_0 = f_0(x)$ defined by equation (3.29).

the set of rearrangements, and thus contains all the weak limits of sequences of rearrangements. Douglas (1994) gave the following characterisation for a non-negative square-integrable function f_0;

$$\begin{aligned}\overline{\operatorname{conv} R(f_0)} &= \Big\{ f \geq 0 \mid \int_\Omega (f-\alpha)_+ d\mu \leq \int_\Omega (f_0 - \alpha)_+ d\mu \\ &\qquad \text{for each } \alpha > 0, \int_\Omega f d\mu = \int_\Omega f_0 d\mu \Big\} \qquad (3.31)\end{aligned}$$

where the + subscript denotes taking the positive part of the function. If we define f_0 as in (3.29) it can be shown that any integrable function $\varphi : [0,1] \to \mathbb{R}$ satisfying $0 \leq \varphi(x) \leq 1$ for each $x \in [0,1]$, and $\int_0^1 \varphi d\mu = 1/2$, belongs to $\overline{\operatorname{conv} R(f_0)}$. This illustrates that $\overline{\operatorname{conv} R(f_0)}$ may be a large class of functions, in particular it includes the constant value 1/2 which is certainly not a rearrangement of f_0. In general, all rearrangements are included, as are functions derived by smoothing a rearrangement while preserving the value of the integral. The limit of a sequence of fine grained rearrangements will be a smoothed 'average' function. Physically, including these limit functions can be thought of as allowing for a small but finite viscosity.

4 Basic equations of motion and approximation by balanced models

Equations (2.1) and (2.3) describe all possible motions of the atmosphere and ocean. In order to understand large scale weather systems, and equivalent ocean circulations, a generic approach is:

(i) Identify an asymptotic regime, and thus small parameter, corresponding to the flows of interest.

(ii) Identify a system of equations, based on scale analysis, that is appropriate to that regime. We refer to this as a 'reduced' system.

(iii) Prove that this system has solutions, which only contain the flows of interest.

(iv) Prove that the solutions of the full equations stay close to that of the 'reduced' system in some norm. The estimate will depend on the small parameter used in the original asymptotic analysis.

This is the approach followed in the chapter by Babin *et al.* They used it to extend existence results for reduced systems to those for the full fluid equations. Note that it is very difficult to show how a simple solution can stay close to a complicated one, while the converse as in step (iv) may be much more practicable. This procedure has been used, for instance, to show that in appropriate regimes, solutions of the 3-dimensional Euler equations stay close to those of the 2-dimensional Euler equations, which are known to be well behaved. One method of doing this is discussed by Babin *et al.*, another approach is described by Marsden *et al.* (1995).

In applications to weather and climate forecasting, it is also important to recognise situations in which the total flow is not well approximated by a reduced system, but the interactions between the motions described by the reduced system and other motions are weak. Such a situation arises in the internal structure of the ocean. This contains large internal gravity waves, Garrett and Munk (1979), which can be shown to couple weakly to the large scale geostrophic circulation, e.g. Gjaja and Holm (1996). We can describe this situation as follows. Let M be the solution space of equations (2.1) or (2.3). Let M_0 be the subspace of M which contains solutions of a reduced system of equations. Define a projection P which maps M to M_0. Represent the solution of (2.1) or (2.3) as $U(t)$ and that of a reduced system as $u(t)$. Then, as shown in Figure 11, if we initialise the reduced model with data $u(0) = PU(0)$, the effective forecast error of the reduced system at time t is $u(t) - PU(t)$. This may be much smaller than $u(t) - U(t)$, and it is important to estimate it.

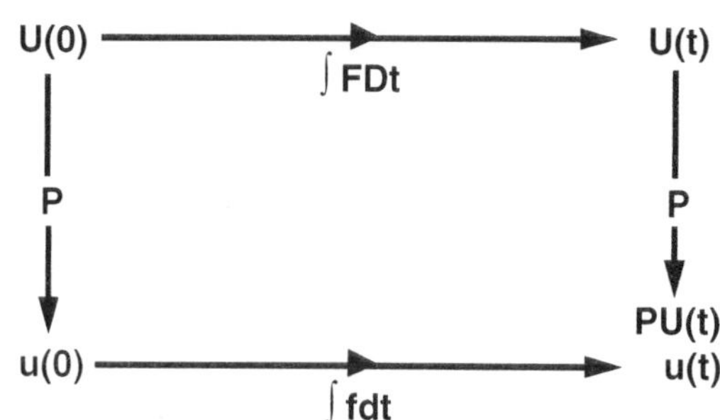

Figure 11: Illustration of the comparison between the error of a prediction $u(t)$ using a reduced set of equations, and the projection P of the exact solution $U(t)$ onto a state consistent with the reduced system.

We illustrate in the following subsections the approach to constructing reduced systems using semi-geostrophic theory. The chapters by Holm *et al.*, Bokhove, McIntyre and Roulstone, and Allen *et al.* describe alternative approaches. The chapter by Baigent and Norbury describes an alternative approach to the semi-geostrophic definition using maximum entropy ideas.

4.1 Approximation by balanced states

We show how a particular reduced system can be defined as a sequence of 'balanced' states. A balanced state is defined for this purpose as a state which is in geostrophic and hydrostatic balance, and is statically and inertially stable (so satisfies (3.19) with S convex). As shown in section 2, this is the natural lowest order approximation for small Rossby number. A similar analysis is possible for axisymmetric states (Shutts *et al.* (1988)).

In order to allow a more complete analysis, we start from the simplified problem (3.5), rather than the original equations (2.1) or (2.3). This avoids the complications caused by compressibility. The procedure is set out in more detail in Cullen (2000). Given a general solution of (3.5), we wish to show how it can be approximated by balanced states in the above sense. For such a solution, calculate $\mathbf{X} = (X, Y, Z)$ using (3.18). Define a projection Π of a general state onto a balanced state by minimising the energy subject to particle displacements satisfying (3.14), as discussed in section 3.5. This gives a state

$$\Pi(X, Y, Z) = (X_b, Y_b, Z_b) \tag{4.1}$$

satisfying (3.19) with $(X_b, Y_b, Z_b) = (x + f^{-1}v_b, y - f^{-1}u_b, g\theta_b/f^2\theta_0)$ and S convex. Write the minimising displacement as $\Xi \equiv (\xi, \eta, \zeta)$. Since $\mathbf{X}$ is preserved on particles by (3.14), we have $\mathbf{X}_b(\mathbf{x} + \Xi) = \mathbf{X}(\mathbf{x})$. Following (3.13), the energy associated with this state is

$$e = \int_\Gamma \left\{ \frac{1}{2}(u_b^2 + v_b^2) - g\theta_b z/\theta_0 \right\} d\tau. \tag{4.2}$$

Preservation of $\mathbf{X}$ under the displacement also means that

$$\frac{\partial(x, y, z)}{\partial(X, Y, Z)} = \frac{\partial(x_b, y_b, z_b)}{\partial(X_b, Y_b, Z_b)}. \tag{4.3}$$

If we define the potential vorticity Q of a general state to be $\frac{\partial(X,Y,Z)}{\partial(x,y,z)}$, then Q is preserved by the projection if it is regarded as a function of (X, Y, Z). Note that smooth solutions of (3.5) preserve the 'Ertel' potential vorticity

$$q = (f + \zeta).\nabla\theta \tag{4.4}$$

on particles, where $\zeta = \nabla \times \mathbf{u}$. Q and q agree to first order in ζ/f. Other choices of Π which are based on preserving (4.4) or closer approximations to it are discussed in the chapters by McIntyre and Roulstone, Holm *et al.* and Allen *et al.* They all provide a generalisation of normal mode projections, which depend on decomposing general fields into a geostrophic and hydrostatic basic state with linear unbalanced waves superposed on it, see Daley (1997). The latter require linearisation of the governing equations about a uniform reference state at rest.

The evolution equations (3.5) can be written as evolution equations for $\mathbf{X} = (X, Y, Z)$:

$$\begin{aligned} \frac{D\mathbf{X}}{Dt} + f^{-1}\mathbf{k} \times \nabla p' &= 0 \\ \frac{Dw}{Dt} + \frac{\partial p'}{\partial z} &= g\theta/\theta_0 \\ \frac{D\alpha}{Dt} &= 0 \\ \mathbf{v}.\mathbf{n} &= 0 \text{ on } \partial\Gamma. \end{aligned} \tag{4.5}$$

Here the continuity equation has been written in Lagrangian form in terms of the specific volume α. Imagine that we have computed a solution of (4.5), with particle positions $\mathbf{x}(t)$, and particle values $\mathbf{X}(t)$. At each time t, we project to the minimum energy state using (4.1). This involves displacing the particles to $\mathbf{x}(t) + \Xi(t)$ while preserving their values of X, Y and Z. Write $\frac{D^*}{Dt}$ to express a derivative following the 'minimum energy' particle positions $\mathbf{x} + \Xi$, and write the 'velocity' that achieves this as $\mathbf{V} = (U, V, W)$, so $\mathbf{V} = \frac{D}{Dt}(\mathbf{x} + \Xi)$ and

$\frac{D^*}{Dt} \equiv \frac{\partial}{\partial t} + \mathbf{V}\cdot\nabla$. Then the equations expressing the evolution of this balanced state are

$$\begin{aligned} \frac{D^*\mathbf{X_b}}{Dt} + (f^{-1}\mathbf{k}\times\nabla p')^* &= 0 \\ \frac{D^*\alpha}{Dt} &= 0 \\ (X_b, Y_b, Z_b) &= \nabla S. \end{aligned} \tag{4.6}$$

Here $(f^{-1}\mathbf{k}\times\nabla p')^*(\mathbf{x}+\Xi) \equiv f^{-1}\mathbf{k}\times\nabla p'(\mathbf{x})$. This means that (4.6) cannot be solved without prior knowledge of the solution of (4.5). $(\nabla p')^*$ is not in general the gradient of a scalar. The second equation in (4.6) implies $\nabla.\mathbf{V} = 0$. The boundary conditions on (3.14) imply that $\mathbf{V}.\mathbf{n} = 0$ on $\partial\Gamma$. The solution of (4.6) has the same 'balanced potential vorticity' Q as (4.5) if Q is regarded as a function of $\mathbf{X}$. However, Q is not conserved on particles under (4.5) and therefore not under (4.6) either.

The evolution of e can be calculated by first rewriting (4.6) in terms of the original variables

$$\begin{aligned} \frac{D^*}{Dt}(u_b, v_b) + (-fV, fU) + (\nabla_h p')^* &= 0 \\ \frac{D^*}{Dt} &\equiv \frac{\partial}{\partial t} + \mathbf{V}\cdot\nabla \\ \frac{D^*\theta_b}{Dt} &= 0 \\ \frac{D^*\alpha}{Dt} &= 0 \\ (fv_b, -fu_b, g\theta_b/\theta_0) &= \nabla s. \end{aligned} \tag{4.7}$$

Here $s = S - \frac{1}{2}f^2(x^2+y^2)$. Multiply the first of these equations by (u_b, v_b) and the third by z and add. Using $W = \frac{D^*z}{Dt}$, and the fact that $\mathbf{V}.\nabla s = fUv_b - fVu_b + gW\theta/\theta_0$ integrates to zero because $\nabla.\mathbf{V} = 0$, we find that

$$\frac{D^*e}{Dt} = -\int_\Gamma (u_b, v_b, 0).(\nabla_h p')^* d\tau. \tag{4.8}$$

This is zero if $(\nabla_h p')^*$ can be written as $\nabla_h\pi$ for some scalar π, because $\nabla.(u_b, v_b, 0) = 0$. For flows close to balance, so that $|\Xi|$ is small, we can write

$$(\nabla_h p')^* = \nabla_h p' - \Xi\cdot\nabla(\nabla p') + O(|\Xi|^2). \tag{4.9}$$

Using (4.9), the condition that $(\nabla_h p')^* = \nabla_h\pi$ becomes

$$\frac{\partial\xi}{\partial y}\frac{\partial^2 p'}{\partial x^2} + \frac{\partial\eta}{\partial y}\frac{\partial^2 p'}{\partial x\partial y} - \frac{\partial\xi}{\partial x}\frac{\partial^2 p'}{\partial y\partial x} - \frac{\partial\eta}{\partial x}\frac{\partial^2 p'}{\partial y^2} = 0. \tag{4.10}$$

This is satisfied for flows independent of one horizontal coordinate. It is of interest to seek other cases where (4.10) is either zero, or smaller than would be expected from general estimates. In such cases $\frac{\partial e}{\partial t}$ will be small, and if $E - e$ is initially small, it will only grow slowly, so that the flow stays close to balance. In the general case, (4.8) and (4.9) allow $E - e$ to be estimated in terms of Ξ and p'. We will exploit this in the next subsection.

4.2 Approximation of evolution by a reduced set of equations

We now illustrate how a reduced set of equations can be constructed, which only uses knowledge of balanced states, and how estimates of the difference between the solution of a reduced system and the full system can be made.

A reduced set of equations for predicting the evolution of a balanced state, approximating the real state of the atmosphere, can be obtained by replacing the term $(f^{-2}\mathbf{k}\times\nabla p')^*$ in (4.6) by a term calculated directly from the balanced state variables. The geostrophic and hydrostatic pressure in the balanced state is given by (3.15) as s. In order to calculate $(\nabla p')^*$ from the balanced variables, we first need to estimate the true pressure p' from s, and then estimate $(\nabla p')^*$ from (4.9). However, the latter estimate depends on Ξ, which is arbitrary and unrelated to the balanced state. It can only be estimated by first estimating the energy difference $E - e$ between the true state and the balanced state, and then using (3.10) to estimate Ξ. We illustrate two estimates, appropriate for different asymptotic regimes.

Consider first the case $Ro \leq O(1), B = NH/fL \gg 1$ which implies $Fr \ll 1$. As discussed in section 2, we then have $W/H \ll U/L$ and hydrostatic balance which for equations (3.5) means $\frac{\partial p'}{\partial z} = g\theta/\theta_0$. The definition of N^2, (2.10), means that $g/\theta_0\frac{\partial\theta}{\partial z} \gg (fL/H)^2$, so that there is a large vertical variation of θ. First calculate what vertical θ conserving displacement is needed to change the pressure, assumed hydrostatic, from p' to s. This requires a change of θ of magnitude $(p' - s)\theta_0/(gH)$ and thus a vertical displacement of magnitude $(p' - s)\theta_0/(gH\frac{\partial\theta}{\partial z}) = (p' - s)/(N^2H)$. The condition $\nabla\cdot\Xi = 0$ means that the associated horizontal displacements will have to be of magnitude $L(p' - s)/(N^2H^2)$. Condition (3.14) then shows that the horizontal velocity components will be changed by an amount $\delta u \simeq fL(p' - s)/(N^2H^2)$. If, instead, the horizontal velocity components were changed by a horizontal displacement so as to be in geostrophic balance with the exact pressure p, the change required would be $(p' - s)/(fL) \simeq B^2\delta u$, which is larger by the factor B^2. Therefore, in this regime, the projection Π will change the pressure field to match the horizontal velocity components, rather than vice versa. This is consistent with geostrophic adjustment theory, see Haltiner and Williams (1980). In particular it will preserve the horizontal non-divergent velocity components to $O(B^{-2})$.

We can therefore estimate p' to $O(B^{-2})$ by assuming $\hat{\mathbf{k}}.\nabla \times (u, v, 0) = \hat{\mathbf{k}}.\nabla \times (u_b, v_b, 0)$, where $\hat{\mathbf{k}} = (0, 0, 1)$. We write $\mathbf{U}$ for the fluid trajectory $\dot{\mathbf{x}}$ which will be determined by the balance condition (4.12). This is to be distinguished from the trajectory $\mathbf{V}$ defined diagnostically from the solution of (4.5). The reduced equations then take the form

$$
\begin{aligned}
\frac{D}{Dt}(u_b, v_b) + (-fV, fU) + \nabla_h \pi &= 0 \\
\frac{D}{Dt} &\equiv \frac{\partial}{\partial t} + \mathbf{U} \cdot \nabla \\
\frac{D\theta}{Dt} &= 0 \\
\frac{D\alpha}{Dt} &= 0 \\
(fv_b, -fu_b, g\theta/\theta_0) &= \nabla s.
\end{aligned}
\tag{4.11}
$$

Substituting $\hat{k}.\nabla \times (u, v, 0) = \hat{k}.\nabla \times (u_b, v_b, 0)$, and enforcing consistency between (3.5) and (4.11), gives

$$
\hat{\mathbf{k}}.\nabla \times (U.\nabla u_b) + \hat{\mathbf{k}}.\nabla \times (-fV, fU, 0) = \hat{\mathbf{k}}.\nabla \times (u_b.\nabla u_b). \tag{4.12}
$$

We determine π by the condition $\nabla_h.(u_b, v_b, 0) = 0$, which follows from the last equation of (4.11), which with (4.12), forms the appropriate reduced set of equations. Since the pressure gradient term is approximated by the gradient of a scalar, these equations conserve the balanced energy e as defined by (4.2). In the special case where the initial data for (3.5) satisfies $\nabla.(u, v, 0) = 0$ and has $\theta = \theta(z)$ only, (4.11) reproduces the solution of (3.5) exactly.

The argument above also shows that in the case $B \ll 1$, the projection preserves the pressure field to $O(B^2)$. In this case, we simply set $\pi = s$, so that the appropriate set of reduced equations is

$$
\begin{aligned}
\frac{D}{Dt}(u_b, v_b) + (-fV, fU) + \nabla_h s &= 0 \\
\frac{D}{Dt} &\equiv \frac{\partial}{\partial t} + \mathbf{U} \cdot \nabla \\
\frac{D\theta}{Dt} &= 0 \\
\frac{D\alpha}{Dt} &= 0 \\
(fv_b, -fu_b, g\theta/\theta_0) &= \nabla s.
\end{aligned}
\tag{4.13}
$$

These are precisely the semi-geostrophic equations studied by Cullen *et al.* (1987), Cullen and Purser (1989), Cullen *et al.* (1991) and many others.

This analysis suggests that the semi-geostrophic model is a good model of balanced flow if $B \ll 1$, but (4.11) is superior if $B \gg 1$, and is exact in the special case of non-divergent flow with stratification independent of the horizontal coordinates. This is consistent with results in the literature. The case $B \gg 1$ corresponds to most published integrations with the shallow water equations. Allen *et al.* (1990) and others have shown that the semi-geostrophic model is not very accurate in such cases, and a number of other reduced models are better. The chapter by McIntyre and Roulstone shows how the replacement of the pressure gradient by an approximation is a generic method of constructing reduced systems of equations.

As shown in the previous subsection, a complete estimate of the difference between the solution of the full equations and reduced equations also requires an estimate of the magnitude of the displacements Ξ. This can be deduced from the energy difference $E - e$. Using (3.10), we obtain

$$E - e \geq \int_{\Gamma} \lambda_{\min} |\xi|^2 d\tau \tag{4.14}$$

where $\lambda_{\min}$ is the smallest eigenvalue of Λ as defined by (3.11). We then have to estimate $E - e$ in terms of Ξ and p'. Estimates of p' can be obtained in principle from estimates of s and $\mathbf{u}_b$ as in the derivation of (4.11) and (4.13). Estimates of s have to be obtained from an existence theory for the chosen reduced system of equations. This programme of work has not yet been carried through.

4.3 Existence of solutions to reduced equations

In this subsection we discuss the properties of the solutions of the two characteristic reduced systems (4.13) and (4.11). Equations (4.13) can be written in terms of $\mathbf{X}$ variables as

$$\begin{aligned} \frac{D\mathbf{X}}{Dt} + f^{-1}\mathbf{k} \times \nabla s &= 0 \\ \frac{D\alpha}{Dt} &= 0 \\ (X, Y, Z) &= \nabla S. \end{aligned} \tag{4.15}$$

Using (3.16) and (3.19), the last term in the first equation of (4.15) has components

$$f^{-1}\left(-\frac{\partial s}{\partial y}, \frac{\partial s}{\partial x}, 0\right) = (y - Y, X - x, 0). \tag{4.16}$$

It was shown by Cullen and Purser (1989) that ρ as defined by (3.20) is conserved under (4.15) following fluid particles. We can thus rewrite equations

(4.15) and 4.16), using (3.21) and (3.22), as

$$\begin{aligned}\frac{D\rho}{Dt} &= 0 \\ \frac{D}{Dt} &\equiv \frac{\partial}{\partial t} + \mathbf{V}.\nabla_{\mathbf{X}} \\ \mathbf{V} &= (y - Y, X - x, 0) \\ (x, y, z) &= \nabla_{\mathbf{X}} R \\ \rho &= \det \frac{\partial^2 R}{\partial (X, Y, Z)^2}.\end{aligned} \tag{4.17}$$

These equations are to be solved in $\mathbf{X}$ coordinates for all $\mathbf{X} \in \mathbb{R}^3$, with boundary condition

$$\mathbf{x} = \nabla_{\mathbf{X}} R \in \Gamma. \tag{4.18}$$

Equations (4.17) and (4.18) can be interpreted as describing a motion of particles in (X, Y, Z) space, with velocity $\mathbf{V} = (y - Y, X - x, 0)$, conserving ρ. The solution procedure is therefore as follows. Given $\rho(X, Y, Z)$ as defined by (3.20), satisfying the compatibility conditions $\rho \geq 0, \int \rho = \mu(\Gamma)$, use the solution procedure discussed in section 3.5 to find a convex $R(X, Y, Z)$. This gives $\mathbf{x}$ as a function of $\mathbf{X}$, and allows $\mathbf{V}$ to be computed. The solution can then be advanced in time. Further details are given in Cullen and Purser (1989).

The 'velocity' $\mathbf{V}$ makes sense provided R is differentiable. The derivative of a convex function, however, may be multi-valued. If so, further work is required to allow the evolution equations (4.15) to make sense. Proving good behaviour of such 'transport' equations requires one of the two following properties to hold (Brenier, private communication). Either

$$|\mathbf{V}(t, \mathbf{X}) - \mathbf{V}(t, \tilde{\mathbf{X}})| \leq C(t)\eta(|\mathbf{X} - \tilde{\mathbf{X}}|) \tag{4.19}$$

where $C \in L^1_{\text{loc}}(\mathbb{R}^+)$ and $\int_0^1 \frac{ds}{\eta(s)} = \infty$; this is Cauchy-Lipschitz theory with Osgood's condition (see Gerard (1992)). Or

$$\int_{|\mathbf{X}| \leq R} |\nabla \mathbf{V}(t, \mathbf{X})| dx \leq C_R(t), \forall R < \infty \text{ with } C_R \in L^1_{\text{loc}}(\mathbb{R}^+). \tag{4.20}$$

This corresponds to the limiting case of the transport theory of DiPerna and Lions (1989). Benamou and Brenier (1998) have shown that weak solutions of (4.15) exist for initial data in $L^p(\mathbb{R}^3)$ for $p > 3$. For the periodic problem in two dimensions, the results of Caffarelli (1996) suggest that strong solutions may exist, since regularity of ρ can then be used to prove that R has two more derivatives than ρ. Cullen and Gangbo (2001) prove that weak solutions of the 2-dimensional shallow water version of (4.15) exist in $L^\infty((0, T); L^r(\mathbb{R}^2))$

for initial data in $L^r(B)$, where $r > 1$ and B is an open bounded ball in $\mathbb{R}^2$.

Now consider equations (4.11). These can be written in terms of $\mathbf{X}$, giving equations similar to (4.15) apart from the replacement of the last terms in the first equation by $f^{-1}\mathbf{k} \times \nabla\pi$. This equation no longer conserves ρ following particles, since in general $\nabla\pi$ cannot be written as a gradient in $\mathbf{X}$ space. It will, however, satisfy conservation of a form of potential vorticity closer to (4.4). In the case $w = 0$, the first equation of (4.11) is exactly the equation for the evolution of 2-dimensional incompressible flow. The theory for these equations is well established, giving existence and regularity under suitable conditions for infinite time, see e.g. Kato and Ponce (1986). In the special case where θ is a function of z only and $\mathbf{u}_b$ a function of (x, y) only, the solution of (4.11) satisfies $w = 0$ and so exists for all time, and is also a solution of (3.5). In general, it is necessary to control w, as in the work of Babin *et al.* (1996) and Marsden *et al.* (1995). In the other special case where (4.11) is vertically averaged, to give a reduced form of the shallow water equations, there is a good chance that $\nabla_h.\mathbf{v}$ can be controlled and a theory obtained based on the results for the case $\nabla_h.\mathbf{v} = 0$. See the chapter by Babin *et al.*, Babin *et al.* (1997) or Embid and Majda (1996).

4.4 Calculation of solutions to reduced equations

We illustrate first a solution procedure for equations (4.13). Write them in the form

$$
\begin{aligned}
\mathbf{Q}\begin{pmatrix} U \\ V \\ W \end{pmatrix} + \frac{\partial}{\partial t}\nabla s &= \mathbf{H} \\
\nabla.\mathbf{U} &= 0 \\
(fv_b, -fu_b, g\theta/\theta_0) &= \nabla s
\end{aligned}
\tag{4.21}
$$

where

$$
\mathbf{Q} = \begin{pmatrix} fv_{bx} + f^2 & fv_{by} & fv_{bz} \\ -fu_{bx} & f^2 - fu_{by} & -fu_{bz} \\ g\theta_x/\theta_0 & g\theta_y/\theta_0 & g\theta_z/\theta_0 \end{pmatrix}
\tag{4.22}
$$

and

$$
\mathbf{H} = \begin{pmatrix} f^2 u_b \\ f^2 v_b \\ 0 \end{pmatrix}.
\tag{4.23}
$$

This formulation is essentially due to Schubert (1985). Equations (4.11) take a similar form, with (4.23) replaced by

$$
\mathbf{H} = \begin{pmatrix} -f\frac{\partial\pi}{\partial y} \\ f\frac{\partial\pi}{\partial x} \\ 0 \end{pmatrix}.
\tag{4.24}
$$

Either of these equations can be solved by forming a single equation for $\frac{\partial \nabla s}{\partial t}$:

$$\begin{aligned} \nabla.\mathbf{Q}^{-1}\frac{\partial}{\partial t}\nabla s &= \nabla.\mathbf{Q}^{-1}\mathbf{H} \\ \left(\mathbf{Q}^{-1}\frac{\partial}{\partial t}\nabla s\right)_n &= (\mathbf{Q}^{-1}\mathbf{H})_n \text{ on } \partial\Gamma. \end{aligned} \tag{4.25}$$

These form an elliptic equation for $\frac{\partial \nabla s}{\partial t}$ if $\mathbf{Q}$ has no negative eigenvalues. For the semi-geostrophic equations, Shutts and Cullen (1987) show that $\mathbf{Q}$ is exactly the matrix that appears in (3.16), and negative eigenvalues correspond to unstable states not describable by semi-geostrophic theory. The existence results discussed in the previous subsection all prove existence of solutions with non-negative eigenvalues.

4.5 Generic forms of reduced equations

Equations (4.17) are an example of a generic form of reduced equations for large scale atmospheric and oceanic flow which take the form

$$\begin{aligned} \frac{Dq}{Dt} &= 0 \\ \frac{D}{Dt} &\equiv \frac{\partial}{\partial t} + \mathbf{V}.\nabla \\ \mathcal{H}(\mathbf{V},\theta) &= q \end{aligned} \tag{4.26}$$

where q is a form of potential vorticity such as in (4.3) or (4.4). Here $\mathcal{H}$ contains an elliptic operator which allows all the other fields to be derived from q. In the case of (4.17), this is the Monge–Ampère equation for R, and the calculation of $\mathbf{V}$ from R. Boundary conditions have to be chosen to allow solution of the elliptic problem. (4.18) are the appropriate conditions for (4.17), and do not imply any physical restriction other then no flow through rigid boundaries. Other examples where the Ertel form of potential vorticity (4.4) is used are given in the chapters by Allen *et al.*, Holm *et al.*, and Roulstone and McIntyre. Boundary conditions are more of an issue in these cases.

The conservation law $\frac{D\theta}{Dt} = 0$ is implied by (4.26). The statement is often then made that potential vorticity is advected quasi-2-dimensionally along θ surfaces. However, this ignores the nontrivial motion of the θ surfaces themselves, and may be misleading. In particular, the θ surfaces will usually intersect the lower and, in the ocean, upper boundary at positions which change in time.

This formulation is very convenient for proving existence of solutions, as shown in section 4.3. However, it also allows information about the properties of the solution to be deduced. The conservation of potential vorticity makes the equations similar to the equations for 2-dimensional incompressible flow, which conserve vorticity. The equations for 3-dimensional incompressible flow do not conserve any vorticity-like quantity. Hence the large scale flow of the atmosphere and ocean is often described in terms of 2-dimensional turbulence, e.g. Leith (1983).

Equations (4.26) also allow deductions about the sensitivity of the solutions to small perturbations. If the fluid is displaced by a field Ξ, retaining potential vorticity conservation, then we have

$$\mathcal{H}(\mathbf{V}+\delta\mathbf{V},\theta+\delta\theta) = q - \Xi.\nabla q. \tag{4.27}$$

Subtracting the final equation of (4.26) from (4.27) and linearising gives $\hat{\mathcal{H}}(\delta\mathbf{V},\delta\theta) = -\Xi.\nabla q$. This can be solved for $(\delta\mathbf{V},\delta\theta)$ in terms of Ξ. The key quantities will be the eigenfunctions and eigenvalues of $\hat{\mathcal{H}}^{-1}$. Small eigenvalues will result in large sensitivity of the solutions to perturbations in the direction of the associated eigenfunction, unless $\Xi.\nabla q$ happens to be small. These properties are exploited in section 8.5.

The most useful reduced systems of the form (4.26) also conserve energy. Both (4.11) and (4.13) conserve the balanced energy e. As discussed in the chapters by Holm *et al.*, Bokhove, and Roulstone and McIntyre, such systems can often be written in Hamiltonian form, with the potential vorticity equation expressing the Liouville theorem or conservation of symplectic structure. Where the conserved potential vorticity takes the form (4.3), the conservation law is clearly seen as a conservation of phase space volume, with $\mathbf{X}$ being the phase space coordinates.

5 Sub-grid models and their interaction with resolved dynamics

5.1 Introduction

Equations (2.1) and (2.3) have to be solved in practical applications with a sub-grid model which expresses the effects of unresolved motions. It is also necessary to include forcing terms. In the atmosphere the effects of phase changes of water are particularly important. This section is primarily written from an atmospheric viewpoint, though the purely dynamical aspects apply to the ocean as well. In the ocean, the effects of salinity have also to be included in the sub-grid model.

The sub-grid model in operational atmospheric models includes

(a) Actual forcing terms, such as radiative heating.

(b) Explicit terms describing phase changes of water.

(c) Interactions with the lower boundary.

(d) Averaged effects of sub-grid dynamics.

A generalisation of (3.5) to include these effects can be written

$$\begin{aligned}
\frac{D\mathbf{u}_h}{Dt} + (-fv, fu) + \nabla_h p' &= (F_u, F_v) \\
\frac{Dw}{Dt} - g\theta/\theta_0 + \frac{\partial p'}{\partial z} &= 0 \\
\nabla.\mathbf{v} &= 0 \\
\frac{D\theta}{Dt} &= F_h + H - LP + S_h \\
\frac{Dr}{Dt} &= F_r + P + S_r \\
\mathbf{v}.\mathbf{n} &= 0 \text{ on } \partial\Gamma.
\end{aligned} \tag{5.1}$$

Here r represents the water vapour content, (F_u, F_v, F_h, F_r) represent sub-grid increments to (u, v, θ, r) respectively. S_h, S_r are source terms for heat and moisture at the lower boundary. H is a source term for heat in the free atmosphere (typically radiation). P is a source/sink term for water vapour, resulting in a term LP in the thermodynamic equation where L is the latent heat. We only consider very simple versions of these terms, ignoring such issues as the difference between condensed water and ice.

If equations (3.5) were being solved exactly, only the source terms would have to be included and (F_u, F_v, F_h, F_r) could be omitted. The viscous and conduction terms from (2.1) and (2.3) would then be resolved and have to be included. In the realistic situation where the equations have to be averaged in space and time before being solved, it is of interest to study how the sub-grid model interacts with the larger scale dynamics. One way of doing this is to include the same effects in a reduced set of equations, and study the effect on the solutions. Observations and computations both show that reduced systems of equations can be quite accurate in describing weather systems even where there are strong effects due to latent heat release and there are significant regions of instability, such as in the mesoscale convective systems which are common over continents in summer. A recent such study is described by Olsson and Cotton (1997). Similarly, observations show that, if a simple boundary layer model is incorporated in the reduced dynamics, the interactions of fronts with the atmospheric boundary layer can be described well, e.g. Ostdiek and Blumen (1997).

5.2 Effect on energy minimisation

Geostrophic and hydrostatic balance can be generalised for equation (5.1) to

$$\begin{aligned}
(-fv, fu) + \nabla_h p' &= (F_u, F_v) \\
-g\theta/\theta_0 + \frac{\partial p'}{\partial z} &= 0.
\end{aligned} \tag{5.2}$$

The first equation in (5.2) is sometimes called 'geotriptic' balance (Johnson (1966)), and we will use this term forthwith. If (F_u, F_v) can be linearised to a form $-c_D(u, v)$, then we can show that (5.2) represents the condition for the energy (3.13) to be made stationary subject to variations

$$\begin{aligned}
\delta\mathbf{v} &= f(\eta, -\xi) - c_D(\xi, \eta) \\
\delta\theta &= 0 \\
\Xi &= (\xi, \eta, \zeta) \\
\nabla.\Xi &= 0 \\
\Xi.\mathbf{n} &= 0 \text{ on } \partial\Gamma.
\end{aligned} \tag{5.3}$$

With this linearisation of F, the geotriptic and hydrostatic relations (5.2) can be written

$$(fv - c_D u, -fu - c_D v, g\theta/\theta_0) = \nabla s. \tag{5.4}$$

The condition for a minimum is that the matrix $\mathbf{Q}$ whose components are

$$\mathbf{Q} = \begin{pmatrix} fv_x - c_D u_x + f^2 + c_D^2 & fv_y - c_D u_y & fv_z - c_D u_z \\ -fu_x - c_D v_x & f^2 + c_D^2 - fu_y - c_D v_y & -fu_z - c_D v_z \\ g\theta_x/\theta_0 & g\theta_y/\theta_0 & g\theta_z/\theta_0 \end{pmatrix} \tag{5.5}$$

is positive definite. Equation (5.5) shows that the effect of the drag term is to increase the effective stability of the flow and make it easier to satisfy the condition for energy minimisation. Knox (1997) reviews several studies which show that sub-grid scale effects lead to modifications of the inertial stability criteria, and hence allow the existence of regions in the atmosphere that appear to violate the large scale inertial stability condition.

The further manipulations of section 3.5 and those in section 5.3 can only be carried out if we assume c_D is independent of x and y. This is not very realistic, though it would be a viable approximation for process studies, where we study the effect of surface friction on weather systems over a homogeneous surface (e.g. all ocean, or all pine forest).

5.3 Effect on maintenance of balance in the flow

We seek a reduced system of equations based on equations (5.1). To do this, we make some simple but qualitatively realistic choices for the various sub-grid terms. Greater sophistication is obviously possible.

As in the previous subsection, let $(F_u, F_v) = -c_D(u, v)$. Set $F_h = 0$, and treat S_h as a specified function of position. The moisture source term that affects the thermodynamic equation depends on the rate of change of the difference between the saturation vapour pressure and the actual pressure following a fluid particle. This depends strongly on pressure, which has a large vertical variation. A useful simple approximation is to set $LP = w\frac{\partial}{\partial z}(\theta - \theta_E)$, where θ_E is the 'effective' potential temperature. This should only be used in saturated regions, where $r \geq r_{\mathrm{SAT}}(p)$. Much more information is given in standard meteorological textbooks, such as Haltiner and Williams (1980). These simplifications decouple the moisture variable r from the equations, except at the boundary between saturated and unsaturated regions. This choice of LP allows the most important effect of moisture on the large scale dynamics to be discussed.

We now write down a reduced system of equations analogous to the semi-geostrophic equations (4.13). In principle, other sets of reduced equations, such as (4.11), could be used as a starting point. (4.13) can be written

$$\begin{aligned} \frac{D}{Dt}(u_b, v_b) + f(v_b - V, U - u_b) &= 0 \\ \frac{D\theta}{Dt} &= 0 \\ \frac{D\alpha}{Dt} &= 0 \\ (fv_b, -fu_b, g\theta/\theta_0) &= \nabla s. \end{aligned} \tag{5.6}$$

An extension of these to include the simplified sub-grid terms is

$$\begin{aligned} \frac{D}{Dt}(u_c, v_c) + f(v_c - V, U - u_c) &= -c_D(u_c - U, v_c - V) \\ \frac{D\theta_E}{Dt} &= S_h; r \geq r_{SAT} \\ \frac{D\theta}{Dt} &= S_h; r < r_{SAT} \\ \frac{Dr}{Dt} &= 0 \\ \frac{D\alpha}{Dt} &= 0 \\ (fv_c - c_D u_c, -fu_c - c_D v_c, g\theta/\theta_0) &= \nabla s. \end{aligned} \tag{5.7}$$

The particular form of the first equation is designed to ensure the drag acts as a sink of kinetic energy. It can be shown that a term $-c_D(u_c^2 + v_c^2)$ is added to the kinetic energy equation. (5.7) can be written in the form (4.21), where now

$$\mathbf{Q} = \begin{pmatrix} fv_{cx} - c_D u_{cx} + f^2 + c_D^2 & fv_{cy} - c_D u_{cy} & fv_z - c_D u_{cz} \\ -fu_{cx} - c_D v_{cx} & f^2 + c_D^2 - fu_{cy} - c_D v_{cy} & -fu_{cz} - c_D v_{cz} \\ g\Theta_x/\theta_0 & g\Theta_y/\theta_0 & g\Theta_z/\theta_0 \end{pmatrix} \tag{5.8}$$

$$\mathbf{H} = \begin{pmatrix} (f^2 + c_D^2)u_c \\ (f^2 + c_D^2)v_c \\ gS_h/\theta_0 \end{pmatrix} \tag{5.9}$$

with $\Theta = \theta$ if $r < r_{\mathrm{SAT}}$, $\Theta = \theta_E$ otherwise. The form of $\mathbf{Q}$ given by (5.8) is a simple generalisation of the form (5.5) which appears in the energy minimisation problem. This simple form of equation is not valid at saturation boundaries.

We can see by comparing (5.8) and (5.9) to (4.21) that the effect of the sub-grid terms is divided between altering the $\mathbf{Q}$ matrix, which determines the response to the forcing, and altering the forcing itself. Thus friction makes the model less responsive to forcing, and latent heating makes it more responsive. The friction also contributes to the forcing directly, while latent heating does not. It is therefore rather misleading to think of latent heating driving large scale atmospheric circulations, it is better to think of it as changing the characteristic motions that result. See, for instance, Thorpe and Emanuel (1985).

By analogy with the solutions of (4.21), we can expect the existence of solutions to (5.7) where the eigenvalues of $\mathbf{Q}$ are non-negative. The case where a region of fluid has zero eigenvalues, corresponding to zero potential vorticity, means that the fluid is well-mixed in one or more of $(fv_c - c_D u_c, -fu_c - c_D v_c, g\theta/\theta_0)$. An example is a well-mixed boundary layer where θ is independent of height. In such situations the implicit transport (U, V, W) becomes instantaneous diffusion, which spreads the effect of forcing uniformly through the well-mixed region. In the boundary layer example, the mixing is immediate through the depth of the boundary layer, see Cullen *et al.* (1987).

A particular effect of latent heating is through the effect of saturation boundaries, which is not directly treatable in (5.8). If the air is either saturated or unsaturated everywhere in a region, observations show that negative eigenvalues of $\mathbf{Q}$ can only be sustained where the forcing is very strong, such as in the lower layers of the atmosphere over sub-tropical deserts. The approximation $\det \mathbf{Q} = 0$ is quite good in frontal zones with active precipitation, Emanuel (1983). The initiation of major convective events, such as thunderstorms, is associated with the rapid transition of a significant mass of air from unsaturated to saturated, which results in a discontinuous loss of stability.

A simplified model of this process has been constructed by Shutts (1987). He solved a finite-dimensional version of (5.7) in a vertical cross section without frictional drag. The data was represented by piecewise constant values of $v_c + fx$, θ and r. All variables were considered as independent of y. The solution procedure described in section 3.5 was used to construct solutions. The effects of latent heat release were included by carrying out standard parcel thermodynamic calculations. An iterative procedure was used, first constructing the solution ignoring latent heating, then calculating the latent heating and solving for new parcel positions. When there is such a discontinuous loss of stability, the parcels jump in the vertical to a new equilibrium position and energy is lost. This can be interpreted as there being a mass sink at low levels and a mass source at high levels, Shutts (1995). The solutions are well-defined without including the effects of evaporative cooling, which according to Emanuel *et al.* (1994) are needed to make the large scale flow stable. The convective transport stabilises the atmosphere by redistributing the heating away from the source region through the jump in the parcel position and the compensating subsidence of other parcels. The lost energy is assumed to be converted into unbalanced waves which disperse and dissipate.

This vertical jump corresponds to a mass rearrangement which cannot be generated by a smooth trajectory, and thus represents a 'generalised flow', as discussed by Brenier (1990). It should be possible to extend the mathematical theory to cover this case and prove that the above iteration converges. This would involve characterising the solution of (4.21) with a matrix $\mathbf{Q}$ which is not positive definite as having a solution for (u, v, w) which is a generalised flow, rather than a continuous velocity field. However, the 'pressure tendency' $\frac{\partial s}{\partial t}$ will remain well-defined and bounded everywhere.

Standard procedures for representing moist convection in operational weather forecasting models use this type of iteration, applied in the vertical only. The theory discussed above may allow these procedures to be put on a sound basis. A similar model has been used by Shutts (1987) to discuss the interaction of balanced flow with orography.

5.4 Maintenance of large scale balances by sub-grid transport

In this subsection we consider a more general approach to representing the effects of the sub-grid dynamical terms (F_u, F_v, F_h, F_r) in (5.1). This is based on methods developed for the ocean by Gent and McWilliams (1996). A similar analysis has been carried out by Buhler and McIntyre (1998).

The sub-grid dynamical terms come from averaging the nonlinear terms in the basic equations. If we take these to be (3.5), together with a moisture conservation equation, then

$$\begin{aligned}
\frac{\overline{D}\overline{\mathbf{v}}_h}{Dt} + (-f\overline{v}, f\overline{u}) + \nabla_h \overline{p}' &= \overline{\mathbf{v}}.\nabla\overline{\mathbf{v}}_h - \overline{\mathbf{v}.\nabla\mathbf{v}_h} \\
\frac{\overline{D}\overline{w}}{Dt} - g\overline{\theta}/\theta_0 + \frac{\partial \overline{p}'}{\partial z} &= \overline{\mathbf{v}}.\nabla\overline{w} - \overline{\mathbf{v}.\nabla w} \\
\nabla.\overline{\mathbf{v}} &= 0 \\
\frac{\overline{D}}{Dt}\overline{\theta} &= \overline{\mathbf{v}}.\nabla\overline{\theta} - \overline{\mathbf{v}.\nabla\theta} \\
\frac{\overline{D}\overline{r}}{Dt} &= \overline{\mathbf{v}}.\nabla\overline{r} - \overline{\mathbf{v}.\nabla r} \\
\overline{\mathbf{v}}.\mathbf{n} &= 0 \text{ on } \partial\Gamma \\
\frac{\overline{D}}{Dt} &\equiv \frac{\partial}{\partial t} + \overline{\mathbf{v}}.\nabla.
\end{aligned} \tag{5.10}$$

The overbar denotes the averaging operator, which can be any form of low pass filter in space and time. The sub-grid correlation terms are written in exact form on the right hand sides. It can be considered as a bilinear operator acting on the prognostic quantities $(\overline{\mathbf{v}}, \overline{\theta}, \overline{r})$. Gent and McWilliams propose a specific way of representing the sub-grid correlations in terms of the averaged quantities. They first split the operator into antisymmetric and symmetric parts. The antisymmetric part applied to a single variable ϕ can always be written as a pseudo-advection $\mathbf{W}.\nabla\overline{\phi}$. If the only nonlinearity in the equations were a set of terms $\mathbf{v}.\nabla\overline{\phi_n}$, for a set of scalars ϕ_n, then the antisymmetric part of the sub-grid model could be written as $\mathbf{W}.\nabla\overline{\phi_n}$ with the same pseudo-advection velocity $\mathbf{W}$ for each n. As an example of this, consider the case of 2-dimensional motion in hydrostatic balance. Assuming there is no variation in the y direction, the y momentum and thermodynamic equations from (5.10) become

$$\begin{aligned}
\frac{\overline{D(v+fx)}}{Dt} &= \overline{\mathbf{v}}.\nabla\overline{(v+fx)} - \overline{\mathbf{v}.\nabla(v+fx)} \\
\frac{\overline{D}}{Dt}\overline{\theta} &= \overline{\mathbf{v}}.\nabla\overline{\theta} - \overline{\mathbf{v}.\nabla\theta} \\
\frac{\overline{D}\overline{r}}{Dt} &= \overline{\mathbf{v}}.\nabla\overline{r} - \overline{\mathbf{v}.\nabla r}.
\end{aligned} \tag{5.11}$$

Gent and McWilliams formally do this for general 3-dimensional flow, by adding or subtracting appropriate terms from each side of the momentum equations. They thus write (5.10), with the hydrostatic approximation, as

$$\begin{aligned}
\frac{D^*\overline{\mathbf{v}}_h}{Dt} + (-fV, fU) + \nabla_h\overline{p}' &= \nabla.\mathbf{E}_h \\
-g\overline{\theta}/\theta_0 + \frac{\partial\overline{p}'}{\partial z} &= 0 \\
\nabla.\mathbf{V} &= 0 \\
\frac{D^*\overline{\theta}}{Dt} &= 0 \\
\frac{D^*\overline{r}}{Dt} &= F_r \\
\mathbf{V}.\mathbf{n} &= 0 \text{ on } \partial\Gamma.
\end{aligned} \tag{5.12}$$

They write out the full forms of $\mathbf{E}_h$ and F_r in their paper. Here, D^*/Dt represents advection by a velocity $\mathbf{V} = \overline{\mathbf{v}} + \mathbf{v}'$, where $\mathbf{v}'$ is the 'sub-grid velocity', which they choose an explicit formula for in their paper. These equations are identical to (5.10), apart from the use of hydrostatic balance. The usefulness of the form (5.12) depends on being able to model $\mathbf{E}_h$ in terms of known averaged quantities, which will be easier if it can be argued that $\mathbf{E}_h$ is small compared with other terms in the momentum equation. The right hand side of the θ equation is set to zero because of their assumption that there are no sub-grid fluxes across isentropic surfaces (isopycnal surfaces in the ocean). The use of the 'total' velocity $\mathbf{V}$ in the Coriolis term and the choice that $\mathbf{V}$ satisfies the continuity equation and boundary conditions means that the right hand side of the momentum equation in (5.12) takes the form of a divergence as shown, and means that non-acceleration theorems are respected.

We can now make a link between (5.12) and reduced systems of equations such as (4.13). If $\mathbf{E}_h$ is either zero, or can be represented explicitly in terms of the mean quantities, and the averaging scale is sufficiently large for $\overline{\mathbf{v}}_h$ and $\overline{\theta}$ to be in geostrophic and hydrostatic balance, then (5.12) and (4.13) become identical, with a forcing term added to the right hand side of the momentum equation in (4.13). The existence theory for (4.13) shows that $\mathbf{V}$ is uniquely determined. Thus the part of the sub-grid model represented by the pseudo-advection plays the role of maintaining large scale balance and is completely determined by the large scale balance requirement. In addition, the equations can be solved with $\mathbf{E}_h = 0$, in which case exactly the semi-geostrophic solution is obtained. Since this solution can be discontinuous, it implies the averaging operator must be a parcel average, rather than an Eulerian average.

The analysis by Buhler and McIntyre (1998) is specifically in terms of Lagrangian averaging. They derive equations of the same form as (5.12) in which the 'total' velocity $\mathbf{V}$ is the Lagrangian mean velocity, and the transported momentum $\mathbf{v}_h$ is $\mathbf{V} + \mathbf{p}$, where p is the 'pseudo-momentum' of the waves that have been filtered out. There is an additional term in the equations representing the perturbation pressure associated with the waves. If the perturbation pressure is negligible, and the averaging scale can be assumed to be large enough that the total momentum $\mathbf{V} + \mathbf{p}$ is in geostrophic balance, then again (5.12) and (4.13) become identical. The neglect of perturbation pressures was noted in section 3.4 to be the basic assumption made in deriving the semi-geostrophic equations. This interpretation allows us to give a physical meaning to the two velocities appearing in the theory, and to connect the semi-geostrophic equations with a formal averaging of the primitive equations.

If this type of formulation is used for the moist convectively unstable case discussed in the previous subsection, then $\mathbf{V}$ becomes a generalised flow, and its effect cannot be represented as simple advection, but as an antisymmetric operator acting on a full vertical column of values of temperature and moisture. Similarly, it may be possible to represent the transport in an unstable or well-mixed boundary layer in this formalism.

The link between the specific reduced system (4.13) and (5.12) is natural because both are likely to be most accurate when the flow is approximately

2-dimensional. A similar identification should be possible for systems describing approximately axisymmetric flows. More generally, Hamiltonian reduced systems which represent the fluid trajectory separately from the momentum, such as some of those described in the chapters by McIntyre and Roulstone and by Holm, can be given a physical interpretation in terms of a Lagrangian mean velocity that includes sub-grid transport and a transported momentum that includes a wave component.

The practical implications of this link in designing sub-grid models for operational atmospheric models are discussed in section 7. The physical implications are also important. For instance, a fundamental part of the general circulation of the atmosphere is the average ascending motion in the tropics. Much of this represents the overall effect of a large number of individual convective events, on quite small scales. Emanuel *et al.* (1994, p.1125) state that 'nearly all the upward motion associated with ensemble averaged ascent must appear as increased mass flux in cumulus clouds'. In this formulation, the increased upward motion would appear as an increased generalised flow $\mathbf{V}$.

5.5 Treatment of unstable regions

The previous sections have emphasised the importance of understanding how the sub-grid model interacts with flows that can be described by a reduced set of equations. Such equations can typically only describe stable states of the atmosphere. Observations, however, indicate that, on average, it is possible to have long-lived unstable regions where there is strong forcing. The best example is the near-surface layers of the atmosphere over deserts strongly heated by the sun. This situation can also occur where air is being cooled by precipitation at a rate too fast for the air to be assumed to remain in a stable position. This can happen in large scale convective systems. Shutts *et al.* (1988) include this effect in a reduced system of equations, but the requirement of large scale balance is probably unreasonable, and the resulting solution may exaggerate the effect of the precipitation.

It is simplest to consider the first example. In (5.10), assume that $\frac{\partial \overline{\theta}}{\partial z}$ is negative, then the mean fields are statically unstable and N^2 as given by (2.10) is negative. The left hand side of (5.10) has solutions which grow exponentially. Assuming a 2-dimensional wave-like perturbation $\exp(\omega t + i(kx + mz))$, the growth rate is given by

$$\omega = \frac{k}{m}\sqrt{(-N^2)}. \tag{5.13}$$

The growth rate is largest on the smallest horizontal scale permitted by the averaging, and the largest vertical scale permitted by the atmosphere. A sub-grid model which is correct in the sense of allowing quasi-steady solutions to persist has to exclude these growing solutions, or reduce the growth rate to be comparable to observed time-scales of variability. For example, suppose we choose a sub-grid model which diffuses in the x direction, where the maximum growth rate is on the smallest scale. A linearised version of (5.10) with such a model included is then

$$\begin{aligned}\frac{\partial \overline{u}}{\partial t} + \frac{\partial \overline{p}'}{\partial x} &= K\frac{\partial^2 \overline{u}}{\partial x^2} \\ -g\overline{\theta}/\theta_0 + \frac{\partial \overline{p}'}{\partial z} &= 0 \\ \frac{\partial \overline{u}}{\partial x} + \frac{\partial \overline{w}}{\partial z} &= 0 \\ \frac{\partial \overline{\theta}}{\partial t} + w\frac{\partial \overline{\theta}}{\partial z} &= K\frac{\partial^2 \overline{\theta}}{\partial x^2}.\end{aligned} \tag{5.14}$$

Assuming solutions proportional to $\exp(\omega t + i(kx + mz))$ gives

$$(\omega + k^2 K)^2 m^2 + k^2 N^2 = 0. \tag{5.15}$$

This has no growing solutions if $K > (\sqrt{(-N^2)})/mk$. If K is chosen to be greater than this value where m and k are the smallest wave numbers permitted by the domain, or by the region over which N^2 is negative, the sub-grid model will be consistent with the persistence of the observed unstable region. The values of diffusion required will typically be large, since it will have to damp on a time-scale comparable to the growth rate of the instabilities multiplied by k/m. Following (5.13), the largest value of k/m permitted by the size of the unstable region corresponds to the maximum growth rate, and must therefore be used in calculating the damping time-scale required. This diffusion should only be used in unstable regions of the flow.

5.6 Example of modelling of the response to thermal forcing by a reduced set of equations

In this subsection we give an example of how the large scale response to atmospheric forcing can be described by the solution of a reduced set of equations with the appropriate forcing terms included. We thus solve (5.7) on the sphere, with a given heat source in the Northern hemisphere subtropics and a mountain ridge crossing the equator confining the response to the forcing. The experiment is described by Mawson and Cullen (1992).

The last equation in (5.7) shows that, outside the boundary layer, $\nabla_h s$ must be small near the equator. Differentiating with respect to z shows that $\nabla_h \theta$ will be small. The total velocity $\mathbf{u}$ can be found from (4.21) with (5.8) and (5.9). If the heating function S_h is localised, a steady solution for $\mathbf{u}$ will be approximately given by

$$w\frac{\partial \Theta}{\partial z} = S_h. \tag{5.16}$$

This suggests that there will be ascending motion in the region of the heat source, with inflow from other regions at low levels to balance it. (5.8) shows that the largest response in the horizontal wind will be in directions where $f^2 + c_D^2$ is smallest, Schubert *et al.* (1991), so that the greatest response will be from the equatorial side of the heat source above the boundary layer. In the boundary layer, (5.4) shows that a pressure gradient will be set up parallel to the inflow.

The experiment includes a mountain ridge crossing the equator to the west of the region where the heat source is applied. This should have the effect of confining the inflow towards the heated region, and locally inducing a maximum cross-equatorial wind on the eastern side of the ridge.

The results are illustrated in Figure 12. This shows the low level wind induced by the heat source. The coastlines are included for illustrative purposes only. The heat source is centred over Northern India and the mountain ridge crosses the equator in east Africa. The solutions of (5.7) show a maximum cross-equatorial flow on the eastern side of the ridge, and a circulation around the heat source. Solutions of a similar model based directly on (5.1) are also shown. It is clear that the bulk of the solution is represented by the reduced equations (5.7), though there are detailed differences. Observations, Findlater (1969), confirm the existence of a low level jet in the area shown. Thus these computations and comparisons with observations suggest the value of the various modelling schemes proposed in this section.

6 Evolutionary properties of atmosphere/ocean circulations

6.1 General remarks

In this section we seek to derive a picture of the large scale dynamics of the atmosphere or ocean as a dynamical system. The basic circulation is symmetric about the poles, with higher temperatures at the equator, as a result of the pole/equator radiation difference. There are permanent asymmetries in the circulation, which are most likely to be a response to the asymmetric land-sea and mountain distribution, particularly in the Northern hemisphere. The Southern hemispheric circulation is much more symmetric. The total circulation also has large transient asymmetries. It is permanently unsteady, but with occasional episodes when there are large quasi-steady circulations. These are called blocking patterns, and can give anomalous weather over periods from weeks to a few months. Qualitatively, the weather map looks much the same from day to day (Figure 13). The characteristic scales of weather systems are always essentially the same. Deterministic prediction is, however, limited to a few life-cycles of the individual weather systems.

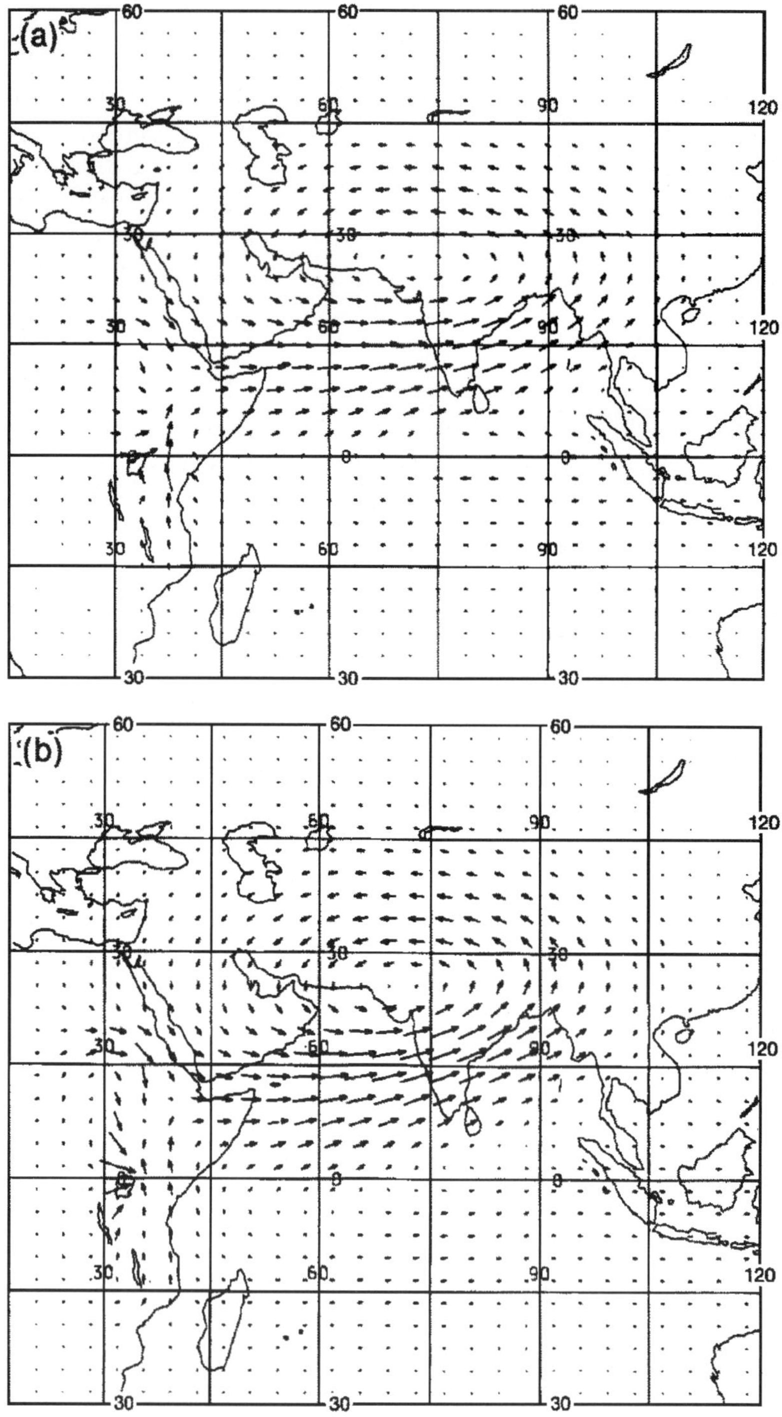

Figure 12: Forecast wind vectors at surface between longitudes 10°E and 180°E and latitudes 30°S to 60°N from idealised simulation of cross-equatorial flow using (a) equations (5.7), (b) equations (5.1). After Mawson and Cullen (1992).

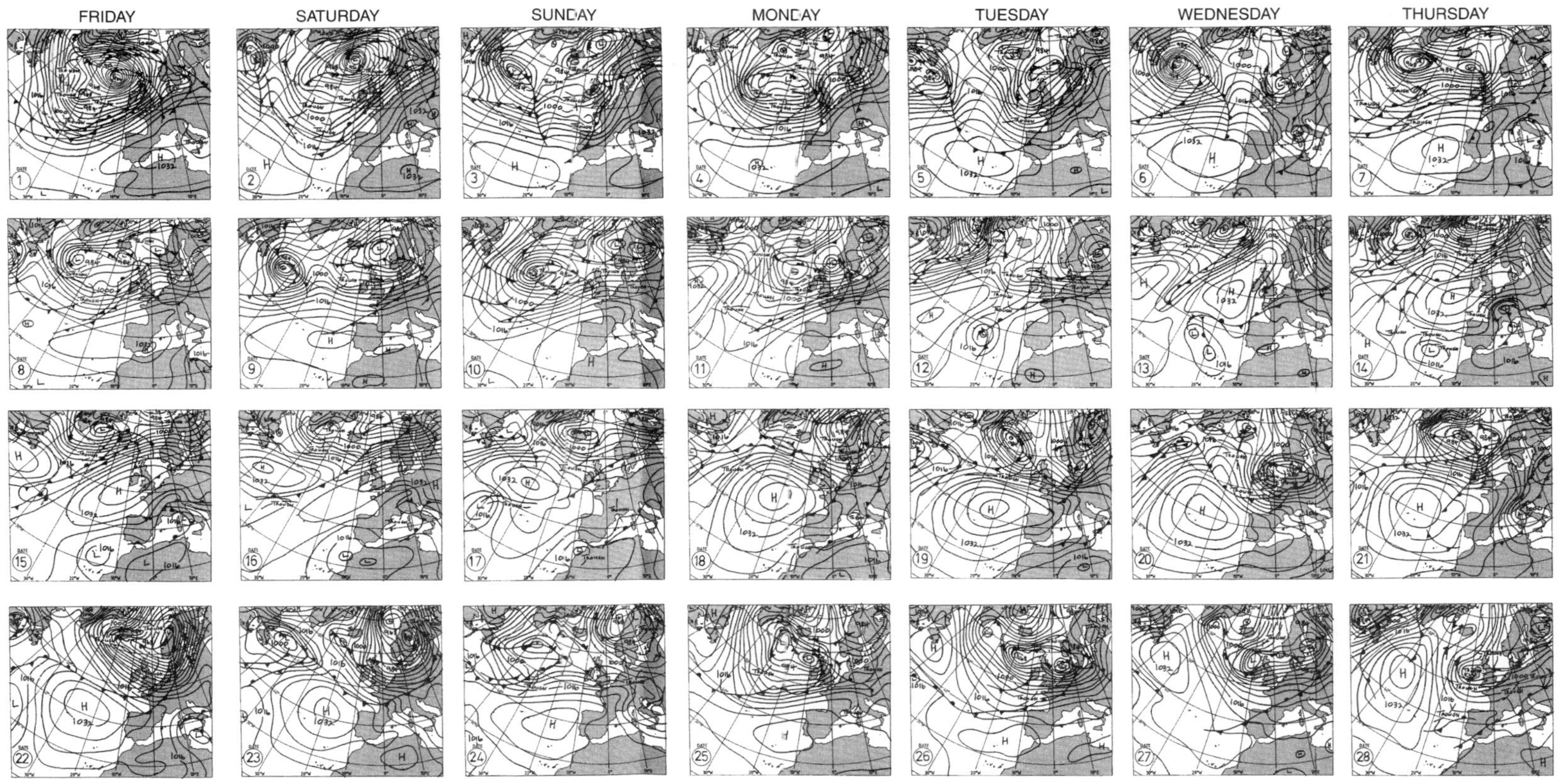

Figure 13: Daily weather maps from 1–28 February 2002. Dates are ringed at lower left hand corner of each map. Maps are of sea level pressure with contour interval of 4 hpa. Source *Weather Log*, Royal Meteorological Society.

It is natural, therefore, to seek a model of the underlying large scale dynamics which has stationary statistics and gives quasi-periodic solutions which can be continued indefinitely in time. The asymmetries and seasonal variations would then follow when the boundary conditions and forcing terms were included. Existing theory does not achieve this, because no argument has been found which prevents a systematic generation of small scales in the flow. The stationary statistics of the observed flow have to be explained by a balance between forcing and dissipation.

Since the pioneering work of Lorenz (1963), there have been many attempts to describe the qualitative features of the large scale dynamics discussed above using low order systems of equations. In this section we, instead, exploit the property, discussed in section 4, that the solutions to the general equation stay close to those of a reduced system. As discussed in section 4.5, we use a reduced system which is Hamiltonian, so in particular has a conserved energy and potential vorticity. The Hamiltonian property means there can be no attractors.

The general method is to exploit the simultaneous conservation of energy, potential vorticity and potential temperature. We first show that the condition for the flow to be steady is that it is a stationary point of the energy with respect to perturbations which simultaneously rearrange the potential vorticity and potential temperature. This type of condition is a general characterisation of steady states of Hamiltonian systems, see Abarbanel and Holm (1987). We then use the principle of Kelvin (Thomson (1910)) to say that stable steady states correspond to strict maxima or minima of the energy. A typical result would be that there are only two globally stable steady states associated with a given potential vorticity and potential temperature distribution, one achieved by maximising and one achieved by minimising the energy. The actual result in a particular case will depend on the boundary conditions and physical restrictions on the permitted rearrangements.

Given an initial distribution of potential vorticity and potential temperature, construct the minimum energy state. If the initial data has greater energy, the evolution will then remain a fixed distance in energy above this basic state. A Hamiltonian system has no attractors. Thus if the initial data represents an unsteady state, it cannot evolve to a steady state or a periodic solution as these would be attractors. If the initial data is a steady state, it cannot be guaranteed to be nonlinearly stable unless the energy is equal to a strictly minimising or maximising value.

More information can be obtained by seeking locally stable states, which are extrema of the energy with respect to continuous displacements, a subset of the global rearrangements. These can be used to explain the long-lived anomalies in atmospheric or oceanic flows. However, a general solution cannot be attracted to the neighbourhood of such a steady state, as that would make it an attractor. Non-Hamiltonian forcing terms must be included to make this happen.

The other results we discuss concern the preservation of the statistics of the flow. As shown in section 4.5, the reduced systems of equations commonly used to describe large scale atmospheric motions have a structure similar to the equations governing 2-dimensional incompressible flow. The solutions of the 2-dimensional Euler equations describe '2-dimensional turbulence', Leith (1983), with a k^{-3} energy spectrum. Theory and computation indicates that such flows exhibit a migration of the energy to the largest available scale, and the enstrophy to the smallest scales, the 'enstrophy cascade'. It then has to be argued that the observed preservation of the mean scale of disturbances is due to the systematic injection of energy at particular scales by non-conservative effects.

We discuss this behaviour first in terms of bounds on the L_2 norm of the velocity and its gradient. If both these are separately conserved, or bounded uniformly in time, then the mean scale is also bounded in time. (This argument was first advanced by Fjortoft (1953) based on energy and enstrophy conservation.) We also discuss the issue in terms of local estimates of the velocity gradients. Control of these would prevent systematic generation of small scales even in small regions of the flow, and thus give greater control over the flow statistics.

6.2 Model equations

Many of the results available can be simply illustrated by the equations for 2-dimensional incompressible flow. These can be written on a region Γ as

$$\begin{aligned} \frac{D\zeta}{Dt} &= 0 \\ \frac{D}{Dt} &\equiv \frac{\partial}{\partial t} + \mathbf{U}.\nabla \\ \mathbf{U} &= \left(-\frac{\partial\psi}{\partial y}, \frac{\partial\psi}{\partial x}\right) \\ \zeta &= \nabla^2\psi \\ \mathbf{U.n} &= 0 \text{ on } \delta\Gamma. \end{aligned} \tag{6.1}$$

Burton and McLeod (1991) proved that if the region Γ is a disk, the energy is maximised by a symmetrical rearrangement of the vorticity ζ which is monotonically increasing with stream-function. The energy can only be minimised, in general, over the 'weak closure' of the rearrangements of vorticity, in the sense of section 3.6. These results and others are described in the chapter by Douglas earlier in this volume.

Applications of this type of theory to large scale atmospheric flow requires inclusion of the variation with latitude of the vorticity associated with the

Earth's rotation. The equations for 2-dimensional incompressible flow on a rotating spherical surface are

$$\begin{aligned} \frac{D}{Dt}(\zeta + f) &= 0 \\ \frac{D}{Dt} &\equiv \frac{\partial}{\partial t} + \mathbf{U}.\nabla\zeta \\ \mathbf{U} &= \left(-\frac{\partial\psi}{\partial y}, \frac{\partial\psi}{\partial x}\right) \\ \zeta &= \nabla^2\psi \\ f &= 2\Omega\sin\phi \\ \mathbf{U}.\mathbf{n} &= 0 \text{ on } \partial\Gamma. \end{aligned} \tag{6.2}$$

Solutions of these equations are rearrangements of the initial absolute vorticity $\zeta + f$. The energy is conserved, but depends on the stream-function which is calculated only from the relative vorticity ζ. The angular momentum is conserved. The planetary vorticity f is given by $2\Omega\sin\phi$, which is a smooth monotonic function of latitude. The natural conjecture is that the rearrangements of absolute vorticity which extremise the energy are functions of latitude only, and are monotonically increasing or decreasing with latitude. Only the minimum energy state, corresponding to the absolute vorticity increasing with latitude, as f does, is likely to be physically realisable.

The difficulty is again the possibility of 'mixing' rearrangements, as discussed in section 3.6. If the given distribution of $\zeta + f$ can be mixed, subject to angular momentum conservation, to give a distribution equal to f, the minimum energy reachable will be zero and there will be no non-trivial stable steady state. It is clear that this cannot happen if the given values of $\zeta + f$ do not cover the full range of values of f. Thus there will be stable steady states with negative relative vorticity at the North pole, and positive at the South pole. This corresponds to a basic easterly flow, which is opposite to the basic westerly flow required to balance the equator-pole temperature difference. The stability of westerly basic flows against this type of mixing has still to be examined.

A more complete reduced model is the semi-geostrophic system (4.13). This can be written as an evolution equation for an inverse potential vorticity in the form (4.17), (4.18). We first illustrate stability results for it where the solution region Γ is a channel of width $2D$ and height H, with periodicity $2L$ in the x direction. The evolution equations (4.17) take the form

$$\begin{aligned} \frac{D\rho}{Dt} &= 0 \\ \frac{D}{Dt} &\equiv \frac{\partial}{\partial t} + U\frac{\partial}{\partial X} + V\frac{\partial}{\partial Y} + W\frac{\partial}{\partial Z} \end{aligned}$$

$$
\begin{aligned}
U &= f\left(\frac{\partial R}{\partial Y} - Y\right) \\
V &= f\left(X - \frac{\partial R}{\partial X}\right) \\
W &= 0.
\end{aligned} \tag{6.3}
$$

R is determined from the Monge–Ampère equation (3.22), with $\nabla R_{\mathbf{X}} = (x, y, z)$, and the boundary conditions (4.18) become

$$
\begin{aligned}
\frac{\partial R}{\partial X}(X,Y,Z) &= \frac{\partial R}{\partial X}(X+2L,Y,Z) \\
-D &\le \frac{\partial R}{\partial Y} \le D \\
0 &\le \frac{\partial R}{\partial Z} \le H.
\end{aligned} \tag{6.4}
$$

It is convenient to define

$$
\Psi = R - \frac{1}{2}(X^2 + Y^2) \tag{6.5}
$$

where Ψ acts as a stream-function for the flow defined in (6.3). The problem is to be solved in (X, Y, Z) space $(-L, L) \times (-\infty, \infty) \times (-\infty, \infty)$. The compatibility conditions (3.23) for the Monge–Ampère equation become

$$
\begin{aligned}
\rho(X+L,Y,Z) &= \rho(X-L,Y,Z) \\
\int_{-L}^{L}\int_{-\infty}^{\infty}\int_{-\infty}^{\infty} \rho\, dX dY dZ &= 4LDH.
\end{aligned} \tag{6.6}
$$

We now seek to identify stable steady states. In order to apply Kelvin's principle we seek a class of perturbations which is dynamically consistent with (6.3). We first show that this class can be written as $\overline{\mathrm{conv}\,\mathcal{R}_h(\rho_0)}$, defined by

$$
\rho \in \overline{\mathrm{conv}\,\mathcal{R}_h(\rho_0)} \text{ if } \begin{cases} \rho(.,Z) \in \overline{\mathrm{conv}\,\mathcal{R}(\rho_0(.,Z))} \text{ for almost all } Z \\ \int Y\rho dX\, dY\, dZ = \int Y\rho_0 dX dY dZ. \end{cases} \tag{6.7}
$$

The first condition restricts the rearrangements of ρ to the (X, Y) variables only, and includes the weak limits. This is similar to the space of 'stratified' rearrangements used by Burton and Nycander (1999). The additional condition is that the mean Y over the particles cannot be changed. This corresponds to angular momentum conservation. The latter is implied by the periodic boundary conditions, which mean, using (4.13), that $\int_{-L}^{L} v_g dx = \int_{-L}^{L} f^{-1}\frac{\partial \phi}{\partial x} dx = 0$.

6.3 Steady states

We next demonstrate the characterisation of steady states of the semi-geostrophic system in terms of stationary points of the energy with respect to rearrangements. The energy for the problem (6.3) is given by (4.2), which can be rewritten in the form (3.24) as

$$E = \int_{-L}^{L}\int_{-D}^{D}\int_{0}^{H}\frac{1}{2}f^2\left\{(X-x)^2+(Y-y)^2-zZ\right\}dxdydz \qquad (6.8)$$
$$= \int_{-L}^{L}\int_{-\infty}^{\infty}\int_{-\infty}^{\infty}\frac{1}{2}f^2\left\{(X-x)^2+(Y-y)^2-zZ\right\}\rho dXdYdZ.$$

Given $\rho = \rho_0(X,Y,Z)$ satisfying (4.24), seek the conditions under which $\delta E = 0$ for perturbations to ρ satisfying $\rho + \delta\rho \in \mathcal{R}(\rho_0)$. These can be generated by keeping ρ fixed on particles in $\mathbf{X}$ space and perturbing X and Y with a displacement field χ. The displacement must be non-divergent, so can be written as $(-\frac{\partial\alpha}{\partial Y}, \frac{\partial\alpha}{\partial X}, 0)$ for an arbitrary function $\alpha(X,Y,Z)$, and must satisfy the periodicity condition so that $\alpha(X-L,Y,Z) = \alpha(X+L,Y,Z)$. The restriction that the mean Y cannot be changed is automatically enforced by the periodicity condition. We then have

$$\delta E = \int_{-L}^{L}\int_{-\infty}^{\infty}\int_{-\infty}^{\infty} f^2\{(X-x)\delta X + (Y-y)\delta Y - X\delta x - Y\delta y - Z\delta z\}\rho dXdYdZ \qquad (6.9)$$

where the integration is taken over particles, so that there is no $\delta\rho$, and we have used the invariance of $\int_{-L}^{L}\int_{-D}^{D}\int_{0}^{H}(x^2+y^2)dxdydz$. Cullen and Purser (1989) show that solutions of (6.3) minimise the energy at each time instant in the sense that

$$\int_{-L}^{L}\int_{-\infty}^{\infty}\int_{-\infty}^{\infty}(-X\delta x - Y\delta y - Z\delta z)\rho dXdYdZ = 0. \qquad (6.10)$$

Substituting the definitions of Ψ and $\delta\mathbf{X}$ gives after some manipulations

$$\int_{-L}^{L}\int_{-\infty}^{\infty}\int_{-\infty}^{\infty}\left(\nabla.\left(\alpha\rho\frac{\partial\Psi}{\partial Y}, -\alpha\rho\frac{\partial\Psi}{\partial X}\right) - \alpha\left(-\frac{\partial\Psi}{\partial X}\frac{\partial\rho}{\partial Y} + \frac{\partial\Psi}{\partial Y}\frac{\partial\rho}{\partial X}\right)\right) dXdYdZ = 0. \qquad (6.11)$$

Assuming that ρ vanishes at a sufficiently large $|Y|$ and requiring (6.11) to hold for arbitrary α gives

$$-\frac{\partial\Psi}{\partial X}\frac{\partial\rho}{\partial Y} + \frac{\partial\Psi}{\partial Y}\frac{\partial\rho}{\partial X} = 0. \qquad (6.12)$$

This condition is precisely that for the flow to be steady as we can see from (6.3). Note that the linearity of (6.11) in α means that $\delta E = 0$ for a perturbation obtained as the limit of a sequence of perturbations defined by displacements α_n, and thus for any perturbation to ρ within $\overline{\mathrm{conv}\,\mathcal{R}_h(\rho_0)}$.

6.4 Stable steady states — barotropic case

We next characterise steady states which are stable by requiring the stationary point of the energy to be an extremum. It is first clear that the maximum energy attainable under these conditions is infinite. Generate a rearrangement by a displacement $\alpha = A\sin(\pi kX/L)$, implying $\delta X = 0, \delta Y = \frac{A\pi k}{L}\cos\left(\frac{\pi kX}{L}\right)$. Since $|y| < D$ for all particles, (3.6) shows that $E \to \infty$ as $A \to \infty$. It is therefore only meaningful to seek minimum energy states. In the stability problem for the barotropic vorticity equation, however, the maximum energy state is well defined. This is because the evolution equation is written in physical space and so the displacements χ have to be within the physical domain. Therefore $\alpha = 0$ on the domain boundaries. This problem was treated by Burton and McLeod (1991).

We then seek to minimise E for $\rho \in \overline{\mathrm{conv}\,\mathcal{R}_h(\rho_0)}$. (6.8) shows that the minimum energy is attained by making the map from (X,Y) to (x,y) as close as possible to the identity map $x = X, y = Y$ and maximising the correlation between z and Z, thus minimising $\int -zZdXdYdZ$. The identity map corresponds to having $\rho = 1$ for $|Y| \leq D, 0 \leq Z \leq H$ and $\rho = 0$ elsewhere. If we are given data where all the non zero values of ρ are greater than 1, it can be expected that 'mixing' these values with the zero values taken by ρ for sufficiently large $|Y|$ will allow a zero energy to be obtained.

First consider distributions independent of Z. In general, if the identity map is in $\overline{\mathrm{conv}\,\mathcal{R}_h(\rho_0)}$, the minimum energy will be zero. If the support of ρ has size greater than $4DL$, it is easy to show that the identity map is not in $\overline{\mathrm{conv}\,\mathcal{R}_h(\rho_0)}$ and there will be a non-trivial stable state. If ρ takes values greater than 1 anywhere, the energy can be reduced by mixing.

This argument shows that the only candidates for a minimum energy state will be a distribution with $\rho_1(Y) \leq 1$ everywhere. Since ρ is an inverse potential vorticity, this condition excludes values of potential vorticity less than 1, which correspond to anticyclonic relative vorticity. This agrees with the result of Kushner and Shepherd (1995) that there were no stable shear flows with anticyclonic shear. This argument would also show that no steady states with anticyclonic relative vorticity were stable in a limited domain with rigid boundary conditions under semi-geostrophic dynamics. They could be stable in doubly periodic flows, because there is then no region of $\rho = 0$ in the (X,Y) planes to mix with the non-zero values.

6.5 Stable steady states — baroclinic case

When ρ_0 depends on Z, let $2LS(Z)$ be the area over which $\rho_0(X,Y,Z)$ is non-zero for each value of Z. The size of the set for which ρ is non-zero cannot be reduced. However, zero values can be mixed in to increase the size of the set to $2D$ for each Z. If $S(Z) \leq 2D$ for all Z, then the resulting state has ρ a function of Z only, and the energy will be the minimum rest state potential

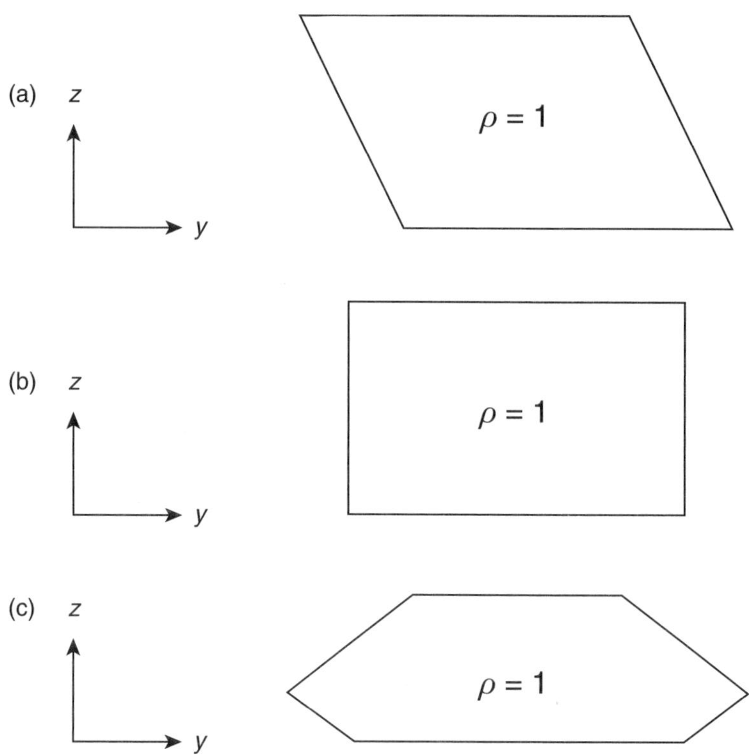

Figure 14: (a) ρ distribution corresponding to a steady baroclinic shear flow. (b) The minimum energy rearrangement of this distribution. (c) The minimum energy configuration for a ρ distribution which cannot be made independent of Z.

energy. However, if $S(Z) > 2D$ for some Z, this minimum energy state cannot be reached by an allowable rearrangement of ρ_0. In this situation there will be kinetic energy in the minimum energy state, and thus a nonlinearly stable flow. The angular momentum constraint is then satisfied by choosing the mean Y for which ρ_1 is non-zero. In general, this flow will depend on Z.

Figure 14(a) shows the distribution of ρ corresponding to a baroclinic shear flow with U increasing with z. The minimum energy rearrangement in that case is a flow which has no vertical shear, Figure 14(b). In general, Figure 14(c), the minimum energy state will have vertical shear, but aligned such that the Z variations of the ρ distribution is minimised. Burton and Nycander (1996) prove a similar result for the 3-dimensional quasi-geostrophic case. The identification of minimum energy states which are baroclinic allows more control over possible dynamic evolution, since only the excess energy above this minimum value is available for transient motion.

6.6 Locally stable states

The global stability results discussed above require extremisation of the energy with respect to any rearrangement of the potential vorticity, even if the rearrangement cannot be generated by a physically reasonable velocity field. Burton and Nycander (1999) showed that a local patch of anomalous potential vorticity in a background flow with uniform potential vorticity can be nonlinearly stable as a maximum energy state of the quasi-geostrophic equations.

Blocking patterns in the atmosphere have a horizontal scale comparable to the Earth's radius. Explaining them as local stable steady states requires inclusion of the variation with latitude of the vorticity associated with the Earth's rotation. However, this then allows dispersive Rossby wave solutions, which makes it difficult to maintain a steady localised perturbation to a zonally symmetric flow. The simplest type of solution for a finite amplitude steady state is a Rossby wave superposed on a mean flow which cancels out the wave speed. A solution of (6.2) with modified boundary conditions which illustrates this is as follows:

$$
\begin{aligned}
\Gamma &= (-L, L) \times (0, D)\\
f &= \beta y\\
\zeta &= A \sin\left(\frac{\pi k x}{L}\right) \sin\left(\frac{\pi l y}{D}\right) \qquad (6.13)\\
\psi &= -\frac{A}{\pi^2 k^2/L^2 + \pi^2 l^2/D^2} \sin\left(\frac{\pi k x}{L}\right) \sin\left(\frac{\pi l y}{D}\right) - \frac{\beta y}{\pi^2 k^2/L^2 + \pi^2 l^2/D^2}.
\end{aligned}
$$

This is a steady solution for any given k and l. It satisfies the boundary conditions that ψ and $\mathbf{u}$ are periodic in x with period $2L$, and $v = 0$ on $y = 0, D$. The dispersive nature of Rossby waves means that this type of solution can only be obtained for monochromatic waves. This has been exploited by Neven (1994) who constructed a set of localised 'modon' steady solutions by patching together monochromatic solutions.

6.7 Regularity estimates and the enstrophy cascade

The generation of small scales in the enstrophy in 2-dimensional incompressible flow governed by (6.1) or (6.2) can be understood using the analytical tools which are used to prove existence and regularity of these equations. The rigorous theory we draw on is set out in Gerard (1992). The equations state that the absolute vorticity $\zeta + f$ is a rearrangement of the initial distribution. Existence proofs by the standard methods require that the solutions are contained in a compact set. However, as discussed in section 3.6, the set of rearrangements is not compact. We therefore have to identify a 'reachable' compact subset of the set of rearrangements — rearrangements that can actually be reached in the time evolution. Given initial absolute vorticity which has bounded gradients, we can do this by estimating the rate of growth of the vorticity gradients. (6.2) implies that

$$\frac{D}{Dt}\nabla(\zeta + f) + \begin{pmatrix} \frac{\partial u}{\partial x} & \frac{\partial v}{\partial x} \\ \frac{\partial u}{\partial y} & \frac{\partial v}{\partial y} \end{pmatrix} \nabla(\zeta + f) = 0. \tag{6.14}$$

This can be written in terms of the stream-function ψ as

$$\frac{D}{Dt}\nabla(\zeta + f) + \begin{pmatrix} -\frac{\partial^2 \psi}{\partial x \partial y} & \frac{\partial^2 \psi}{\partial x^2} \\ -\frac{\partial^2 \psi}{\partial y^2} & \frac{\partial^2 \psi}{\partial y \partial x} \end{pmatrix} \nabla(\zeta + f) = 0. \tag{6.15}$$

These equations can be used to estimate the rate of increase of vorticity gradients, using a bound on the velocity gradients in terms of the vorticity and its gradients (Gerard (1992),p 424):

$$\| \nabla \mathbf{U}(t) \| \leq C \log \left(2 + \| \nabla(\zeta + f)(t) \|\right). \tag{6.16}$$

Exact definitions of the norms used are given in Gerard. However, they are essentially maximum norms. The bound is derived from the solution procedure for the Poisson equation for ψ in terms of ζ. Because of the dependence of the bound in (6.16) on the vorticity gradients, the estimate of vorticity gradients obtained from (6.14) allows doubly exponential growth in time. This does not prevent regularity being proved for all time, but allows the accumulation of enstrophy at small scales. This can be expressed as a statement that

$$\| (\nabla \mathbf{U})^2 \| \leq C \| \mathbf{U}^2 \| \tag{6.17}$$

where C grows exponentially in time. If the L_2 norm is used instead of the maximum norm, then an estimate of the form (6.17) holds with C independent of time, but dependent on the domain size. Thus the mean scale of the flow is bounded, but local regions where small scales are generated are permitted. This agrees with widespread computational experience.

Now consider the quasi-geostrophic equations. For simplicity, we consider the shallow water version of these equations, where the depth is close to a mean value h_0, and f is constant. For more details, see the chapter by White earlier in this volume.

$$\begin{aligned} \frac{Dq}{Dt} &= 0 \\ \frac{D}{Dt} &\equiv \frac{\partial}{\partial t} + \mathbf{U}.\nabla \\ \mathbf{U} &= \left(-\frac{\partial \psi}{\partial y}, \frac{\partial \psi}{\partial x}\right) \\ q &= g h_0 \nabla^2 \psi - f^2 \psi \\ \mathbf{U}.\mathbf{n} &= 0 \text{ on } \delta\Gamma. \end{aligned} \tag{6.18}$$

q is the quasi-geostrophic potential vorticity. In the case $L \ll L_R = \sqrt{(gh_0)}/f$, (6.18) behaves similarly to the incompressible equations (6.1). In the case $L \gg L_R$, the equations become the 'equivalent barotropic' equations:

$$\begin{aligned} -f^2\frac{\partial \psi}{\partial t} + \mathbf{U}.\nabla q &= 0 \\ \mathbf{U} &= \left(-\frac{\partial \psi}{\partial y}, \frac{\partial \psi}{\partial x}\right) \\ q &= gh_0\nabla^2\psi - f^2\psi \\ \mathbf{U}.\mathbf{n} &= 0 \text{ on } \delta\Gamma. \end{aligned} \tag{6.19}$$

Farge and Sadourny (1989) and Larichev and McWilliams (1991) present results which suggest that the enstrophy cascade is substantially suppressed in the equivalent barotropic model (6.19) if the initial data satisfies $L > L_R$. Thus there appears to be a range of scales greater than L_R which cannot really be regarded as turbulent. In the atmosphere, baroclinic development occurs at the scale $L = L_R$. Energy migrating to larger scales will thus be in the equivalent barotropic regime, where the shape of the spectrum only changes slowly in time, with a gradual migration of energy to large scales. When combined with the intermittent injection of energy at the scale L_R, this is consistent with the observed stationary statistics of weather maps. Motions on smaller scales than L_R dissipate quickly through the enstrophy cascade. In the ocean, the barotropic mode has a large L_R, about 3000 km, but the internal Rossby radius is only about 30 km. Thus internal oceanic structures, which often have a scale larger than 30 km, may not behave turbulently either.

Now consider a 2-dimensional version of the semi-geostrophic model, as expressed by (4.17) and (4.18). As before, these equations describe a sequence of minimum energy states, where the energy is

$$E = \frac{1}{2}f^2\int_\Gamma \{(x-X)^2 + (y-Y)^2\}d\tau. \tag{6.20}$$

An equation of the form (6.15) still governs the rate of growth of inverse potential vorticity gradients. Using (6.5), we obtain

$$\frac{D}{Dt}(\nabla\rho) + \begin{pmatrix} -\frac{\partial^2\Psi}{\partial X\partial Y} & \frac{\partial^2\Psi}{\partial X^2} \\ -\frac{\partial^2\Psi}{\partial Y^2} & \frac{\partial^2\Psi}{\partial Y\partial X} \end{pmatrix}\nabla\rho = 0. \tag{6.21}$$

The equation that determines Ψ from ρ is a Monge–Ampère equation for R, followed by addition of the fixed function $\frac{1}{2}(X^2+Y^2)$ to obtain Ψ. Thus, in effect, ρ is transported by a convex stream-function. If the flow is steady, convexity prevents exponential growth of gradients of ρ, except for time periods short compared with that needed for a particle to make a complete circuit of a streamline. Only algebraic growth is permitted.

In general, we can write that the energy minimisation property means that (6.20) is minimised against area preserving displacements of $\mathbf{x}$ for fixed $\mathbf{X}$. Consider a closed loop S in the fluid, with line element $d\mathbf{s}$. Then we can cyclically displace $\mathbf{x}$ by a distance ds around the loop, which changes the local value by an amount $(d\mathbf{x}/ds)ds$, giving

$$\delta E = f^2 \int_S (\mathbf{x} - \mathbf{X}) \cdot \frac{\delta \mathbf{x}}{ds} ds = 0. \tag{6.22}$$

Writing $\mathbf{U} = f(y - Y, X - x)$, (6.22) can be written

$$\delta E = \int_S \frac{\delta \mathbf{U}}{ds} \cdot (\mathbf{U} ds) = 0. \tag{6.23}$$

The rate of extension of the loop can be written

$$\frac{d|S|}{dt} = \int_S \frac{\delta \mathbf{U}}{ds} \cdot \mathbf{ds} = 0. \tag{6.24}$$

Comparing (6.23) with (6.24) shows that there can be no growth of the line element coming from the parts with $\mathbf{ds}$ correlated with $\mathbf{U}$. Thus the normal straining mechanism leading to the enstrophy cascade is excluded. Growth is possible from the parts of the line element uncorrelated with $\mathbf{U}$. This will normally be algebraic growth only, as in the steady case. Transient exponential growth is possible if the time evolution changes the velocity field in a way that the line element does not align to it as it grows.

This argument also excludes the enstrophy cascade in the solutions of more general reduced systems of equations where the Monge–Ampère equation is used to generate other fields from the potential vorticity. Some of these systems are described in the chapter by McIntyre and Roulstone.

7 Applications to numerical model design

7.1 General considerations

We can summarise the results reviewed in the previous subsections that are important for the design of practical atmospheric numerical models as follows. Most of the points apply equally to ocean models.

(i) In a given asymptotic regime, the solution of the full equations stays close to that of an appropriate reduced system (section 4).

(ii) Most useful simple models appropriate for the weather and climate forecasting problem can be described in terms of advecting a conserved potential vorticity, from which the other fields can be derived (section 4.5).

(iii) An accurate representation of potential vorticity conservation is therefore needed to give accurate predictions of the adiabatic evolution of the atmosphere.

(iv) The solution procedure for the full equations needs to respect the link between the potential vorticity and the other fields in the appropriate reduced system of equations.

(v) The sub-grid model forms part of the method by which the overall solution stays close to that of the reduced system. (section 5.4)

(vi) The solution procedure should control the distance by which the solution departs from that of the reduced system to the theoretically predicted level.

7.2 Semi-implicit methods and relation to schemes for reduced equations

We illustrate these points by using (5.1) as the model, thus including the sub-grid model, but avoiding other complications present in (2.1). A standard algorithm for solving these equations, and for solving (2.1), uses 'semi-Lagrangian' methods for advection terms and an implicit scheme for calculating the pressure, Staniforth and Cote (1991). A simple form of such a scheme can be written as

$$\begin{aligned}
\mathbf{u}_h^{n+1} &= \mathbf{u}_{hd}^n - \delta t\left((-fv, fu)^{n+1} + \nabla_h p'^{n+1} - (F_u, F_v)^{n+1}\right)\\
w^{n+1} &= w_d^n + \delta t\left(g\theta^{n+1}/\theta_0 - \frac{\partial p'^{n+1}}{\partial z}\right)\\
\nabla.\mathbf{v^{n+1}} &= 0 \\
\theta^{n+1} &= \theta_d^n + \delta t\left(F_h^{n+1} + H - LP + S_h\right)\\
r^{n+1} &= r_d^n + \delta t\left(F_r^{n+1} + P + S_r\right)\\
\mathbf{v^{n+1}}.\mathbf{n} &= 0 \text{ on } \partial\Gamma.
\end{aligned} \tag{7.1}$$

Superscripts n and $n+1$ denote discrete time levels. δt is the time-step. The calculation is assumed to be performed at discrete grid-points. Suffix d denotes a value calculated at the 'departure point' of a trajectory which finishes at the grid-point being calculated at the end of the time-step. Values at departure points have to be obtained by interpolation from the nearest grid-points. In practice second order time accuracy is used, rather than the first order scheme written out in (7.1).

In (7.1) the sub-grid terms (F_u, F_v, F_h, F_r) are all computed at time level $n+1$. This is because the momentum terms (F_u, F_v) form part of the lowest order balance as discussed in section 5.2. The time integration is designed so that the lowest order balance will be satisfied at time level $n+1$ if the acceleration terms are zero. It was shown in section 5.5 that the sub-grid term in F_h has to provide damping on a faster time scale than the growth of any instability that would be generated by the explicitly resolved dynamics. If the time-step is long compared with this time scale, as is typically true, then the sub-grid diffusion has to be calculated using implicit time differencing, so that the term F_h has to be evaluated at time level $n+1$. A similar argument for moist statically unstable regions shows that the term F_r also has to be evaluated at time level $n+1$.

Equation (5.1) contains equations for the explicit evolution of all the variables except the pressure. A standard solution procedure is to calculate the pressure at time level $n+1$ by using the condition that the continuity equation and boundary condition are satisfied at time level $n+1$. This is closely analogous to the analytical solution procedure described in section 3.3. It gives

$$\begin{aligned} \delta t \nabla^2 p'^{n+1} &= \nabla.\mathbf{u}_d^n - \delta t \left(\nabla.(-fv, fu, -g\theta/\theta_0)^{n+1} - \nabla.(F_u, F_v, 0)^{n+1} \right) \qquad (7.2) \\ \delta t \frac{\partial p'^{n+1}}{\partial n} &= \mathbf{u}_d^n.\mathbf{n} - \delta t \left((-fv, fu, -g\theta/\theta_0)^{n+1}.\mathbf{n} - (F_u, F_v, 0)^{n+1}.\mathbf{n} \right) \text{ on } \partial\Gamma. \end{aligned}$$

In order to solve (7.2), we have to make a linearised estimate of all the terms on the right hand side evaluated at time level $n+1$. This can be done by using the reduced set of evolution equations (5.7), which predict the evolution of $\nabla s = (fv_c - c_D u_c, -fu_c - c_D v_c, g\theta/\theta_0)$, precisely the terms that are required apart from F_h. The solution procedure for (5.7) is based on the form (4.21), with the matrices $\mathbf{Q}$ and $\mathbf{H}$ given by (5.8) and (5.9). Therefore

$$\begin{aligned} (\nabla s)^{n+1} &= (fv_c - c_D u_c, -fu_c - c_D v_c, g\theta/\theta_0)^{n+1} \qquad (7.3) \\ &= (\nabla s)^n + \mathbf{H}^n \delta t - \mathbf{Q}\delta t \begin{pmatrix} u \\ v \\ w \end{pmatrix}^{n+1}. \end{aligned}$$

In order to obtain a single scalar equation for p'^{n+1}, we substitute this into (7.1) rather than (7.2). This gives equations for the velocity components at time $n+1$ of the form

$$(\mathbf{I}+\mathbf{Q}\delta t^2)\mathbf{u}^{n+1} = \mathbf{u}_d^n - \delta t \left(\nabla p'^{n+1} - (fv, -fu, g\theta/\theta_0)^n - (F_u, F_v, 0)^n - \mathbf{H}^n \delta t \right). \qquad (7.4)$$

This can be reduced to a single equation for p^{n+1} similar to (7.2) by using the condition $\nabla.\mathbf{u}^{n+1} = 0$.

Successful use of this algorithm depends on the ellipticity of the operator $\mathbf{I} + \mathbf{Q}\delta t^2$. If $\mathbf{Q}$ has no negative eigenvalues, so that the flow is statically and inertially stable, this is assured. Otherwise there is a time-step limit

$$\delta t^2 \leq 1/\lambda \qquad (7.5)$$

where λ is the largest negative eigenvalue of $\mathbf{Q}$. This is a statement that the time-step must be short enough to resolve the physical instabilities that will occur for such data. In operational models, this causes a difficulty, because such regions are small in extent, and it is undesirable to restrict the time-step for the complete model to that required in the most unstable region. Thus the sub-grid model should be designed assuming a time as well as space averaging, and reduce the maximum growth rates of instabilities to be on the time-scale required, as shown in section 5.5.

If the time-step is large, and the time level used in the implicit calculations is entirely at time level $n+1$, as illustrated in (7.4), then as the time-step becomes large, the solution of (7.4) tends towards

$$\mathbf{Q}\delta t\mathbf{u}^{n+1} = -\left(\nabla p'^{n+1} - (fv, -fu, g\theta/\theta_0)^n - (F_u, F_v, 0)^n - \mathbf{H}^n\delta t\right). \quad (7.6)$$

Imposing the condition $\nabla.\mathbf{u}^{n+1} = 0$ gives an equation exactly of the form (4.25), with the identification $p = s$. Thus the solution of (7.4) with a long time-step will tend exactly to the solution of (5.7), in the case where $(F_u, F_v) = -(c_D u, c_D v)$. This achieves the aim of a solution of the general equations close to that of a reduced system if the time averaging scale is large, and also allows the sub-grid model to be more completely accounted for than in (5.7). Thus a linearisation of the sub-grid model only has to be included in the definition of the time tendency of the balanced state, not in that of the balanced state itself.

The identification of the long time-step solution with the specific semi-geostrophic reduced system (5.7) is a result of the explicit treatment of the Du/Dt term by the semi-Lagrangian algorithm. If the trajectory computation is made implicit, other reduced systems can be obtained as the long time-step limit. If the implicit calculation is simplified to use only a fixed reference θ profile which is independent of x and y, the quasi-geostrophic equations are obtained in the long time-step limit.

A further development of this algorithm would use the ideas of section 5.4 to identify the velocity calculated at time level $n+1$ with a total transport velocity $\mathbf{V}$ including sub-grid effects. This would require, for instance, allowing an implicit correction to be made to the convective mass transport. This can be achieved by writing the convection scheme as a nonlinear first estimate based on the values at the beginning of the time-step, together with a linearised correction.

A scheme based on these principles for the full compressible equations has been developed, the formulation and tests of various aspects are set out in Cullen *et al.* (1997).

7.3 Potential vorticity conservation

Now consider the accurate treatment of potential vorticity advection by this algorithm. The potential vorticity det $\frac{\partial(X,Y,Z)}{\partial(x,y,z)}$ as given by (4.3) is conserved by the dynamical reduced equations (4.13), though not when the sub-grid model is included. (X, Y, Z) are given by (3.18). Exact conservation of the potential vorticity following the motion is not practical using a conventional grid-point method. Contour dynamics algorithms, which use the potential vorticity as a variable and advect contours of fixed potential vorticity, can achieve exact conservation, e.g. Dritschel and Ambaum (1997). It is possible to ensure conservation in the trajectory calculation by ensuring that the control volume associated with the departure points corresponding to each grid-box has the same problem as the grid-box. This is illustrated in two dimensions in Figure 15. We require in that case

$$\text{Area}(x_{1d}, z_{1d}, x_{2d}, z_{2d}) = (x_2 - x_1)(z_2 - z_1). \tag{7.7}$$

Two ways of achieving this type of conservation are described by Leslie and Purser (1995) and Lin and Rood (1996). Conservation can also be improved by using the variables (X, Y, Z) directly in the integration scheme. Noting that Z is simply proportional to θ, (7.1) becomes

$$\begin{aligned}
(-Y, X)^{n+1} &= (-Y, X)^n_d - f^{-1}\delta t(\nabla_h p'^{n+1} - (F_u, F_v)^{n+1}) \\
w^{n+1} &= w^n_d + \delta t\left(g\theta^{n+1}/\theta_0 - \frac{\partial p'^{n+1}}{\partial z}\right) \\
\nabla.\mathbf{v}^{\mathbf{n+1}} &= 0 \\
\theta^{n+1} &= \theta^n_d + \delta t\left(F_h^{n+1} + H - LP + S_h\right) \\
r^{n+1} &= r^n_d + \delta t\left(F_r^{n+1} + P + S_r\right) \\
\mathbf{v}^{\mathbf{n+1}}.\mathbf{n} &= 0 \text{ on } \partial\Gamma.
\end{aligned} \tag{7.8}$$

Note that the implicit treatment of the Coriolis term is preserved. A similar procedure is analysed by Bates *et al.* (1995), where the implicit treatment of the Coriolis term is shown to be essential. Conservation is, however, lost in the interpolation to the departure points. In practice, this interpolation has to be done to higher order accuracy than the interpolation used in finding the departure point positions, so may be a less serious source of error. An explicit form of the semi-Lagrangian scheme which, in effect, uses (X, Y) as variables has been developed by Rochas, see Temperton (1997).

Schemes which aim to achieve potential vorticity conservation are examples of 'geometric' integration schemes. Reviews of such schemes are given by Budd and Iserles (1999).

7.4 Grid design

The first design requirement stated in subsection 7.1 is to respect the fact that solutions of the full equations stay close to those of appropriate reduced

Modified semi-Lagrangian method

Preserve the volume in the departure point calculation

Figure 15: Control volumes for ensuring conservation of potential vorticity using semi-Lagrangian advection.

systems. It is thus important that other motions, which are not part of the reduced solution, are not generated purely computationally, and that the numerical method can represent the solution of the reduced system accurately. If finite difference methods are used to implement the numerical procedure outlined above, it is important to ensure that the finite difference grid is suitable for representing the geostrophic and hydrostatic balance that defines the reduced system. Cullen (1989) describes a set of experiments to establish the best finite difference strategy for solving the semi-geostrophic system (4.13).

It is also important to ensure that implicit equations such as (7.4) can be solved easily in finite difference form. One consideration is the removal of zero eigenvalues from the discretised version of the Laplacian operator in (7.4). Another is to remove zero eigenvalues from the discretised version of $\mathbf{Q}$. It is always necessary to derive the discrete form of (7.4) from discrete approximations to (7.1). This can result in large stencils for approximating terms in (7.4), which lead to zero eigenvalues corresponding to 'checkerboard' modes, see Haltiner and Williams (1980, p. 149, eq. 5-113).

The optimum grid for solving (7.2) is staggered so that u is a half grid length from p' in the x direction, v in the y direction and w in the z direction (Figure 16(a)). The horizontal part of this arrangement is often referred to as the Arakawa C grid. This leads to the minimum possible stencil for representing $\nabla^2 p'$. The optimum grid for dealing with the $\mathbf{Q}$ term in (7.4) can be deduced from the optimum grid for solving (4.25). This requires the components of ∇s to be a half grid-length from s in the appropriate direction. When applied to (7.4), the requirement is that v is a half grid length from p' in the x direction, u in the y direction and θ in the z direction, illustrated in Figure 16(b). The requirements for positions of u and v are thus contradictory, actual choices are discussed below. The staggering of θ from p' in the vertical is known as

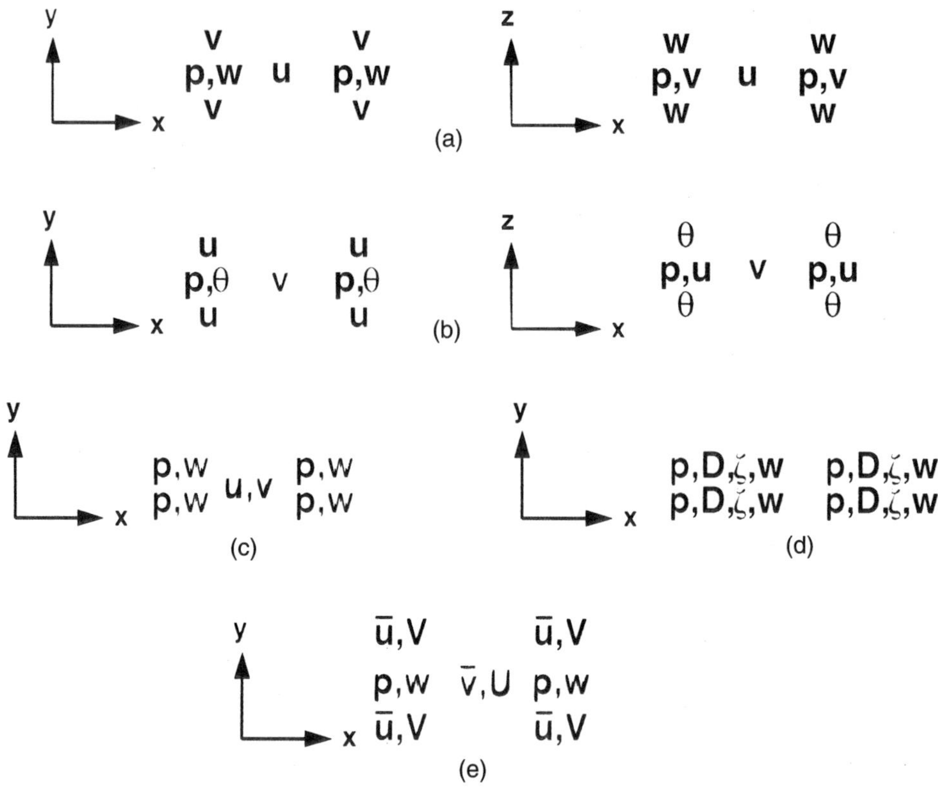

Figure 16: Various arrangements of variables on the grid for solution of (7.4). (a) Natural staggering to give best stencil for $\nabla^2 p$. (b) Natural staggering for representing ∇s. (c) Arakawa B grid. (d) Natural arrangement using D and ζ as variables. (e) Natural arrangement using $\overline{\mathbf{u}}$ and $\mathbf{U}$ as variables.

the 'Charney–Phillips' grid. The advantages of this were demonstrated by Arakawa and Moorthi (1987).

In most atmospheric models using this type of integration scheme, the grid is chosen to optimise the representation of $\nabla^2 p'$. However, in ocean models, where the resolution is often coarser compared with the scales of the motions being studied, the advantages of the two representations of u and v are more equal, and a compromise arrangement, the Arakawa 'B' grid is used (see Figure 16(c) and Bryan (1989)). There are two other possible resolutions of the difficulty. The equivalence of u and v to ∇s only applies to the geostrophic part of the wind, which has no horizontal divergence. Instead of using u and v as variables, the horizontal divergence D and vertical component of vorticity ζ can be used. An ideal arrangement then positions both of these at the same point as p', Figure 16(d). This is discussed by Bates *et al.* (1995). The other possibility is based on the formulation of section 5.4 where there is an averaged velocity $\overline{\mathbf{u}}$

which is different from the total effective transport velocity $\mathbf{U}$. The solution procedure would be ideal if these are held at the positions shown in Figure 16(e). This arrangement is used by Lin and Rood (1997), where the transport velocity is calculated at a different time level from the transported variables, as well as on a different grid.

7.5 Example of implementation

We illustrate some of the above considerations by showing a comparison of potential vorticity forecasts from two versions of the UK Meteorological Office atmospheric model. The important features of the integration schemes are listed below.

Scheme A (Cullen and Davies (1991))

(i) Variables held on Arakawa B grid in horizontal, θ held at same level as p.

(ii) Explicit time integration.

(iii) Advection using standard two step finite difference scheme with fourth order accuracy in space.

(iv) Sub-grid model in all equations includes horizontal eddy mixing term.

Scheme B (Cullen *et al.* (1997))

(i) Variables held on Arakawa C grid in horizontal, Charney–Phillips grid used in vertical.

(ii) Semi-implicit time integration as discussed above.

(iii) Semi-Lagrangian advection, with monotonic interpolation for thermodynamic variables, but not for winds.

(iv) No horizontal mixing in sub-grid model.

Scheme B includes several, but not all, of the desirable features discussed above. Comparative results after 10 days of integration are shown in Figure 17. A latitude–longitude grid with 96×72 points was used, with 19 levels in the vertical. The output is for the level $\theta = 315K$. The potential vorticity distribution for scheme B shows much greater spatial coherence, even though the spatial smoothing has been removed from the sub-grid model. This is consistent with the discussion in section 6.7, and suggests that the design principles discussed in this section are important.

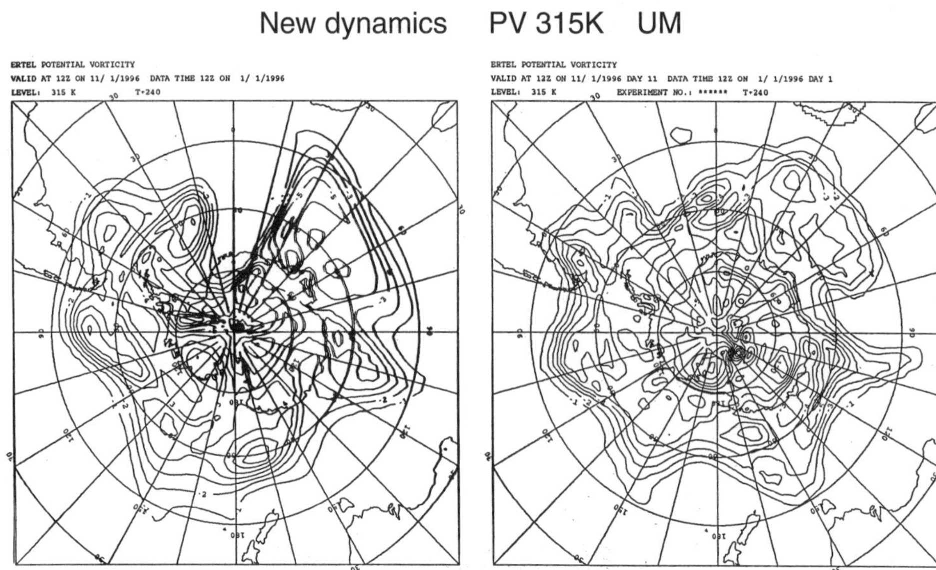

Figure 17: 10-day forecasts of potential vorticity at 315K using the Meteorological Office Unified Model in a climate model configuration and integration scheme A (right) and the same with integration scheme B (left).

8 Application to data assimilation and predictability studies

8.1 Basic data assimilation techniques

Data assimilation is the process of finding a model state which best fits incomplete or imperfect data. The basic methods used in modern assimilation systems are reviewed by Daley (1997). We follow his definitions and notation in this section. The standard theory is developed for assimilation of data into a defined discrete model. The assumptions made about model behaviour and error structure are typically quite crude. In this section we show how some of the dynamical knowledge set out in earlier sections can be brought to bear on the problem.

8.2 Projection of observed information onto balanced states

Atmospheric observations are usually designed to sample the large scale low frequency motions associated with weather patterns. In section 4.5 of Daley (1997), it is pointed out that it is necessary to ensure that the data is projected onto these motions, and is not used to initialise small scale high frequency motions which can also be described by the equations. In practice, it is often necessary to use explicit procedures to damp unwanted high frequency motions, and force the data to be fitted by the low frequency motions, as illustrated in the chapter by Lynch.

The characterisation of the solution of a set of reduced equations as a sequence of minimum energy states described in sections 4.1 and 5.2 can be used to give a method of explicitly nudging a general state of a model solving equations (3.5) or (5.1) towards a state of geostrophic or geotriptic and hydrostatic balance. Given a general state $(\mathbf{u}, \theta, p)$, not satisfying (5.4), carry out a virtual displacement Ξ of the fluid defined by

$$\begin{aligned} \nabla^2 s &= \nabla.(fv - c_D u, -fu - c_D v, g\theta/\theta_0) \\ \frac{\partial s}{\partial n} &= (fv - c_D u, -fu - c_D v, g\theta/\theta_0).\mathbf{n} \text{ on } \partial\Gamma \\ \Xi \equiv (\xi, \eta, \zeta) &= \alpha((fv - c_D u, -fu - c_D v, g\theta/\theta_0) - \nabla s). \end{aligned} \tag{8.1}$$

This ensures that $\nabla.\Xi = 0$ and $\Xi.\mathbf{n} = 0$ on $\partial\Gamma$. Use this displacement to change $(\mathbf{u}, \theta)$ on particles according to (5.3). Then the change to the balanced energy (4.2) can be shown to be

$$\delta e = \int_\Gamma \left(-(f^2 + c_D^2)u^2 - (f^2 + c_D^2)v^2 - (g\theta/\theta_0)^2 - (\nabla s)^2\right) d\tau \tag{8.2}$$

which is negative definite. Thus, with suitable choice of α, these displacements can be applied iteratively to give convergence to an energy minimising state satisfying (5.2). In practice, the vertical variation of s from (8.1) will be very close to the hydrostatic value $g\theta/\theta_0$. Assuming this, the components ξ and η of the displacement can be calculated explicitly from (8.1), and ζ from the condition $\nabla.\Xi = 0$. This procedure can be generalised to the free surface case. In the shallow water case, s is replaced by the fluid depth h, and the first two equations in (8.1) are not needed. The procedure can also formally be applied using the full sub-grid terms (F_u, F_v) in (8.1), though convergence cannot then be proved. The procedure can also be carried out on the sphere using the proper value of f, convergence is assured by the theorem of McCann quoted in section 3.5.

It can be shown that this iteration will converge fastest on the scales characterised by high frequency gravity waves. It will thus do a similar job to Lynch's digital filter, see the chapter by Lynch, while ensuring that the potential vorticity is preserved on particles.

In order to try and project the observed information more directly onto the balanced part of the flow, it is convenient to change variables so that the primitive variables (u, v, θ) are replaced by three new variables, one of which contains all the balanced information. These are the potential vorticity, the geostrophic departure, and the horizontal divergence. This will give a decomposition of the form

$$
\begin{aligned}
u &= u_1 + u_2 + u_3 \\
v &= v_1 + v_2 + v_3 \\
\theta &= \theta_1 + \theta_2 + \theta_3
\end{aligned}
\tag{8.3}
$$

where components $1, 2$ and 3 are derived from the potential vorticity, geostrophic departure, and horizontal divergence respectively. An appropriate projection can be derived by a standard linear normal mode analysis of the hydrostatic version of (3.5), as in Bartello (1995). θ has to be linearised about a basic state $\theta = \Theta(z)$ and f is assumed constant. The velocities are linearised about a state of rest. Write the basic state Brunt-Väisälla frequency $\frac{g}{\theta_0}\frac{\partial \Theta}{\partial z}$ as N^2. The basic state potential vorticity is $f\frac{\theta_0}{g}N^2$. The linearised perturbation potential vorticity is proportional to

$$
q = f\frac{g}{\theta_0}\frac{\partial \theta_1}{\partial z} + N^2\left(\frac{\partial v_1}{\partial x} - \frac{\partial u_1}{\partial y}\right). \tag{8.4}
$$

(u_1, v_1, θ_1) are derived from q by assuming geostrophic and hydrostatic balance, which take the form

$$
(u_1, v_1, \theta_1) = \left(-f^{-1}\frac{\partial p'}{\partial y}, f^{-1}\frac{\partial p'}{\partial x}, \frac{\theta_0}{g}\frac{\partial p'}{\partial z}\right). \tag{8.5}
$$

The derivation of (u_1, v_1, θ_1) is achieved by first substituting (8.5) into (8.4) to obtain

$$
N^2\nabla_h^2 p' + f^2\frac{\partial^2 p'}{\partial z^2} = fq. \tag{8.6}
$$

This is solved for p', using appropriate boundary conditions, and (u_1, v_1, θ_1) are then derived from (8.5).

The second variable is chosen to have the same geostrophic departure as the given fields. The divergence of the geostrophic departure can be written, using (8.5), as $f\left(\frac{\partial v}{\partial x} - \frac{\partial u}{\partial y}\right) - \nabla_h^2 p'$. Using the hydrostatic relation, the condition that the vertical derivative of the geostrophic departure given by (u_2, v_2, θ_2) is the same as the original is

$$
f\frac{\partial}{\partial z}\left(\frac{\partial v_2}{\partial x} - \frac{\partial u_2}{\partial y}\right) - \frac{g}{\theta_0}\nabla_h^2\theta_2 = f\frac{\partial}{\partial z}\left(\frac{\partial v}{\partial x} - \frac{\partial u}{\partial y}\right) - \frac{g}{\theta_0}\nabla_h^2\theta. \tag{8.7}
$$

We choose the second variable such that the potential vorticity is zero. This can be done by defining a stream-function Ψ satisfying

$$
f^2\frac{\partial^2}{\partial z^2}\nabla_h^2\Psi + N^2\nabla_h^4\Psi = f\frac{\partial}{\partial z}\left(\frac{\partial v_2}{\partial x} - \frac{\partial u_2}{\partial y}\right) - \frac{g}{\theta_0}\nabla_h^2\theta_2 \tag{8.8}
$$

and setting

$$
(u_2, v_2, \theta_2) = \left(-f\frac{\partial^2\Psi}{\partial y\partial z}, f\frac{\partial^2\Psi}{\partial x\partial z}, -\nabla_h^2\Psi\frac{\partial\Theta}{\partial z}\right). \tag{8.9}
$$

Substitution of (u_2, v_2, θ_2) into the first equation of (8.4) confirms that this mode has $q = 0$. The third variable is the horizontal divergence, which is represented by a velocity potential χ. We require that (u_3, v_3) have no vertical component of vorticity and that $\theta_3 = 0$, giving

$$\nabla^2 \chi = \frac{\partial u_3}{\partial x} + \frac{\partial v_3}{\partial y} \tag{8.10}$$
$$(u_3, v_3, \theta_3) = \left(\frac{\partial \chi}{\partial x}, \frac{\partial \chi}{\partial y}, 0\right).$$

The geostrophic departure and horizontal divergence are not normal modes of the linearisation of equations (3.5). The normal modes can only conveniently be written in Fourier space, as in Bartello (1995). The form written here is sometimes called the Craya–Herring cyclic basis. Methods similar to this have been implemented by Mohebalhojeh and Dritschel (2000).

The projection can be generalised to spherical geometry with $\frac{\partial \Theta}{\partial z}$ a general function of z by performing a spherical normal mode analysis, which cannot be written as a simple analytic procedure in real space. It is also possible to derive equivalent equations to (8.4), (8.5) and (8.6) directly with f and $\frac{\partial \Theta}{\partial z}$ allowed to vary. This gives so-called 'implicit normal mode' analysis, Temperton (1989).

We can alternatively use the energy minimisation procedure described above to decompose a general state. This can be used in spherical geometry or in limited domains, and does not require any linearisations or restrictions on $\frac{\partial \theta}{\partial z}$. Given general (u, v, θ, p'), use (8.1) to find a balanced state (u_b, v_b, θ_b, s), where s is the geostrophic and hydrostatic pressure, with the same 'balanced' potential vorticity, (4.3), as the original data. The original data can be reconstructed from (u_b, v_b, θ_b, s), given as functions of $\mathbf{x}$, and the integrated displacements Ξ, also given as functions of $\mathbf{x}$. The latter can be split into two components

$$\Xi = \Xi_1 + \Xi_2 = (\xi_1, \eta_1, \zeta_1) + (\xi_2, \eta_2, \zeta_2)$$
$$\nabla_h \Xi_1 = 0, \zeta_1 = 0 \tag{8.11}$$
$$\frac{\partial \eta_2}{\partial x} - \frac{\partial \xi_2}{\partial y} = 0.$$

Ξ_1 generates the horizontally divergent winds as in (8.10), and Ξ_2 generates the ageostrophic vorticity and temperature changes as in (8.8).

8.3 Initialisation of ageostrophic winds

In sections 4.4 and 5.3, we set out equations for the evolution of balanced states in the form (4.21). These include an equation for the total velocity $\mathbf{u}$, which does not satisfy the geostrophic relation, and acts as the response to dynamical and physical forcing. In section 5.4, it is shown that this total velocity can be interpreted as including the effect of sub-grid transport. It includes, for example, the vertical motion balancing radiative heating or cooling. In order to analyse or initialise numerical forecasts, it is necessary to calculate this total velocity in a way which is consistent with the evolution of the balanced state.

The linear decomposition (8.3) of the previous subsection has one component (u_1, v_1, θ_1) which describes the geostrophic and hydrostatic part of the data. If the total velocity $\mathbf{u}$ describing the evolution of the balanced state is decomposed according to (8.3), it will also include components from the other modes $\mathbf{u}_2, \mathbf{u}_3$. These components can be interpreted as being 'slaved' to the geostrophic state, see Warn *et al.* (1994). They can be calculated by nonlinear normal mode initialisation, see Daley (1997), or other related methods. Nonlinear normal mode initialisation, including the physical forcing terms, is widely used in operational forecast models. The link between this procedure and solving an equation of the form (4.21) is made by Leith (1980). He shows that adiabatic nonlinear normal mode initialisation is equivalent to solving the quasi-geostrophic form of (4.21), with a fixed reference profile replacing the actual $\partial\theta/\partial z$. Use of (4.21) with (5.8) and (5.9) would then correspond to a variable coefficient generalisation of normal mode initialisation, with physical forcing included.

It should be noted that a balance assumption is likely to be most accurate where the ageostrophic circulation required is small, and vice-versa. If forcing is applied to a state with very small potential vorticity, (4.21) shows that the ageostrophic circulation required will be large, and possibly quite unrealistic. The real flow would simply be unbalanced under such conditions. Thus the obvious natural procedure of finding the balanced state that best fits the observations, and then calculating the associated total velocity from (4.21), is likely to be unsafe. The integration scheme (7.4) for the full equations avoids this problem by solving for $(\mathbf{I} + \mathbf{Q}\delta t^2)\mathbf{u}$, so that balance is not enforced when $\mathbf{Q}$ has a small eigenvalue. (7.4) should also thus be suitable as an initialisation scheme. This is consistent with the incremental nudging method of Lorenc *et al.* (1991), which effectively trusts the forecast model to initialise the total velocity field from information about the balanced state.

For the same reason, it is unwise to use the balanced ageostrophic motion as an error measure or in a cost function. The balanced pressure tendency is a safe variable to use instead, because the solution procedures discussed in section 5.3 will ensure that this evolves smoothly in time even when, as in deep convection, the balance requirement forces discontinuous mass transfers.

Many observations, such as satellite cloud observations, would be most useful as measures of the vertical motion. Using this data requires deducing the balanced state consistent with a given vertical motion. This is much harder, and leads to the use of the 4-dimensional variational analysis method discussed below. The problem can be addressed by perturbing (4.21):

$$\mathbf{Q}'\begin{pmatrix} u \\ v \\ w \end{pmatrix} + \mathbf{Q}\begin{pmatrix} u' \\ v' \\ w' \end{pmatrix} + \frac{\partial}{\partial t}\nabla s' = \mathbf{H}'. \tag{8.12}$$

Taking the curl of (8.12) gives

$$\nabla \times \mathbf{Q}' \begin{pmatrix} u \\ v \\ w \end{pmatrix} + \nabla \times \mathbf{Q} \begin{pmatrix} u' \\ v' \\ w' \end{pmatrix} = \nabla \times \mathbf{H}'. \tag{8.13}$$

If $\mathbf{u}'$ is given, $\mathbf{H}$ is given by (5.9) and the radiative forcing is assumed to be correct, $\mathbf{H}'$ and $\mathbf{Q}'$ represent a correction to the geostrophic flow and can be written in terms of a scalar s'. (8.13) thus becomes an equation for s' in terms of the geostrophic flow and $\mathbf{u}'$.

8.4 The use of Lagrangian methods to optimise the fit between two meteorological fields

One of the difficult issues in data assimilation is that the differences between two meteorological fields may not be well expressed in terms of a standard Eulerian norm such as the L^2 norm. This is because, at least away from equatorial regions, the evolution of the flow is dominated by advection of potential vorticity (section 4.5), and thus typical errors are likely to be displacements of the potential vorticity and hence the other fields derived from them. This is illustrated in Figures 18 and 19. The root mean square position errors of both extra-tropical and tropical cyclones grow approximately linearly with forecast time. The L_2 error of upper level height (equivalent to pressure) fields, which are dominated by large-scale and usually well forecast patterns, also grows fairly linearly with time. However, the L_2 error of upper level winds, which are dominated by smaller scales features, increases most rapidly early in the forecast period and then saturates. The L_2 error of surface pressure, whose patterns are typically of smaller scale than upper air patterns, behaves in an intermediate fashion. Hoffman *et al.* (1995) decomposed forecast errors of 500 hpa height into displacement errors, amplitude errors, and a residual. They found that nearly 80 percent of the error in a particular 500 hpa forecast could be explained as displacement error. The procedures of section 3.1 and the chapter by Douglas earlier in this volume using rearrangements should also be applicable to this form of analysis.

Daley (1997) discusses the choice of norm to use in the analysis procedure in terms of expected distributions of errors. His equation (15) gives the total cost function to be minimised in the analysis as

$$J = 0.5\{[\mathbf{y}^o - H(\mathbf{x}^a)]^T \mathbf{R}^{-1}[\mathbf{y}^o - H(\mathbf{x}^a)] + [\mathbf{x}^f - \mathbf{x}^a]^T[\mathbf{P}^b]^{-1}][\mathbf{x}^f - \mathbf{x}^a]\}. \tag{8.14}$$

In (8.14), superscripts a, f, and o refer to the analysis, the first-guess forecast, and the observations respectively. $\mathbf{x}$ represents a model state, and $\mathbf{y}$ a set of observed values. The operator H calculates pseudo-observed values from forecast fields. The error measure to be used for the observations has to be related to the expected error structure of the observations, encompassed by the

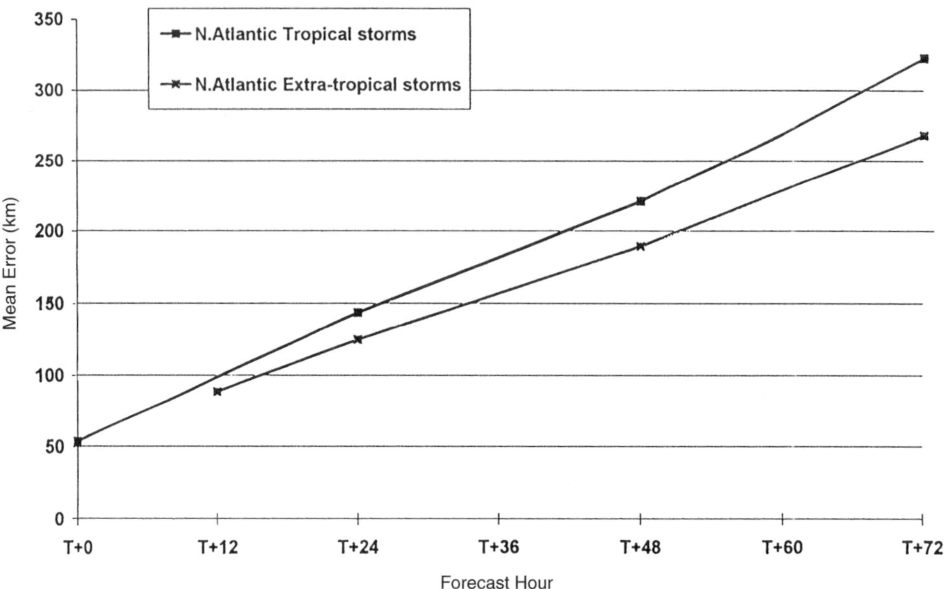

Figure 18: Growth of RMS position error with forecast time for extra-tropical and tropical cyclones in the North Atlantic during 1995.

error covariance matrix **R**. This is discussed by Daley. The error measure to be used for the forecast minus analysis term has to be related to the expected forecast error structure, and so it is in this term that alternative norms such as measures of displacements could be used. The formulation (8.14) allows for the use of different norms in measuring observation error and forecast error, and produces a maximum likelihood estimate.

Hoffman *et al.* (1995) set out a procedure for minimising (8.14) where displacements are used in calculating the cost function. They also suggest using a displacement as a control variable in the minimisation, thus setting $\mathbf{x}_a$ equal to a displacement applied to $\mathbf{x}_f$ plus a residual. Given this approach, it is natural to seek a balanced fit to the data by using an additional displacement satisfying (5.3) to reduce the energy while preserving the potential vorticity, thus moving towards a geostrophic and hydrostatic state.

8.5 Sensitivity of model evolution to small perturbations — applications to data assimilation and ensemble generation

A working assumption for optimising the analysis procedure for numerical weather prediction is that the best analysis is that which leads to the best forecast. Therefore it is particularly important to analyse accurately the fea-

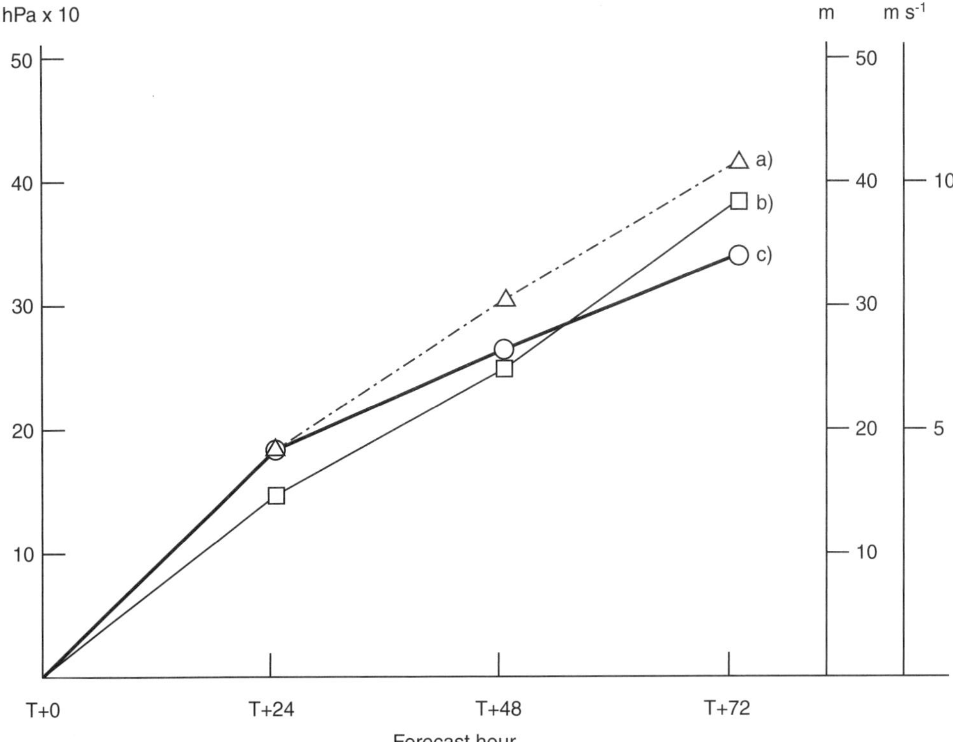

Figure 19: Growth of L_2 error of surface pressure (solid), 500 hpa height (dashed), and 200 hpa wind over the Northern hemisphere with forecast time during 1995.

tures which could lead to the fastest error growth in the subsequent forecast, see Courtier (1997). If we use the idea set out in section 4.5 that the atmosphere stays close to an evolution given by advection of potential vorticity, then information about the sensitive regions can be deduced from a linear perturbation of (4.26):

$$\begin{aligned} \frac{\partial q'}{\partial t} + \mathbf{U}.\nabla q' + \mathbf{U}'.\nabla q &= 0 \\ \hat{\mathcal{H}}(\mathbf{U}', \theta') &= q' \end{aligned} \qquad (8.15)$$

where $\hat{\mathcal{H}}$ is the operator that appears following (4.27). Maximum sensitivity is likely to occur when $\mathbf{U}$ is normal to ∇q and both are large. Then a fixed perturbation to either will have the largest effect on $\mathbf{U}.\nabla q$. According to the principles reviewed in section 6.2, these will be steady states corresponding to stationary but non-extremal points of the energy under potential vorticity rearrangements. In a barotropic flow, these are characterised locally by points where q is non-monotonic as a function of stream-function. The degree of

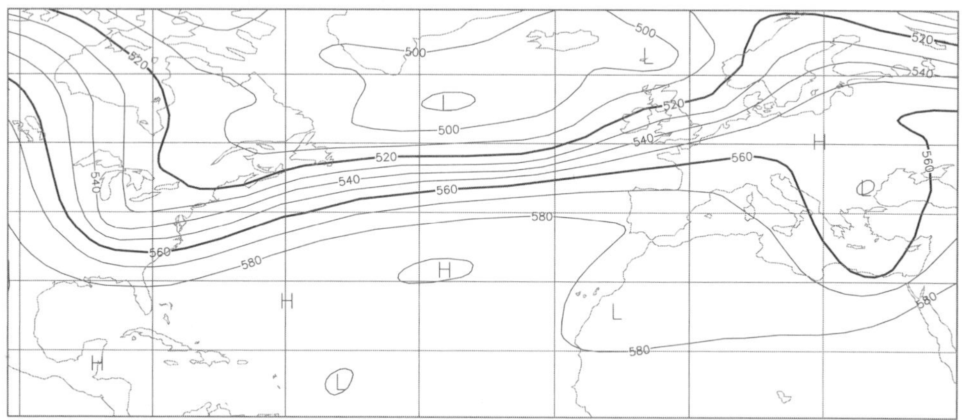

Figure 20: 500 hpa height over part of the Northern hemisphere at 1200UTC on 24 December 1999.

sensitivity can be inferred from the energy difference between the actual profile and that that would be obtained from a local monotone rearrangement of q. Baroclinic steady flows will additionally correspond to non-extremal points of the energy if the vertical correlation of the potential vorticity gradient on isentropic surfaces can be increased. Buizza and Palmer (1995) illustrate a correlation between sensitive regions and the local value of du/dz, which is a crude measure of the potential for baroclinic instability. An example of a non-extremal stationary state is a straight flow with a non-monotonic potential vorticity gradient across it. Such states preceded the 'October storm' which hit the UK on 16 October 1987, and the 'French' storm which hit Paris on 26 December 1999. The pressure field, in effect the stream-function, preceding the French storm is shown in Figure 20. There is an intense band of strong winds across the Atlantic. The associated relative vorticity, which dominates the potential vorticity, has large positive values on the northern edge of the strip and negative values on the southern edge. Elsewhere it is small. The north-south profile of potential vorticity is thus highly non-monotonic.

Another contributing factor will be a small eigenvalue of $\hat{\mathcal{H}}^{-1}$, which will give a large $\mathbf{U}'$ for a given q'. If the potential vorticity inversion is given by (4.21) and (4.25), then (4.25) can be perturbed to give

$$\begin{aligned} &\nabla.\mathbf{Q}^{-1}\frac{\partial}{\partial t}\nabla s' - \nabla.\mathbf{Q}^{-1}\mathbf{Q}'\mathbf{Q}^{-1}\frac{\partial}{\partial t}\nabla s = \\ &\qquad \nabla.\mathbf{Q}^{-1}\mathbf{H}' - \nabla.\mathbf{Q}^{-1}\mathbf{Q}'\mathbf{Q}^{-1}\mathbf{H} \\ &\left(\mathbf{Q}^{-1}\frac{\partial}{\partial t}\nabla s' - \nabla.\mathbf{Q}^{-1}\mathbf{Q}'\mathbf{Q}^{-1}\frac{\partial}{\partial t}\nabla s\right)_n = \\ &\qquad \left(\mathbf{Q}^{-1}\mathbf{H}' - \nabla.\mathbf{Q}^{-1}\mathbf{Q}'\mathbf{Q}^{-1}\mathbf{H}\right)_n \text{ on } \partial\Gamma. \end{aligned} \tag{8.16}$$

It can be seen from these equations that small eigenvalues of the potential vorticity matrix $\mathbf{Q}$ will give the greatest sensitivity. The perturbations are included as values of $\mathbf{Q}'$ and $\mathbf{H}'$. The way they are spread in space depends on the structure of the $\mathbf{Q}^{-1}$ matrix. This has been applied in designing data assimilation systems by stating that the region of influence of an observation should be isotropic in geostrophic and isentropic coordinates, see Desroziers and Lafore (1993). The combination of low potential vorticity and a highly unstable basic state flow was the precursor to both the October storm and the French storm referred to above, Shutts (1991).

There are two methods commonly used in operational centres of calculating sensitivity directly from a numerical model solving the evolution equations. The first is the 'singular vector' method. This is expensive to use, and is thus normally used with either a reduced set of equations or a very low resolution model. A set of equations for the evolution of a perturbation to the original forecast is derived by linearising the governing equations. If these take the form

$$\frac{\partial \mathbf{u}}{\partial t} = N\mathbf{u} \tag{8.17}$$

the 'tangent linear' equations for a perturbation to the solution takes the form

$$\frac{\partial \mathbf{u}'}{\partial t} = L(t)\mathbf{u}' \tag{8.18}$$

where the operator $L(t)$ depends on the control forecast. This system is written in discrete form, when it becomes a matrix equation with time-dependent coefficients. If this is integrated over a fixed number of time-steps, it generates a single matrix evolution operator

$$\mathbf{u}'(t) = E(0,t)\mathbf{u}'(0). \tag{8.19}$$

The eigenvalues of $E(0,t)$ are calculated, and that giving the largest growth of $\mathbf{u}'$ in a suitable norm, usually energy, is chosen. The corresponding eigenfunction gives the structure of the most sensitive perturbation to the initial state. The procedure is fully described by Buizza and Palmer (1995).

The second method is the 'error-breeding method' of Toth and Kalnay (1993). This consists of finding the structure of the fastest growing perturbation to a given initial state iteratively by integrating a perturbed forecast for a few hours, calculating the difference from the control forecast, and then repeating the exercise with an initial perturbation equal to the rescaled final perturbation from the previous iteration. This procedure converges to the mode with the largest Liapunov exponent. It would be interesting to compare the results of both these methods with the dynamically based methods discussed at the beginning of this subsection.

It remains to be explored whether the areas of maximum sensitivity identified by the two latter methods can be characterised in terms of properties of the potential vorticity distribution. If so, local potential vorticity diagnostics could be used to change the assumed error in the first guess forecast, and/or to change the overall weighting of the cost function in (8.14) so that the analysis procedure is focussed towards doing a good job in critical areas.

8.6 Fitting observations by a model trajectory over a time interval

As noted above, 4D variational data assimilation is the most promising avenue for using observations relating to the ageostrophic circulation. In addition, as discussed by Eyre (1997), it allows use of observations with a high time frequency, such as data from aircraft, and gives a framework for use of remotely sensed data which is only indirectly related to the variables predicted directly by the model. Daley (1997) shows that the theoretical basis for 4DVAR comes from the need to predict the evolution of model error, and is a compromise from the more complete method based on the Kalman filter forced by lack of computer power. However, even the Kalman filter method assumes that model errors are uncorrelated in time, which is extremely unlikely. The hope is that more dynamical knowledge can be used to replace these statistical assumptions.

The standard 4DVAR strategy is to minimise the cost function (8.14) over a fixed time interval. The output is a model trajectory which satisfies the (discrete) equations and minimises (8.14). Since the model is deterministic, the control variables have to be the initial values of the model variables. In practice, this minimisation has to be done iteratively using the linearisation (8.18) of the model equations, and their adjoint (see Daley (1997)). A number of additional possibilities are suggested by the preceding sections:

(a) Use of a model integration scheme based on the principles of section 7 to ensure that the model trajectory stays close to a balanced state. In particular, use of (7.4) for initialisation, and use in the tangent linear model (8.18) of the same linearised corrections to the sub-grid model as are needed for the implicit time integration.

(b) Use of potential vorticity, divergence and geostrophic departure as control variables, so that more use can be made of dynamical information in estimating the error structure of the potential vorticity. Overall estimates of the statistics of the unbalanced motion can be made by the methods of section 4.2.

(c) Use of displacement error in calculating the difference between two forecast fields, or between a forecast field and an analysed field derived from a set of observations. Use of displacements in the iteration to minimise the cost function, incorporating the energy minimisation method in the iterative procedure to aid convergence to a near balanced state.

(d) Use of the potential vorticity matrix in estimating how to spread information, or the likely structure of the error matrix.

(e) Use of dynamical estimates of sensitivity in estimating model error, or in the overall weighting of the cost function to be minimised.

In addition, a basic and well recognised limitation of the current 4DVAR approach is the need to treat the model as perfect, and hence use the initial values as control variables. An alternative would be to use the model as a weak constraint. The simplest way of doing this would be to have a complete set of model variables as control variables at a number of times during the assimilation period. This is very expensive and not practical. Courtier (1997) and Daley (1997) review a number of ways of achieving this in an affordable way. Griffith and Nichols (1996) provide examples of how to achieve it in a simple test problem.

9 Summary and conclusions

We have illustrated that considerable insight into the dynamics of the atmosphere and ocean can be obtained by solving 'reduced' systems of equations, and that these solutions can be related to the real flow by rigorously estimating how close the exact solution stays to the solution of the reduced model. The approximations made in deriving the reduced equations are regime dependent, and so it is not possible to find a single reduced model that describes all flows of interest in weather and climate forecasting or ocean modelling. However, the behaviour of the reduced models gives considerable assistance in designing integration schemes, sub-grid models, and data assimilation methods to be used for the more general models using averaged equations which are employed operationally.

In addition, the properties of the reduced models give considerable understanding of the atmosphere and ocean circulation. In particular, they show why there is a permanent unsteady, but statistically stationary circulation, show that there can be locally stable states which may persist for extended periods, and identify regions of maximum instability and thus uncertainty in how the subsequent evolution will go. They also show that the basic dynamics can be thought of as advection of a scalar, and thus typical forecast error growth will be linear rather than exponential when measured in a suitable way.

10 References

Abarbanel, H.D.I. and Holm, D.D. (1987) Nonlinear stability analysis of inviscid flows in three dimensions: Incompressible fluids and barotropic fluids. *Phys. Fluids*, **30**, 3369–3382.

Allen, J.S., Barth, J.A. and Newberger, P. (1990) On intermediate models for barotropic continental shelf and slope flow fields. Part I: Formulation and comparison of exact solutions.. *J. Phys. Oceanog.*, **20**, 1017–1043.

Arakawa, A. and Moorthi, S. (1987) Baroclinic instability in vertically discrete systems. *J. Atmos. Sci.* **45**, 1688–1707.

Babin, A., Mahalov, A. and Nicolaenko, B. (1996) Global splitting, integrability and regularity of 3D Euler and Navier–Stokes equations for uniformly rotating fluids. *Eur. J. Mech., B/Fluids*, **15**, 291–300.

Babin, A., Mahalov, A. and Nicolaenko, B. (1997) Regularity and integrability of rotating shallow water equations. *C.R. Acad. Sci. Paris*, **324**, 593–598.

Bartello, P. (1995) Geostrophic adjustment and inverse cascades in rotating stratified turbulence. *J. Atmos. Sci.*, **52**, 4410–4428.

Bartello, P. and Thomas, S.J. (1996) The cost-effectiveness of semi-Lagrangian advection. *Mon. Weather Rev.*, **124**, 2883–2897.

Bates, J.R., Li, Y., Brandt, A., McCormick, S.F. and Ruge, J. (1995) A global shallow water numerical model based on the semi-Lagrangian advection of potential vorticity. *Quart. J. Roy. Meteor. Soc.*, **121**, 1981–2006.

Batt, K.L. and Leslie, L.M. (1998) Verification of output from a very high resolution numerical weather prediction model: the 1996 Sydney to Hobart yacht race. *Met. Apps.*, **5**, 321–328.

Benamou, J.-D. and Brenier, Y. (1998) Weak existence for the semi-geostrophic equations formulated as a coupled Monge–Ampère/transport problem. *SIAM J. Appl. Math.*, **58**, 1450–1461.

Brenier, Y. (1990) The least action principle and the related concept of generalised flows for incompressible inviscid fluids. *J. Amer. Math. Soc.*, **2**, 225–255.

Brenier, Y. (1991) Polar factorization and monotone rearrangement of vector-valued functions. *Commun. Pure Appl. Math.*, **44**, 375–417.

Bryan, K. (1989) The design of numerical models of the ocean circulation in Oceanic circulation models: combining data and dynamics. In *Proceedings of the NATO Advanced Study Institute, Les Houches, France*, February 15–26, 1988, 465–500.

Budd, C.J. and Iserles, A. (1999) Geometric Integration. *Phil. Trans. Roy. Soc.*, **357**, 943–1133.

Buhler, O. and McIntyre, M.E. (1998) On non-dissipative wave-mean interactions in the atmosphere or oceans. *J. Fluid Mech.*, **354**, 301–343.

Buizza, R. and Palmer, T.N. (1995) The singular-vector structure of the atmospheric global circulation. *J. Atmos. Sci.*, **52**, 1434–1456.

Burton, G.R. and Douglas, R.J. (1998) Rearrangements and polar factorisation of countably degenerate functions. *Proc. Roy. Soc. Edin.* (A), **128**, 671–681.

Burton, G.R. and McLeod, J.B. (1991) Maximisation and minimisation on classes of rearrangements. *Proc. Roy. Soc. Edin.* (A), **119**, 287–300.

Burton, G.R. and Nycander, J. (1999) Stationary vortices in three-dimensional quasi-geostrophic shear flow. *J. Fluid Mech.*, **389**, 255–274.

Caffarelli, L.A. (1996) Boundary regularity of maps with convex potentials II. *Annals of Math.*, **144**, 453–496.

Charney, J.G., Fjortoft, R. and von Neumann, J. (1950) Numerical integration of the barotropic vorticity equation. *Tellus*, **2**, 237–254.

Courtier, P. (1997) Variational methods. *J. Met. Soc. Japan*, **75**, 211–218.

Cullen, M.J.P. (1989) Implicit finite difference methods for modelling discontinuous atmospheric flows. *J. Comp. Phys.*, **81**, 319–348.

Cullen, M.J.P. (2000) On the accuracy of the semi-geostrophic approximation. *Quart. J. Roy. Meteor. Soc*, **126**, 1099–1116.

Cullen, M.J.P. and Davies, T. (1991) A conservative split-explicit integration scheme with fourth order horizontal advection. *Quart. J. Roy. Meteor. Soc.*, **117**, 993–1002.

Cullen, M.J.P., Davies, T., Mawson, M.H., James, J.A. and Coulter, S. (1997) An overview of numerical methods for the next generation UK NWP and climate model. In *Numerical Methods in Atmosphere and Ocean Modelling. The Andre Robert Memorial Volume*, C. Lin, R. Laprise, H. Ritchie, (eds.), Canadian Meteorological and Oceanographic Society, Ottawa, Canada, 425–444.

Cullen, M.J.P. and Douglas, R.J. (1998) Applications of the Monge–Ampère equation and Monge transport problem to meteorology and oceanography. In *Monge–Ampère Equation: Applications to Geometry and Optimisation*, Contemporary Mathematics **226**, American Mathematical Society, 33–53.

Cullen, M.J.P. and Gangbo, W. (2001) A variational approach for the 2–D semi-geostrophic shallow water equations. *Arch. Rat. Mech. Anal.*, **156**, 241–273.

Cullen, M.J.P., Norbury, J., Purser, R.J. and Shutts, G.J. (1987) Modelling the quasi-equilibrium dynamics of the atmosphere. *Quart. J. Roy. Meteor. Soc.*, **113**, 735–757.

Cullen, M.J.P., Norbury, J. and Purser, R.J. (1991) Generalised Lagrangian solutions for atmospheric and oceanic flows. *S.I.A.M. J. Appl. Math.*, **51**, 20–31.

Cullen, M.J.P. and Purser, R.J. (1984) An extended Lagrangian theory of semi-geostrophic frontogenesis. *J. Atmos. Sci.*, **41**, 1477–1497.

Cullen, M.J.P. and Purser, R.J. (1989) Properties of the Lagrangian semi-geostrophic equations. *J. Atmos. Sci.*, **46**, 2684–2697.

Daley, R. (1997) Atmospheric data assimilation. *J. Met. Soc. Japan*, **75**, 319–329.

Desroziers, G. and Lafore, J.-P. (1993) A coordinate transformation for objective frontal analysis. *Mon. Weather Rev.*, **121**, 1531–1553.

DiPerna, R.J. and Lions, P-L.L. (1989) Ordinary differential equations, transport theory and Sobolev spaces. *Invent. Math.*, **98**, 511–547.

Douglas, R.J. (1994) Rearrangements of functions on unbounded domains. *Proc. Roy. Soc. Edin.*(A), **124**, 621–644.

Dritschel, D.G. and Ambaum, M.H.P. (1997) A contour-advective semi-Lagrangian numerical algorithm for simulating fine-scale conservative dynamical fields. *Quart. J. Roy. Meteor. Soc.*, **123**, 1097–1130.

Emanuel, K.A. (1983) The Lagrangian parcel dynamics of moist symmetric instability. *J. Atmos. Sci.*, **40**, 2369–2376.

Emanuel, K.A., Neelin, J.D. and Bretherton, C.S. (1994) On large-scale circulations in convecting atmospheres. *Quart. J. Roy. Meteor. Soc.*, **120**, 1111–1144.

Embid, P.F. and Majda, A.J. (1996) Averaging over fast waves for geophysical flows with arbitrary potential vorticity. *Comm. Partial Diff. Eqs.*, **21**, 619–658.

Eydeland, A., Spruck, J. and Turkington, B. (1990) Multiconstrained variational problems of nonlinear eigenvalue type: new formulations and algorithms. *Math. Comp.* **55**, 509–535.

Eyre, J.R. (1997) Variational assimilation of remotely-sensed observations of the atmosphere. *J. Met. Soc. Japan*, **75**, 331–338.

Farge, M. and Sadourny, R. (1989) Wave-vortex dynamics in rotating shallow water. *J. Fluid Mech.*, **206** , 433–462.

Findlater, J. (1969) Interhemispheric transport of air in the lower troposphere over the western Indian Ocean. *Quart. J. Roy. Meteor. Soc.*, **95**, 400–403.

Fjortoft (1953) On the changes in the spectral distribution of kinetic energy for two-dimensional, non-divergent flow. *Tellus*, **5**, 225–230.

Gage, K.S. and Nastrom, G.D. (1986) Theoretical interpretation of atmospheric wavenumber spectra of wind and temperature observed by commercial aircraft during GASP. *J. Atmos. Sci.*, **43**, 729-740.

Gangbo, W. and McCann, R.J. (1996) The geometry of optimal transportation. *Acta Math.*, **177:2**, 113–161.

Garrett, C and Munk, W. (1979) Internal waves in the ocean. *Ann. Rev. Fluid Mech.*, **11**, 339–369.

Gent, P.R. and McWilliams, J.C. (1996) Eliassen–Palm fluxes and the momentum equation in non-eddy-resolving ocean circulation models. *J. Phys. Oceanog.*, **26**, 2539–2546.

Gerard, P. (1992) Resultats recents sur les fluides parfaits incompressibles bidimensionnels. *Seminaire Bourbaki*, **757**, 411–444.

Gill, A.E. (1982) *Atmosphere-Ocean Dynamics.* Academic Press. 662pp.

Gjaja, I. and Holm, D.D. (1996) Self-consistent Hamiltonian dynamics of wave, mean-flow interaction for a rotating stratified incompressible fluid. *Physica D*, **98**, 343–378.

Griffith, A.K. and Nichols, N.K. (1996) Accounting for model error in data assimilation using adjoint methods. In *Computational Differentiation: Techniques, Applications and Tools*, SIAM, 195–204.

Halmos, P.R. (1950) *Measure Theory.* Van Nostrand.

Haltiner, G.J. and Williams, R.T. (1980) *Numerical prediction and Dynamic Meteorology*, 2nd ed., Wiley, 477pp.

Hoffman, R.N., Liu, Z., Louis, J.-F. and Grassotti, C. (1995) Distortion representation of forecast errors. *Mon. Wea. Rev.*, **123**, 2758–2770.

Holton, J.R. (1992) *An introduction to dynamic meteorology.* Academic Press, 511pp.

Hoskins, B.J. (1975). The geostrophic momentum approximation and the semi-geostrophic equations. *J. Atmos. Sci.*, **32**, 233–242.

Hoskins, B.J., McIntyre, M.E. and Robertson, A.W. (1985) On the use and significance of isentropic potential vorticity maps. *Quart. J. Roy. Meteor. Soc.*, **111**, 887–946.

Johnson, W.B. (1966) The geotriptic wind. *Bull. Amer. meteor. Soc.*, **47**, 982.

Kato, T. and Ponce, G. (1986) Wellposed-ness of the Euler and Navier–Stokes equations in Lebesgue spaces $L_s^p(\mathbb{R}^2)$. *Rev. Mat. Iberoamerica*, **2**, 73–88.

Katz, E.J. (1975) Tow spectra from MODE. *J. Geophys. Res.*, **80**, 1163–1167.

Knox, J.A. (1997) Generalized nonlinear balance criteria and inertial stability. *J. Atmos. Sci.*, **54**, 967–985.

Kushner, P.J. and Shepherd, T.G. (1995) Wave-activity conservation laws and stability theorems for semi-geostrophic dynamics. Part 2. Pseudo-energy based theory. *J. Fluid Mech.*, **290**, 105–129.

Larichev, V.D. and McWilliams, J.C. (1991) Weakly decaying turbulence in an equivalent-barotropic fluid. *Phys. Fluids*, **A 3(5)**, 938–950.

LeBlond, P.H. and Mysak, L.A. (1978) *Waves in the Ocean*, Elsevier, 602pp.

Leith, C.E. (1980) Nonlinear normal mode initialisation and quasi-geostrophic theory. *J.Atmos.Sci.*, **37**, 958–968.

Leith, C.E. (1983) Predictability in theory and practice. *Large-scale dynamical processes in the atmosphere*, B.J.Hoskins and R.P. Pearce (eds.), Academic Press, 65–383.

Leslie, L.M. and Purser, R.J. (1995) Three-dimensional mass-conserving semi-Lagrangian scheme employing forward trajectories. *Mon. Weather Rev.*, **123**, 2551–2566.

Lin, S.-J. and Rood, R.B. (1996) Multidimensional flux form semi-Lagrangian transport schemes. *Mon. Weather Rev.*, **124**, 2046–2070.

Lin, S.-J. and Rood, R.B. (1997) An explicit flux-form semi-Lagrangian shallow water model on the sphere. *Quart. J. Roy. Meteor. Soc.*, **123**, 2477–2498.

Lorenc, A.C., Bell, R.S. and MacPherson, B. (1991) The Meteorological Office analysis correction data assimilation scheme. *Quart. J. Roy. Meteor. Soc.*, **117**, 59–90.

Lorenz, E.N. (1963) Deterministic non-periodic flow. *J. Atmos. Sci.*, **20**, 130–141.

Marsden, J.E., Ratiu, T.S. and Raugel, G. (1995) Equations d'Euler dans une coque spherique mince. *C.R. Acad. Sci. Paris*, **321**, 1201–1206.

McCann, R.J. (1995) Existence and uniqueness of monotone measure preserving maps. *Duke Math. J.*, **80**, 309–323.

McCann, R.J. (2001) Polar factorization of maps on Riemannian manifolds. *Geom. Functional Anal.*, **11**, 589–608.

McIntyre, M.E. and Roulstone, I. (1996). Hamiltonian balanced models: constraints, slow manifolds and velocity splitting. UK Met. Office NWP Sci. Paper no. 41, submitted to *J. Fluid Mech.* See also McIntyre and Roulstone [II, 8] (these volumes).

Mawson, M.H. and Cullen, M.J.P. (1992) An idealised simulation of the Indian monsoon using primitive-equation and quasi-equilibrium models. *Quart. J. Roy. Meteor. Soc.*, **118**, 153–164.

Mohebalhojeh, A.R. and Dritschel, D.F. (2000) On the representation of gravity waves in numerical models of the shallow-water equations *Quant. J. Roy. Meteor. Soc.*, **126**, 669–688.

Neven, E.C. (1994) Baroclinic modons on a sphere. *J. Atmos. Sci.*, **51**, 1447–1464.

Olsson, P.Q. and Cotton, W.R. (1997) Balanced and unbalanced circulations in a primitive equation simulation of a mid-latitude MCC. Part II. Analysis of balance. *J. Atmos. Sci.*, **54**, 457–497.

Ostdiek, V. and Blumen, W. (1997) A dynamic trio: inertial oscillation, deformation frontogenesis, and the Ekman–Taylor boundary layer. *J. Atmos. Sci.*, **54**, 1490–1502.

Pielke, R.A. (1984) *Mesoscale Meteorology*, Academic Press, 612pp.

Pogorelov, A.V. (1973) *Extrinsic geometry of convex surfaces*, Israel Programme for Scientific Translations, 669pp.

Purser, R.J. (1999) Legendre-transformable semi-geostrophic theories. *J. Atmos. Sci.*, **56**, 2522–2535.

Richardson, L.F. (1922). *Weather Prediction by Numerical Process.* Cambridge University Press, reprinted by Dover, 1965. 236pp.

Roulstone, I. and Sewell, M.J. (1996). Potential vorticities in semi-geostrophic theory. *Quart. J. Roy. Meteorol. Soc.*, **122**, 983–992.

Salmon, R. (1985) New equations for nearly geostrophic flow. *J.Fluid Mech.*, **153**, 461–477.

Schubert, W.H. (1985) Semi-geostrophic theory. *J. Atmos. Sci.*, **42**, 1770–1772.

Schubert, W.H., Ciesieleski, P.E., Stevens, D.E. and Kuo H.-C. (1991) Potential vorticity modelling of the ITCZ and the Hadley circulation. *J. Atmos. Sci.*, **48**, 1493–1509.

Shutts, G.J. (1987) Balanced flow states resulting from penetrative, slant-wise convection. *J. Atmos. Sci.*, **44**, 3363–3376.

Shutts, G.J. (1987) The semi-geostrophic weir: a simple model of flow over mountain barriers. *J. Atmos. Sci.*, **44**, 2018–2030.

Shutts, G.J. (1991) Dynamical aspects of the October storm 1987: A study of a successful fine-mesh simulation. *Quart. J. Roy. Meteor. Soc.*, **116**, 1315–1348.

Shutts, G.J. (1995) An analytical model of the balanced flow created by localised convective mass transfer in a rotating fluid. *Dyn. Atmos. Oceans*, **22**, 1–17.

Shutts, G.J. and Cullen, M.J.P. (1987) Parcel stability and its relation to semi-geostrophic theory. *J. Atmos. Sci.*, **46**, 2684–2697.

Shutts, G.J., Booth, M. and Norbury, J. (1988) A geometric model of balanced, axisymmetric flow with embedded penetrative convection. *J. Atmos. Sci.*, **45**, 2609–2621.

Smagorinsky, J. (1974) Global atmospheric modelling and the numerical simulation of climate., in *Weather and Climate Modification*, Wiley, 633–686.

Snyder, C., Skamarock, W.C. and Rotunno, R. (1991) A comparison of primitive-equation and semi-geostrophic simulations of baroclinic waves. *J. Atmos. Sci.*, **48**, 2179–2194.

Staniforth, A. and Cote, J. (1991) Semi-Lagrangian integration schemes for atmospheric models — a review. *Mon. Weather Rev.*, **119**, 2206–2223.

Temperton, C. (1989) Implicit normal mode initialisation for spectral models. *Mon. Weather Rev.*, **117**, 436–451.

Temperton, C. (1997) Treatment of the Coriolis terms in semi-Lagrangian spectral models. In *Numerical Methods in Atmosphere and Ocean Modelling. The Andre Robert Memorial Volume*, C. Lin, R. Laprise, H. Ritchie, (eds.), Canadian Meteorological and Oceanographic Society, Ottawa, Canada, 293–302.

Thomson, W. (Lord Kelvin) (1910) Maximum and minimum energy in vortex motion. *Mathematical and Physical Papers*, vol. 4, Cambridge University Press, 172–183.

Thorpe, A.J. and Emanuel, K.A. (1985) Frontogenesis in the presence of small stability to slant-wise convection. *J. Atmos. Sci.*, **42**, 1809–1824.

Toth, Z. and Kalnay E. (1993) Ensemble forecasting at NMC: the generation of perturbations. *Bull Amer. Meteor. Soc*, **74**, 2317–2330.

Vallis, G.K. (1992) Mechanisms and parametrisations of geostrophic adjustment and a variational approach to balanced flow. *J. Atmos. Sci.*, **49**, 1144–1160.

Vallis, G.K. (1996) Potential vorticity inversion and balanced equations of motion for rotating and stratified flows. *Quart. J. Roy. Meteor. Soc.*, **122**, 291–322.

Veitch, G. and Mawson, M.H. (1993) A comparison of inertial stability conditions in the planetary semi-geostrophic and quasi-equilibrium models. U.K. Met. Office Short Range Forecasting Tech. Report no. 60.

Warn, T., Bokhove, O., Shepherd, T.G. and Vallis, G.K. (1994) Rossby number expansions, slaving, and balance dynamics. *Quart. J. Roy. Meteor. Soc.*, **121**, 723–739.

Rearrangements of Functions with Applications to Meteorology and Ideal Fluid Flow

R.J. Douglas

1 Introduction

This article studies rearrangements of functions, and considers applications to ideal fluid flow, decomposition of weather forecast error, and a system of equations which models large scale atmospheric and oceanic flow. We begin by giving an intuitive idea of when a function is a rearrangement of another function. Let f be a function defined on a bounded set $\Omega \subset \mathbb{R}^n$, and imagine that Ω is a continuum of infinitesimal particles. Suppose we exchange the particle positions, with each particle retaining its value of f. This yields a function g, which is a *rearrangement* of f. This intuitive notion makes sense whether we attach a scalar or a vector to a particle: the idea of rearranging a function can be applied to both scalar and vector valued functions. Roughly speaking, the formal definition of two functions f and g being rearrangements is that for any given set of values, the set where f takes those values has the same size as the corresponding set for g. (We give precise definitions in §2 and §5.)

For a prescribed function f_0, we can consider the *set of all rearrangements of* f_0. Sets of rearrangements of some given function arise naturally in applications in the following ways:

(i) Suppose we have a quantity q which is conserved following the flow in an ideal (i.e. incompressible, inviscid) fluid. If we consider a fluid particle, it may move as the fluid evolves, but it retains the same value of q, and the particle does not change size (as the flow is incompressible). It follows that at any two instants of time t_1 and t_2, $q(t_1)$ and $q(t_2)$ are rearrangements. (Compare with the intuitive notion of rearrangement above.) In particular, at any time t, $q(t)$ belongs to the set of rearrangements of $q(0)$.

(ii) Suppose we know the values of some quantity on fluid particles at some fixed time t, but do not know the positions of particles, save that they must satisfy conservation of mass. Then the set of possible configurations of the quantity are described by elements of a set of rearrangements.

Without further information, the knowledge that some quantity belongs to a set of rearrangements is of limited utility. However if this is harnessed to a principle that a particular type of flow results from maximising or minimising a functional (of the quantity), typically an energy, over all possible configurations, we can prove properties of such flows. In §3 we study ideal fluid flow in a bounded planar domain; vorticity is rearrangement preserved in the sense of (i). There is a principle that maximising kinetic energy over a set of rearrangements should yield a steady flow. We review the result that maximising vorticities have corresponding stream functions such that the equation of a steady flow is indeed satisfied. We study the semigeostrophic equations, a standard model for weather front formation, in §6. A vector valued quantity $\mathbf{X}$, from which we can recover the physical quantities velocity, pressure and potential temperature, is predicted on particles at some fixed time t. A constraint on particles is that the flow is incompressible, so possible configurations of $\mathbf{X}$ are given by elements of a set of rearrangements as described in (ii). The *Cullen–Norbury–Purser* principle states that for a solution, the particles are arranged so that geostrophic energy is minimised. We apply this principle at each time t, so we have a constrained minimisation problem, where the constraint changes with t. We review the result that this principle is well posed, and identify such stable solutions with an extra constraint on the system of equations.

This article is organised as follows. Section 2 introduces rearrangements of scalar valued functions. With reference to simple examples, we give a precise definition of two functions being rearrangements, introduce inequalities satisfied by special rearrangements, consider mappings which relate rearrangements to each other, and establish properties of the set of rearrangements. Section 3 presents applications of rearrangement theory to ideal fluid dynamics; in addition to the application mentioned above, we review some related problems. We discuss a new approach to the evaluation of weather forecast error in §4; the error in the forecast of some meteorological quantity is split into a contribution due to displacement, and a part penalising incorrect qualitative features. The decomposition makes use of the results established in §2. Section 5 reviews properties of rearrangements of vector valued functions; as in §2, the various concepts are illustrated by straightforward examples. After introducing the definition of two vector valued functions being rearrangements, and establishing some equivalent properties, we consider inequalities satisfied by a special rearrangement. In general vector valued rearrangement theory is less rich than the scalar valued case. We note the connection between rearrangement inequalities and optimal mass transfer problems. Application of results in §5 to the semigeostrophic equations is the subject of §6. In addition to the identification of stable solutions with an extra constraint that we discussed above, we explain the way in which the semigeostrophic equations model front formation. We review results that have been proved concerning existence of solution for these equations, and indicate the work which remains to achieve a full existence theory. Note that in this article we restrict attention to the semigeostrophic equations with constant rotation. The reader is referred to Cullen and Douglas (1998) for a description of how the methods described in §6 may be applied to the variable rotation case.

This is far from an exhaustive survey of rearrangements of functions and their applications. There is an extensive literature on the use of isoperimetric inequalities, see for example Talenti (1976). Moreover properties of the directional derivative of the increasing rearrangement (see §2 for the definition of the increasing rearrangement of a function) are pertinent to the study of certain differential equations. (See Mossino and Temam 1981.) The reader is referred to Kawohl (1985) for a survey of applications of rearrangements methods to partial differential equations. In this article we focus on those applications studied during the programme *Mathematics of Atmosphere and Ocean Dynamics*, held at the Isaac Newton Institute for Mathematical Sciences in 1996.

2 Rearrangements of scalar valued functions

This section is concerned with rearrangements of scalar valued functions. We introduce the concept via an example in §2.1; we then give a formal definition and note equivalent properties. Applications of rearrangement theory make use of inequalities which hold for special rearrangements: in §2.2 we introduce the increasing and decreasing rearrangements of a function, and discuss inequalities they satisfy. Measure-preserving mappings, which relate rearrangements, are reviewed in §2.3. We recall the result that any real integrable function on a bounded interval can be written as the composition of its increasing rearrangement with a measure-preserving mapping. In applications we often wish to maximise or minimise a functional (which typically represents an energy) over the set of rearrangements of a prescribed function: in §2.4 we establish properties of this set. A sequence of rearrangements of a prescribed function may have a (weak) limit which is not a rearrangement: we define an enlarged set which contains these limits. An extremum of a functional with respect to this set may or may not be a rearrangement: we illustrate with two simple examples. Finally in §2.5 we extend the definition of rearrangement to functions defined on unbounded domains of infinite size.

2.1 Definition and properties of rearrangements of functions

We give a simple example of two scalar functions which are rearrangements of each other. Let $f(x) = x$ for each $x \in [0, 1]$, and let g be defined by

$$g(x) = \begin{cases} 1 - 2x & \text{if } x \in [0, 1/2], \\ 2x - 1 & \text{if } x \in [1/2, 1]. \end{cases} \tag{2.1}$$

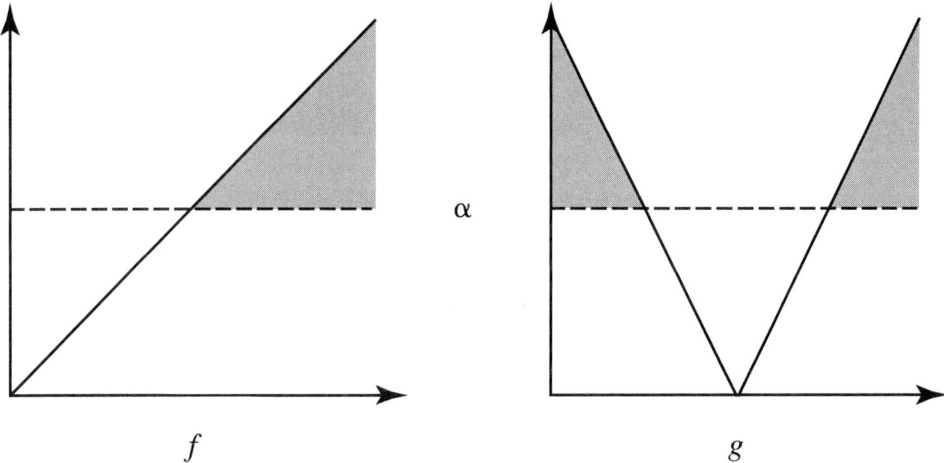

Figure 1: Two rearrangements f and g; the area of the shaded regions is the same for every α

For each $0 < \alpha < 1$,

$$\int_0^1 (f(x) - \alpha)_+ dx = \frac{1}{2}(1-\alpha)^2 = \int_0^1 (g(x) - \alpha)_+ dx \tag{2.2}$$

where the subscript + denotes the positive part of the function, that is $h_+(x) = \max\{h(x), 0\}$. (2.2) says that the area of the region which is bounded above by the graph of f and bounded below by the line $y = \alpha$ is the same as the area of the corresponding regions of g. This is illustrated in Figure 1: the total shaded areas are the same. If $\alpha \geq 1$, then $\int_0^1 (f(x)-\alpha)_+ dx = 0 = \int_0^1 (g(x)-\alpha)_+ dx$, and if $\alpha \leq 0$, $\int_0^1 (f(x) - \alpha)_+ dx = -\alpha + \int_0^1 f(x)dx = -\alpha + \int_0^1 g(x)dx = \int_0^1 (g(x) - \alpha)_+ dx$. Combining the above results, $\int_0^1 (f(x) - \alpha)_+ dx = \int_0^1 (g(x) - \alpha)_+ dx$ for each real α.

We will use this property to make a precise definition of when f and g are *rearrangements* of each other. We follow ideas of Eydeland, Spruck and Turkington (1990). Let Ω be a bounded subset of $\mathbb{R}^n$, and let μ be a measure of the size of subsets of Ω. If $n = 1$, an appropriate choice is length, for $n = 2$, area, and for $n = 3$, volume.

Definition. Let $f, g : \Omega \to \mathbb{R}$ be two integrable functions i.e $\int_\Omega |f(x)|d\mu(x) < \infty$, $\int_\Omega |g(x)|d\mu(x) < \infty$. f is a *rearrangement of* g if

$$\int_\Omega (f(x) - \alpha)_+ d\mu(x) = \int_\Omega (g(x) - \alpha)_+ d\mu(x) \tag{2.3}$$

for every real α.

For most applications, the measure of size we will use is n-dimensional Lebesgue measure, denoted λ_n, which coincides with length, area or volume when $n = 1, 2$ or 3. (Note, however, that there are 'pathological' sets whose size we cannot calculate.) In this article we will restrict attention to measures μ where the size of a set $S \subset \Omega$, denoted $\mu(S)$, is given by $\mu(S) = \int_S h d\lambda_n$, for some non-negative integrable function h. (For a fluid flow problem, such a choice is appropriate if we want the size of the set to be its mass, rather than its volume.) Clearly the choice of h identically equal to 1 recovers n-dimensional Lebesgue measure.

There are other properties which are equivalent to the definition of rearrangement given above.

Theorem 1 *Let $\Omega \subset \mathbb{R}^n$ be a bounded set, and let μ be a measure as above, that is absolutely continuous with respect to n-dimensional Lebesgue measure. Let $f, g : \Omega \to \mathbb{R}$ be integrable functions. Then the following are equivalent.*
(i) f is a rearrangement of g.
(ii) For every real α,

$$\mu(\{x : f(x) \geq \alpha\}) = \mu(\{x : g(x) \geq \alpha\}).$$

(iii) For every (Borel) set $B \subset \mathbb{R}$,

$$\mu\left(\{x : f(x) \in B\}\right) = \mu\left(\{x : g(x) \in B\}\right). \tag{2.4}$$

(iv) For each continuous function $F : \mathbb{R} \to \mathbb{R}$,

$$\int_\Omega F(f(x))d\mu(x) = \int_\Omega F(g(x))d\mu(x). \tag{2.5}$$

(Equation (2.5) *is understood in the sense that if one of the integrals is finite, so is the other and they are equal.)*

Proof. Follows by the methods of, for example, Douglas (1998). □

Property (ii) is used as the definition of rearrangement in some of the literature. (See, for example, Burton 1987.) For $f, g : [0, 1] \to \mathbb{R}$ we can interpret (ii) as follows. Intersect the graphs of f and g with the line $y = \alpha$; for each function, find the length of that part of the line for which the graph lies on or above the line. f and g are rearrangements if the two lengths are equal for every choice of α. For f defined by $f(x) = x$ and g as in (2.1), this length is 0 when $\alpha \geq 1$, $1 - \alpha$ for $0 \leq \alpha \leq 1$, and 1 for $\alpha < 0$. This is illustrated in Figure 2 for a particular choice of α.

Our intuitive notion of f and g being rearrangements is that for any collection of values, the set where f takes those values has the same size as the corresponding set for g. This is stated precisely in Theorem 1 (iii); the restriction to Borel subsets of $\mathbb{R}$ excludes some 'strange' sets, and ensures that we can measure the size of the sets in (2.4). It is immediate from Theorem

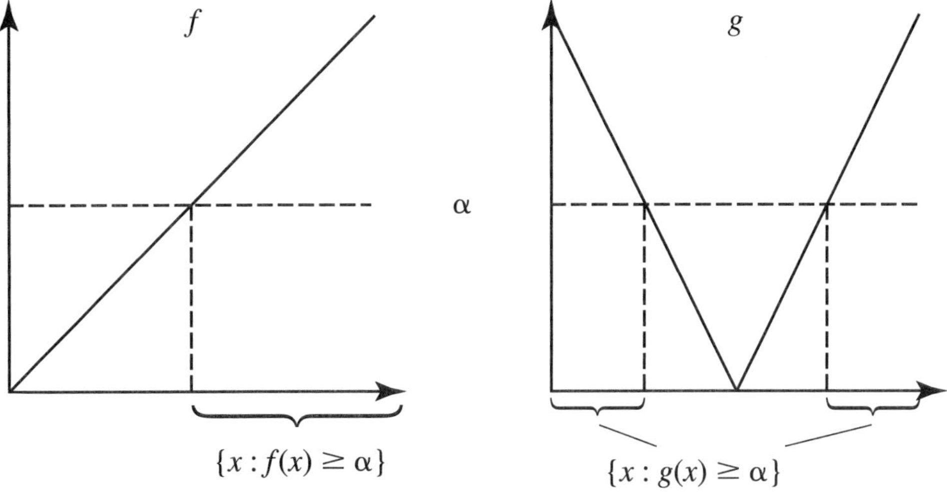

Figure 2: Functions f and g which satisfy Theorem 1 (ii) for every α

1 (iv) that if $\int_\Omega |f|^p d\mu$ is finite for some $p \geq 1$, and g is a rearrangement of f, then $\int_\Omega |f|^p d\mu = \int_\Omega |g|^p d\mu$. More generally any 'reasonable' function F satisfies (2.5).

We now introduce the concept of the *set of rearrangements* of a given function. A function has more than one rearrangement (unless it is constant). For $f : [0,1] \to \mathbb{R}$ defined by $f(x) = x$, the function g given by (2.1) is a rearrangement of f. A different rearrangement h is defined by $h(x) = 1 - x$. In general the set of all the rearrangements of a given function will have infinitely many elements: as all the members of the set are rearrangements of each other, any one can be used to identify the set. (Rearrangement is an equivalence relation.) We write $\mathcal{R}(f_0)$ to denote the set of rearrangements of a prescribed function f_0.

2.2 Special rearrangements and rearrangement inequalities

In this subsection we identify some special elements of the set of rearrangements of a prescribed function, and review inequalities that they satisfy. Such inequalities are pertinent to the study of maximising or minimising functionals (which often represent an energy in physical applications) with respect to a set of rearrangements; we discuss this in §2.4. Our aim is to introduce results that we will require for applications in §3 and §4, and to review some other important rearrangement inequalities; more on this topic can be found in Alvino, Lions and Trombetti (1989) and Burton (1987).

The first special class of rearrangements we shall consider are monotone rearrangements defined on a (bounded) interval. Let $f(x) = x$ for each $x \in [0,1]$, and let g be as in (2.1). f is an increasing function, and it can be proved that it is the only increasing function which is a rearrangement of g. More generally, if $f_0 : [0,1] \to \mathbb{R}$ is an integrable function i.e. $\int_0^1 |f_0(x)| d\lambda_1(x) < \infty$, we can construct an increasing function f_0^* on $[0,1]$ by replacing every set $\{x : f_0(x) \geq \alpha\}$ by an interval of the same size extending leftwards from 1, i.e. $\{x : f_0^*(x) \geq \alpha\} = [1 - \lambda_1(\{x : f_0 \geq \alpha\}), 1]$. It is immediate that $f_0^* \in \mathcal{R}(f_0)$. Two increasing rearrangements of f_0 can differ only at points of discontinuity, and there are only countably many such points. It follows that f_0^* is the (essentially) unique rearrangement of f_0 which is an increasing function; we call f_0^* the *increasing rearrangement of* f_0. If $\Omega \subset \mathbb{R}^d$ is equipped with a measure μ such that $\mu(\Omega) < \infty$, and $f_0 : \Omega \to \mathbb{R}$ is integrable, the above construction can be used to define an increasing function f_0^* on $[0, \mu(\Omega)]$; for simplicity we restrict attention to $\Omega = [0,1]$ with 1-dimensional Lebesgue measure. Using an analogous construction, we can define the *decreasing rearrangement of* f_0, which we write $f_0^\triangle$. It is the (essentially) unique rearrangement of f_0 which is a decreasing function. (See Burton 1987, Lemma 1.) For g as in (2.1), $g^\triangle(x) = 1 - x$.

There are many inequalities involving rearrangements where an extreme value is obtained by a monotone (i.e. increasing or decreasing) rearrangement; we consider some fundamental examples. For scalar valued square integrable functions f, g defined on the unit interval, (that is functions satisfying $\int_0^1 f^2 < \infty, \int_0^1 g^2 < \infty$,)

$$\int_0^1 f(x)g(x)d\lambda_1(x) \leq \int_0^1 f^*(x)g^*(x)d\lambda_1(x) \tag{2.6}$$

where f^*, g^* are the increasing rearrangements of f, g respectively. If we replace the increasing rearrangements by decreasing rearrangements the inequality still holds. Allowing f and g to vary over $\mathcal{R}(f)$ and $\mathcal{R}(g)$ respectively, which pairs of functions maximise $\int_0^1 fg$? From our previous statements it is immediate that the pairs (f^*, g^*), $(f^\triangle, g^\triangle)$ are maximisers; in general there will be others, where the functions are arranged so that the 'big values' of f multiply 'big values' of g. If f and g have rearrangements $\tilde{f}$ and $\tilde{g}$ such that $\tilde{f} = \phi \circ \tilde{g}$ for some increasing function ϕ, then $\tilde{f}$ and $\tilde{g}$ achieve equality in (2.6). If we fix g, there is a unique rearrangement $\hat{f} \in \mathcal{R}(f)$ such that equality holds in (2.6) if and only if there exists an increasing function ϕ such that $\hat{f} = \phi \circ g$. (The reader is referred to Burton 1987 for proofs of these assertions.) In particular, if the sets $\{x : g(x) = \alpha\}$ have zero size for each $\alpha \in \mathbb{R}$,

$$\int_0^1 f(x)g^*(x)d\lambda_1(x) < \int_0^1 f_0^*(x)g^*(x)d\lambda_1(x) \tag{2.7}$$

for every rearrangement f of f_0 which is not equal to f_0^*. We can always find $\tilde{f} \in \mathcal{R}(f_0)$ which maximises $\int_0^1 fg$ for fixed g; we prove this in the next subsection.

For f and g as above, it follows from (2.6) and Theorem 1 (iv) that

$$\left\{\int_0^1 |f^*(x) - g^*(x)|^2 d\lambda_1(x)\right\}^{1/2} \leq \left\{\int_0^1 |f(x) - g(x)|^2 d\lambda_1(x)\right\}^{1/2}. \quad (2.8)$$

We say that the mapping which takes a function to its increasing rearrangement is *non-expansive* on the space of square integrable functions (as it does not increase the L^2-norm). For $1 \leq p < \infty$, if we replace 2 by p and $1/2$ by $1/p$ in (2.8), then the inequality still holds. (See Crowe, Zweibel and Rosenbloom 1986, Corollary 1.) Noting that the left hand side of the inequality is the L^p norm of $f^* - g^*$, and that the right hand side is the L^p norm of $f - g$, we see that mapping a function to its increasing rearrangement is non-expansive on L^p for $1 \leq p < \infty$. More generally it may be shown (by the methods of Lieb and Loss 1997, Theorem 3.5) that

$$\int_0^1 J(f^*(x) - g^*(x))d\lambda_1(x) \leq \int_0^1 J(f(x) - g(x))d\lambda_1(x)$$

where $J : \mathbb{R} \to \mathbb{R}$ is a non-negative convex function satisfying $J(0) = 0$. (All the above inequalities still hold if we replace increasing rearrangements by decreasing rearrangements.)

Another important class of rearrangements are the (Schwarz) symmetric decreasing rearrangements. For an integrable function f defined on a bounded set Ω in the plane, the *Schwarz-symmetrisation* $f^{\triangle}$ is formed by replacing every set $\{x : f(x) \geq \alpha\}$ by a disc centred on the origin of the same area. It follows that $f^{\triangle}$ depends on $|x|$ only, and is a decreasing function of $|x|$. We can extend this concept to three (or indeed n) dimensions: replace discs by balls, and area by volume. For $n \geq 3$, $f^{\triangle}$ is usually referred to as the *spherically symmetric rearrangement of f*. The analogous inequalities to (2.6) and (2.8) extend to spherically symmetric rearrangements, that is

$$\int_{\Omega} f(x)g(x)d\lambda_n(x) \leq \int_B f^{\triangle}(x)g^{\triangle}(x)d\lambda_n(x),$$

$$\left\{\int_B |f^{\triangle}(x) - g^{\triangle}(x)|^2 d\lambda_n(x)\right\}^{1/2} \leq \left\{\int_{\Omega} |f(x) - g(x)|^2 d\lambda_n(x)\right\}^{1/2},$$

where B is a ball (centre the origin) of the same size as Ω. Moreover, for a non-negative function f defined on a bounded domain in $\mathbb{R}^n$, having square integrable first order partial derivatives, and vanishing on the boundary of Ω, the following inequality is satisfied:

$$\int_B |\nabla f^{\triangle}(x)|^2 d\lambda_n(x) \leq \int_{\Omega} |\nabla f(x)|^2 d\lambda_n(x).$$

The set B is as above. (See Brothers and Ziemer 1988 for related inequalities satisfied by the spherically symmetric rearrangement.)

In the next subsection we introduce the concept of a measure-preserving mapping; these maps relate rearrangements to each other. In particular we will see that any integrable function can be written as the composition of its increasing rearrangement with a measure-preserving mapping.

2.3 Measure-preserving mappings and polar factorisation of scalar valued functions

This subsection studies measure-preserving mappings and their relationship with rearrangements. By way of motivation, we return to our statement in the introduction that for an incompressible flow, any Lagrangian conserved quantity is rearrangement preserved. Let $t \to \chi(t,x)$ be the trajectory of the fluid particle which is at x initially. For any given subset U of the fluid at time t, for each particle in U we can find the original position of the particle at time zero; noting that the flow is incompressible, this set of points has the same size as U. It follows that the trajectory mapping is a measure-preserving mapping, the definition of which we give below. Let q be a quantity which is conserved on fluid particles, and let q_0, q_t be its values at times $0, t$ respectively. Now $q_t \circ \chi(t,.) = q_0$, where $\chi(t,.)$ is a measure-preserving mapping (and $\circ$ denotes composition of functions); we see later that this implies q_t and q_0 are rearrangements, which justifies the claim made in the introduction.

Essentially a measure-preserving mapping between two sets U, V is a mapping which satisfies the following property: given any (measurable) set $W \subset V$, the set of points in U which are mapped to W has the same size as W. More precisely:

Definition Let $U \subset \mathbb{R}^n$, $V \subset \mathbb{R}^d$, and let μ, ν be measures (of size) on U, V respectively, with $\mu(U) = \nu(V)$. A *measure-preserving mapping* $s : U \to V$ satisfies $\mu(\{x : s(x) \in W\}) = \nu(W)$ for each (ν-measurable) set $W \subset V$. Halmos (1950) shows that this is equivalent to requiring that for every ν-integrable function f,

$$\int_U f \circ s d\mu = \int_V f d\nu. \tag{2.9}$$

We consider some elementary examples when $U = V = [0,1]$, and μ and ν are length. It is easily seen that $s_1(x) = x$ for $x \in [0,1]$ is a measure-preserving mapping. However g as defined in (2.1) is a measure-preserving mapping which is two to one; it follows that measure-preserving mappings need not be one to one. Such mappings need not be smooth: let $s_2 : [0,1] \to [0,1]$ be defined by

$$s_2(x) = \begin{cases} x + 1/2 & \text{if } x \in [0, 1/2], \\ x - 1/2 & \text{if } x \in (1/2, 1), \\ 0 & \text{if } x = 1. \end{cases}$$

Then s_2 is a measure-preserving-mapping from $[0,1]$ to itself that is discontinuous at $x = 1/2$.

If a measure-preserving mapping s is one to one, and maps measurable sets to measurable sets, then the inverse of s exists and is also a measure-preserving mapping. Such an s is called a *measure-preserving transformation.* We show that measure-preserving mappings and transformations preserve rearrangements. Let bounded sets $\Omega \subset \mathbb{R}^n$, $\Omega' \subset \mathbb{R}^d$ be such that $\mu(\Omega) = \nu(\Omega')$, where ν and μ are measures satisfying the hypotheses of Theorem 1. Let $f, g : \Omega \to \mathbb{R}$ be integrable functions satisfying $f \in \mathcal{R}(g)$, and let $s : \Omega' \to \Omega$ be a measure-preserving mapping. Using the notation $h^{-1}(S) = \{x : h(x) \in S\}$, for each (Borel) set $B \subset \mathbb{R}$ we have

$$\begin{aligned}\nu((f \circ s)^{-1}(B)) &= \nu(s^{-1} \circ f^{-1}(B)) = \mu(f^{-1}(B)) \\ &= \mu(g^{-1}(B)) = \nu(s^{-1} \circ g^{-1}(B)) = \nu((g \circ s)^{-1}(B)).\end{aligned}$$

Thus $f \circ s \in \mathcal{R}(g \circ s)$. If s is a measure-preserving transformation, then the reverse implication holds. Furthermore, if (Ω', ν) is the same space as (Ω, μ), then $f \circ s \in \mathcal{R}(f)$, and in particular, for a quantity q which remains constant on fluid particles following the motion of an incompressible flow, $q(t) \in \mathcal{R}(q(0))$.

If there exists a measure-preserving transformation τ between two measure spaces (U, μ) and (V, ν), they have the same measure theoretic structure; we call such spaces *isomorphic.* It may be shown that bounded sets $\Omega \subset \mathbb{R}^n$ equipped with measures μ satisfying the hypotheses of Theorem 1 are isomorphic to an interval equipped with length (i.e. one-dimensional Lebesgue measure). We can construct the increasing rearrangement of an integrable function $f : \Omega \to \mathbb{R}$ by finding $(f \circ \tau)^*$, where $\tau : [0, \mu(\Omega)] \to \Omega$ is a measure-preserving transformation. Moreover we can apply the inequalities of the previous subsection to $f \circ \tau$.

Any other rearrangement may be obtained from the increasing rearrangement by composition with a suitable measure-preserving mapping. Ryff (1970) proved that for any integrable function f defined on $[0,1]$, there exists a measure-preserving mapping $s : [0,1] \to [0,1]$ such that $f = f^* \circ s$. For integrable functions f defined on Ω as above, we can apply this result to $f \circ \tau$, where $\tau : [0, \mu(\Omega)] \to \Omega$ is a measure-preserving transformation. The existence of a measure-preserving mapping $s : \Omega \to [0, \mu(\Omega)]$ satisfying $f = f^* \circ s$ follows easily. The decomposition into the composition of the increasing rearrangement with a measure-preserving mapping is called a *polar factorisation of* f. The methods of Burton and Douglas (1998) demonstrate that the polar factorisation is unique (i.e. there does not exist a measure-preserving mapping t, with $t \neq s$, such that $f = f^* \circ t$,) if and only if f^* is injective. An immediate consequence of the existence of a polar factorisation is the following result (which we stated without proof in the previous subsection):

Proposition 1 *Let Ω, μ be as in Theorem 1. Let $f, g : \Omega \to \mathbb{R}$ be square integrable functions. Then there exists $\hat{f} \in \mathcal{R}(f)$ such that*

$$\int_\Omega \hat{f}(x)g(x)d\mu(x) = \int_0^{\mu(\Omega)} f^*(x)g^*(x)d\lambda_1(x)$$

i.e. $\sup_{\tilde{f}\in\mathcal{R}(f)} \int_\Omega \tilde{f}g$ *is attained by* $\hat{f}$.

Proof. From the discussion above, g has a polar factorisation $g = g^* \circ s$ for some measure-preserving mapping $s : \Omega \to [0, \mu(\Omega)]$. Define $\hat{f} = f^* \circ s$. It is immediate that $\hat{f} \in \mathcal{R}(f)$, and

$$\begin{aligned} \int_\Omega \hat{f}(x)g(x)d\mu(x) &= \int_\Omega f^*(s(x))g^*(s(x))d\mu(x) \\ &= \int_0^{\mu(\Omega)} f^*(x)g^*(x)d\lambda_1(x) \qquad (2.10) \end{aligned}$$

where (2.10) follows because s is measure-preserving. □

Note that it is not always possible to obtain one rearrangement from another by composing with a measure-preserving mapping. If $f(x) = x$ for $x \in [0, 1]$, and g is as in (2.2), no measure-preserving mapping $s : [0, 1] \to [0, 1]$ such that $f = g \circ s$ exists.

2.4 Maximising or minimising functionals with respect to sets of rearrangements

We introduced the concept of the set of rearrangements of a prescribed function in §2.1. In this subsection we study properties of this set, in anticipation of later applications where we will maximise or minimise a functional over a set of rearrangements. A conventional approach is to take a maximising sequence which converges to some limit, and demonstrate that the limit is a maximiser. However we will show by example that a sequence of rearrangements of a fixed function (or any subsequence thereof) need not have a limit which is a rearrangement; in fact this sequence has a (weak) limit which is not a rearrangement of the original function. This motivates our study of the closed convex hull of the set of rearrangements, a set which contains all weak limits of sequences of rearrangements. We consider two simple examples of extremising a functional with respect to a set of rearrangements, then with respect to the closed convex hull of that set. Our examples demonstrate that the extreme value over the latter set may or may not be attained by a rearrangement.

For Ω and μ as in Theorem 1, we work with square integrable functions $f_0 : \Omega \to \mathbb{R}$, that is f_0 which satisfy $\int_\Omega f_0^2 d\mu < \infty$. Functions in this space (which we denote $L^2(\Omega, \mu)$) have a finite 'energy', but can have arbitrarily small scale oscillations: therefore it is a natural setting for a physical problem.

We define a norm on $L^2(\Omega,\mu)$ by $\|f\|_2 = \{\int_\Omega f^2 d\mu\}^{1/2}$. Elements of this space are integrable.

We now recall some standard definitions, which will enable us to say whether a sequence of functions converges to a limit. For more detail see, for example, Friedman (1982, sections 3.2, 4.10, 4.14).

Definitions. A sequence $(f_n) \subset L^2(\Omega,\mu)$ converges strongly to $f \in L^2(\Omega,\mu)$ if $\|f_n - f\|_2 \to 0$ as $n \to \infty$. A set $\mathcal{F} \subset L^2(\Omega,\mu)$ is *strongly closed* if for every sequence (f_n) in $\mathcal{F}$ which converges strongly to a $f \in L^2(\Omega,\mu)$, say, then the limit function f is actually in $\mathcal{F}$. The *closure* of a set $\mathcal{F}$, denoted $\overline{\mathcal{F}}$, is the smallest closed set which contains $\mathcal{F}$.

A set $\mathcal{F} \subset L^2(\Omega,\mu)$ is *strongly compact* if every sequence in $\mathcal{F}$ has a subsequence which converges strongly to an element of $\mathcal{F}$. A strongly compact set is strongly closed.

It is useful to have another notion of convergence and compactness which can 'ignore' small scale oscillations, and thus be more realistically applied to physical data. We give an example. For each $n \in \mathbb{N}$, let $f_n \in L^2([0,2\pi],\lambda_1)$ be defined by $f_n(x) = \sin nx$. The sequence (f_n) does not converge in $L^2([0,2\pi],\lambda_1)$, nor does any subsequence. However the f_n approach the constant function with value 0 in an averaged sense, as $\int_0^{2\pi} f_n g \to 0$ as $n \to \infty$ for any $g \in L^2([0,2\pi],\lambda_1)$. We say that f_n *converges weakly* to 0. We give the formal definition.

Definitions. For all positive integers n, let $f_n \in L^2(\Omega,\mu)$, and further suppose that $f \in L^2(\Omega,\mu)$. We say that f_n *converges weakly* to f if $\int_\Omega f_n g \to \int_\Omega fg$ as $n \to \infty$ for every $g \in L^2(\Omega,\mu)$. Strong convergence implies weak convergence, but not vice versa.

A set $\mathcal{F} \subset L^2(\Omega,\mu)$ is *weakly (sequentially) closed* if the weak limit of every weakly convergent sequence in $\mathcal{F}$ belongs to $\mathcal{F}$. Such sets are strongly closed.

A set $\mathcal{F} \subset L^2(\Omega)$ is *weakly (sequentially) compact* if every sequence in $\mathcal{F}$ has a subsequence converging weakly to an element of $\mathcal{F}$. For any sequence in $\mathcal{F}$ which has a weak limit, the weak limit belongs to $\mathcal{F}$. Strong compactness implies weak (sequential) compactness.

When minimising a functional over some set, a key property that the set may possess is convexity. In $\mathbb{R}^n$ a set is convex if the line joining any two points of the set lies in the set. In the context of square integrable functions we have:

Definition. A set $\mathcal{F} \subset L^2(\Omega,\mu)$ is *convex* if for every $f, g \in \mathcal{F}$, and $\lambda \in [0,1]$, then $(1-\lambda)f + \lambda g \in \mathcal{F}$. The *convex hull* of a set $\mathcal{F}$, denoted conv $\mathcal{F}$, is the smallest convex set which contains $\mathcal{F}$.

If a set is closed and convex, then it is weakly closed.

An element f of a convex set $\mathcal{F}$ is an *extreme point* if it does not lie in the interior of any line segment in the set; that is if $f = (1-\lambda)f_1 + \lambda f_2$ for some

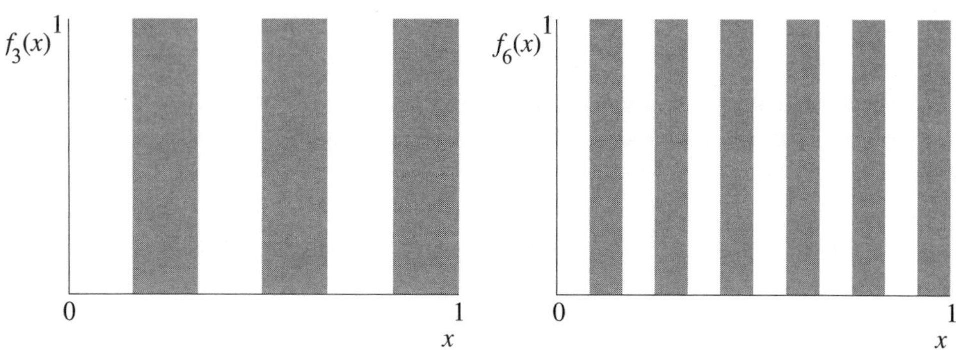

Figure 3: f_3 and f_6 as defined by (2.12)

$f_1, f_2 \in \mathcal{F}$ and $\lambda \in (0,1)$, then $f = f_1 = f_2$. This notion generalises the idea of a vertex of a convex polyhedron.

We show that the set of rearrangements of a non-negative function $f_0 \in L^2(\Omega, \mu)$ is (strongly) closed. Let $(f_n) \subset \mathcal{R}(f_0)$ be a sequence converging to f, say. Then (2.8) yields that for each $n \in \mathbb{N}$,

$$\|f_0^* - f^*\|_2 = \|f_n^* - f^*\|_2 \leq \|f_n - f\|_2.$$

It follows that $\|f_0^* - f^*\|_2 = 0$, that is $f_0^* = f^*$. Thus $f \in \mathcal{R}(f_0)$.

However $\mathcal{R}(f_0)$ is not strongly compact, weakly compact, nor convex unless f_0 is a constant function. For example, there are sequences of rearrangements which converge weakly to limits which are not rearrangements of the original function (from which it follows that the set is not weakly compact). Let $f_0 : [0,1] \to \mathbb{R}$ be defined by

$$f_0(x) = \begin{cases} 0 & \text{if } x \in [0, 1/2), \\ 1 & \text{if } x \in [1/2, 1]. \end{cases} \tag{2.11}$$

Define, for $n \in \mathbb{N}$,

$$f_n(x) = \begin{cases} 0 & \text{if } x = 0, \\ 0 & \text{if } x \in (m/n, (2m+1)/2n], \\ 1 & \text{if } x \in ((2m+1)/2n, (m+1)/n]. \end{cases} \tag{2.12}$$

where $m = 0, 1, \ldots, n-1$. The functions f_3 and f_6 are illustrated in Figure 3. For each $n \in \mathbb{N}$, f_n is equal to zero on a set of length $1/2$, and equal to 1 on a set of length $1/2$, therefore $f_n \in \mathcal{R}(f_0)$. However, given any $g \in L^2(0,1)$, it may be shown that $\int_\Omega f_n g \, d\mu \to 1/2 \int_\Omega g \, d\mu$ as $n \to \infty$, that is f_n converges weakly to the constant function with value $1/2$, which is not a rearrangement of f_0.

This behaviour can make it difficult to maximise or minimise a functional with respect to a set of rearrangements. It is quite possible that there will be a sequence of rearrangements giving progressively less energy, for instance, but the limit function will not be a rearrangement. To compensate for the lack of compactness (and convexity) we hope to use rearrangement inequalities, such as those described in §2.2, to show the existence of maximisers or minimisers. Otherwise we can work with the weak closure of the set of rearrangements of a given function, the smallest weakly closed set that contains the set of rearrangements. It is a weakly compact set: we can extract a weakly convergent subsequence from any extremising sequence, and we may be able to demonstrate that the limit is an extremum. Given a non–negative function $f_0 \in L^2(\Omega)$, the weak closure is equal to the closed convex hull of the set of rearrangements. It is immediate that this set is convex, and it may be shown that its set of extreme points is $\mathcal{R}(f_0)$. Douglas (1994) gave the following characterisation

$$\overline{\mathrm{conv} R(f_0)} = \left\{ f \geq 0 : \int_\Omega (f-\alpha)_+ d\mu \leq \int_\Omega (f_0 - \alpha)_+ d\mu \ \forall \alpha > 0, \int_\Omega f d\mu = \int_\Omega f_0 d\mu \right\},$$

where the + subscript denotes taking the positive part of the function. If we choose f_0 as in (2.11) it can be shown that any integrable function $\varphi : [0,1] \to \mathbb{R}$ satisfying $0 \leq \varphi(x) \leq 1$ for each $x \in [0,1]$, and $\int_0^1 \varphi d\mu = 1/2$, belongs to $\overline{\mathrm{conv}\ R(f_0)}$. This illustrates that $\overline{\mathrm{conv}\ R(f_0)}$ may be a large class of functions, in particular it includes the constant value 1/2 which is certainly not a rearrangement of f_0. In general, all rearrangements are included, as are functions derived by smoothing a rearrangement, if the smoothing preserves the value of the integral, and does not introduce new extreme values.

We conclude this subsection by considering two simple minimisation problems. Let f_0 be as in (2.11). Suppose we minimise

$$\text{(i)} \int_0^1 (f(x) - x)^2 d\lambda_1(x), \ \text{(ii)} \ -\int_0^1 x f(x) d\lambda_1(x)$$

over $f \in \mathcal{R}(f_0)$, then over $f \in \overline{\mathrm{conv}\ \mathcal{R}(f_0)}$. Noting that $\int_0^1 f^2$ is conserved over $f \in \mathcal{R}(f_0)$, the maximiser of $\int_0^1 x f(x)$ over $f \in \mathcal{R}(f_0)$ will be the minimiser of (i). Results in §2.2 yield that f_0 is the (unique) minimiser, and this gives a value 1/12 for the integral. However the identity function belongs to $\overline{\mathrm{conv}\ \mathcal{R}(f_0)}$, therefore the minimum of (i) with respect to that set will be zero, that is we obtain a lower value by minimising with respect to the weak closure. The minimum in (ii) with respect to $\mathcal{R}(f_0)$ is again attained by f_0, giving a value $-3/8$. However in this case no lower value may be obtained by minimising with respect to $\overline{\mathrm{conv}\ \mathcal{R}(f_0)}$; indeed it can be shown that f_0 is the unique minimiser.

The difference between these examples follows from the form of the functionals. In the case of (i), $\int f^2$ is conserved under rearrangements of f_0, but is not necessarily conserved when the weak limit is taken. For example, take a sequence (f_n) of rearrangements of f_0 which converges weakly to the identity function *id*. For each n, $\int_0^1 f_n^2 = 1/2$, but $\int_0^1 id^2 = 1/3$. On the other hand, if (f_n) converges weakly to f, then $\int x f_n \to \int x f$ as $n \to \infty$. In physical applications one has to consider whether it is appropriate to extremise with respect to the set of rearrangements of a given function, or consider a relaxed formulation, extremising with respect to the weak closure.

2.5 Rearrangements of functions on unbounded domains of infinite size

In this article we will only study rearrangements of functions defined on bounded subsets of $\mathbb{R}^n$. However some applications of rearrangement theory are naturally posed on unbounded sets of infinite size. (See for example Douglas 1994.) The properties listed in Theorem 1 are no longer equivalent for functions defined on such domains: we illustrate by an example. Let $f_1, f_2 : [0, \infty) \to \mathbb{R}$ be defined by

$$f_1(x) = \begin{cases} 0 & \text{if } x \in [0,1], \\ 1/x^2 & \text{if } x > 1, \end{cases}$$

$$f_2(x) = \begin{cases} 0 & \text{if } x \in [0,2], \\ 1/(x-1)^2 & \text{if } x > 2. \end{cases}$$

Now f_1 and f_2 satisfy Theorem 1 (i) and (ii), but not (iii) and (iv). We fix the properties that two functions which are rearrangements satisfy by the following definition.

Definition. Let Ω be an unbounded subset of $\mathbb{R}^n$ of infinite size, that is $\lambda_n(\Omega) = \infty$ where λ_n denotes n-dimensional Lebesgue measure. Two non-negative integrable functions $f, g : \Omega \to \mathbb{R}$ are *rearrangements* if

$$\lambda_n(\{x : f(x) \geq \alpha\}) = \lambda_n(\{x : g(x) \geq \alpha\})$$

for every $\alpha > 0$.

With this definition, f_1 and f_2 defined above are rearrangements. Roughly speaking we are imposing the following condition; for any given set of values not including zero, the set where f takes those values has the same size as the corresponding set for g. However the value zero may be taken on sets of different size. (The concept of a decreasing rearrangement makes sense in this context, but the notion of an increasing rearrangement does not in general.) The key restriction is to functions f such that $\lambda_n(\{x : f(x) \geq \alpha\}) < \infty$ for each $\alpha > 0$. Similar assumptions in the literature are that the functions *vanish at infinity* (see Lieb and Loss 1997, §3.2) or that they are *rearrangeable* (see Simon 1994). Non-negative integrable functions satisfy these conditions.

For non-negative, integrable $f, g : [0, \infty) \to \mathbb{R}$, inequalities (2.6) and (2.8) still hold, (replacing the unit interval with the half line,) where f^*, g^* are the decreasing rearrangements of f, g respectively. (See for example Lieb and Loss 1997, Theorems 3.4 and 3.5.) We note an important rearrangement inequality for non-negative integrable functions defined on $\mathbb{R}$. The *symmetric decreasing rearrangement* of a non-negative integrable function $f : \mathbb{R} \to \mathbb{R}$ is constructed by moving each set $\{x : f(x) \geq \alpha\}$ to an interval of the same size, symmetric about zero. We denote this function $f^{\triangle}$. For non-negative integrable functions f, g and h defined on $\mathbb{R}$, *Riesz's inequality* is satisfied:

$$\int_{-\infty}^{\infty}\int_{-\infty}^{\infty} f(x)g(x-y)h(y)dxdy \leq \int_{-\infty}^{\infty}\int_{-\infty}^{\infty} f^{\triangle}(x)g^{\triangle}(x-y)h^{\triangle}(y)dxdy.$$

Riesz's inequality has analogues in higher dimensions. (See Lieb and Loss 1997, Theorem 3.7.) The case when g is the Newtonian potential is of particular interest for kinetic energy minimisation in fluid mechanics.

3 Application to steady vortices in ideal fluid flow

This section discusses applications of rearrangements of functions to ideal fluid flow. For an energy functional written in terms of some Lagrangian conserved quantity, we maximise relative to the set of rearrangements of a prescribed function; maximisers correspond to special flows. We illustrate this variational method by a specific example. Vorticity is preserved following an ideal fluid flow in a bounded planar domain. We show that maximisers of the kinetic energy yield stream functions which satisfy the equation of a steady flow in §3.1. Section 3.2 is concerned with demonstrating that maximisers exist. We review some similar constrained energy maximisation problems in §3.3.

3.1 A variational principle for ideal fluid flow

In this subsection we seek steady flows of an ideal fluid by maximising a functional over a set of rearrangements. Consider an ideal (i.e. incompressible and inviscid) fluid of unit density flowing without body forces in a planar domain Ω bounded by a simple closed curve $\partial\Omega$. The fluid velocity $\mathbf{u}$ and pressure p satisfy the incompressible Euler equations

$$\begin{aligned} \frac{\partial \mathbf{u}}{\partial t} + \mathbf{u}.\nabla\mathbf{u} &= -\nabla p, \\ \nabla.\mathbf{u} &= 0. \end{aligned} \tag{3.1}$$

We impose the boundary condition $\mathbf{u}.\mathbf{n} = 0$ on $\partial\Omega$. The vorticity is given by $\nabla \wedge \mathbf{u}$, and it may be written $\omega\mathbf{k}$ where $\mathbf{k}$ is the unit vector perpendicular to the plane of Ω. Taking the curl of (3.1), we see that the time derivative following the flow of ω is zero, that is ω is preserved on particles. Moreover the flow is incompressible, therefore (as described in §2.3) for each time t we have $\omega(t) \in \mathcal{R}(\omega(0))$. Noting Theorem 1 (iv), an alternative way to express this fact is that the infinite family of Casimir integrals

$$C_F = \int_\Omega F(\omega) d\lambda_2$$

is conserved (for all t), where F is an arbitrary function (which can be approximated by continuous functions). Therefore we are restricted to a particular *isovortical surface*, or *symplectic leaf.* A *stream function* ψ exists which satisfies

$$\mathbf{u} = (\partial\psi/\partial y, -\partial\psi/\partial x)$$

at each time t; it follows that $\omega = -\triangle\psi$. The boundary condition $\mathbf{u}.\mathbf{n} = 0$ implies that ψ is constant on $\partial\Omega$; we can take the constant to be zero. We determine ψ from ω by solving the boundary value problem

$$-\triangle\psi = \omega \text{ in } \Omega, \qquad (3.2)$$

$$\psi = 0 \text{ on } \partial\Omega. \qquad (3.3)$$

For a given square integrable function ω, we can find ψ which satisfies (3.2) and (3.3) in the weak sense. (Roughly speaking this means that ψ is zero on $\partial\Omega$ and satisfies

$$\int_\Omega \psi(-\triangle F) d\lambda_2 = \int_\Omega \omega F d\lambda_2$$

for every smooth function F which vanishes on $\partial\Omega$. As in the previous section, λ_2 denotes 2-dimensional Lebesgue measure.) Call this function $K\omega$. This defines a mapping K, which we think of as the inverse of $-\triangle$ (with zero Dirichlet boundary conditions).

We seek solutions of

$$-\triangle\psi = \phi \circ \psi \qquad (3.4)$$

for some stream function ψ, and some function ϕ, where $\circ$ denotes composition. (3.4) is the equation for the stream function of a steady flow; the steady Euler equation is satisfied with $-p = |\nabla\psi|^2/2 + \Phi \circ \psi$ where $\Phi' = \phi$. We maximise kinetic energy over a family of flows whose vorticities are rearrangements of each other; there is a principle that a maximiser should yield a steady flow. The origins of this idea can be found in the work of Kelvin (1910); the modern formulation using rearrangements of functions is due to Benjamin (1976). (However this was in the context of steady vortex rings in three-dimensional ideal fluid flow - see §3.3 for a discussion of this problem.) The kinetic energy E of the fluid is given by

$$E = \frac{1}{2}\int_\Omega |\mathbf{u}|^2 d\lambda_2 = \frac{1}{2}\int_\Omega |\nabla\psi|^2 d\lambda_2 = \frac{1}{2}\int_\Omega \omega K\omega d\lambda_2.$$

We have the following result (due to Burton 1987) which justifies the above principle.

Theorem 2 *Let $\omega_0 : \Omega \to \mathbb{R}$ be square integrable (i.e. $\omega_0 \in L^2(\Omega)$). Suppose that $\bar{\omega} \in \mathcal{R}(\omega_0)$ is such that $E(\bar{\omega}) \geq E(\omega)$ for each $\omega \in \mathcal{R}(\omega_0)$. Then*

$$-\Delta\bar{\psi} = \phi \circ \bar{\psi} \tag{3.5}$$

for some increasing function ϕ, where $\bar{\psi} = K\bar{\omega}$.

Proof. (Sketch only.) It is easily seen that K is a linear operator; it may also be shown that

$$\int_\Omega uKv d\lambda_2 = \int_\Omega vKu d\lambda_2 \tag{3.6}$$

for $u, v \in L^2(\Omega)$, and furthermore that $E(\omega) > 0$ if $\omega \neq 0$. Now for $\omega \in \mathcal{R}(\omega_0)$, $\omega \neq \bar{\omega}$,

$$\begin{aligned}
E(\bar{\omega}) &\geq E(\omega) \\
&= E(\omega - \bar{\omega}) + \frac{1}{2}\int_\Omega \omega K\bar{\omega} d\lambda_2 + \frac{1}{2}\int_\Omega \bar{\omega} K\omega d\lambda_2 - \frac{1}{2}\int_\Omega \bar{\omega} K\bar{\omega} d\lambda_2 \quad (3.7) \\
&= E(\omega - \bar{\omega}) + \int_\Omega (\omega - \bar{\omega}) K\bar{\omega} d\lambda_2 + E(\bar{\omega}) \quad (3.8) \\
&> \int_\Omega (\omega - \bar{\omega}) K\bar{\omega} d\lambda_2 + E(\bar{\omega}).
\end{aligned}$$

By way of explanation, we have used linearity of K and (3.6) to obtain (3.7) and (3.8) respectively. Writing $\bar{\psi} = K\bar{\omega}$ we have

$$\int_\Omega \omega\bar{\psi} d\lambda_2 < \int_\Omega \bar{\omega}\bar{\psi} d\lambda_2$$

for every $\omega \in \mathcal{R}(\omega_0)$, $\omega \neq \bar{\omega}$. Burton (1987, Theorem 5) yields that $\bar{\omega} = \phi \circ \bar{\psi}$ for some increasing function ϕ. (This result was discussed in §2.2.) We identify $\bar{\omega}$ with $-\Delta\bar{\psi}$ and obtain (3.5). □

3.2 Existence of steady vortices

In the previous subsection we saw that a maximiser of the kinetic energy with respect to the set of rearrangements of a given function has a corresponding stream function which satisfies the equation of a steady flow. We demonstrate that such maximisers, and hence such steady flows, exist; the fact that E possesses at least one maximiser relative to $\mathcal{R}(\omega_0)$ is due to Burton (1987, 1989).

Theorem 3 *Let $\omega_0 \in L^2(\Omega)$. Then there exists $\bar{\omega} \in \mathcal{R}(\omega_0)$ such that $E(\bar{\omega}) \geq E(\omega)$ for every $\omega \in \mathcal{R}(\omega_0)$.*

Proof. Firstly we note that for $\omega \in \mathcal{R}(\omega_0)$,

$$E(\omega) \leq \|\omega\|_2 \|K\omega\|_2 \leq \|K\| \, \|\omega\|_2^2 = \|K\| \, \|\omega_0\|_2^2, \tag{3.9}$$

where $\|.\|_2$ denotes the L^2 norm, and $\|K\|$ is the (finite) operator norm of $K : L^2(\Omega) \to L^2(\Omega)$. By way of explanation, the first inequality in (3.9) follows by the Cauchy-Schwarz inequality, and the equality by Theorem 1. It follows that $\sup_{\omega \in \mathcal{R}(\omega_0)} E(\omega)$ is finite.

Let (ω_n) be a maximising sequence for E relative to the weak closure of the set of rearrangements of ω_0, which we denote $\overline{\mathcal{R}}(\omega_0)$. (This set, and its properties, were discussed in §2.4.) $\overline{\mathcal{R}}(\omega_0)$ is weakly sequentially compact therefore (ω_n) has a subsequence, which we again denote (ω_n), which converges weakly in $L^2(\Omega)$ to $\hat{\omega}$, say, in $\overline{\mathcal{R}}(\omega_0)$. The linear operator $K : L^2(\Omega) \to L^2(\Omega)$ is compact: consequently $K\omega_n \to K\hat{\omega}$ strongly in $L^2(\Omega)$ as $n \to \infty$. It follows that $E(\omega_n) \to E(\hat{\omega})$ as $n \to \infty$.

Write $K\hat{\omega} = \psi$. Proposition 1 yields the existence of $\bar{\omega} \in \mathcal{R}(\omega_0)$ such that

$$\int_\Omega \bar{\omega}\psi d\lambda_2 = \int_0^{\lambda_2(\Omega)} \omega_0^* \psi^* d\lambda_1.$$

Moreover $\overline{\mathcal{R}}(\omega_0)$ is equal to the closed convex hull of $\mathcal{R}(\omega_0)$, so no larger value can be obtained by maximising $\int_\Omega \omega\psi$ over $\omega \in \overline{\mathcal{R}}(\omega_0)$: in particular,

$$\int_\Omega \bar{\omega}\psi d\lambda_2 \geq \int_\Omega \hat{\omega}\psi d\lambda_2. \tag{3.10}$$

Repeating a calculation from the proof of Theorem 2, we have

$$E(\bar{\omega}) \geq \int_\Omega (\bar{\omega} - \hat{\omega})\psi d\lambda_2 + E(\hat{\omega}) \geq E(\hat{\omega}),$$

where the second inequality follows from (3.10). It follows that $E(\bar{\omega}) \geq E(\omega)$ for every $\omega \in \mathcal{R}(\omega_0)$. □

An alternative principle could be proposed; minimise the kinetic energy over a family of vorticities which are rearrangements of each other. If ω_0 is one-signed, then Burton (1989) yields the existence of a unique minimiser $\bar{\omega}$ of E relative to $\mathcal{R}(\omega_0)$ which satisfies $\bar{\omega} = \phi \circ K\bar{\omega}$ for a decreasing function ϕ. However Burton and McLeod (1991) demonstrate that if ω_0 takes both positive and negative values, no minimiser of E relative to $\mathcal{R}(\omega_0)$ exists. (For such ω_0, a unique minimiser $\hat{\omega}$ for E relative to $\overline{\mathcal{R}}(\omega_0)$ does exist; moreover $\hat{\omega} = \phi \circ K\hat{\omega}$ for some decreasing ϕ. Minimisers are characterised in Burton and McLeod,Theorem 2.1.)

3.3 A survey of some related ideal fluid flow problems

Benjamin (1976) proposed a theory of steady vortex rings in an ideal fluid involving maximising a functional over a set of rearrangements of a given function. We describe his original formulation. Let (r, θ, z) denote cylindrical coordinates. Consider an axisymmetric three-dimensional ideal fluid flow having zero velocity in the θ-direction, and whose far-field behaviour is a uniform flow in the z-direction. Subject to suitable regularity assumptions, a *Stokes stream function* Ψ exists which is related to the velocity $\mathbf{u}$ by

$$\mathbf{u} = \left(-\frac{1}{r}\frac{\partial \Psi}{\partial z}, 0, \frac{1}{r}\frac{\partial \Psi}{\partial r} \right)$$

at each time t. The vorticity $\nabla \times \mathbf{u}$ is solely in the θ-direction; we denote its magnitude by ω. Define a differential operator $\mathcal{L}$ by

$$\mathcal{L}(.) \equiv -\frac{1}{r}\frac{\partial}{\partial r}\left(\frac{1}{r}\frac{\partial(.)}{\partial r}\right) - \frac{1}{r^2}\frac{\partial^2(.)}{\partial z^2}.$$

It may be shown that

$$\mathcal{L}\Psi = \frac{\omega}{r}.$$

Moreover $\xi = \omega/r$ is a Lagrangian conserved quantity; noting that the flow is incompressible yields that ξ remains a rearrangement of its initial value. The *impulse* defined by

$$\mathcal{I} = \int_{\mathbb{R}^3} \xi r^2$$

is also a Lagrangian conserved quantity. We define $\mathcal{K}$ to be an inverse operator for $\mathcal{L}$ with suitable asymptotic conditions at infinity. (We omit the precise details.) The kinetic energy is given by

$$E = \frac{1}{2}\int_{\mathbb{R}^3} |\mathbf{u}|^2 = \frac{1}{2}\int_{\mathbb{R}^3} \frac{1}{r^2}|\nabla \psi|^2 = \frac{1}{2}\int_{\mathbb{R}^3} \xi \mathcal{K} \xi$$

where ψ is the difference between Ψ and the far-field stream function. Benjamin's proposal was to maximise E over $\xi \in \mathcal{R}(\xi_0)$, where ξ_0 is a prescribed non-negative function, subject to the constraint that $\mathcal{I}$ has some given value. He conjectured that maximisers would yield steady vortex rings. For a maximiser ξ, write $\psi = \mathcal{K}\xi$; one seeks to demonstrate that

$$\mathcal{L}\psi = \phi \circ (\psi - \lambda r^2/2) \tag{3.11}$$

where λ is the Lagrange multiplier corresponding to the constraint on the impulse, and ϕ is some function. λ represents the far-field velocity. The Stokes stream function has the form $\Psi = \psi - \lambda r^2/2$.

Recent results have justified Benjamin's variational approach to some extent. In view of the symmetry, one works on the half-plane $\Pi = \{(r, z) \in \mathbb{R}^2 : r > 0\}$. Burton (1999) has proved the existence of maximisers for a relaxed formulation of the problem, where ξ is constrained to lie in the weak closure of the set of rearrangements of ξ_0, for a wide class of non-negative functions

ξ_0. If $\hat{\xi}$ is a maximiser, then an equation corresponding to (3.11) holds. (Note, however, that the problem is now posed on Π.) The maximisers may be rearrangements of ξ_0; however this is not always the case. (See Burton 1999.)

Benjamin (1976) also proposed a second variational formulation; in this case $\lambda > 0$, which represents the velocity of the vortex ring relative to the fluid velocity at infinity, is prescribed. One maximises $E - \lambda\mathcal{I}$ over the set of rearrangements of a given non-negative function. This problem is studied in, for example, Badiani and Burton (1999).

The strategy of maximising an energy relative to the set of rearrangements of a given function has been successfully applied to other problems. Nycander (1995) demonstrated the existence of a localised, stationary, stable vortex in a background flow of constant shear. The flow was modelled by the incompressible two-dimensional Euler equations. The three-dimensional quasi-geostrophic equations are a more realistic model of geophysical flows; for such a flow in an external shear flow, Burton and Nycander (1999) proved the existence of a vortex having the same properties as in the simpler problem. In this case the constraint set was a set of stratified rearrangements, that is functions obtained by rearranging in the plane at each fixed z, where z is the vertical coordinate. This paper is a good introduction to problems of the type discussed in this section.

4 Decomposition of weather forecast error using rearrangements of functions

In this section we consider an application of rearrangements of functions to the evaluation of weather forecast error: the problem is how to compare the forecast of a physical quantity to the values actually observed. We seek a notion of error which favours what a meteorologist would call a good forecast. In §4.1 we define a forecast error which is composed of a contribution due to differences in qualitative features between the forecast and the actual distribution, and a contribution which measures the error due to displacement. To calculate the latter we need to define what we mean by the length of the shortest path joining two rearrangements: approaches to this are discussed in §4.2. Qualitative features error is evaluated by minimising a function over a set of rearrangements: the minimising rearrangements are characterised in §4.3. In §4.4 we note limitations in the simple formulation outlined in §4.1: we discuss a more sophisticated strategy.

4.1 A simple formulation of forecast error decomposition

There have been many attempts to quantify weather forecast error. In this section we consider a new approach to the problem of evaluating the error of a forecast quantity compared to the actual distribution. Most weather forecast

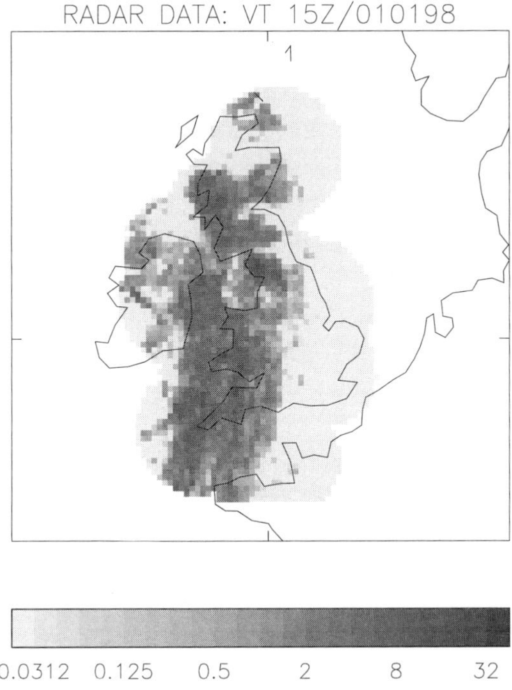

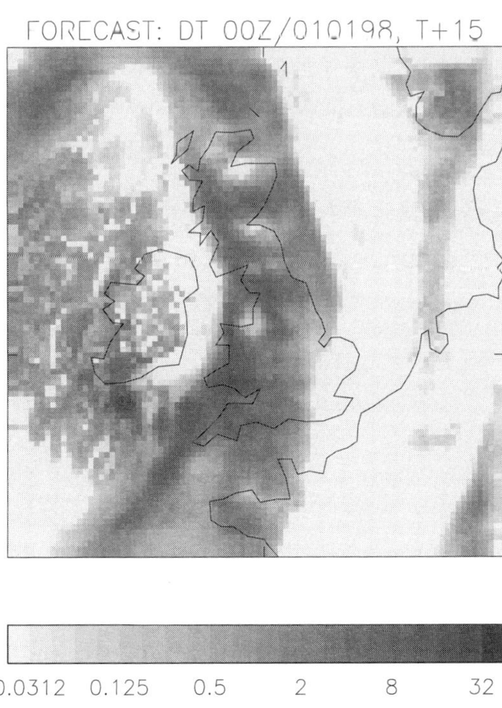

Figure 4: A forecast where the error is predominantly due to displacement. Picture courtesy of U.K. Meteorological Office Forecast Calibration Alignment project.

errors are simply displacements of significant weather in space or time, such as a rainband arriving a few hours later than expected. Figure 4 is an example (for real meteorological data) of a forecast which has captured the essential features of the true distribution as measured by the radar data, but has a displacement error. Conventional forecast error scores, such as integrating the square of the difference of the two functions and taking the square root, have limited use as they reward conservative forecasts. For example a forecast which is very good except for misplacing the location of a small intense storm by only a few grid lengths may have a poorer score than one that only forecasts a weak storm in that area. A more informative assessment of the two forecasts is that both have some positioning errors, but the latter has failed to capture the correct qualitative features. Following the pioneering work of Hoffman *et al.* (1995), we describe a forecast error which is split into two parts: the error due to displacement, and the error due to differences in qualitative features. This measure of error would favour the former forecast in our example above; moreover the two error values (one corresponding to displacement and one to difference in qualitative features) would be descriptive of how the forecasts had failed.

A simple formulation of forecast error decomposition is as follows. Find a displaced version of the forecast which is a best fit to the actual distribution. Use a conventional forecast error score (such as the L^2 difference described above) to evaluate the error between the displaced forecast and the true distribution: this value represents the error due to difference in qualitative features. Now calculate 'the length of the shortest path' which connects the forecast and the displaced forecast (in the space of allowable displacements). This value measures the displacement error. The total error is a weighted sum of the two. Figure 5 illustrates the idea of finding a displaced forecast which most closely matches the true distribution. Essentially we identify similar qualitative features in the forecast and actual distribution, and adjust the forecast so that corresponding features are in the same position.

Consider an important quantity q in weather forecasting which is a function of spatial variables, e.g. rainfall, and suppose we have a distribution predicted by a model, and a distribution derived from observations, which we assume to be the true distribution. Suppose $q : \Omega \to \mathbb{R}^d$, where $\Omega \subset \mathbb{R}^n$ is bounded. We will be interested in the cases $n = 2$ or 3, and as we have rainfall in mind as the physical quantity, we restrict attention to $d = 1$ and q non-negative. (Note however that the idea of forecast error decomposition still makes sense if $d > 1$: in this case we work with rearrangements of vector valued functions, which are described in §5.) Let q_1 be the observed (true) distribution, q_2 the distribution predicted by the model. We make the physically reasonable assumption that q_1 and q_2 are square integrable.

In a conventional approach the forecast error is $\|q_1 - q_2\|_2 = \{\int_\Omega (q_1(x) - q_2(x))^2 d\lambda_n(x)\}^{1/2}$, where λ_n denotes n-dimensional Lebesgue measure. (For

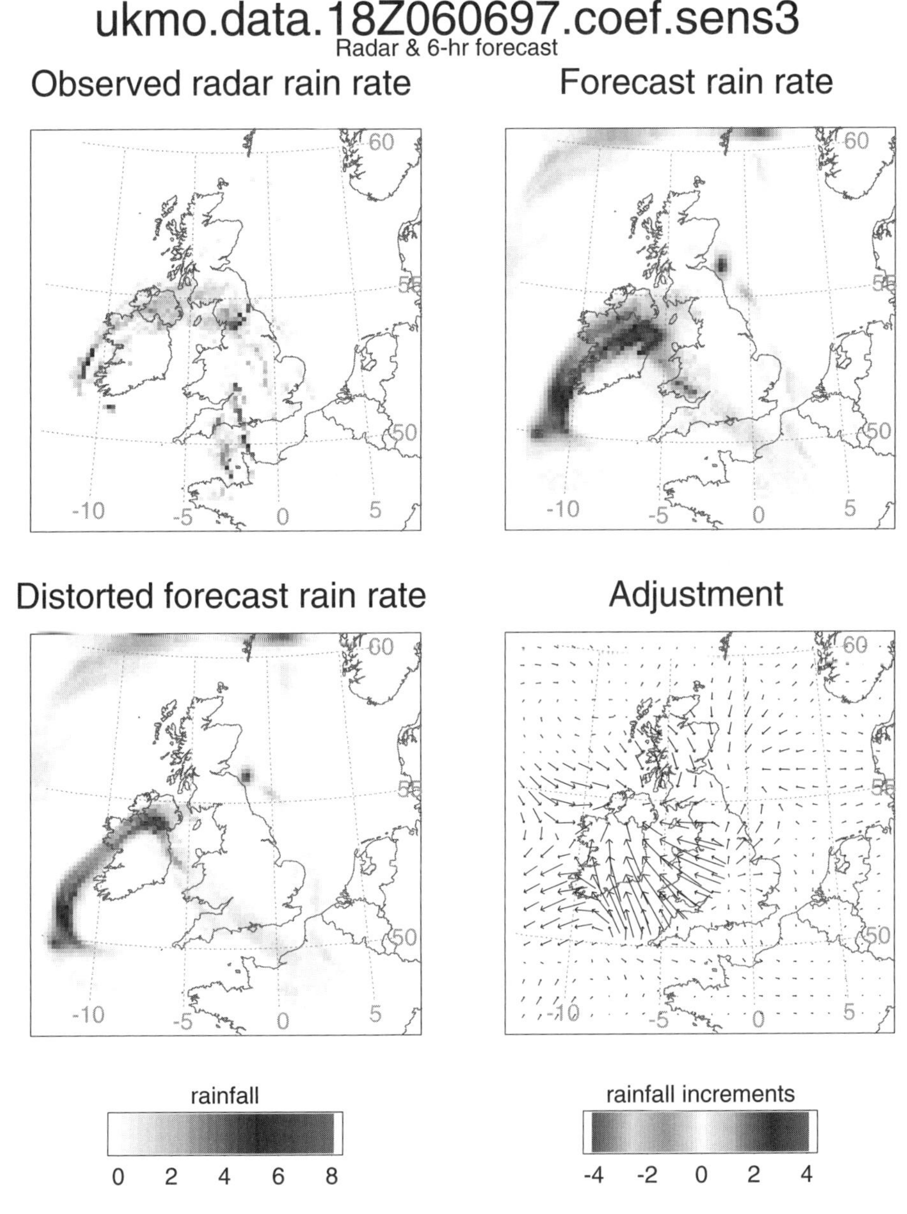

Figure 5: Displacing the forecast to find a best fit to the true distribution. Picture courtesy of UK Meteorological Office Forecast Calibration Alignment project.

$n = 2$ think of this as area, for $n = 3$, volume.) In our proposed approach we split the error as follows:

(i) Find $\hat{q} \in \mathcal{R}(q_2)$ such that $\hat{q}$ attains

$$\inf_{q \in \mathcal{R}(q_2)} \|q_1 - q\|_2. \tag{4.1}$$

(ii) Calculate the 'length of the shortest path' between q_2 and $\hat{q}$ in $\mathcal{R}(q_2)$. We discuss approaches to defining this quantity, which we write dist $(q_2, \hat{q})$, in the next subsection.

Equation(4.1) represents the error due to difference in qualitative features, whilst dist $(q_2, \hat{q})$ represents the error due to displacement. The total error is a weighted sum of the two, that is

$$TE(q_1, q_2) = (1 - \theta) \inf_{q \in \mathcal{R}(q_2)} \|q_1 - q\|_2 + \theta \inf_{\hat{q} \in \mathcal{M}} \text{dist } (\hat{q}, q_2) \tag{4.2}$$

where $0 < \theta < 1$, and $\mathcal{M}$ denotes the set of $q \in \mathcal{R}(q_2)$ which attain the infimum in (4.1). θ is chosen so that one component of error does not dominate the other: to determine a value, one would run a number of test cases.

For our definition of total error to make sense we need to show that (4.1) is attained by some $\hat{q} \in \mathcal{R}(q_2)$; we demonstrate that this is the case (and establish conditions for the minimiser to be unique) in §4.3.

4.2 Approaches to calculating length of a path in the set of rearrangements

In this subsection we give two possible definitions of the 'length of the shortest path' connecting two rearrangements. Firstly we deal with the question of what we mean by a path g which links q_2 and $\hat{q}$ in $\mathcal{R}(q_2)$; we will require that $g : [0, 1] \times \Omega \to \mathbb{R}$ satisfies $g(0, .) = q_2$, $g(1, .) = \hat{q}$, $g(t, .) \in \mathcal{R}(q_2)$ for each $t \in [0, 1]$, and that g is smooth in some sense. We say that the pair (g, v) defines a path if it is a solution of

$$\frac{\partial g}{\partial t} + \nabla.(gv) = 0, \text{ with } \nabla.v = 0, \tag{4.3}$$

subject to

$$g(0, .) = q_2, g(1, .) = \hat{q}.$$

(We assume the velocity field is sufficiently smooth to apply the transport theory of Diperna and Lions 1989.) Denote the set of paths by G. The distance between q_2 and $\hat{q}$ in $\mathcal{R}(q_2)$ is defined via a Least Action Principle, that is

$$\text{dist } (q_2, \hat{q}) = \inf_{(g,v) \in G} A(g, v), \tag{4.4}$$

for some Action integral $A = A(g, v)$. Two possible choices are as follows;

$$A_1(g,v) = \int_0^1 \int_\Omega |v(t,x)|^2 dxdt, \tag{4.5}$$

$$A_2(g,v) = \int_0^1 \int_\Omega g(t,x)|v(t,x)|^2 dxdt. \tag{4.6}$$

For a given path, (4.5) penalises the magnitude of the velocity which transports q_2 to $\hat{q}$: thinking of $g(t,.)$ as a density, (4.6) costs the 'kinetic energy' of the transportation. The choice of Action integral will depend on the type of displacement error one wishes to penalise most heavily; this may vary depending on the requirements of the customer using the forecast. Consider a patch of Ω which we move using a specified velocity field: if the values on the patch are high, this will give a bigger value for (4.6) than if the values are low. In contrast (4.5) is independent of the values we are transporting. Therefore the choice $A = A_2$ in (4.4) is more suitable for an application where one is primarily concerned with the position of any heavy rainfall.

If we discard the condition in (4.3) that the velocity is divergence free, Benamou and Brenier (1998b) show that (4.4) with $A = A_2$ is a time continuous formulation of an optimal mass transfer problem. (We discuss optimal mass transfer problems in §5.3.) They also describe an augmented Lagrangian numerical technique to compute solutions; one could hope to modify this scheme to calculate (4.4).

4.3 A characterisation of minimising rearrangements

We demonstrate that the formulation of forecast error decomposition given in §4.1 is well posed: we show that the infimum in (4.1) is attained, and establish a necessary and sufficient condition for the minimiser to be unique. Next we note a special case when this condition is always satisfied; then we conjecture the extent to which minimisers are determined in the general case. There may be infinitely many minimisers: we give an example. Finally we discuss numerical computation of the relevant quantities.

We have the following result.

Proposition 2 *Let q_1, $q_2 : \Omega \to \mathbb{R}$ be non-negative square integrable functions, where $\Omega \subset \mathbb{R}^n$ is bounded. Then*

(i) there exists $\hat{q} \in \mathcal{R}(q_2)$ such that $\hat{q}$ attains $\inf_{q \in \mathcal{R}(q_2)} \|q_1 - q\|_2$, and

(ii) $\hat{q}$ is the unique minimiser of $\|q_1 - q\|_2$ over $q \in \mathcal{R}(q_2)$ if and only if $\hat{q} = \varphi \circ q_1$ for some increasing function φ.

Proof. (i) Recall from §2.3 that q_1 has a polar factorisation $q_1 = q_1^* \circ s$ for some measure-preserving mapping $s : \Omega \to [0, \lambda_n(\Omega)]$, where q_1^* is the increasing rearrangement of q_1. Let $q \in \mathcal{R}(q_2)$. Now

$$\begin{aligned} \|q_1 - q\|_2 &\geq \|q_1^* - q_2^*\|_2 & (4.7)\\ &= \|q_1^* \circ s - q_2^* \circ s\|_2 & (4.8)\\ &= \|q_1 - q_2^* \circ s\|_2. \end{aligned}$$

By way of explanation, (4.7) follows from Crowe, Zweibel and Rosenbloom (1986, Corollary 1): this inequality was discussed in §2. The alternative characterisation of a measure-preserving mapping given in §2.3 yields (4.8). It follows that $\hat{q} = q_2^* \circ s$ attains $\inf_{q\in\mathcal{R}(q_2)} \|q_1 - q_2\|_2$.

(ii) From above we see that $\hat{q}$ attains $\inf_{q\in\mathcal{R}(q_2)} \|q_1 - q\|_2$ if and only if $\|q_1 - \hat{q}\|_2 = \|q_1^* - q_2^*\|_2$. Noting from Theorem 1 that the L^2 norm is preserved under rearrangement, it follows that $\hat{q}$ attains $\inf_{q\in\mathcal{R}(q_2)} \|q_1 - q\|_2$ if and only if

$$\int_\Omega q_1 \hat{q} d\lambda_n = \int_0^{\lambda_n(\Omega)} q_1^* q_2^* d\lambda_1. \tag{4.9}$$

Burton (1987, Theorems 3 and 5) yields that $\hat{q}$ is the unique element of $\mathcal{R}(q_2)$ where equality holds in (4.9) if and only if $\hat{q} = \varphi \circ q_1$ for some increasing function φ. (This result was discussed in §2.) □

Remark If q_1 has no level sets of positive measure, that is the sets $L_\alpha = \{x \in \Omega : q_1(x) = \alpha\}$ have zero size for each $\alpha \in \mathbb{R}$, then we can find an increasing function φ such that $\varphi \circ q_1$ is a rearrangement of q_2. Proposition 2 (ii) yields that this is the unique minimiser of $\|q_1 - q\|_2$ over $q \in \mathcal{R}(q_2)$.

Conjecture Minimisers of $\|q_1 - q\|_2$ over $q \in \mathcal{R}(q_2)$ are uniquely determined up to level sets of q_1 with positive measure, and when two minimisers are restricted to any level set of q_1 with positive measure, they are rearrangements on that level set.

We show by example that the minimiser need not be unique. Let $q_1, q_2 : [-1,1]^2 \to \mathbb{R}$ be respectively the actual distribution and forecast of some meteorological quantity, where

$$q_1(x,y) = \begin{cases} 3 & \text{if } x^2 + y^2 \leq 1/16,\\ 2 & \text{if } 1/16 < x^2 + y^2 \leq 9/16,\\ 0 & \text{otherwise,} \end{cases}$$

and

$$q_2(x,y) = \begin{cases} 3 & \text{if } x^2 + y^2 \leq 1/4,\\ 0 & \text{otherwise.} \end{cases}$$

Further define

$$q_3(x,y) = \begin{cases} 3 & \text{if } x^2 + y^2 \leq 1/16, \text{ or } 6/16 \leq x^2 + y^2 \leq 9/16,\\ 0 & \text{otherwise.} \end{cases}$$

It is easily seen that $q_3 \in \mathcal{R}(q_2)$, and moreover

$$\|q_1 - q_2\|_2 = \|q_1 - q_3\|_2 = \inf_{q \in \mathcal{R}(q_2)} \|q_1 - q\|_2.$$

It may be shown that any $\tilde{q} \in \mathcal{R}(q_2)$ which satisfies

$$\{(x,y) : x^2 + y^2 \leq 1/16\} \subset \{(x,y) : \tilde{q}(x,y) = 3\} \subset \{(x,y) : x^2 + y^2 \leq 9/16\}$$

is a minimiser. In fact for most choices of piecewise constant functions q_1 and q_2, there will be infinitely many $\tilde{q} \in \mathcal{R}(q_2)$ which attain $\inf_{q \in \mathcal{R}(q_2)} \|q_1 - q\|_2$. We discuss this defect in our formulation in the next subsection.

Finally we note that to calculate (4.1), the proof of Proposition 2 (i) yields that it suffices to evaluate $\|q_1^* - q_2^*\|_2$. Furthermore if we calculate a polar factorisation of q_1, we have a formula for a rearrangement $\hat{q}$ of q_2 which attains (4.1). Therefore numerical computation of the relevant quantities consists of finding increasing rearrangements and polar factorisations.

4.4 Limitations in the proposed strategy, and an alternative formulation

We begin by highlighting two limitations in the formulation of forecast error decomposition which we introduced in §4.1: firstly a numerical implementation may require many computations, and secondly errors in qualitative features may be penalised as though they were displacement errors. We describe an alternative formulation which may overcome these difficulties.

As noted in the previous subsection, if we are dealing with piecewise constant data, there may be infinitely many rearrangements of q_2 which attain (4.1). When we calculate the displacement error, we have to find dist $(q_2, \hat{q})$ for each minimiser $\hat{q}$, taking the least value (if such is attained) as the displacement error. A discretised version of this problem may be computationally expensive.

We illustrate the second difficulty by supposing that we wish to decompose rainfall forecast error over the United Kingdom. It is possible that $\hat{q}$, the rearrangement of the forecast rainfall q_2 which is nearest to the actual distribution q_1, is achieved by rearranging rainfall from a wet area (e.g. Scotland) to a dry area a long distance away (e.g. the South East of England). This could happen if the forecast overestimated the extent of a storm in the wet area, and failed to predict a small amount of rainfall in the dry area. The displacement error term will be large; however the error is really one of qualitative features. We can lessen the impact of this problem by restricting the region over which we perform the forecast error decomposition to those which are 'meteorologically similar'.

A possible solution to these difficulties is to minimise the combined quantity of qualitative difference and displacement error over $q \in \mathcal{R}(q_2)$, that is consider the problem

$$\inf_{q \in \mathcal{R}(q_2)} \{(1-\theta)\, \|q_1 - q\|_2 + \theta \text{ dist } (q, q_2)\}, \tag{4.10}$$

for some $0 < \theta < 1$. If (4.10) is uniquely attained by $\hat{q} \in \mathcal{R}(q_2)$, then by evaluating $\|q_1 - \hat{q}\|_2$ and dist $(\hat{q}, q_2)$ we recover respectively the difference in qualitative features, and displacement error. (Note, however, that these values need not be the same as the quantities calculated according to the scheme outlined in §4.1.) If we consider the example in §4.3, then q_2 is the unique minimiser of (4.10), so the above formulation may not suffer from non-uniqueness problems to the extent the formulation in §4.1 does. Benamou and Brenier (1998b) discuss how their numerical scheme (mentioned in §4.2) may be adapted to calculate (4.10).

5 Rearrangements of vector valued functions

The intuitive idea of when two functions are rearrangements given in the introduction (in terms of exchanging values on particles) is equally applicable to both scalar and vector valued functions. In this section we consider rearrangements of vector valued functions; we seek analogues of the special rearrangements which exist and inequalities which hold in the scalar valued case. We introduce the concept via an example in §5.1, and state some equivalent formulations. Then we consider the generalisation of the increasing rearrangement, the monotone rearrangement, in §5.2: this is the unique rearrangement (of a prescribed vector valued function) equal to the gradient of a convex function. We can establish an inequality satisfied by the monotone rearrangement by considering an appropriate class of optimal mass transfer problems: this is discussed in §5.3. Section 5.4 deals with the regularity of the optimal mappings, in anticipation of an application to a system of equations in §6. Finally we discuss the polar factorisation of a vector valued function, that is writing the function as the composition of its monotone rearrangement with a measure-preserving mapping. We give conditions for when such a decomposition is known to exist, and when it is unique, in §5.5.

5.1 Definition and properties of rearrangements of vector valued functions

We introduce the concept of rearrangement of vector valued functions by an example. Define a function f on the unit square $[0,1] \times [0,1]$, which we write $[0,1]^2$, by

$$f(\mathbf{x} = (x,y)) = \begin{cases} (1,1) & \text{if } 1/2 \leq x \leq 1, 1/2 \leq y \leq 1, \\ (0,0) & \text{otherwise}, \end{cases} \tag{5.1}$$

and another function g by

$$g(\mathbf{x} = (x,y)) = \begin{cases} (0,0) & \text{if } 1/8 < y < 7/8, \\ (1,1) & \text{if } y \leq 1/8 \text{ or } y \geq 7/8. \end{cases}$$

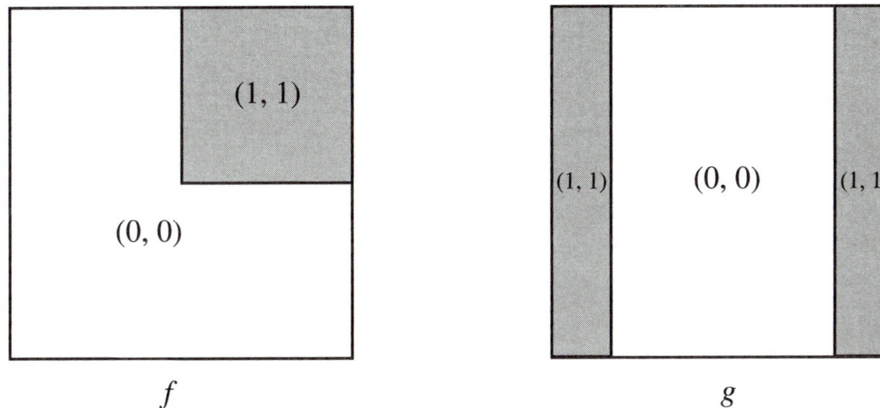

Figure 6: Functions f and g

f and g are illustrated in Figure 6. Consider the set of points where f takes the value $(1,1)$, that is $\{\mathbf{x} \in [0,1]^2 : f(\mathbf{x}) = (1,1)\}$, and the corresponding set for g, $\{\mathbf{x} \in [0,1]^2 : g(\mathbf{x}) = (1,1)\}$. These sets are equal to $[1/2,1] \times [1/2,1]$ and $[0,1] \times [0,1/8] \bigcup [0,1] \times [7/8,1]$ respectively, and both have area $1/4$. Similarly the sets $\{\mathbf{x} : f(\mathbf{x}) = (0,0)\}$ and $\{\mathbf{x} : g(\mathbf{x}) = (0,0)\}$ have equal area (of $3/4$). As f and g take no other values, it follows that $\{\mathbf{x} : f(\mathbf{x}) \in B\}$ has the same area, or 2–dimensional Lebesgue measure, as $\{\mathbf{x} : g(\mathbf{x}) \in B\}$ for every (Borel) set $B \subset \mathbb{R}^2$. We will say that two vector valued functions are rearrangements when this property holds.

More generally we have the following definition.

Definition Let Ω be a bounded set in $\mathbb{R}^n$, and let $f, g : \Omega \to \mathbb{R}^d$ be integrable functions. Then f and g are *rearrangements* if

$$\mu(\{\mathbf{x} : f(\mathbf{x}) \in B\}) = \mu(\{\mathbf{x} : g(\mathbf{x}) \in B\}) \tag{5.2}$$

for every Borel subset of $\mathbb{R}^d$, where μ is the 'size' (or measure) of the set. As before an appropriate choice of 'size' for n = 1,2, or 3 would be length, area or volume respectively. For the purposes of this article we restrict to measures that are of the form $\mu(E) = \int_E f d\lambda_n$ where f is a non-negative integrable function, and λ_n denotes n-dimensional Lebesgue measure. (In the language of measure theory, μ is absolutely continuous with respect to n-dimensional Lebesgue measure. There is no need to be so restrictive for the results that follow; see Douglas 1998 and Burton and Douglas 1998 for appropriate choices of measure space.) As in the scalar case, the restriction to Borel sets is to ensure that (5.2) always makes sense.

Definition (5.2) is identical to (2.4) for scalar valued rearrangements, except that d will now be greater than 1. Other characterisations of two vector valued functions being rearrangements are given in the following result:

Theorem 4 *Let $\Omega \subset \mathbb{R}^n$ be bounded and let μ be absolutely continuous with respect to n-dimensional Lebesgue measure. Then, for integrable functions $f, g : \Omega \to \mathbb{R}^d$, the following are equivalent.*

(i) f is a rearrangement of g.

(ii) For each $F \in C(\mathbb{R}^d)$ (the set of continuous functions from $\mathbb{R}^d \to \mathbb{R}$)

$$\int_\Omega F(f(x))d\mu(x) = \int_\Omega F(g(x))d\mu(x). \tag{5.3}$$

((5.3) is understood in the sense that if one of the integrals is finite, then so is the other and they are equal.)

(iii) For each $c \in \mathbb{R}^d$,

$$\mu(\{x : f(x) \geq c\}) = \mu(\{x : g(x) \geq c\}),$$

where the inequalities are calculated component by component.

Proof. See Douglas (1998, Theorem 2.2). □

Brenier (1991) defined rearrangement of vector valued functions via (ii). (He restricted to a subclass of $C(\mathbb{R}^d)$ so that the integrals in (5.3) are guaranteed to be finite.) Cullen, Norbury and Purser (1991), following Baigent (1988), used (iii) as a definition; this is a direct extension of the characterisation of scalar valued rearrangements given in Theorem 1 (ii). We can unify these definitions as follows: each may be reduced to the requirement that (5.3) holds for each $F : \mathbb{R}^d \to \mathbb{R}$ in a class of functions $\mathcal{U}$. The different definitions above correspond to different choices of $\mathcal{U}$. $\mathcal{U}$ must be a sufficiently large class so that for each $n \in \mathbb{N}$ such that $\int_\Omega |f|^n$ is finite, we have $\int_\Omega |f|^n = \int_\Omega |g|^n$, where $|.|$ denotes Euclidean distance on $\mathbb{R}^d$. (We note that the characterisation of scalar valued rearrangements based on ideas of Eydeland, Spruck and Turkington 1990 can also be generalised to the vector valued case. See Douglas 1998 for details.)

For an integrable function f, we can define the *set of rearrangements of* f, denoted $\mathcal{R}(f)$. We will introduce a special element of this set, the monotone rearrangement of f, in the next subsection, and demonstrate that the set is closed (in L^1). However in general it is neither convex nor compact in the space of integrable functions. Consequently when we minimise a functional with respect to the set of rearrangements of a prescribed integrable function, we cannot assume that every minimising sequence has a subsequence which converges to some rearrangement. Instead we use rearrangement inequalities. (See §6.2.)

A vector valued function $f : \Omega \to \mathbb{R}^d$ may be written in terms of real valued components $f_i : \Omega \to \mathbb{R}$, where $f = (f_1, \ldots, f_d)$. If f and g are vector valued functions which are rearrangements then the corresponding components of f

and g are rearrangements in the scalar valued sense. However it is important to note that the condition that each component of a vector valued function f is a rearrangement (in the scalar sense) of the corresponding component of a vector valued function g, is not sufficient for f to be a rearrangement of g in the vector valued sense. We give an example. Let f be as in the previous example, then

$$f_1(\mathbf{x}) = f_2(\mathbf{x}) = \begin{cases} 1 & \text{if } \mathbf{x} \in [1/2, 1] \times [1/2, 1], \\ 0 & \text{if } \mathbf{x} \notin [1/2, 1] \times [1/2, 1]. \end{cases}$$

Define

$$g(\mathbf{x}) = \begin{cases} (1,0) & \text{if } \mathbf{x} \in [1/2, 1] \times [1/2, 1], \\ (0,1) & \text{if } \mathbf{x} \in [0, 1/2] \times [1/2, 1], \\ (0,0) & \text{otherwise.} \end{cases}$$

Then $g_1 = f_1$ and

$$g_2(\mathbf{x}) = \begin{cases} 1 & \text{if } \mathbf{x} \in [0, 1/2] \times [1/2, 1], \\ 0 & \text{if } \mathbf{x} \notin [0, 1/2] \times [1/2, 1]. \end{cases}$$

It is easily seen that $f_1 \in \mathcal{R}(g_1)$, and $f_2 \in \mathcal{R}(g_2)$. However, $f \notin \mathcal{R}(g)$, because f takes the value (1,1) on a region of area 1/4, while g never takes this value. Consequently in general we cannot apply scalar valued rearrangement results to the components of vector valued functions and hope to obtain results about vector valued rearrangements. Indeed, the theory of rearrangements of vector valued functions is less rich than that of rearrangements of scalar valued functions. We will see an example of this in the next subsection.

5.2 Monotone rearrangement of vector valued functions

For a real function f defined on an interval, there is an (essentially) unique $f^* \in \mathcal{R}(f)$ which is an increasing function. (See §2.2.) There is an analogous rearrangement for vector valued functions. The concept of increasing does not make sense for vector valued functions, because $\mathbb{R}^d$ is not well–ordered for $d \geq 2$. Instead we note that an increasing scalar valued function is the derivative of a convex function. Replacing derivative with gradient, this is a property which is well defined for vector valued functions, therefore we seek a rearrangement of a vector valued function equal to the gradient of a convex function. Let f be as in (5.1), and define $f^\# : [0,1]^2 \to \mathbb{R}^2$ by

$$f^\#(\mathbf{x} = (x, y)) = \begin{cases} (1,1) & \text{if } x + y \geq 2 - \frac{1}{\sqrt{2}}, \\ (0,0) & \text{otherwise.} \end{cases} \tag{5.4}$$

$f^\#$ is easily seen to be a rearrangement of f, and moreover $f^\# = \nabla\Psi$ (at the points where Ψ is differentiable), where $\Psi : [0,1]^2 \to \mathbb{R}$ is a convex function defined by

$$\Psi(\mathbf{x}=(x,y)) = \begin{cases} x+y-\left(2-\frac{1}{\sqrt{2}}\right) & \text{if } x+y \geq 2-\frac{1}{\sqrt{2}}, \\ 0 & \text{otherwise.} \end{cases}$$

It can be shown that $f^{\#}$ is the unique rearrangement of f equal to the gradient of a convex function. To generalise this concept to integrable functions $u : \Omega \subset \mathbb{R}^n \to \mathbb{R}^d$, it is clear we require $n = d$. We have the following result.

Theorem 5 *For $n = d$, let Ω and μ be as in Theorem 4. Suppose $u : \Omega \to \mathbb{R}^d$ is an integrable function. Then*

(i) There exists $u^{\#} \in \mathcal{R}(u)$, such that $u^{\#} = \nabla\psi$ (at the points of differentiability of ψ), where $\psi : \mathbb{R}^d \to \mathbb{R} \cup \{+\infty\}$ is convex. Moreover $u^{\#}$ is unique in the sense that if $\phi : \mathbb{R}^d \to \mathbb{R} \cup \{+\infty\}$ is another convex function, and $\nabla\phi \in \mathcal{R}(u)$, then $u^{\#} = \nabla\phi$ almost everywhere.

(ii) If in addition Ω is open and connected with smooth boundary, and μ is equivalent to d-dimensional Lebesgue measure, then the mapping $u \to u^{\#}$ is continuous (in $L^1(\Omega, \mu, \mathbb{R}^d)$, the space of integrable functions from $\Omega \to \mathbb{R}^d$).

Proof. (i) follows from the main theorem of McCann (1995). (ii) is part of Brenier (1991, Theorem 1.1). □

We call $u^{\#}$ the *monotone rearrangement of u* — it is a *cyclically monotone* function, that is it satisfies

$$\sum_{i=1}^{n} u^{\#}(x_i).(x_i - x_{i-1}) \geq 0$$

for (almost) all finite sequences $x_1, \ldots, x_n = x_0$ of points in Ω. (For two points, this collapses to the usual definition of monotone.)

For functions f as in Theorem 4, with $n \neq d$, we can still define a monotone rearrangement if we work on a different domain in $\mathbb{R}^d$ which has the same 'size' as Ω. Let $\mathcal{B}$ be the open ball in $\mathbb{R}^d$, centre the origin, such that $\mu(\Omega) = \lambda_d(\mathcal{B})$, where λ_d denotes d-dimensional Lebesgue measure. We 'move Ω onto $\mathcal{B}$', and work with the resulting function. There exists a measure-preserving transformation (see §2.3) $\tau : (\Omega, \mu) \to (\mathcal{B}, \lambda_d)$. The composition of f with τ satisfies the hypotheses of Theorem 5, and we say that $(f \circ \tau)^{\#}$ is the monotone rearrangement of f. Using this construction we have a simple proof of the following result.

Proposition 3 *Let Ω, μ, and f be as in Theorem 4. Then $\mathcal{R}(f)$ is closed (in $L^1(\Omega, \mu, \mathbb{R}^d)$).*

Proof. Let $(f_n) \subset \mathcal{R}(f)$, and suppose that $f_n \to g$ in $L^1(\Omega, \mu, \mathbb{R}^d)$ as $n \to \infty$. (Recall that this means that $\int_\Omega |f_n - g| d\mu \to 0$ as $n \to \infty$, where $|.|$ denotes Euclidean distance on $\mathbb{R}^d$.) Let $\mathcal{B}$ be as above, and for each $n \in \mathbb{N}$, let $f_n^\#$ denote the monotone rearrangement of f_n. Then Theorem 5 yields that $f_n^\# \to g^\#$ as $n \to \infty$, and we note that $f_n^\# = f^\#$ for each $n \in \mathbb{N}$. It follows that $g^\# = f^\#$, whence $g \in \mathcal{R}(f)$. □

Recall from §2 that the increasing rearrangement satisfies the following inequalities for two square integrable real valued functions f and g defined on the unit interval:

$$\begin{aligned} \int_0^1 f^*(x) g^*(x) d\lambda_1(x) &\geq \int_0^1 f(x) g(x) d\lambda_1(x), \\ \|f^* - g^*\|_2 &\leq \|f - g\|_2, \end{aligned}$$

where h^* denotes the increasing rearrangement of h, and $\|.\|_2$ the L^2 norm, that is $\|h\|_2 = \{\int_0^1 |h|^2 d\lambda_1\}^{1/2}$. For Ω and μ as in Theorem 5, square integrable functions $u_1, u_2 : \Omega \to \mathbb{R}^d$ do not in general satisfy

$$\begin{aligned} \int_\Omega u_1^\#(x).u_2^\#(x) d\mu(x) &\geq \int_\Omega u_1(x).u_2(x) d\mu(x), \qquad (5.5) \\ \|u_1^\# - u_2^\#\|_2 &\leq \|u_1 - u_2\|_2. \end{aligned}$$

In the next subsection we show that (5.5) holds in the special case when u_2 is the identity function, and demonstrate that the inequality is strict when $u_1 \neq u_1^\#$. In fact (5.5) is known to hold when $u_2 : \Omega \to \Omega$ is a measure-preserving mapping. (This result is due to Burton and Douglas 1998, Proposition 2.8.)

5.3 Rearrangement inequalities from optimal mass transfer problems

In this subsection we prove a rearrangement inequality via a reformulation as an optimal mass transfer problem. We begin with a brief review of the latter. The prototype optimal mass transfer problem is the following: given two sets U, V of equal volume, find the optimal volume-preserving mapping between them, where optimality is measured against a non-negative cost function $c = c(x, y)$. One interprets $c(x, y)$ as being the cost per unit mass for transporting material from $x \in U$ to $y \in V$; the optimal map minimises the total cost of redistributing the mass of U through V. Optimal mass transfer problems have a wide range of applications, particularly in economics; in the original problem of Monge (1781), the question was how best to move a pile of soil to an excavation, minimising the work done.

We consider a mathematical formulation of such a problem. Let $U, V \subset \mathbb{R}^d$ be such that $\mu(U) = \nu(V)$, where μ, ν measure 'size' on U, V respectively. (More precisely, let μ and ν be Borel measures.) The set of possible strategies for redistributing the mass of U onto V is the set S of measure-preserving mappings between (U, μ) and (V, ν). The problem of finding a strategy which minimises the cost becomes the following; is there a measure-preserving mapping s which attains

$$\inf_{s \in S} \int_U c(x, s(x)) d\mu(x), \tag{5.6}$$

and if so, what is it? This question has been solved for a certain class of cost functions when μ vanishes on sufficiently small sets; the reader is referred to the lucid account of Gangbo and McCann (1996) for details. In what follows we restrict attention to the cost function $c(x, y) = |x - y|^2/2$, where $|.|$ denotes Euclidean distance.

Suppose we are considering the following minimisation problem:

$$\inf_{f \in \mathcal{R}(f_0)} \frac{1}{2} \int_\Omega |f(x) - x|^2 d\mu(x), \tag{5.7}$$

where Ω and μ are as in Theorem 5, and $f_0 : \Omega \to \mathbb{R}^d$ is a square integrable function. Define a measure ν for (Borel) subsets $B \subset \mathbb{R}^d$ by $\nu(B) = \mu(f_0^{-1}(B))$. Then $\nu(B)$ measures how much of Ω is mapped to the set B. The set of all measure-preserving mappings S from (Ω, μ) to $(\mathbb{R}^d, \nu)$ is exactly $\mathcal{R}(f_0)$, therefore (5.7) is equivalent to

$$\inf_{f \in S} \frac{1}{2} \int_\Omega |f(x) - x|^2 d\mu(x), \tag{5.8}$$

which is of the form (5.6). (To link with notation used elsewhere in the literature, the set S is sometimes referred to as the set of mappings s which push the measure μ forward to the measure ν, which is denoted $s_{\#}\mu = \nu$.) We have the following result.

Theorem 6 *Let Ω and μ be as in Theorem* 5*. Suppose $f_0 : \Omega \to \mathbb{R}^d$ is a square integrable function. Then* (5.7) *is uniquely attained by $f_0^{\#}$, the monotone rearrangement of f_0, or equivalently*

$$\int_\Omega f_0^{\#}(x).x d\mu(x) > \int_\Omega f(x).x d\mu(x) \tag{5.9}$$

for each $f \in \mathcal{R}(f_0) \backslash \{f_0^{\#}\}$.

Proof. (5.8) is uniquely attained by $f_0^{\#}$. (See for example Gangbo and McCann 1996, Theorem 1.2.) The above discussion yields that (5.7) is uniquely attained by $f_0^{\#}$. This result is easily seen to be equivalent to (5.9) for every $f \in \mathcal{R}(f_0) \backslash \{f_0^{\#}\}$; write $|f(x) - x|^2$ as $(f(x) - x).(f(x) - x)$, multiply out, and apply Theorem 4. □

In the special case when Ω is open, connected, and has smooth boundary, and μ is equivalent to λ_d, (5.9) can be deduced from Brenier (1991, Theorem

3.1 and Proposition 3.1 (iii) and (v)). Douglas (1998, Theorem 3.1) showed this result could also be obtained from Brenier (1991, Theorems 1.1, 1.2 and Proposition 2.1).

5.4 Regularity of optimal mass transfers

For two subsets $U, V \subset \mathbb{R}^d$, we considered the problem of optimally transferring the mass of (U, μ) to (V, ν) in the previous subsection, where optimality was measured against the cost function $c(x, y) = |x - y|^2/2$. If there exists some non-negative integrable function g such that $\mu(B) = \int_B g\lambda_d$ for each (Borel) set $B \subset \mathbb{R}^d$ (where λ_d denotes d-dimensional Lebesgue measure), then a unique optimal mapping exists, and it is equal to the gradient of a convex function. In this subsection we are interested in the regularity of the optimal mapping; in §6.6 we study an application of this theory to existence of solutions for a system of partial differential equations. We repeat some of Evans (1997, Section 4); however we quote some additional results for $d = 2$.

Our review will of necessity use technical language; we give a brief guide to the notation used, and refer the reader to Adams (1975) for more details. For $0 < \beta < 1$, $u \in C^\beta(\overline{U})$ if u is continuous on $\overline{U}$, and if there exists $\gamma > 0$ such that for every $x, y \in U$ we have

$$|u(x) - u(y)| \leq \gamma|x - y|^\beta. \tag{5.10}$$

(This is similar to, but weaker than, a Lipschitz condition.) The Hölder space $C^{1,\beta}(\overline{U})$ consists of $C^1(\overline{U})$ functions (i.e. continuously differentiable up to the boundary of U) for which there exists a $\gamma > 0$ such that (5.10) holds for every first order derivative (of the function), and every $x, y \in U$. Replacing statements about first order derivatives with second order derivatives, we define $C^{2,\beta}(\overline{U})$.

Denote the optimal mapping between (U, μ) and (V, ν) by $\nabla\Psi$, where Ψ is a convex function. Without further assumptions, little is known about the regularity of Ψ. We will assume that U and V are bounded, connected, open sets in $\mathbb{R}^d$; further suppose that there is a non-negative integrable function h such that $\nu(B) = \int_B h\lambda_d$ for each (Borel) subset B of $\mathbb{R}^d$, and that g and h are bounded above and below, away from zero. If the *target set* V is not convex, then $\nabla\Psi$ can have a singular part: see Caffarelli (1992a) for an example. However if the target set is convex, then Ψ is strictly convex, and belongs to $C^{1,\alpha}_{loc}(U)$ for some $0 < \alpha < 1$. (The subscript *loc* means that the function is $C^{1,\alpha}$ on all compact subsets of U.) If f and g are smoother functions, then so is Ψ. Suppose $g \in C^\beta(U)$ and $h \in C^\beta(V)$ for some $0 < \beta < 1$; then $\Psi \in C^{2,\alpha}_{loc}(U)$ for each $0 < \alpha < \beta$. These interior regularity results are due to Caffarelli (1992a). If U is convex as well, we obtain regularity up to the boundary. For g and h bounded away from 0 and $+\infty$ we have $\Psi \in C^{1,\alpha}(\overline{U})$ for some $0 < \alpha < 1$; if the boundaries of U and V are smooth, and $g \in C^\beta(U)$ and $h \in C^\beta(V)$ for

$0 < \beta < 1$, then $\Psi \in C^{2,\alpha}(\overline{U})$ for every $0 < \alpha < \beta$. These assertions were proved (independently) by Urbas (1997) and Caffarelli (1992b, 1996).

For some applications, the hypothesis that both U and V are convex is too restrictive: see §6.3, 6.6. One might hope to make only the assumptions needed for the interior regularity theory, and then demonstrate (partial) regularity up to the boundary. If $d \geq 3$ there is a class of counterexamples due to Pogorelov (1964) which make it difficult to prove further results. However if we restrict attention to $d = 2$, and make the additional assumption that the target set V is strictly convex, then Douglas and McCann (1998) have proved that the $\Psi \in C^1(\overline{U})$ (so we have C^1 regularity up to the boundary). Wolfson (1997) establishes conditions for the existence of smooth area-preserving mappings (which coincide with the optimal mapping) between two domains with smooth boundaries. If the domains satisfy a *pseudo-convexity* condition, that is if

$$\min_{\partial U} \kappa_1 + \min_{\partial V} \kappa_2 > 0, \tag{5.11}$$

where κ_1 and κ_2 are the minimum curvature values for the curves which comprise ∂U and ∂V respectively, then there is a smooth area-preserving mapping between U and V. Roughly speaking, the pseudo-convexity condition is a joint convexity condition; if U is a disc (which is 'as convex as possible'), then (5.11) may still be satisfied even when V fails to be convex.

5.5 Polar factorisation of vector valued functions

For a real valued integrable function defined on a bounded interval, Ryff (1970) showed that it could be written as the composition of its increasing rearrangement with a measure-preserving mapping. This decomposition is the *polar factorisation* introduced in §2.3. Our purpose in this subsection is to extend this concept to vector valued functions. The monotone rearrangement, introduced in §5.2, plays the role of the increasing rearrangement. The question we address is the following: for a given integrable vector valued function $u : \Omega \subset \mathbb{R}^d \to \mathbb{R}^d$, when does a measure-preserving mapping $s : \Omega \to \Omega$ such that $u = u^{\#} \circ s$ exist, and when is it unique (i.e. when is it impossible to find a different measure-preserving mapping $t : \Omega \to \Omega$ such that $u = u^{\#} \circ t$)? The expression $u = u^{\#} \circ s$ is called a polar factorisation of u; this term was introduced by Brenier (1991).

We begin with a simple example; let f be as (5.1). The monotone rearrangement of f, $f^{\#}$, is defined by (5.4). We seek an area-preserving mapping which maps the set where f takes the value $(1,1)$ to the set where $f^{\#}$ does, and similarly for $(0,0)$. Denote the set where f takes the value $(1,1)$ by A: the corresponding set for $f^{\#}$ by $A^{\#}$. Split the set $A \backslash A^{\#}$ into two triangles of equal size by bisecting with the line $x = y$; we call the triangle above the line B_1, the one below B_2. Denote the part of $A^{\#} \backslash A$ above the line $x = y$ by $B_1^{\#}$, the remainder by $B_2^{\#}$. Figure 7 illustrates this notation. We can map B_1 onto $B_1^{\#}$ by a rotation and a translation - we write τ_1 for this map, and τ_2 for the

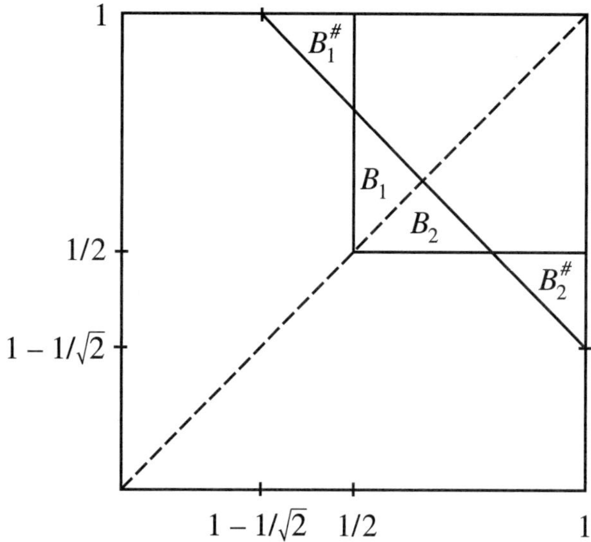

Figure 7: Construction of a polar factorisation

analogous map which moves B_2 onto $B_2^\#$. Define $s : [0,1]^2 \to [0,1]^2$ by

$$s(\mathbf{x}) = \begin{cases} \tau_1(\mathbf{x}) & \text{if } \mathbf{x} \in B_1, \\ \tau_2(\mathbf{x}) & \text{if } \mathbf{x} \in B_2, \\ \tau_1^{-1}(\mathbf{x}) & \text{if } \mathbf{x} \in B_1^\#, \\ \tau_2^{-1}(\mathbf{x}) & \text{if } \mathbf{x} \in B_2^\#, \\ \mathbf{x} & \text{otherwise.} \end{cases}$$

Now s is an area-preserving mapping which satisfies $f = f^\# \circ s$: thus f has a polar factorisation. (It is easily seen that it is not unique.)

To prove a general result on the existence and uniqueness of polar factorisations, we require some definitions and notation.

Definitions. Let $\mu = \lambda_d$, and let Ω and u be as in Theorem 5. u is *nondegenerate* if u does not map a set of positive size to a set of zero size, or more precisely if $\mu(\{x : u(x) \in E\}) = 0$ for every set $E \subset \mathbb{R}^d$ of zero Lebesgue measure. Otherwise u is *degenerate.*

The function f defined by (5.1) is degenerate: it maps a set of area $1/4$ to the point $(1,1)$, which has zero area.

u is *countably degenerate* if we can make u nondegenerate by removing countably many level sets. (Level sets of a function are sets of the form $\{x : u(x) = c\}$.)

If we remove the level sets corresponding to $(1,1)$ and $(0,0)$, the function f defined by (5.1) is nondegenerate: therefore f is countably degenerate.

A function is *almost injective* if it is an injective function except possibly on a set of measure zero. If u is nondegenerate, $u^{\#}$ is almost injective (see Burton and Douglas 1998, Lemma 2.4).

We have the following result on the existence and uniqueness of a polar factorisation (due to Burton and Douglas 1998).

Theorem 7 *Let $\mu = \lambda_d$, and let Ω and u be as in Theorem 5. Then*

(i) If u is countably degenerate, then u has a polar factorisation $u = u^{\#} \circ s$, where $s : \Omega \to \Omega$ is a measure-preserving mapping.

(ii) If $u^{\#}$ is almost injective, then the polar factorisation exists and is unique.

From our earlier remarks, (ii) applies to nondegenerate functions. (The fact that a nondegenerate function has a unique polar factorisation was first established by Brenier 1991, Theorem 1.2.) The idea of the proof of (ii) is as follows. Suppose $u^{\#}$ is injective - we show that $(u^{\#})^{-1} \circ u$ is a measure-preserving mapping. To show uniqueness, suppose $u = u^{\#} \circ s = u^{\#} \circ t$ for two measure-preserving mappings $s, t : \Omega \to \Omega$; then injectivity of $u^{\#}$ implies that s and t must be equal (except possibly on a set of measure zero). For (i), the essence of the proof is to map the level sets of positive measure of u to the corresponding level sets of $u^{\#}$, and apply (ii) to the nondegenerate function which remains after these sets have been removed.

Can we extend this result? It is easy to see that the polar factorisation cannot be unique if u has a level set of positive measure. $u^{\#}$ has a corresponding level set of positive measure, and we can choose a non-trivial measure-preserving mapping $\tau : \Omega \to \Omega$ which leaves points other than those in the level set fixed. If $u = u^{\#} \circ s$ is a polar factorisation of u, then so is $u = u^{\#} \circ (\tau \circ s)$: s is not equal to $\tau \circ s$. If $d \geq 2$, even when a function fails to have any level sets of positive measure, the polar factorisation, if it exists, may not be unique. (See Burton and Douglas 1998 for an example.) Burton and Douglas (2001) proved that the polar factorisation is unique if and only if the monotone rearrangement is almost injective. (They also introduced a class of integrable vector-valued functions which do not have polar factorisations.)

Finally we address the following problem: given a square integrable vector valued function $u : \Omega \subset \mathbb{R}^d \to \mathbb{R}^d$, what is the projection of u onto the set of measure-preserving mappings from Ω to Ω? This question is linked to polar factorisation by the following proposition.

Proposition 4 *Let Ω and μ be as in Theorem 7, and suppose $u : \Omega \to \mathbb{R}^d$ is square integrable. Then*

(i) Measure-preserving mappings $s : \Omega \to \Omega$ which satisfy $u = u^{\#} \circ s$ maximise $\int_\Omega u(x).s(x)d\mu(x)$ over the set S of measure-preserving mappings from Ω to Ω.

(ii) Suppose that u has a polar factorisation. Then every $s \in S$ which maximises $\int_\Omega u(x).s(x)d\mu(x)$ satisfies $u = u^\# \circ s$.

Proof. (i) Burton and Douglas (1998, Proposition 2.8). (ii) Burton and Douglas (2001). □

We say that a measure-preserving mapping $s : \Omega \to \Omega$ belongs to the projection of u onto S if it is a closest element of S to u where distance is measured by the L^2-norm i.e. it attains

$$\inf_{s \in S} \|u - s\|_2^2 = \inf_{s \in S} \int_\Omega |u(x) - s(x)|^2 d\mu(x).$$

In general there will be many 'closest' elements of S to u, therefore the projection will be multi valued. It is easily shown that minimising $\|u - s\|_2^2$ over $s \in S$ is equivalent to maximising $\int_\Omega u(x).s(x)d\mu(x)$ over $s \in S$. If u has a polar factorisation, then Proposition 4 says that s is in the projection of u onto S if and only if s arises from a polar factorisation of u. It follows that if $u^\#$ is almost injective, then the projection of u onto S will be unique. In particular the projection of a nondegenerate function u onto S is single valued. (This result was first proved by Brenier 1991.)

6 Applications of rearrangements of functions to atmospheric and oceanic flow

This section studies the semigeostrophic equations, a model for slowly varying flows constrained by rotation and stratification. They have, in particular, been used to study front formation in meteorology. After stating the equations in §6.1, we show how stable solutions can be interpreted as a sequence of minimum geostrophic energy states in §6.2; the energy minimisation is carried out over a set of rearrangements. At each fixed time t, the minimising rearrangement is the gradient of a convex function; as we discuss in §6.3, tracking singularities of this potential is thought of as weather fronts forming and evolving. In the special case when the minimising rearrangement is nondegenerate (see §5.5), we show in §6.4 that we can find the mapping which relates the Lagrangian and Eulerian variables. Retaining the nondegeneracy assumption, in §6.5 we exploit the so-called *duality structure* of the semigeostrophic equations to reformulate the system as a coupled Monge–Ampère/ transport problem. We discuss existence and uniqueness of solution for this coupled system with reference to relevant results in the literature in §6.6.

6.1 The semigeostrophic equations

Weather systems and equivalent large scale flows in the ocean can be characterised as slowly varying flows constrained by rotation and stratification.

A standard model for such flows are the three-dimensional Boussinesq equations of semigeostrophic theory on an f plane. The derivation and validity of the semigeostrophic equations are discussed in greater detail elsewhere in this volume; we note that the system is valid when the timescale for a change in velocity following a fluid particle is much greater than f^{-1}. (See Shutts and Cullen 1987.) In particular this is true in the case of front formation in meteorology, where the flow is largely parallel to the front. We state the equations in the form used by Hoskins (1975).

$$\frac{Du_g}{Dt} - fv_{ag} = 0, \frac{Dv_g}{Dt} + fu_{ag} = 0, \tag{6.1}$$

$$\frac{D\theta}{Dt} = 0, \tag{6.2}$$

$$\nabla.\mathbf{u} = 0, \tag{6.3}$$

$$\nabla\phi = \left(fv_g, -fu_g, \frac{g\theta}{\theta_0}\right) \tag{6.4}$$

where

$$\begin{aligned} \mathbf{u} &\equiv (u, v, w) \equiv \mathbf{u}_g + \mathbf{u}_{ag}, \\ \mathbf{u}_g &\equiv (u_g, v_g, 0), \\ \frac{D}{Dt} &\equiv \frac{\partial}{\partial t} + \mathbf{u}.\nabla \end{aligned}$$

The term f is the Coriolis parameter, assumed constant (so we are considering constant rotation), g is the acceleration due to gravity, θ_0 is a reference value of the potential temperature θ, and ϕ is a pressure variable. Subscripts g and ag denote geostrophic and ageostrophic velocity (or wind) components respectively, where the geostrophic velocity is defined to be the horizontal component of velocity in balance with the pressure gradient. This definition is included in equation (6.4), as is the statement of hydrostatic balance.

We solve the equations (for the velocities $\mathbf{u}$, $\mathbf{u}_g$, potential temperature θ, and pressure variable ϕ) in an open, bounded, connected set $\Omega \subset \mathbb{R}^3$ which has smooth boundary, with normal velocity $\mathbf{u}.\mathbf{n} = 0$ on $\partial\Omega$. (In the meteorological literature the equations are sometimes posed on $\overline{\Omega}$.) For $\mathbf{x} = (x, y, z) \in \Omega$, by making the substitution

$$\mathbf{X} \equiv (X, Y, Z) \equiv (x + v_g/f, y - u_g/f, (g/f^2\theta_0)\theta) \tag{6.5}$$

it is shown in Purser and Cullen (1987) that we may replace (6.1) and (6.2) by

$$\frac{D\mathbf{X}}{Dt} = \mathbf{u_g}.$$

It is immediate from (6.4) and (6.5) that

$$\mathbf{X} = \nabla P \text{ where } P = \left(\frac{\phi}{f^2} + \frac{1}{2}(x^2 + y^2)\right). \tag{6.6}$$

(6.5) is known as the *geostrophic transformation* and is due to Hoskins (1975).

We think of $\mathbf{X}$ as a function of the physical space co-ordinates $\mathbf{x}$. Rewriting in terms of $\mathbf{X}$ and $\mathbf{x}$, we have

$$\frac{D\mathbf{X}}{Dt} = fJ(\mathbf{X} - \mathbf{x}) \tag{6.7}$$

where

$$J = \begin{pmatrix} 0 & -1 & 0 \\ 1 & 0 & 0 \\ 0 & 0 & 0 \end{pmatrix}.$$

We now have a system of equations (6.6), (6.7) and (6.3) for unknowns $\mathbf{X}$, $\mathbf{u}$ and ϕ which we solve on $[0, t_1) \times \Omega$, for some $t_1 > 0$. We seek solutions in a restricted class - solutions that are stable in a sense we make precise in the next subsection.

We introduce the Lagrangian coordinates

$$\tilde{\mathbf{X}}(t, \mathbf{x}) = \mathbf{X}(t, \chi(t, \mathbf{x}))$$

where

$$t \to \chi(t, \mathbf{x})$$

is the trajectory of the fluid particle which is at $\mathbf{x}$ initially. The Lagrangian form of (6.6), (6.7) and (6.3) is

$$\begin{aligned} \tilde{\mathbf{X}}(t, \mathbf{x}) &= \nabla P(t, \chi(t, \mathbf{x})), && (6.8) \\ \frac{\partial \tilde{\mathbf{X}}}{\partial t} &= f(J(\tilde{\mathbf{X}} - \chi)), && (6.9) \\ \det D\chi(t, \mathbf{x}) &= 1, && (6.10) \end{aligned}$$

where $D\chi$ denotes the Jacobian matrix of the mapping χ.

6.2 Stable solutions of the semigeostrophic equations as a sequence of minimum energy states

Cullen, Norbury and Purser (1991) sought solutions of (6.6), (6.7) and (6.3) which can be interpreted as a sequence of constrained minimum energy states, where the constraints evolve with time. They considered solutions obtained by the following procedure. At each fixed time t, predict $\mathbf{X}$ on particles using (6.7). (In the Lagrangian variables defined above, we predict $\tilde{\mathbf{X}}(t, .)$ but not $\chi(t, .)$.) Now apply the *Cullen–Norbury–Purser* principle which states that for a solution, the particles are arranged to minimise geostrophic energy. We claim that this determines $\mathbf{X}(t, .)$, (that is a minimiser exists and is unique,) and that $\mathbf{X}(t, .) = \nabla\psi(t, .)$ for some convex function $\psi(t, .)$.

For a solution obtained by the Cullen–Norbury–Purser methods we identify $P(t, .)$ with $\psi(t, .)$ at each time t. Therefore we have a solution of (6.6),

(6.7) and (6.3) with $P(t,.)$ convex at each time t. Conversely, if we apply the Cullen–Norbury–Purser methods to a solution of (6.6), (6.7) and (6.3) with $P(t,.)$ convex at each time t, we obtain the given solution. (Recall that the monotone rearrangement is essentially unique.) It follows that seeking solutions that can be interpreted as a sequence of (constrained) minimum energy states is equivalent to solving (6.6), (6.7) and (6.3) with the extra constraint that $P(t,.)$ is convex at each time t. As convexity corresponds to a minimum energy principle, we expect such solutions to be stable (in some sense). In Shutts and Cullen (1987) it is shown that they are stable with respect to small displacements in a frozen pressure field.

We justify our earlier claim that the Cullen–Norbury–Purser principle is well posed i.e. demonstrate that an energy minimiser exists and is unique. The geostrophic energy E is defined as

$$\begin{aligned} E &= \int_\Omega \frac{1}{2}u_g^2 + \frac{1}{2}v_g^2 + \frac{g\theta z}{\theta_0} d\lambda_3(\mathbf{x}) \\ &= f^2\frac{1}{2}\int_\Omega X^2(\mathbf{x}) + x^2 + Y^2(\mathbf{x}) + y^2 d\lambda_3(\mathbf{x}) - f^2\int_\Omega \mathbf{x}.\mathbf{X}(\mathbf{x})d\lambda_3(\mathbf{x}), \end{aligned}$$

where λ_3 denotes volume (or more precisely 3-dimensional Lebesgue measure). Suppose one possible state of the fluid at a fixed time t is described by a function $\mathbf{X}_t = (X_t, Y_t, Z_t)$. (One way to calculate a candidate state of the fluid would be to solve equation (6.7) on particles, i.e. solve (6.9) for $\tilde{\mathbf{X}}(t,.)$, and then assume the particles at time t are in the same configuration that they are in at time 0.) Values are fixed on particles, but the particles may take up any configuration (subject to the incompressibility constraint that they do not change volume), therefore possible states of the fluid are described by elements of $\mathcal{R}(\mathbf{X}_t)$. The Cullen–Norbury–Purser principle states that we should minimise the geostrophic energy over all possible states, therefore we study the minimisation problem

$$\inf_{\mathbf{X}\in\mathcal{R}(\mathbf{X}_t)} E(\mathbf{X}), \tag{6.11}$$

where the energy minimiser (if it exists and is unique) gives the actual state of the fluid. The following theorem is due to Douglas (1998, Theorem 3.1) (in which we make an additional physically reasonable assumption which ensures (6.11) is finite):

Theorem 8 *Suppose that $\mathbf{X}_t \in L^p(\Omega, \lambda_3, \mathbb{R}^3)$, where $2 \le p < \infty$. Then* (6.11) *is uniquely attained by the monotone rearrangement of $\mathbf{X}_t$, that is the unique element of $\mathcal{R}(\mathbf{X}_t)$ equal to the gradient of a convex function.*

Proof. Applying Theorem 4 yields that $\int_\Omega X^2(\mathbf{x}) + Y^2(\mathbf{x}) d\lambda_3(\mathbf{x})$ has the same value for each $\mathbf{X} \in \mathcal{R}(\mathbf{X}_t)$: therefore it is sufficient to study

$$\sup_{\mathbf{X}\in\mathcal{R}(\mathbf{X}_t)} \int_\Omega \mathbf{x}.\mathbf{X}(\mathbf{x})d\lambda_3(\mathbf{x}). \tag{6.12}$$

Theorem 6 yields that (6.12) is uniquely attained by $\mathbf{X}_t^{\#}$, the monotone rearrangement of $\mathbf{X}_t$. □

We note that in simple examples equation (6.7) may be easy to solve analytically, and in such cases the methods of Cullen, Norbury and Purser described above have been used to calculate solutions to the semigeostrophic equations. In general, however, this is not the case: to prove the existence of solutions, the equations are reformulated in $\mathbf{X}$ (dual) space as a coupled Monge–Ampère/transport problem. We describe how this is done in §6.5 and discuss existence results in this setting in §6.6.

Benamou (1992) interprets the minimum energy principle in the following way: assume that the configuration of the fluid at each time t is an energy minimum with respect to permutation of the fluid particles. Recalling from (6.6) that $\mathbf{X} = \nabla P$, and using the methods of Brenier (1991, Proposition 2.1), it can be shown that $P(t,.)$ must be convex. The approach used here differs in that we prove that an energy minimiser exists (and is unique), rather than assuming it exists.

6.3 The semigeostrophic equations as a model for frontogenesis

The energy minimisation problem of the previous subsection yielded that stable solutions of the semigeostrophic equations satisfy $\mathbf{X}(t,.) = \nabla P(t,.)$ at each time t for some convex function $P(t,.)$. In this subsection, following (for example) Cullen (1983) and Evans (1997, Section 7.3), we interpret singularities of $P(t,.)$ as weather fronts, that is regions across which there are large variations in wind and temperature. As time evolves, we track the singularities of $P(t,.)$; this is thought of as weather fronts forming and moving. This is the sense in which the semigeostrophic equations are considered a model for frontogenesis. In this subsection we merely outline this idea, giving references to some relevant literature.

To study the regularity of $P(t,.)$, we note that $\mathbf{X}(t,.)$ is the unique minimiser of

$$\inf_{\mathbf{X}\in\mathcal{R}(\mathbf{X}_t)} \int_\Omega |\mathbf{X}(\mathbf{x}) - \mathbf{x}|_2^2 d\lambda_3(\mathbf{x}),$$

and as described in §5 this may be rewritten as an optimal mass transfer problem, which has an associated regularity theory. (See §5.4.) For each time t, define a (Borel) measure ν_t by $\nu_t(B) = \lambda_3(\mathbf{X}(t,.))^{-1}(B))$ for Borel subsets B of $\mathbb{R}^3$. The image space of $\mathbf{X}$ is known as *dual space*; ν_t measures what volume of physical space is mapped to a given set in dual space. We noted in §5 that the set of measure-preserving mappings between (Ω, λ_3) and $(\mathbb{R}^3, \nu_t)$ is exactly $\mathcal{R}(\mathbf{X}(t,.))$. Our regularity theory holds in the special case when ν_t satisfies some additional properties. For a given t, suppose that $\mathbf{X}(t,.)$ is nondegenerate in the sense of Brenier (1991), that is $\mathbf{X}(t,.)$ does not map a set of positive volume to a set of zero volume. Then $\nu_t(E)$ is well defined for Lebesgue measurable subsets of $\mathbb{R}^3$, and moreover ν_t is absolutely continuous with respect to Lebesgue measure; the crucial consequence of this is that we may write ν_t as

$$\nu_t(E) = \int_E \rho(t, \mathbf{X}) d\lambda_3(\mathbf{X})$$

for Lebesgue measurable subsets $E \subset \mathbb{R}^3$, where $\rho(t, .)$ is a non-negative integrable function which we call the *pseudo-density.* (In the language of measure theory, $\rho(t, .)$ is the Radon–Nikodym derivative of ν_t with respect to Lebesgue measure.) Define the *support* of $\rho(t, .)$, denoted supp $\rho(t, .)$, to be the smallest closed subset $F \subset \mathbb{R}^3$ such that $\lambda_3(\Omega) = \nu_t(F)$. To study the singularities of $P(t, .)$, we consider the regularity of $\nabla P(t, .)$, the optimal mass transfer from (Ω, λ_3) to (supp $\rho(t, .), \nu_t)$ with respect to the cost $c(\mathbf{x}, \mathbf{y}) = |\mathbf{x} - \mathbf{y}|^2/2$.

Recall from §5 that regularity of $\nabla P(t, .)$ cannot be guaranteed in the case where supp $\rho(t, .)$ is not convex. Therefore loss of convexity of supp $\rho(t, .)$ as time evolves can be seen as a mechanism for introducing singularities of $P(t, .)$. A computationally appealing example is the following. Suppose Ω is convex, and let $\rho(0, .)$ be a convex pseudo-density patch, that is 1 on some bounded convex set and 0 otherwise; now let time evolve. If the geostrophic velocity is sufficiently regular, we see in §6.5 that $\rho(t, .) \in \mathcal{R}(\rho(0, .))$ for each t, therefore $\rho(t, .)$ will continue to be the characteristic function of some set (i.e. 1 on some set, 0 elsewhere). However supp $\rho(t, .)$ may lose convexity as time evolves, and singularities of the optimal mapping may form and move. The author is not aware of any such numerical computations.

The 2-dimensional semigeostrophic equations, that is (6.1), and the appropriate forms of (6.3) and (6.4), are still a physically relevant system in certain regimes. They can be studied in exactly analogous fashion to the 3-dimensional equations. As explained in §5, additional regularity results for the optimal mapping are available when we restrict to two dimensions. In particular we can interpret Wolfson's result (1997) on the existence of an area-preserving diffeomorphism as showing that fronts do not form when the pseudo-density is a characteristic function, and Ω and supp $\rho(t, .)$ satisfy a 'joint convexity condition'. (See §5 for details, but note that if Ω is a ball, and supp $\rho(t, .)$ is 'nearly convex', then fronts will not form.) Computations have been performed for two-dimensional semigeostrophic models: Cullen and Roulstone (1993) numerically simulated the evolution of an unstable wave as it passes through frontal collapse.

As stated in Evans (1997, Section 7.3), 'There are extremely interesting mathematical problems here, which have only in part been studied.'

6.4 Identification of the trajectory mapping via polar factorisation

At each fixed time t, stable solutions (in the sense of §6.2) of the semigeostrophic equations satisfy $\mathbf{X}(t,.) = \nabla\psi(t,.)$ for some convex function $\psi(t,.)$ which we identify with $\phi/f^2 + (x^2 + y^2)/2$. In this subsection we study the Lagrangian version of the problem, that is equations (6.8), (6.9) and (6.10), using the techniques of Benamou and Brenier (1992, 1998a). Our aim is to find the trajectory mapping χ.

For each t, assume $\tilde{\mathbf{X}}(t,.)$ is square integrable and nondegenerate (as defined in §5.5, that is $\tilde{\mathbf{X}}(t,.)$ does not map a set of positive size to a set of zero size). Then, for each t, Theorem 7 yields that $\tilde{\mathbf{X}}(t,.)$ has a unique polar factorisation into the composition of its monotone rearrangement with a measure-preserving mapping, that is there exists a convex function $\psi(t,.)$, and a unique measure-preserving mapping $s(t,.)$ such that

$$\tilde{\mathbf{X}}(t,.) = \nabla\psi(t, s(t,.)). \tag{6.13}$$

We identify $\psi(t,.)$ with $P(t,.)$ (as before), and $s(t,.)$ with the trajectory mapping $\chi(t,.)$ at each fixed time t. For each t, Proposition 4 yields that $\chi(t,.)$ is the projection of $\tilde{\mathbf{X}}(t,.)$ onto the set of measure-preserving mappings from Ω to Ω. Denoting this projection mapping by Π_S, (6.9) yields that

$$\frac{\partial\tilde{\mathbf{X}}}{\partial t} = f(J(Id - \Pi_S)\tilde{\mathbf{X}}) \tag{6.14}$$

where Id denotes the identity map on Ω. Benamou (1992) refers to (6.14) as the *dynamic rearrangement equation.*

The nondegeneracy assumption implies that $\tilde{\mathbf{X}}(t,.)$ has no level sets of positive measure: this excludes some physically reasonable solutions. As discussed in §5.5, if $\tilde{\mathbf{X}}(t,.)$ has at least one level set of positive measure, then the polar factorisation (6.13) (if such exists) is not unique: hence in this case we cannot define the trajectory mapping uniquely without further information. A question of current interest is how to choose the trajectory mapping when the polar factorisation (6.13) is known to exist but is not unique.

6.5 Reformulation of the semigeostrophic equations as a coupled Monge–Ampère/transport problem

In §6.2 we showed that solutions of the semigeostrophic equations (6.6), (6.7) and (6.3) which can be interpreted as a sequence of minimum energy states are those for which $P(t,.)$ is a convex function at each time t. However we did not address the problem of showing that such solutions exist. Our aim in this subsection is to rewrite the semigeostrophic equations as a coupled Monge–Ampère/ transport problem in $\mathbf{X}$ (or dual) space; we discuss existence of solution for this formulation in the next subsection. We assume that $\mathbf{X}(t,.)$ is nondegenerate for each t, and show that the (Legendre–Fenchel) conjugate convex function of $P(t,.)$ (which we denote $R(t,.)$) satisfies a Monge–Ampère problem in a generalised sense. The right-hand side of the Monge–Ampère equation is the pseudo-density (at time t). The Monge–Ampère problem replaces (6.6), the convexity condition and the nondegeneracy assumption. (This reformulation uses the so-called duality structure of the semigeostrophic equations.) We can use (6.7) to derive an evolution equation for the pseudo-density; this yields a coupled system which is a closed form of the semigeostrophic equations. The advecting velocity is the geostrophic velocity (written in dual space coordinates); it may be derived from R.

To establish the coupled problem in dual space and to discuss existence results we need to use technical language: the reader is referred to Friedman (1982) for a more detailed description of the terms used. We first address the question of why we reformulate the semigeostrophic equations in dual space, rather than establish an existence theory in the physical space Ω. To use the results of Diperna and Lions (1989) to show existence of solution for the evolution equation (6.7) we require the advecting velocity $\mathbf{u}(t,.) \in W^{1,1}(\Omega)$ (or better) at each time t, that is $\mathbf{u}(t,.)$ is integrable, and has a derivative in the sense of integration by parts against smooth functions which vanish on $\partial\Omega$, which is also integrable. However the convexity of $P(t,.)$ at each time t gives information about the regularity of $\mathbf{u}_g$ (via (6.6)) rather than $\mathbf{u}$. To avoid this difficulty we work in dual space.

The methods of §6.2 yielded $\mathbf{X}$ as a function of $\mathbf{x}$ at each fixed time t; now we wish to reformulate the semigeostrophic equations in $\mathbf{X}$ variables, therefore we find a mapping from $\mathbf{X}$ to $\mathbf{x}$. For each t, given $\mathbf{X}(t,.) = \nabla P(t,.)$, essentially we restrict attention to the case where $\nabla P(t,.)$ is injective, so that we have $\mathbf{x}(t,.) = (\nabla P(t,.))^{-1}$. For a convex function there is a natural notion of generalised derivative, the subdifferential: the *subdifferential of f at a point* $\mathbf{x}$, which we denote $\partial f(\mathbf{x})$, is the set of hyperplanes to the graph of f at the point $(\mathbf{x}, f(\mathbf{x}))$. A convex function may have 'kinks', but it is differentiable at all other points on the interior of the set where it is finite: the subdifferential of a convex function f is the singleton set $\{\nabla f(\mathbf{x})\}$ for almost every $\mathbf{x}$ where $f(\mathbf{x})$ is finite. Without loss of generality we can choose $P(t,.)$ to be lower semicontinuous (which, roughly speaking, is a one-sided version of continuity), and proper, that is $P(t,.)$ never takes the value $-\infty$ and is finite at some point. We define the (Legendre–Fenchel) *conjugate convex* function of $P(t,.)$, which we denote $R(t,.)$, by

$$R(t,\mathbf{X}) = \sup\{\mathbf{x}.\mathbf{X} - P(t,\mathbf{x}) : \mathbf{x} \in \mathbb{R}^3\}. \tag{6.15}$$

We have the following result.

Proposition 5 *Suppose* $\mathbf{X}(t,.) : \Omega \to \mathbb{R}^3$ *is an integrable function which satisfies* $\mathbf{X}(t,.) = \nabla P(t,.)$ *almost everywhere for some proper lower semicon-*

tinuous convex function $P(t,.) : \mathbb{R}^3 \to \mathbb{R} \cup \{+\infty\}$. *Then, for almost every* $\mathbf{x} \in \Omega$,

$$P(t,\mathbf{x}) + R(t,\mathbf{X}(t,\mathbf{x})) = \mathbf{x}.\mathbf{X}(\mathbf{x},t). \tag{6.16}$$

Moreover if $\mathbf{X}(t,.)$ *does not map a set of positive size to a set of zero size (i.e.* $\mathbf{X}(t,.)$ *is nondegenerate), then*

$$\mathbf{x} = \nabla R(t,\mathbf{X}(\mathbf{x},t)) \tag{6.17}$$

for almost every $\mathbf{x} \in \Omega$.

Proof. From our remarks above we have that $\partial P(t,.)(\mathbf{x}) = \{\mathbf{X}(t,\mathbf{x})\}$ for almost every $\mathbf{x} \in \Omega$. Standard convex analysis (see Rockafellar 1970, Theorem 23.5) demonstrates that (6.16) holds for almost every $\mathbf{x} \in \Omega$. Noting that $P(t,.)$ is convex, lower semicontinuous and proper, and adopting the notation that f^* denotes the conjugate convex function of f, Rockafellar (1970, Theorem 12.2) yields that $P^{**}(t,.) = P(t,.)$. Rewriting (6.16) we have

$$R^*(t,\mathbf{x}) + R(t,\mathbf{X}(t,\mathbf{x})) = \mathbf{x}.\mathbf{X}(t,\mathbf{x})$$

for a.e. $\mathbf{x} \in \Omega$. It follows that $\mathbf{x} \in \partial R(t,.)(\mathbf{X}(t,\mathbf{x}))$ for almost every $x \in \Omega$. If $\mathbf{X}(t,.)$ is nondegenerate, then $\partial R(t,.)(\mathbf{X}(t,\mathbf{x})) = \{\nabla R(t,\mathbf{X}(t,\mathbf{x}))\}$ for a.e. $\mathbf{x} \in \Omega$, and we obtain (6.17). □

The nondegeneracy assumption ensures that $\mathbf{X}(t,.)$ does not map a set of positive size to the set where $R(t,.)$ is not differentiable, therefore (6.17) holds. For two points $\mathbf{x},\mathbf{y} \in \Omega$, $\mathbf{x} \neq \mathbf{y}$, with $\mathbf{X}(t,\mathbf{x}) = \mathbf{X}(t,\mathbf{y}) = \mathbf{k}$, say, the methods of the above proof yield that $\mathbf{x},\mathbf{y} \in \partial R(t,.)(\mathbf{k})$, therefore $R(t,.)$ is not differentiable at $\mathbf{k}$. So if we cannot make $\nabla P(t,.)$ injective by ignoring a set of measure zero, (6.17) is replaced by a differential inclusion. We do not study this problem in this article; however we note that it is the appropriate generalisation to the case where we do not assume the (physically unjustified) nondegeneracy condition.

At each fixed t, if we restrict $\mathbf{X}(t,.)$ to $\Omega_t \subset \Omega$, where Ω_t is chosen so that (6.17) holds for every $\mathbf{x} \in \Omega_t$ and $\lambda_3(\Omega_t) = \lambda_3(\Omega)$, then $\mathbf{X}(t,.)$ is injective on Ω_t. Now we can define the inverse mapping $\mathbf{x} = \mathbf{x}(t,\mathbf{X})$, and $\mathbf{x}(t,.) = \nabla R(t,.)$ on $\mathbf{X}(\Omega_t)$. $\mathbf{X}(t,.) = \nabla P(t,.)$, (6.16), and $\mathbf{x}(t,.) = \nabla R(t,.)$ are sometimes referred to as the *duality structure* of the semigeostrophic equations. There is a considerable literature which has identified and exploited this structure for particular solutions (see for example Cullen and Purser 1984, Chynoweth, Porter and Sewell 1988).

For each t let ν_t be the measure introduced in §6.3. $\mathbf{X}(t,.)$ is nondegenerate by assumption, therefore (as discussed in §6.3) ν_t may be written

$$\nu_t(E) = \int_E \rho(t,\mathbf{X})d\lambda_3(\mathbf{X})$$

for every (Lebesgue measurable) subset E of $\mathbb{R}^3$, where (integrable non-negative) $\rho(t,.)$ is the pseudo-density. In particular at each time t we have the constraint

$$\int_{\mathbb{R}^3} \rho(t,\mathbf{X})d\lambda_3(\mathbf{X}) = \int_{\mathbb{R}^3} d\nu_t = \lambda_3(\Omega).$$

Moreover the methods of Burton and Douglas (1998, Lemma 6) yield that $\mathbf{x}(t,\mathbf{X}) = \nabla R(t,\mathbf{X})$ for ν_t a.e. $\mathbf{X} \in \mathbb{R}^3$, or equivalently, for a.e. $\mathbf{X} \in \text{supp}\ \rho(t,.)$ (where supp $\rho(t,.)$ is the smallest closed set which contains all the mass of ν_t).

We rewrite (6.6) using the methods of Brenier (1991). Fixing t, for each $F \in C_c(\mathbb{R}^3)$ (where $C_c(\mathbb{R}^3)$ denotes continuous functions with compact support in $\mathbb{R}^3$) it follows from (6.6) that

$$\int_{\Omega} F(\nabla P(t,\mathbf{x}))d\lambda_3(\mathbf{x}) = \int_{\mathbb{R}^3} F(\mathbf{X})\rho(t,\mathbf{X})d\lambda_3(\mathbf{X}). \tag{6.18}$$

If we make the substitution $\mathbf{X} = \nabla P(t,\mathbf{x})$ in the left hand integral of (6.18), a formal calculation, using the fact that $\mathbf{x}(t,\mathbf{X}) = \nabla R(t,\mathbf{X})$ on supp $\rho(t,.)$, yields

$$\int_{\text{supp}\ \rho(t,.)} F(\mathbf{X})\det\ H(R(t,\mathbf{X}))d\lambda_3(\mathbf{X}) = \int_{\mathbb{R}^3} F(\mathbf{X})\rho(t,\mathbf{X})d\lambda_3(\mathbf{X})$$

for each $F \in C_c(\mathbb{R}^3)$, where H denotes the Hessian matrix. This gives the following Monge–Ampère problem for each t:

$$\begin{gathered} \det\ H(R(t,\mathbf{X})) = \rho(t,\mathbf{X}), \\ \nabla R(t,\mathbf{X}) \text{ maps the support of } \rho(t,\mathbf{X}) \text{ into } \overline{\Omega}. \end{gathered} \tag{6.19}$$

It follows that (6.18) is a weak formulation of (6.19).

We can use the evolution equation (6.7) to derive the following transport equation for ρ, which holds in the classical weak sense (see Benamou and Brenier 1992, 1998a for details):

$$\frac{\partial \rho}{\partial t} + \nabla.(\rho\mathbf{U}) = 0 \tag{6.20}$$

where $\mathbf{U} = f(y - Y, X - x, 0)$ and $\nabla \equiv (\partial/\partial X, \partial/\partial Y, \partial/\partial Z)$. Noting that $\mathbf{x}(t,.) = \nabla R(t,.)$ ν_t a.e. in $\mathbb{R}^3$ for each t, we have

$$\mathbf{U} = f(J(Id - \nabla R)) \tag{6.21}$$

where Id denotes the identity mapping on $\mathbb{R}^3$. The transport equation (6.20) is coupled to the Monge–Ampère problem (6.18) by (6.15) and (6.21). It follows from (6.21) that if R is sufficiently smooth, $\nabla.\mathbf{U} = 0$. When this holds, the solution of (6.20) can be viewed as rearranging the initial values of ρ i.e. $\rho(t,.)$ is a rearrangement of $\rho(0,.)$ for each positive time t.

6.6 Weak solutions of the semigeostrophic equations

The previous section yielded equations (6.15), (6.18), (6.20) and (6.21) for unknowns R, P, ρ and $\mathbf{U}$. We discuss existence of solution for the system (6.18) and (6.20) (which holds in the classical weak sense), coupled by (6.15) and (6.21), posed on $[0, t_1) \times \mathbb{R}^3$ for some $t_1 > 0$, with initial condition $\rho(0, .) = \rho_0$ given. The difficulty is showing that $\mathbf{U}$ is sufficiently regular to solve the evolution equation (6.20) in some suitable sense: in view of (6.21), what we have to show is that given $\rho(t, .)$, the function $R(t, .)$ obtained from (6.18) and (6.15) is sufficiently smooth. One possible strategy to prove existence of solutions is to use a fixed point argument. From our earlier assumptions we have that $\rho(t, .)$ is non-negative and integrable; if we could show that the solution to the Monge–Ampère problem $R(t, .) \in W^{2,1}(\mathbb{R}^3)$, then (6.21) yields that $\mathbf{U}(t, .) \in W^{1,1}(\mathbb{R}^3)$, and we could use the transport theory of Diperna and Lions (1989) to solve (6.20). The author is not aware of any proof of this result. Given the additional hypothesis that $\rho_0 \in L^p(\mathbb{R}^3)$ for $p > 3$, and is compactly supported, Benamou and Brenier (1992, 1998a) proved a result of weak existence - they showed existence of solutions for an approximate problem and then passed to the limit. The smoother approximate problem was obtained by discretising time and mollifying; the fact that a limiting weak solution exists is reliant on the continuity of the mapping $u \to u^\#$ (where $u^\#$ denotes the monotone rearrangement of u). Similar results were obtained by Otto (1997). The author is not aware of any uniqueness results.

For a given t, let S_t denote the set of measure-preserving mappings from $(\text{supp}\ \rho(t, .), \nu_t) \to (\Omega, \lambda_3)$; it is easily seen that $\nabla R(t, .)$ is the optimal mapping in S_t against the cost $c(W, Z) = |W - Z|^2/2$. We can choose the 'target set' Ω convex, so can we use the regularity theory for $R(t, .)$ stated in §5.4 to prove existence of strong solutions (at least for short times)? This question is open to the best of the author's knowledge. For notational convenience we write $Y(t)$ for the interior of supp $\rho(t, .)$. If we assume that Ω and supp $\rho(t, .)$ are bounded, convex and have smooth boundaries, Caffarelli (1996) shows that if $\rho(t, .) \in C^\alpha(Y(t))$ (for some $0 < \alpha < 1$), and is bounded above and below, away from zero, then $R(t, .) \in C^{2,\alpha}(\overline{Y(t)})$: in this case we have regularity up to the boundary, and classical transport theory (using characteristics) may be applied. However the same regularity theory yields $P(t, .) \in C^{2,\alpha}(\overline{\Omega})$; in the light of §6.3, restricting to solutions where the support of $\rho(t, .)$ is convex and has smooth boundary is limiting the theory to the case when no fronts form! If we relax the assumptions to Ω convex, supp $\rho(t, .)$ bounded and connected, (with the same restrictions on $\rho(t, .)$,) then there is an interior regularity theory due to Caffarelli (1992a). If $\rho(t, .) \in C^{1,\beta}(Y(t))$ for some $0 < \beta < 1$, then $R(t, .) \in C^{3,\alpha}_{loc}(Y(t))$ for each $0 < \alpha < \beta$. The challenge is to demonstrate regularity up to the boundary (e.g. $R(t, .) \in C^{2,\alpha}(\overline{Y(t)})$) in this case.

We conclude this section with an observation of Evans (1997). The system (6.20), (6.21) and (6.19), posed on $[0, t_1) \times Y(t)$, is similar in structure to the vorticity formulation of the 2-dimensional Euler equations:

$$\frac{\partial \omega}{\partial t} + \nabla.(\omega \mathbf{u}) = 0, \tag{6.22}$$

$$\mathbf{u} = \left(\frac{\partial \psi}{\partial y}, -\frac{\partial \psi}{\partial x}\right), \tag{6.23}$$

$$-\Delta \psi = \omega. \tag{6.24}$$

Here $\mathbf{u}$ is the velocity, ω the vorticity, ψ the streamfunction. For the semigeostrophic equations, we have a Monge–Ampère problem rather than (6.24): roughly speaking, (6.22), (6.23) and (6.24) is a linearisation of (6.20), (6.21) and (6.19). There is an existence theory (for strong global solutions) for the 2-dimensional Euler equations, (see for example Lions 1996,) so we can hope that a parallel theory can be developed for the semigeostrophic equations.

Acknowledgements

The author was grateful for the hospitality of the Isaac Newton Institute for Mathematical Sciences during the programme *Mathematics of Atmosphere and Ocean Dynamics*. The author's research was supported by EPSRC Research Grant reference GL/L43220.

References

Adams, R.A. (1975) *Sobolev Spaces*. New York, Academic Press.

Alvino, A., Lions, P-L. & Trombetti, G. (1989) On optimisation problems with prescribed rearrangements. *Nonlinear Anal. Theory Methods Appl.* **13**, 185–220.

Badiani, T.V. & Burton, G.R. (2001) Steady vortex rings in $\mathbb{R}^3$ and rearrangements. *Proc. R. Soc. Lond.* A **457**, 1115–1135.

Baigent, S. (1987) *Applications of vector valued rearrangements to modelling the weather*. MSc Dissertation, Oxford University.

Benamou, J.-D. (1992) *Transformation conservant la measure, mécanique des fluides incompressibles et modèle semi-géostrophique en météorologie*. Thèse de 3eme cycle, Université de Paris IX-Dauphine.

Benamou, J.-D. & Brenier, Y. (1998a) Weak existence for the semigeostrophic equations formulated as a coupled Monge–Ampere/transport problem. *SIAM J. Appl. Math.* **58**, 1450–1461.

Benamou, J-D. & Brenier, Y. (1998b) A numerical method for the Optimal Time-Continuous Mass Transport Problem and related problems. *Monge Ampère Equation: Applications to Geometry and Optimization, Contemporary Mathematics* **226**, American Mathematical Society, 1–11.

Benjamin, T.B. (1976) The alliance of practical and analytical insights into the nonlinear problems of fluid mechanics. *Applications of methods of functional analysis to problems in mechanics. Lecture notes in mathematics*, **503**, Springer, 8–29.

Brenier, Y. (1991) Polar factorisation and monotone rearrangement of vector-valued functions. *Comm. Pure Appl. Math.* **44**, 375–417.

Brothers, J.E. & Ziemer, W.P. (1988) Minimal rearrangements of Sobolev functions. *J. Reine Angew. Math.* **384**, 153–179.

Burton, G.R. (1987) Rearrangement of functions, maximisation of convex functionals, and vortex rings. *Math. Ann.* **276**, 225–253.

Burton, G.R. (1989) Variational problems on classes of rearrangements and multiple configurations for steady vortices. *Ann. Inst. H. Poincaré Anal. Non Linéaire* **6**, 295–319.

Burton, G.R. (1999) Vortex-rings of prescribed impulse. *Submitted.*

Burton, G.R. & Douglas, R.J. (1998) Rearrangements and polar factorisation of countably degenerate functions. *Proc. Roy. Soc. Edin.* **128A**, 671–681.

Burton, G.R. & Douglas, R.J. (2001) Uniqueness of the polar factorisation and projection of a vector-valued mapping. *Submitted.*

Burton, G.R. & McLeod, J.B. (1991) Maximisation and minimisation on classes of rearrangements. *Proc. Roy. Soc. Edin.* **119A**, 287–300.

Burton, G.R. & Nycander, J. (1999) Stationary vortices in three-dimensional quasi-geostrophic shear flow. *J. Fluid Mech.* **389**, 255–274.

Caffarelli, L. (1992a) The regularity of mappings with a convex potential. *J.A.M.S.* **5**, 99–104.

Caffarelli, L. (1992b) Boundary regularity of maps with convex potentials. *Comm. Pure Appl. Math.* **45**, 1141–1151.

Caffarelli, L. (1996) Boundary regularity of maps with convex potentials – II. *Ann. of Math.* **144**, 453–496.

Chynoweth, S., Porter, D. & Sewell, M.J. (1988) The parabolic umbilic and atmospheric fronts. *Proc. R. Soc. Lond.* A**419**, 337–362.

Crowe, J.A., Zweibel, J.A. & Rosenbloom, P.C. (1986) Rearrangements of functions. *J. Funct. Anal.* **66**, 432–438.

Cullen, M.J.P. (1983) Solutions to a model of a front forced by deformation. *Q.J. Roy. Met. Soc.* **109**, 565–573.

Cullen, M.J.P. & Douglas, R.J. (1998) Applications of the Monge–Ampère equations and Monge transport problem to meteorology and oceanography. *Monge Ampère Equation: Applications to Geometry and Optimization, Contemporary Mathematics* **226**, American Mathematical Society, 33–53.

Cullen, M.J.P., Norbury, J. & Purser, R.J. (1991) Generalised Lagrangian solutions for atmospheric and oceanic flows. *SIAM J. Appl. Math.* **51**, 20–31.

Cullen, M.J.P. & Purser, R.J. (1984) An extended Lagrangian theory of semigeostrophic frontogenesis. *J. Atmos. Sci.* **41**, 1477–1497.

Cullen, M.J.P. & Roulstone, I. (1993) A Geometric Model of the Nonlinear Equilibration of Two-Dimensional Eady Waves. *J. Atmos. Sci.* **50**, 328–332.

Diperna, R.J. & Lions, P.L. (1989) Ordinary differential equations, transport theory and Sobolev spaces. *Invent. Math.* **98**, 511–547.

Douglas, R.J. (1994) Rearrangements of functions on unbounded domains. *Proc. Roy. Soc. Edin.* **124A**, 621–644.

Douglas, R.J. (1998) Rearrangements of vector valued functions, with application to atmospheric and oceanic flows. *SIAM J. Math. Anal.* **29**, 891–902.

Douglas R.J. & McCann, R.J. (1998) *Unpublished note.*

Evans, L.C. (1997) Partial Differential Equations and Monge–Kantorovich Mass Transfer. *Current Developments in Mathematics*, International Press, Cambridge, 26–78.

Eydeland, A., Spruck, J. & Turkington, B. (1990) Multiconstrained variational problems of nonlinear eigenvalue type: new formulations and algorithms. *Math. Comp.* **55**, 509–535.

Friedman, A. (1982) *Foundations of Modern Analysis.* Dover, New York.

Gangbo, W. & McCann, R.J. (1996) The geometry of optimal transportation. *Acta Math.* **177**, 113–161.

Halmos, P.R. (1950) *Measure Theory.* Van Nostrand.

Hoffman, R.N., Liu, Z., Louis, J.-F. & Grassotti, C. (1995) Distortion representation of forecast errors. *Mon. Wea. Rev.* **123**, 2758–2770.

Hoskins, B.J. (1975) The geostrophic momentum approximation and the semigeostrophic equations. *J. Atmos. Sci.* **32**, 233–242.

Kawohl, B. (1985) *Rearrangements and convexity of level sets in PDE*, Springer Lecture Notes in Mathematics **1150**.

Lieb, E.H. & Loss, M. (1997) *Analysis.* American Mathematical Society.

Lions, P.L. (1996) *Mathematical Topics in Fluid Mechanics, Volume 1, Incompressible Models.* Oxford, Clarendon Press.

McCann, R.J. (1995) Existence and uniqueness of monotone measure preserving maps. *Duke Math. J.* **80**, 309–323.

Monge, G. 1781 *Mémoire sur la théorie des déblais et de remblais.* Mémoires de l'Académie des Sciences.

Mossino, J. & Temam, R. (1981) Directional Derivative of the increasing rearrangement mapping and application to a queer differential equation in plasma physics. *Duke Math. J.* **48**, 475–495.

Nycander, J. (1995) Existence and stability of stationary vortices in a uniform shear flow. *J. Fluid Mech.* **287**, 119–132.

Otto, F. (1997) *Private communication.*

Pogorelov, A.V. (1964) *Monge–Ampère Equations of Elliptic Type.* Noordhoff, Groningen.

Purser, R.J. & Cullen, M.J.P. (1987) A duality principle in semigeostrophic theory. *J. Atmos. Sci.* **44**, 3449–3468.

Rockafellar, R.T. (1970) *Convex Analysis.* Princeton University Press.

Ryff, J.V. (1970) Measure preserving transformations and rearrangements. *J. Math. Anal. and Applics.* **30**, 431–437.

Shutts, G.J. & Cullen, M.J.P. (1987) Parcel stability and its relation to semigeostrophic theory. *J. Atmos. Sci.* **46**, 2684–2697.

Simon, B. (1994) *Réarrangment relatif sur un espace mesuré et applications.* Thèse de l'Université de Poitiers.

Talenti, G. (1976) Elliptic Equations and Rearrangements. *Ann. Scuola Norm. Sup. Pisa.* **3**, 697–718.

Thomson, Sir William (Lord Kelvin) (1910) Maximum and minimum energy in vortex motion. *Mathematical and Physical Papers*, vol. **4**, Cambridge University Press, 172–183.

Urbas, J. (1997) On the second boundary value problem for equations of Monge–Ampère type. *J. reine angew. Math.* **487**, 115–124.

Wolfson, J.G. (1997) Minimal Lagrangian diffeomorphisms and the Monge–Ampère equation. **46**, 335–373.

Statistical Methods in Atmospheric Dynamics: Probability Metrics and Discrepancy Measures as a means of defining Balance

Stephen Baigent and John Norbury

1 Introduction

The aim of this chapter is to introduce a probabilistic approach to defining balance for geophysical flows. Geophysical flows are split into fast and slow components where the slow component describes some 'average' or 'macroscopic' large scale evolution, whereas the fast component describes the more rapid fluctuations of particle positions on a 'microscopic' fine or very local scale. We will treat the fast variables as random variables and, in the spirit of statistical physics, we will identify fluid microstates as measure-preserving maps, and fluid macrostates as probability distributions on the set of microstates. A balanced flow will then be characterised by a particular macrostate (i.e. probability distribution), which is, in a sense to be defined later, the most likely for the observed macroscopic properties of the fluid. We will characterise semigeostrophic flow as the most likely evolution of minimum energy states consistent with the large-scale constraints of the system.

Like many other physical systems, the atmosphere has both fast and slow dynamics. The slow dynamics describe the macroscopic or averaged evolution of the air 'parcels' (say approximately 10 km by 10 km in the horizontal, by 100m in the vertical, or more), whereas the fast dynamics describes the microscopic or fine scale rapid movement of air 'particles' (of sizes about 10 – 100 m^3) that are 10^8–10^9 $\times$ smaller. To the synoptic large scale modeller these fast microscale motions are mostly irrelevant and a nuisance. First, these (often unstable) motions do not usually provide information relevant to the large-scale model; secondly they are often associated with unwanted phenomena, such as gravity waves; and thirdly they often lead to instabilities that plague numerical calculations. However, the small scale vertical movement of thermal energy is a daily process that is necessary in maintaining the large scale vertical balance represented by the hydrostatic law, and so small-scale motions may, over longer time-scales, significantly affect the large-scale dynamics.

There has been considerable research into devising models for which these microscopic features are eliminated. Typically this is achieved in two ways. First by the reduction of the full model to a lower dimensional system by projecting the original system onto a lower dimensional manifold. The quasigeostrophic [20, 12] and semigeostrophic [13, 24] models are examples of this. In the second approach, for instance nonlinear normal mode initialisation, the primitive equations are retained and the initial conditions etc. are filtered in order to eliminate fast components (see, for example, [18, 2]). The filtering process actually constitutes projection onto a manifold, which is supposed to correspond to an invariant manifold of the full model, so that by projecting initial conditions onto this manifold it is hoped that the evolution remains on this manifold and the fast components remain eliminated. Usually it is assumed that the invariant manifold is locally attracting, or at least that any small amount of noise containing a fast component does not grow.

These two approaches may be related, along the lines shown by Leith [17]: the full model projected onto various iterative approximations to the invariant slow manifold corresponds to making increasingly higher order Rossby number approximations to the dynamics. In particular the first iterate corresponds to the quasigeostrophic approximation, and defines the manifold of *geostrophic balance*, where the horizontal pressure forces balance the coriolis forces induced by the planetary rotation to the accuracy of the basic approximation. In the vertical direction we have the hydrostatic equation where the vertical pressure gradient is balanced by the buoyancy force.

Geostrophic balance is obtained from the equations for horizontal momentum in the full model by setting the Rossby number to zero. Both quasi- and semigeostrophic theories use geostrophic balance as the basic equilibrium state of the atmosphere at mid-latitudes, around which the actual dynamics fluctuates. Geostrophic and hydrostatic equilibrium can also be obtained via the Cullen, Norbury and Purser minimisation principle (CNP) [10], which defines a balanced state in terms of minimisers of an energy functional over a class of virtual displacements of air parcels that conserve horizontal angular momentum (and fluid mass) in the rotating system.

In this paper we propose a statistical approach to characterising equilibrium or balanced states. We seek to define a basic equilibrium state of the atmosphere as that state which is statistically most likely, given all the information we have, such as, for example, the system energy. We develop a stochastic description for the fast dynamics through a probability distribution on the set of all fluid particle configurations, and obtain the balanced state as a mapping from the slow to the fast variables derived from a special, 'most likely' probability distribution on the fluid configurations. We consider an *ensemble* of independently evolving replica systems, and define the equilibrium macrostate at a time t as the most likely probability distribution of system configurations for the ensemble at time t.

We explore the use of the Kullback–Leibler discrepancy measure to determine the 'distance' of one fluid macrostate from some fixed reference macrostate, namely that corresponding to uniformly rotating flow, and define a balanced state to be that macrostate which minimises the Kullback–Leibler discrepancy measure subject to the known statistics of the system. Note that this does not involve making approximations in expansions of the dynamical equations in the Rossby number, although smallness of the Rossby number is needed to motivate the separated time scales. Furthermore, the constraints may be time-dependent on the slow time scale, so that non-conservative effects such as heating and dissipation could easily be included.

We show that the most likely state of a system chosen at random from the ensemble is geostrophic balance, and this is the state (with probability one) of all systems in the ensemble when the statistical 'temperature' tends to zero, i.e. defines the *ground state* of the system. The CNP principle thus defines the system ground state, and geostrophic balance physically results from the continual dissipation of energy until the ground state is achieved. Semigeostrophic flow is an evolution through the set of ground states.

Section road map

We begin in §2 by reviewing the Boussinesq equations for 3D flow on an f-plane, rescaling and making a special transformation of coordinates to span phase space with fast and slow variables. Our aim will be to define balanced manifolds as mappings from the slow to the fast variables.

Next in section §3 we review the CNP minimisation principle and rearrangement theory, and examine how these ideas are applied to define geostrophic balance. A simple illustrative Lattice model for geostropic balance is discussed in §4, and problems with defining unique solutions are highlighted. In §5 the principle is also reformulated as the problem of finding the minimum L^2 distance (or discrepancy) between a given probability distribution and a uniform distribution, so that geostrophically-balanced flow is that flow which is closest in this metric to uniformly rotating flow, subject to the applied constraints. Next in §6 the geostrophic balance manifold is used to derive a quasi-steady state system known as the semigeostrophic equations.

Motivated by the success of statistical physics in providing a model for equilibrium thermodynamics, we briefly review in §7 the maximum entropy principle. This principle is then applied in §8 to define balance for the Boussinesq equations in 3D. A new Lattice model incorporating the MEP for defining balance is introduced in §9.

Finally, the Appendix outlines some basic convexity theory and related ideas for Banach spaces that we use in constructing mappings between our phase space variables and the original physical configuration space variables.

2 Boussinesq equations for 3D f-plane flow

The starting point for most simple models of the atmosphere is the set of primitive equations for the inviscid, adiabatic motion of an incompressible fluid in a rotating frame, the fluid occupying a fixed domain $D \subset \mathbb{R}^3$. In addition the fluid is assumed to be in hydrostatic balance, that is, subject to a gravitational force $-g\boldsymbol{z}$, where $\boldsymbol{z}$ is (measured) parallel to the axis of rotation, that causes the vertical pressure gradient. Potential temperature is taken as given on fluid particles by defining mass/density conserving coordinates $\boldsymbol{x} = (x, y, z)$ (the context indicates when $\boldsymbol{x} = \boldsymbol{x}(\boldsymbol{a}, t)$ or $\boldsymbol{x}$ is a coordinate system). The (x, y, z) are chosen such that Lebesgue measure $d\mu$ in this coordinate system is equal to the mass measure, so that the fluid density is unity. In terms of these mass coordinates, the fluid occupies a closed, bounded label set $\mathcal{L} \subset \mathbb{R}^3$ with $d\mu(\boldsymbol{a}) = d\mu(\boldsymbol{x})$ for $\boldsymbol{a} \in \mathcal{L}, \boldsymbol{x} \in D$. Let the velocity field[1] of the fluid at each label point $\boldsymbol{a} \in \mathcal{L}$ and time $t > 0$ be $\boldsymbol{u} = (u, v, w)$, the potential temperature field be θ, and the geopotential pressure field be ϕ. The material derivative d/dt is defined by the rate of change following a particle with velocity $\boldsymbol{u} = d\boldsymbol{x}/dt$ as

$$\frac{d}{dt} = \frac{\partial}{\partial t} + (\boldsymbol{u} \cdot \nabla_{\boldsymbol{x}}). \tag{2.1}$$

(Here the subscript $\boldsymbol{x}$ indicates the choice of coordinates for the gradient ∇.) With these definitions, the scaled horizontal force-acceleration equations read, for $\boldsymbol{x} \in \mathcal{L}$, where henceforth we identify the label set $\mathcal{L}$ with the initial particle locations, so that $\mathcal{L} = D$, and $t \geq 0$, with no source terms for horizontal momentum or potential temperature,

$$\frac{du}{dt} - fv = -\frac{\partial \phi}{\partial x} = -f\frac{\partial \hat{\phi}}{\partial x} \tag{2.2}$$

$$\frac{dv}{dt} + fu = -\frac{\partial \phi}{\partial y} = -f\frac{\partial \hat{\phi}}{\partial y}. \tag{2.3}$$

For simplicity we consider θ constant on particles (no heating or rainfall) so that

$$\frac{d\theta}{dt} = 0. \tag{2.4}$$

For hydrostatic balance in the vertical we have

$$\frac{g\theta}{\theta_0} = \frac{\partial \phi}{\partial z} = f\frac{\partial \hat{\phi}}{\partial z}. \tag{2.5}$$

Here we have scaled the geopotential by the coriolis parameter f to give a new potential $\hat{\phi} = \phi/f$.

[1]At present, we assume that all defined functions lie in some appropriate function space such that the indicated derivatives exist.

Finally, since $\boldsymbol{x} = (x, y, z)$ correspond to mass coordinates, we have unit fluid density and conservation of mass is represented by

$$\text{div}\,(u, v, w) = 0. \tag{2.6}$$

The function θ is measured relative to some reference value θ_0, g is the gravitational constant, and f is the coriolis rotation parameter (assumed constant).

Since $\boldsymbol{u} = d\boldsymbol{x}/dt$ and f is constant, we can rewrite the equations for horizontal momentum as

$$\frac{d}{dt}(u - fy) = -\frac{\partial \phi}{\partial x} = -f\frac{\partial \hat{\phi}}{\partial x} \tag{2.7}$$

$$\frac{d}{dt}(v + fx) = -\frac{\partial \phi}{\partial y} = -f\frac{\partial \hat{\phi}}{\partial y}. \tag{2.8}$$

Now defining new variables X, Y, Z by

$$X = x + \frac{1}{f}v,\ Y = y - \frac{1}{f}u,\ Z = \frac{g\theta}{f^2\theta_0} \tag{2.9}$$

allows us to rewrite (2.7), (2.8), (2.4) together with (2.9) as

$$\frac{dX}{dt} = -\frac{\partial \hat{\phi}}{\partial y}(\boldsymbol{x}) \tag{2.10}$$

$$\frac{dY}{dt} = \frac{\partial \hat{\phi}}{\partial x}(\boldsymbol{x}) \tag{2.11}$$

$$\frac{dZ}{dt} = 0 \tag{2.12}$$

$$\frac{dx}{dt} = f(y - Y(\boldsymbol{x})) \tag{2.13}$$

$$\frac{dy}{dt} = f(X(\boldsymbol{x}) - x). \tag{2.14}$$

The vertical motion is implicit through relations (2.5) and (2.6). Since, in practice with our North Atlantic horizontal space scale and our hourly timescale, $1/f$ is small, (2.10)–(2.14) is a singular perturbation problem in which X, Y are the *slow* variables and x, y are the *fast* variables which are formally slaved to the X, Y variables in the $1/f \to 0$ limit. From (2.13) and (2.14) the horizontal velocity components $(dx/dt, dy/dt)$ are large unless $\|(X, Y) - (x, y)\| = O(1/f)$. Thus if initially (X, Y) and (x, y) are not close, the air parcels move very rapidly, on a time scale $O(1/f)$. On the other hand, the X, Y variables appear to move on an $O(1)$ time scale.

Hence we seek balanced flows where the X, Y components will evolve with small deviations from a mean slow evolution, whereas the x, y components will vary around this slow evolution with faster fluctuations. Singular perturbation problems can be treated with approaches based upon the persistence of invariant manifolds [28], or time averaging methods of perturbation theory [27]. Usually, modellers [13] define a basic equilibrium state by setting $\epsilon \equiv 1/f = 0$ in (2.2) and (2.3) to obtain the zeroth order (in ϵ) approximation to the horizontal velocity field:

$$u_0 = -\frac{\partial \hat{\phi}_0}{\partial y}, v_0 = \frac{\partial \hat{\phi}_0}{\partial x}. \tag{2.15}$$

These are known as the horizontal geostrophic velocities. They are the horizontal velocities for which the horizontal pressure gradient matches the coriolis force on air parcels. Note also that the geopotential $\hat{\phi}_0$ in (2.15) is the zeroth order approximation of $\hat{\phi}$ in a series expansion in powers of ϵ. Combining (2.9) and (2.15) we obtain

$$X = x + \frac{1}{f}\frac{\partial \hat{\phi}_0}{\partial x}, \ Y = y + \frac{1}{f}\frac{\partial \hat{\phi}_0}{\partial y}, \tag{2.16}$$

together with

$$Z = \frac{1}{f}\frac{\partial \hat{\phi}_0}{\partial z}, \tag{2.17}$$

which follows from the zeroth order terms of (2.5). Equations (2.16) and (2.17) can be written concisely using the new potential $P_0(\boldsymbol{x}) = \frac{1}{2}(x^2 + y^2) + \frac{1}{f}\hat{\phi}_0(\boldsymbol{x})$ as

$$\boldsymbol{X}(\boldsymbol{x}) = \nabla_{\boldsymbol{x}} P_0(\boldsymbol{x}). \tag{2.18}$$

For $\epsilon = 0$ this mapping $\nabla P_0 : D \to \mathbb{R}^3$ defines an invariant manifold for equations (2.10)–(2.14) defined by $X = x, Y = y$. Note that $\partial u_0/\partial x + \partial v_0/\partial y = 0$, so that no vertical motion ($\partial w_0/\partial z = 0$ implies $w_0 = 0$) is required for this invariant manifold, and $Z \equiv 0$ is consistent with equation (2.12). However, a leading order stratified model could be allowed by having $f^{-1}\partial\hat{\phi}/\partial z$ tend to $Z(z)$ as $\epsilon = f^{-1} \to 0$; if the basic motion is horizontal then again $dZ/dt \equiv 0$. To establish the persistence or not of this invariant manifold for small $\epsilon = f^{-1} > 0$, one could resort to theories such as Hamiltonian perturbation theory. These approaches are not pursued here; instead we estimate $\boldsymbol{x}$ (treated as a random variable) from the known statistics, such as the conserved fluid energy, to establish a nearby model with an invariant manifold.

Note that the energy of the full flow (2.2)–(2.6) is conserved. Forming the inner product of (2.2), (2.3) and (2.5) with the velocity field $\boldsymbol{u} = (u, v, w)$ we have

$$\left(u\frac{du}{dt} + v\frac{dv}{dt}\right) + \boldsymbol{u}\cdot(-fv, fu, -g\theta/\theta_0) \quad = \quad -\boldsymbol{u}\cdot\nabla\phi \tag{2.19}$$

$$\frac{1}{2}\frac{d}{dt}\left(u^2 + v^2\right) - w\frac{g\theta}{\theta_0} \quad \overset{(A)}{=} \quad -\text{div}\,(\boldsymbol{u}\phi) \tag{2.20}$$

$$\frac{d}{dt}\left\{\frac{1}{2}\left(u^2 + v^2\right) - \frac{gz\theta}{\theta_0}\right\} \quad \overset{(B)}{=} \quad -\text{div}\,(\boldsymbol{u}\phi), \tag{2.21}$$

where (A) follows from the incompressibility condition (2.6) and (B) from (2.4). Integrating this last expression over D and using the zero normal velocity component boundary condition on ∂D yields

$$E = \int_D \left\{ \frac{1}{2}\left(u^2 + v^2\right) - \frac{gz\theta}{\theta_0} \right\} d\mu(\boldsymbol{x}) = \text{constant.} \tag{2.22}$$

In the terms of the transformed coordinates (2.9) and the Lagrangian fluid labels $\boldsymbol{a}$ this energy becomes (using $d\mu(\boldsymbol{x}) = d\mu(\boldsymbol{a})$)

$$E = f^2 \int_{\mathcal{L}} \left\{ \frac{1}{2}\left((X-x)^2 + (Y-y)^2\right) - zZ \right\} d\mu(\boldsymbol{a}). \tag{2.23}$$

For the purposes of the next section on the CNP minimisation principle, we note that the appended integral

$$E' = f^2 \int_{\mathcal{L}} \left\{ \frac{1}{2}\left((X-x)^2 + (Y-y)^2 + (Z-z)^2\right) \right\} d\mu(\boldsymbol{a}) \tag{2.24}$$

is also conserved by the flow. This follows because (i) Z is materially transported by particles from (2.4) and (2.17), and (ii) the incompressible fluid is confined to a fixed domain, so that the integrated Z^2 and z^2 terms in (2.24) are also constants during the motion, that is the integrals are conserved by the flow.

3 The CNP minimisation principle and rearrangements

The CNP minimisation procedure [9, 10, 11] defines geostrophic balance and semigeostrophic flow in terms of minima of the fluid energy (2.24) over virtual displacements of fluid particles that *conserve the particle horizontal momenta.*

When CNP was first introduced in [10], it was shown that for smooth mass-conserving variations of fluid particles (which are equivalent to smooth momentum-conserving virtual displacements) the extrema satisfy $\boldsymbol{X} = \nabla_{\boldsymbol{x}} P_0(\boldsymbol{x})$, where P_0 is some smooth potential function, thus identifying, via (2.18), an extreme energy state with geostrophic balance. To ensure stability, the minimum energy state was chosen, and for such states P_0 is convex. The minimum energy configuration was used to define the semigeostrophic equations in geostrophic momentum coordinates:

$$\frac{dX}{dt} = f\left(\frac{\partial R_0}{\partial Y} - Y\right) \tag{3.1}$$

$$\frac{dY}{dt} = f\left(X - \frac{\partial R_0}{\partial X}\right) \tag{3.2}$$

$$\frac{dZ}{dt} = 0, \tag{3.3}$$

where $R_0(\boldsymbol{X}) := \boldsymbol{X}\cdot\nabla P_0^{-1}(\boldsymbol{X}) - P_0(\nabla P_0^{-1}(\boldsymbol{X}))$ is the Fenchel conjugate of P_0 (see also A.4 in the Appendix for a more general definition of the Fenchel conjugate). On constant Z surfaces, the semigeostrophic flow (3.1)–(3.3) advects the semigeostrophic potential vorticity (see, for example, [13])

$$\rho(\boldsymbol{X}) := \det D^2 R_0(\boldsymbol{X}). \tag{3.4}$$

To see this, we note that, since $\partial(x,y,z)/\partial(a,b,c) = 1$ (which follows from (2.6)), the Eulerian expression for PV

$$\rho(\boldsymbol{X}) = \det D^2 R_0(\boldsymbol{X}) = \frac{\partial(X,Y,Z)}{\partial(x,y,z)} \tag{3.5}$$

can be given in Lagrangian terms, on a particle labelled $\boldsymbol{a}\in D$ as

$$\rho(\boldsymbol{a},t) = \frac{\partial(X(\boldsymbol{a},t),Y(\boldsymbol{a},t),Z(\boldsymbol{a},t))}{\partial(a,b,c)}. \tag{3.6}$$

To see that this is conserved following the particle, we first note from (2.12) that $Z(\boldsymbol{a},t) = Z(\boldsymbol{a},0)$ for all $t\geq 0$. Therefore we have

$$\begin{aligned}
\frac{\partial(X(\boldsymbol{a},t),Y(\boldsymbol{a},t),Z(\boldsymbol{a},t))}{\partial(a,b,c)} &= \frac{\partial(X(\boldsymbol{a},t),Y(\boldsymbol{a},t),Z(\boldsymbol{a},t))}{\partial(X(\boldsymbol{a},0),Y(\boldsymbol{a},0),Z(\boldsymbol{a},0))} \\
&\quad\times\frac{\partial(X(\boldsymbol{a},0),Y(\boldsymbol{a},0),Z(\boldsymbol{a},0))}{\partial(a,b,c)} \\
&= \frac{\partial(X(\boldsymbol{a},t),Y(\boldsymbol{a},t),Z(\boldsymbol{a},0))}{\partial(X(\boldsymbol{a},0),Y(\boldsymbol{a},0),Z(\boldsymbol{a},0))}\times\rho(\boldsymbol{a},0) \\
&= \frac{\partial(X(\boldsymbol{a},t),Y(\boldsymbol{a},t))}{\partial(X(\boldsymbol{a},0),Y(\boldsymbol{a},0))}\times\rho(\boldsymbol{a},0) \\
&= \rho(\boldsymbol{a},0).
\end{aligned}$$

The last line follows from the fact that

$$r(t) = \partial(X(\boldsymbol{a},t),Y(\boldsymbol{a},t))/\partial(X(\boldsymbol{a},0),Y(\boldsymbol{a},0)) = 1,$$

which follows from $r(0)=1$ and

$$\begin{aligned}
\frac{d}{dt}\frac{\partial(X(\boldsymbol{a},t),Y(\boldsymbol{a},t))}{\partial(X(\boldsymbol{a},0),Y(\boldsymbol{a},0))} &= \frac{\partial(\dot X(\boldsymbol{a},t),Y(\boldsymbol{a},t))}{\partial(X(\boldsymbol{a},0),Y(\boldsymbol{a},0))} + \frac{\partial(X(\boldsymbol{a},t),\dot Y(\boldsymbol{a},t))}{\partial(X(\boldsymbol{a},0),Y(\boldsymbol{a},0))} \\
&\stackrel{A}{=} \frac{\partial(\dot X(\boldsymbol{X},t),Y(\boldsymbol{X},t))}{\partial(X,Y)} + \frac{\partial(X(\boldsymbol{X},t),\dot Y(\boldsymbol{X},t))}{\partial(X,Y)} \\
&= \frac{\partial\dot X}{\partial X} + \frac{\partial\dot Y}{\partial Y} \\
&= f\left(\frac{\partial^2 R_0}{\partial X\partial Y} - \frac{\partial^2 R_0}{\partial Y\partial X}\right) = 0.
\end{aligned}$$

Here step 'A' follows under the assumption that $\boldsymbol{a} \mapsto \boldsymbol{X}(\boldsymbol{a}, t)$ is invertible for each t. The conservation of vorticity represents the consistency of the Eulerian and Lagrangian descriptions of this field of fluid mechanics.

In order to extend semigeostrophic theory to more realistic flows and incorporate frontogenesis, the set of variations is enlarged to include all measure-preserving maps of the fluid labels onto themselves, whereby the CNP minimisation becomes:

CNP Principle

Given $\boldsymbol{X}(\cdot) : \mathcal{L} \to \mathbb{R}^3$ find

$$\inf_{\boldsymbol{s} \in \mathcal{S}} \int_{\mathcal{L}} \frac{1}{2} |\boldsymbol{X}(\boldsymbol{a}) - \boldsymbol{s}(\boldsymbol{a})|^2 \, d\mu(\boldsymbol{a}), \tag{3.7}$$

where $\mathcal{S}$ is the set of all measure-preserving transformations of $\mathcal{L}$ into itself, i.e.

$$\mathcal{S} = \{ s : \mathcal{L} \to \mathcal{L} \; : \; \mu(A) = \mu(s^{-1}A) \text{ for all Borel } A \subset \mathcal{L} \}.$$

For this extended problem the convex minimisers (when they exist) need not be smooth, thus allowing for the formation of atmospheric fronts.

In [11] Cullen and Purser treat (3.7) for piecewise constant momenta X, Y and potential temperature θ in the two-dimensional deformation model [8], for which there is a single horizontal coordinate plus the vertical coordinate. $\mathcal{L}$ is now an open convex subset of $\mathbb{R}^2$ and $\{U_i\}_{i=1}^n$ a collection of disjoint open convex sets such that $\cup_{i=1}^n \bar{U}_i = \bar{\mathcal{L}}$.

For a given piecewise constant vector-valued function $\boldsymbol{X}^0 : \mathcal{L} \to \mathbb{R}^2$ satisfying

$$\boldsymbol{X}^0(\boldsymbol{x}) = \begin{cases} \boldsymbol{X}_i & \boldsymbol{x} \in U_i \\ \boldsymbol{0} & \text{otherwise,} \end{cases} \tag{3.8}$$

Cullen and Purser defined $\boldsymbol{X} : \mathcal{L} \to \mathbb{R}^2$ to be a *rearrangement* of $\boldsymbol{X}^0$ if

$$\mu\{\boldsymbol{x} \mid \boldsymbol{X}(\boldsymbol{x}) = \boldsymbol{X}_i\} = \mu\{\boldsymbol{x} \mid \boldsymbol{X}^0(\boldsymbol{x}) = \boldsymbol{X}_i\}, \tag{3.9}$$

and showed that there exists a unique (continuous but not necessarily smooth) convex function $\phi : \mathcal{L} \to \mathbb{R}$ such that the closure of $\nabla\phi(\mathcal{L})$ is a rearrangement of $\boldsymbol{X}^0$. To do this, they defined, for a set of arbitrary real numbers $\{h_i\}_{i=1}^n$, a piecewise linear convex function

$$\phi(\boldsymbol{x}) = \max_i \left\{ \boldsymbol{x}^T \boldsymbol{X}_i - h_i \right\}. \tag{3.10}$$

The function ϕ defines a convex polyhedron with $m(\leq n)$ faces embedded in $\mathbb{R}^3$. The m faces have gradients $\{\boldsymbol{X}_{i_k}\}_{k=1}^m$ for some set of integers $i_k \in [1, n]$, and the faces project onto open convex sets E_{i_k} in $\mathcal{L}$ such that $\cup_{k=1}^m \bar{E}_{i_k} = \bar{D}$. Cullen and Purser showed that there is a unique set of h_i (up to an additive constant) such that $m = n$ and $\mu\{E_{\sigma(i)}\} = \mu\{U_i\}$ for some permutation $\sigma \in S_n$, thus defining the optimal solution (3.10).

4 A lattice model for geostrophic balance

A Lattice problem is studied in Benamou [6] and Baigent [4], whereby $\mathcal{L}$ was approximated by a fixed lattice of points $\Gamma = \{\boldsymbol{a}_1, \ldots, \boldsymbol{a}_n\} \subset \mathbb{R}^3$, and (3.7) replaced by

$$\inf_{\sigma \in S_n} \sum_{i=1}^{n} \frac{1}{2} |\boldsymbol{X}_i - \boldsymbol{a}_{\sigma(i)}|^2, \tag{4.1}$$

where S_n is the permutation group of order n. Clearly, the infimum in (4.1) is attained, so 'inf' can be replaced by 'min'. An n-square matrix $K = ((k_{ij}))$ is *doubly stochastic* when each $k_{ij} \geq 0$ and $\sum_{i=1}^{n} k_{ij} = 1 = \sum_{j=1}^{n} k_{ij}$. According to Birkhoff's theorem, the $n!$ n-square *permutation* matrices $P_\sigma = ((\delta_{i\,\sigma(j)}))$ $(\sigma \in S_n)$ are the *extreme points* of the set of doubly stochastic matrices, and any doubly stochastic matrix K can be written as a convex sum of permutation matrices P_σ, say

$$K = \sum_{\sigma \in S_n} p_\sigma P_\sigma, \tag{4.2}$$

where $0 \leq p_\sigma \leq 1$ for $\sigma \in S_n$ satisfy $\sum_{\sigma \in S_n} p_\sigma = 1$. Hence, the p_σ define a probability distribution over the set of permutations $\sigma \in S_n$. This decomposition of doubly stochastic matrices into a *convex* sum of permutation matrices is central to our statistical approach, since it defines a probability distribution on the set of *system configurations* $\sigma \in S_n$, and therefore furnishes an *ensemble* interpretation. However it is important to note that, while this decomposition holds true when $n \to \infty$, it may break down in the *uncountable* continuum limit [26], since there are extreme points that do not correspond to nonsingular measure-preserving maps.

Returning to (4.1), we now follow [4] and convert (4.1) to a Linear Program. Let $c_{ij} = \frac{1}{2}|\boldsymbol{X}_i - \boldsymbol{a}_j|^2$. Then since any extreme value of a linear function over a closed convex subset of $\mathbb{R}^{n^2}$ occurs at an extreme point (Theorem 2, [29, p 3]), we may *relax* (4.1) to

$$\begin{aligned} &\text{minimise} &&: \; f(k) = \sum_{i,j=1}^{n} k_{ij} c_{ij} \\ &\text{subject to} &&: \; k \in F = \{k \mid \sum_{i=1}^{n} k_{ij} = 1 = \sum_{j=1}^{n} k_{ij},\, 0 \leq k_{ij} \leq 1\}; \end{aligned} \tag{4.3}$$

so explicitly,

$$\min_{\sigma \in S_n} \sum_{i=1}^{n} \frac{1}{2} |\boldsymbol{X}_i - \boldsymbol{a}_{\sigma(i)}|^2 = \min_{k \in \mathcal{D}} \sum_{i,j=1}^{n} \frac{1}{2} |\boldsymbol{X}_i - \boldsymbol{a}_j|^2 k_{ij} \tag{4.4}$$

where $\mathcal{D}$ is the set of n-square doubly stochastic matrices. In the field of Linear Programming (4.3) is known as the *Optimal Assignment Problem.* The corresponding dual canonical maximisation problem can be constructed following

Trustrum [29, pp 33–34] to obtain

$$\begin{aligned} \text{maximise} \quad &: \quad g(u,v) = \sum_{i=1}^{n} u_i + \sum_{j=1}^{n} v_j \\ \text{subject to} \quad &: \quad (u,v) \in G = \{(u_1,\dots,u_n,v_1,\dots,v_n) \in R^{2n} \mid u_i + v_j \le c_{ij}\}. \end{aligned} \tag{4.5}$$

The u_i, v_j are the Lagrange multipliers for the primal program. The Fundamental Duality Theorem (Theorem 3, [29, p 19]) states that if the primal and dual programs have feasible solutions then they each have optimal solutions and their optimal values are the same. This is easy, as the set of $n!$ real (bounded) sums $\{\sum_{i=1}^{n} c_{i\sigma(i)} \mid \sigma \in S_n\}$ clearly has a largest element and therefore the primal has an optimal solution k corresponding to a permutation matrix P^σ (although this solution may not be unique). A feasible solution can easily be constructed for the Dual as follows. Let $\{u_i\}_{i=1}^{n}$ be any set of real numbers. Define $v_j = \min_i\{c_{ij} - u_i\}$. Then the u_i, v_j are real numbers satisfying the constraint $u_i + v_j \le c_{ij}$ for all i and j, and therefore are feasible solutions. Hence we have shown that the optimisation problem (4.1) has an optimal solution. Next we wish to establish the properties of the optimal solution.

Trustrum's canonical equilibrium theorem (Theorem 5,[29, p 23]) then shows that the optimal solution to the assignment problem satisfies, upon writing $\boldsymbol{x}_i = \boldsymbol{a}_{\sigma^*(i)}$, where σ^* is the optimal ordering,

$$\begin{aligned} u_i + v_i &= \tfrac{1}{2}|\boldsymbol{X}_i - \boldsymbol{x}_i|^2 \quad (i = 1,2,\dots,n), \\ u_i + v_j &\le \tfrac{1}{2}|\boldsymbol{X}_i - \boldsymbol{x}_j|^2 \quad i \ne j. \end{aligned} \tag{4.6}$$

It is easy to see, after choosing $j = \sigma(i)$ and then comparing equivalent sums, that in this ordering

$$\sum_{i=1}^{n} \boldsymbol{X}_i^T \boldsymbol{x}_i \ge \sum_{i=1}^{n} \boldsymbol{X}_i^T \boldsymbol{x}_{\sigma(i)}, \tag{4.7}$$

for all $\sigma \in S_n$ not equal to the identity permutation. Hence the Linear Program (4.3) and its dual (4.5) are equivalent to the optimisation problem (4.1).

Now define two functions $\phi, \psi : \mathbb{R}^3 \to \mathbb{R}$ by

$$\phi(\boldsymbol{x}) = \max_i\{\boldsymbol{X}_i^T \boldsymbol{x} - v_i - \frac{1}{2}|\boldsymbol{X}_i|^2\}, \;\; \psi(\boldsymbol{X}) = \max_i\{\boldsymbol{X}^T \boldsymbol{x}_i - u_i - \frac{1}{2}|\boldsymbol{x}_i|^2\}, \tag{4.8}$$

so that ϕ and ψ are a Legendre transform pair and $\phi(\boldsymbol{x}_i) = u_i$, $\psi(\boldsymbol{X}_i) = v_i$. Geometrically, ϕ (and similarly for ψ) represents a convex polyhedron constructed by taking the upper envelope of a set of n planes whose gradients are $\boldsymbol{X}_i$ and whose intercepts with the vertical coordinate axis are $-v_i - \frac{1}{2}|\boldsymbol{X}|^2$. By construction, ϕ, ψ are continuous, piecewise linear, convex functions, differentiable almost everywhere, with

$$\boldsymbol{X} = \frac{\partial \phi}{\partial \boldsymbol{x}}, \;\; \boldsymbol{x} = \frac{\partial \psi}{\partial \boldsymbol{X}} \; (a.e.). \tag{4.9}$$

Hence, we have shown that an optimal solution $\sigma^* \in S_n$ for (4.1) exists, and that $\boldsymbol{x}_i = \boldsymbol{a}_{\sigma^*(i)} = \nabla\psi(\boldsymbol{X}_i)$ where $\psi : \mathbb{R}^3 \to \mathbb{R}$ is a bounded convex function, differentiable almost everywhere.

Now let us consider the full problem (3.7). Baigent [3] generalised the definition (3.9) of a vector-valued rearrangement of $\boldsymbol{X}^0 : \Omega \to \mathbb{R}$ to be any everywhere-bounded vector-valued function $\boldsymbol{X}$ satisfying

$$\mu\{\boldsymbol{x} \mid \boldsymbol{X}(\boldsymbol{x}) \geq \boldsymbol{c}\} = \mu\{\boldsymbol{x} \mid \boldsymbol{X}^0(\boldsymbol{x}) \geq \boldsymbol{c}\} \qquad \forall \boldsymbol{c} \in \mathbb{R}^n, \tag{4.10}$$

and showed that each vector-valued function $\boldsymbol{X}$ lies in an equivalence class of rearrangements generated by a cyclically monotone rearrangement $\boldsymbol{X}^*$, deducing that $\boldsymbol{X}^*$ could be written as the subdifferential of a proper lower semicontinuous convex function. Brenier [7] also gave a number of equivalent definitions for the rearrangement of a vector-valued function and established a beautiful result (the Polar Factorisation Theorem [7]) showing how any vector-valued function can be factored into a measure-preserving map composed with the subgradient of a convex function.

5 CNP and probability metrics

The definition (4.10) of a rearrangement suggests another, more direct link to probability theory: Two vector-valued functions $\boldsymbol{U}, \boldsymbol{V}$ are in the same rearrangement class if and only if, when viewed as vector random variables, they have the same probability distribution function, or in other words they are *identically distributed.* Not surprisingly, the CNP minimisation principle appears in the Statistics literature as we now discuss.

In [15] Knott and Smith considered the following problem:

> given two probability distributions F and G for random variables $\boldsymbol{U}$ and $\boldsymbol{V}$ which take values in $\mathbb{R}^m$, what is the joint distribution function H for $(\boldsymbol{U}, \boldsymbol{V})$ taking values in $\mathbb{R}^{2m}$ which minimises the expected squared Euclidean distance $E_H(|\boldsymbol{U} - \boldsymbol{V}|^2)$?

Thus given a random variable $\boldsymbol{U}$ taking values in $\mathbb{R}^m$ with distribution function F and a second distribution function G, they sought a smooth mapping $S : \mathbb{R}^m \to \mathbb{R}^m$ defined by $\boldsymbol{U} \mapsto s(\boldsymbol{U})$ such that the random variable $s(\boldsymbol{U})$ has distribution function G, and which minimises the expectation $E_F(|\boldsymbol{U} - s(\boldsymbol{U})|^2)$. Knott and Smith showed that a smooth minimiser, *when it exists*, must have the form $s^* = \nabla\phi$ for some smooth convex function $\phi : \mathbb{R}^m \to \mathbb{R}$. Knott and Smith also make the connection with a certain Monge-Ampère equation. Let the probability density functions of $\boldsymbol{U}, \boldsymbol{V}$ be f, g respectively. When $m = 2$, and the coordinates are $\boldsymbol{u} = (u_1, u_2)$, the function $\psi : \mathbb{R}^2 \to \mathbb{R}$ defined by $\psi(\boldsymbol{u}) = \frac{1}{2}|\boldsymbol{u}|^2 - \phi(\boldsymbol{u})$, where $\nabla\phi$ is the minimiser, satisfies the strictly elliptic

Monge-Ampère equation

$$\psi_{u_1u_1}\psi_{u_2u_2} - \psi^2_{u_1u_2} = \frac{dF(u_1, u_2)}{dG(\psi_{u_1}, \psi_{u_2})} = \frac{f(u_1, u_2)}{g(\psi_{u_1}, \psi_{u_2})} > 0. \tag{5.1}$$

Rüschendorf and Rachev [25] generalised the ideas of Knott and Smith to L^p spaces by using a special case of the Monge-Kantorovich problem (MKP) (for a review, see [23]). Here, given two probability distributions P and Q on $\mathbb{R}^k$, let $M(P,Q)$ be the set of all joint probability distributions π on $\mathbb{R}^k \times \mathbb{R}^k$ whose fixed marginals are P and Q, so that

$$dP(\boldsymbol{X}) = \int_{x\in\mathbb{R}^k} d\pi(\boldsymbol{X}, \boldsymbol{x}),\ dQ(\boldsymbol{x}) = \int_{X\in\mathbb{R}^k} d\pi(\boldsymbol{X}, \boldsymbol{x}). \tag{5.2}$$

One interpretation of this definition is that P and Q represent the initial and final distribution of mass and $M(P,Q)$ all possible transfers of mass in P to Q. In particular $M(\mu,\mu)$ (with μ Lebesgue measure) corresponds to the set of doubly stochastic probability measures.

Define the following $\mathcal{L}^2$ metric on $M(P,Q)$:

$$\mathcal{L}^2_\pi(P,Q) = \int_{\mathbb{R}^k\times\mathbb{R}^k} |\boldsymbol{X} - \boldsymbol{x}|^2\, d\pi(\boldsymbol{X}, \boldsymbol{x}), \tag{5.3}$$

and the corresponding *minimal probability metric*, also known as the L^2 Wasserstein distance,

$$\sigma(P,Q) = \inf_{\pi\in M(P,Q)} \mathcal{L}^2_\pi(P,Q). \tag{5.4}$$

The problem can thus be interpreted as carrying out the mass transfer in such a way as to minimise the 'total cost' $\mathcal{L}^2(P,Q)$. Now suppose that $\boldsymbol{X}$ and $\boldsymbol{x}$ are random variables with respective probability distributions P and Q, i.e. $\boldsymbol{X} \sim P$, $\boldsymbol{x} \sim Q$; then according to [25] we have:

Theorem 1 *If* $\int_{\mathbb{R}^k} |\boldsymbol{X}|^2 dP(\boldsymbol{X}) < \infty$, $\int_{\mathbb{R}^k} |\boldsymbol{x}|^2 dQ(\boldsymbol{x}) < \infty$ *then*

1. *There exists a solution* π *of (5.4), or equivalently, there exist random variables* $\boldsymbol{X} \sim P$ *and* $\boldsymbol{x} \sim Q$ *with* $\sigma(P,Q) = \int_{\mathbb{R}^k\times\mathbb{R}^k} |\boldsymbol{X} - \boldsymbol{x}|^2\, d\pi(\boldsymbol{X}, \boldsymbol{x})$.

2. *Let* $\boldsymbol{X} \sim P$ *and* $\boldsymbol{x} \sim Q$; *then* $(\boldsymbol{X}, \boldsymbol{x})$ *is a solution of (5.4) if and only if* $\boldsymbol{x} \in \partial\Phi^*(\boldsymbol{X})$, *where* $\partial\Phi^*$ *denotes the subdifferential of a proper convex function* $\Phi^*(\boldsymbol{X})$.

At the 'total cost' minimum, we have

$$\inf_{s \text{ has distribution } Q} \int_{\mathbb{R}^k} |\boldsymbol{X} - s(\boldsymbol{X})|^2\, dP(\boldsymbol{X}) = \tag{5.5}$$
$$\inf_{\pi\in M(P,Q)} \int_{\mathbb{R}^k\times\mathbb{R}^k} |\boldsymbol{X} - \boldsymbol{x}|^2\, d\pi(\boldsymbol{X}, \boldsymbol{x}).$$

Further, with $\Phi(\boldsymbol{x})$ for fixed $\boldsymbol{x} \in D$, the Fenchel conjugate of $\Phi^*(\boldsymbol{X})$,

$$\sigma(P,Q) = \int_{\mathbb{R}^k} |\boldsymbol{X} - \partial\Phi^*(\boldsymbol{X})|^2 \, dP(\boldsymbol{X}) = \int_{\mathbb{R}^k} |\boldsymbol{x} - \partial\Phi(\boldsymbol{x})|^2 \, dQ(\boldsymbol{x}). \quad (5.6)$$

For example, the choices $dP(\boldsymbol{X}) = \frac{1}{\mu(\Omega)} \sum_{i=1}^n \mu(U_i)\delta(\boldsymbol{X}-\boldsymbol{X}_i)\, d\boldsymbol{X}$ and $dQ(\boldsymbol{x}) = d\boldsymbol{x}/\mu(\mathcal{L})$ (where μ is Lebesgue measure) lead to the piecewise constant momentum model studied by Cullen and Purser. Whereas the choice $dP(\boldsymbol{X}) = \frac{1}{n}\sum_i \delta(\boldsymbol{X}-\boldsymbol{X}_i)\, d\boldsymbol{X}$ and $dQ(\boldsymbol{x}) = \frac{1}{n}\sum_i \delta(\boldsymbol{x}-\boldsymbol{x}_i)\, d\boldsymbol{x}$ leads to the Lattice model and hence the linear program (4.3).

It can also be shown [15, 7] that at the minimum Φ^* satisfies the elliptic Monge-Ampère equation (for D^2 the matrix operator of second partial derivatives w.r.t. $\boldsymbol{X}$):

$$\det D^2\Phi^*(\boldsymbol{X}) = \frac{p(\boldsymbol{X})}{q(\partial\Phi^*(\boldsymbol{X}))} > 0 \;\; (a.e.), \quad (5.7)$$

where $dQ(\boldsymbol{x}) = q(\boldsymbol{x})\, d\boldsymbol{x}$ and $dP(\boldsymbol{X}) = p(\boldsymbol{X})\, d\boldsymbol{X}$.

6 Applications to semigeostrophic theory

As shown in [7], by an appropriate choice of the mass distributions P and Q these results can be applied to solve the CNP minimisation problem (3.7). By (5.6), this is the same as the relaxed problem of finding the minimum distance over all doubly stochastic measures $M(\mu,\mu)$:

$$\inf_{s\in\mathcal{S}} \int_{\mathbb{R}^k} \frac{1}{2}|\boldsymbol{X}(\boldsymbol{a}) - s(\boldsymbol{a})|^2 \, d\mu(\boldsymbol{a}) = \inf_{\gamma\in M(\mu,\mu)} \int_{D^2} \frac{1}{2}|\boldsymbol{X}(\boldsymbol{a}) - \boldsymbol{a}'|^2 \, d\gamma(\boldsymbol{a},\boldsymbol{a}'), \quad (6.1)$$

with $dQ(\boldsymbol{a}) = d\mu(\boldsymbol{a})$ and $P : \mathbb{R}^3 \to \mathbb{R}$ satisfying, for all continuous f,

$$\int_{\mathbb{R}^3} f(\boldsymbol{X}) \, dP(\boldsymbol{X}) = \int_D f(\boldsymbol{X}(\boldsymbol{a})) \, d\mu(\boldsymbol{a}). \quad (6.2)$$

Equation (6.1) compares with (4.4) in the Lattice model. By Theorem 1, there exists a proper convex function Φ^* such that $\boldsymbol{x} \in \partial\Phi^*(\boldsymbol{X})$ and for which

$$\int_{D^2} \frac{1}{2}|\boldsymbol{X}(\boldsymbol{a}) - \boldsymbol{x}(\boldsymbol{a})|^2 \, d\mu(\boldsymbol{a}) = \int_D \frac{1}{2}|\boldsymbol{X}(\boldsymbol{a}) - (\partial\Phi^* \circ \boldsymbol{X})(\boldsymbol{a})|^2 \, d\mu(\boldsymbol{a}). \quad (6.3)$$

Furthermore, there exists a proper convex function $\Phi(\boldsymbol{x})$, the Fenchel conjugate of $\Phi^*(\boldsymbol{X})$, such that

$$\boldsymbol{X} \in \partial\Phi(\boldsymbol{x}). \quad (6.4)$$

Defining $\hat{\phi_0} = f(\Phi - \frac{1}{2}(x^2 + y^2))$ then gives, using the relations (2.9),

$$u_0(\boldsymbol{x}) = f(y - Y(\boldsymbol{x})) = f\left(y - \frac{\partial\Phi}{\partial y}\right),$$

and

$$v_0(\boldsymbol{x}) = f(X(\boldsymbol{x}) - x) = f\left(\frac{\partial \Phi}{\partial x} - x\right),$$

the equations for balance (a.e.):

$$u_0 = -\frac{\partial \hat{\phi}_0}{\partial y},\ v_0 = \frac{\partial \hat{\phi}_0}{\partial x},\ \frac{g\theta}{\theta_0} = f\frac{\partial \hat{\phi}_0}{\partial z}, \tag{6.5}$$

which agree with horizontal geostrophic balance (2.15) and the zeroth order approximation to the hydrostatic balance relation (2.5).

On the balance (slow) manifold $\boldsymbol{x} = \nabla\Phi^*(\boldsymbol{X})$ (a.e.), and

$$\begin{gathered} \frac{\partial \hat{\phi}_0}{\partial x}(\boldsymbol{x}(\boldsymbol{X})) = f\left(X - \frac{\partial \Phi^*}{\partial X}\right),\ \frac{\partial \hat{\phi}_0}{\partial y}(\boldsymbol{x}(\boldsymbol{X})) = f\left(Y - \frac{\partial \Phi^*}{\partial Y}\right), \\ \frac{\partial \hat{\phi}_0}{\partial z}(\boldsymbol{x}(\boldsymbol{X})) = fZ. \end{gathered} \tag{6.6}$$

We now return to the original 3D Boussinesq equations (2.10)–(2.14), and write down the *quasi-steady state* approximation for these equations, where the fast variable $\boldsymbol{x}$ is assumed to evolve sufficiently rapidly to steady state that it can effectively be replaced by its steady state value on the balance manifold. This filters out the fast dynamics and leaves approximate equations for the slow dynamics obtained by projecting the full dynamics onto the balance manifold (see any text on singular perturbation theory, for example, [19]). Hence we substitute the first two equations in (6.6) into equations (2.10) and (2.11) to obtain equations for the evolution of the horizontal momentum components in 3D semigeostrophic flow:

$$\dot{X} = f(y(\boldsymbol{X}) - Y),\ \dot{Y} = f(X - x(\boldsymbol{X})), \tag{6.7}$$

to which we add $\dot{Z} = 0$. By construction, these equations evolve on the manifold of geostrophic balance.

At the minimum of (6.1) we have, from (5.7), since q is the uniform distribution,

$$p(\boldsymbol{X}) = \det D^2\Phi^*(\boldsymbol{X}) \tag{6.8}$$

which can be shown to correspond with the usual semigeostrophic potential vorticity $\rho(\boldsymbol{X})$ (see page 349).

Remark 1 Equations (5.6) and (6.8) show how to define the semi-geostrophic energy in the Eulerian description. We define $\Psi(\boldsymbol{X}) = \frac{1}{2}|\boldsymbol{X} - \partial\Phi^*(\boldsymbol{X})|^2$ so that

$$\frac{1}{2}\sigma(P,\mu)(\boldsymbol{X}) = \int_{\mathbb{R}^3} \Psi(\boldsymbol{X})\, dP(\boldsymbol{X}) = \int_{\mathbb{R}^3} \Psi(\boldsymbol{X})\rho(\boldsymbol{X})\, d\mu(\boldsymbol{X}),$$

which agrees with the form of the energy given in [24].

Remark 2 The above calculations show that one interpretation of the CNP minimisation principle for defining semi-geostrophic balance is to find the density ρ that is closest to a uniform probability density in the $\mathcal{L}^2$ metric. In fact, the minimal metric (5.4) can be written in the form

$$\sigma(P,Q) = \int_0^1 |P^{-1}(\alpha) - Q^{-1}(\alpha)|^2 \, d\alpha, \tag{6.9}$$

where $\boldsymbol{X} \sim P, \boldsymbol{x} \sim Q$ and $dQ = d\mu$, Lebesgue measure (see, for example, [23]).

Remark 3 Suppose a potential $U : \mathcal{L} \to \mathbb{R}$ is added to the $L^2_\pi(P,Q)$ metric, and we consider

$$\sigma_U(P,Q) = \inf_{\pi \in M(P,Q)} \int_{D^2} \{|\boldsymbol{X} - \boldsymbol{x}|^2 + U(\boldsymbol{x})\} \, d\pi(\boldsymbol{X}, \boldsymbol{x}). \tag{6.10}$$

Then the new expression to be minimised can be written as

$$\{E_\pi(|\boldsymbol{X}|^2) + E_\pi(|\boldsymbol{x}|^2) + E_\pi(U(\boldsymbol{x}))\} - 2E_\pi(\boldsymbol{x} \cdot \boldsymbol{X}),$$

where $E_\pi(f) = \int f \, d\pi$ (the expectation of f). The bracketed term is invariant under all $\pi \in M(P,Q)$, but the second term is not. Following through the proof of Theorem 1 in [25], we see that the addition of U does not alter the conclusions of Theorem 1; the Fenchel conjugation duality still holds: $\exists$ convex Ψ, s.t. $\boldsymbol{x} \in \partial\Psi(\boldsymbol{X})$ and $\boldsymbol{X} \in \partial\Psi^*(\boldsymbol{x})$.

Let us now pause to recap on progress. We relaxed the original CNP minimisation problems (3.7) and (4.1) by a mathematical device, i.e. by transforming the nonlinear CNP minimisation to a linear program, as expressed by (4.4) and (6.1). The idea was to introduce doubly stochastic matrices and doubly stochastic measures (see also [5]). How should we interpret this mathematical device physically? Is there any physical meaning to the equalities (4.4) and (6.1) that in hindsight would motivate the relaxation procedure? In the remainder of the chapter we suggest that an ensemble approach and the application of maximum entropy methods provide a framework with which (4.4) and (6.1) have a physical interpretation.

7 The maximum entropy method

Statistical physics provides a framework for calculating the probability that a system is in a particular state. An ideal gas in a 5 m^3 box contains about 10^{28} gas molecules, each of which requires 6 coordinates for description. No feasible experiment can simultaneously measure all the 6×10^{28} coordinates necessary to describe the state of the gas. However, it is possible to give a reasonable and very accurate description of the gas in thermodynamic equilibrium by way of macroscopic variables such as energy, temperature, pressure, etc.

There are several different approaches to classical statistical mechanics, although they share many mathematical similarities, and often lead to the same results. In one approach, largely due to Boltzmann, particles are classified into 'bins'. As a specific example, suppose each bin identifies particles of energy E_k, and that there are n_k particles in the kth bin, m bins and N particles in total. When particles are distributed amongst the bins according to their energies, we obtain a *macrostate* $\{n_1, \ldots, n_m\}$. A *microstate* is a particular distribution of the particles into the m bins that imagines that particles could be distinguished. Thus there are many microstates that give rise to a given macrostate (since in the macrostate the particles are imagined to be indistinguishable). The number of independent microstates for the macrostate $\{n_1, \ldots, n_m\}$ is $W = N!/(n_1! \ldots, n_m!)$. The macrostate must also satisfy

$$\sum_{k=1}^{m} n_k = N \tag{7.1}$$

$$\sum_{k=1}^{m} n_k E_k = \text{const.} N. \tag{7.2}$$

In the limit $N \to \infty$, but m *remaining fixed*, and setting $p_k = \lim_{N\to\infty} n_k/N$, then

$$S(p) = \frac{1}{N} \log W \to -\sum_{k=1}^{m} p_k \log p_k.$$

Thus S is proportional to the number of microstates realising the macrostate $\{n_1, \ldots, n_m\}$ as $N \to \infty$, now written in terms of the fractions of particles (the p_k's) in each bin. Boltzmann argued that the most probable distribution, which he identified as the equilibrium of the particle system, is the macrostate which maximises $S(p)$ subject to the constraints (7.1) and (7.2) rewritten in terms of the p_k's (after taking the limit as $N \to \infty$):

$$\sum_{k=1}^{m} p_k = 1 \tag{7.3}$$

$$\sum_{k=1}^{m} p_k E_k = \text{const.} \tag{7.4}$$

The maximising macrostate p^* has the explicit form:

$$p_k^* = \frac{\exp(-\beta E_k)}{\sum_{l=1}^{m} \exp(-\beta E_l)}, \tag{7.5}$$

where β is a constant which must be determined by satisfying the energy constraint (7.4). (The conditions for existence and uniqueness of such a β are given in Theorem 2 below.)

One of the drawbacks of Boltzmann's approach here is that it requires the identification of microstates for a given model, and there appears to be no hard and fast rule for defining such a microstate. For example, if the particles were classified according to their temperatures rather than energies, the microstates would be defined differently.

A somewhat different approach, that nevertheless often leads to identical mathematics, was put forward by Gibbs. Rather than treating individual systems, Gibbs considered ensembles consisting of multiple copies of the same system evolving independently of one another from a random set of initial configurations, and described the statistical properties of the ensemble in terms of the probabilities of finding a randomly chosen system in a given configuration. In this approach, one simply needs to define a configuration space Ω for all possible configurations ω of an individual system and a probability distribution p on Ω for which $p(\omega)$ is the probability that a randomly chosen system is in the configuration $\omega \in \Omega$. For technical reasons, to be discussed later, we will for now assume that Ω is *finite*. We let $\mathcal{M}$ denote the set of probability vectors on Ω, i.e. if $p \in \mathcal{M}$ then $p \geq 0$ and $\sum_{\Omega} p(\omega) = 1$. For each $p \in \mathcal{M}$, $\sum_{\omega \in A} p(\omega)$ is the fraction of systems in the ensemble whose configurations belong to A. Given a function $f : \Omega \to \mathbb{R}$ and the probability density p, $\mathcal{E}_p(f) = \sum_{\omega \in \Omega} p(\omega) f(\omega)$ will denote the expectation of f.

The *statistical entropy* of the distribution p is then defined to be

$$\eta(p) = -\sum_{\omega \in \Omega} p(\omega) \log p(\omega). \tag{7.6}$$

Now suppose that we are given the following system moments (that is, expectations of the functions f_i):

$$\bar{f}_i = \mathcal{E}_p(f_i) \quad i = 1, \ldots, m, \tag{7.7}$$

where the $\bar{f}_i$ are known real numbers. According to the maximum entropy principle, the *maximum entropy distribution* is any distribution p^* that maximises the statistical entropy $\eta(p)$ over $p \in \mathcal{M}$ subject to the statistics (7.7). It can be shown (e.g. [14]) that all maximisers (they could be nonunique) take the form

$$p_{\boldsymbol{\beta}}(\omega) = \frac{1}{Z(\boldsymbol{\beta})} \exp(-\boldsymbol{\beta} \cdot \boldsymbol{f}(\omega)), \tag{7.8}$$

where $Z(\boldsymbol{\beta}) = \sum_{\omega \in \Omega} \exp(-\boldsymbol{\beta} \cdot \boldsymbol{f}(\omega))$, $\boldsymbol{f} = (f_1, \ldots, f_m)$, and $\boldsymbol{\beta} = (\beta_1, \ldots, \beta_m)$. (Here the β_i can be identified as the Lagrange multipliers associated with each constraint in (7.7).) The remaining problem is then to determine the values of the β_i given the numbers $\bar{f}_i$.

For the case $m = 1$, say where the single known moment is the energy constraint (and, as already stated, Ω is a finite set)

$$\bar{h} = \mathcal{E}_{p_\beta}(h), \tag{7.9}$$

we have [14, Corollary 1.1.5]

Theorem 2 *If $\bar{h}$ is a real number and $\min_{\omega\in\Omega} h(\omega) < \bar{h} < \max_{\omega\in\Omega} h(\omega)$, then there exists a unique parameter $\bar{\beta} \in \mathbb{R}$ such that the distribution $p_{\bar{\beta}}$ has the energy $\mathcal{E}_{p_{\bar{\beta}}}(h) = \bar{h}$ and maximises entropy over all distributions with the same energy $\bar{h}$.*

(Note that here it is important that Ω is finite, so that the extrema of h can easily be shown to exist.) In fact, we may classify the extreme values of the energy h over $\omega \in \Omega$ by the limits

$$\lim_{\beta\to+\infty} \mathcal{E}_{p_\beta}(\omega) = \min_{\omega\in\Omega} h(\omega), \quad \lim_{\beta\to-\infty} \mathcal{E}_{p_\beta}(\omega) = \max_{\omega\in\Omega} h(\omega). \tag{7.10}$$

Usually $T = 1/\bar{\beta}$ is defined as the system temperature, since $\bar{\beta}$ is conjugate to $\bar{h}$. In this sense, the first limit in (7.10) says that as the system temperature $T \downarrow 0$ (i.e. $\beta \uparrow +\infty$) the distribution p_β tends to the uniform density (also known as the *ground state*) on the set of system configurations $\{\omega \in \Omega : h(\omega) = \min_{\omega'\in\Omega} h(\omega')\}$ (e.g. [14, p6]).

8 Geophysical balance defined by the maximum entropy principle

The first challenge of using the MEP (Maximum Entropy Principle) to define a balanced state in the 3D Boussinesq equations (2.10)–(2.14) is dealing with an uncountably infinite configuration space, namely the set $\mathcal{S}$ of all measure-preserving maps of the fluid domain D into itself. To apply the MEP method we first have to be able to define a probability measure on $\mathcal{S}$. The set $\mathcal{S}$ is not a linear space, but we may identify each $s \in \mathcal{S}$ with a linear map of L^1 into itself as follows: Let $\varphi \in L^1$ and s be a measure-preserving transformation of D into itself. Then for each $A \in \mathcal{A}$ (here $\mathcal{A}$ is the set of all Borel subsets of D), consider the finite measure m_φ defined by

$$m_\varphi(A) = \int_A \varphi(\boldsymbol{a})\, d\mu(\boldsymbol{a}). \tag{8.1}$$

Then $m_\varphi \circ s^{-1}$ also defines a finite measure, so that by the Radon-Nikodyn theorem (see, for example, [16]) there exists an L^1 function denoted by $P_s\varphi$ which satisfies

$$\int_A (P_s\varphi)(\boldsymbol{a})\, d\mu(\boldsymbol{a}) = \int_{s^{-1}(A)} \varphi(\boldsymbol{a}')\, d\mu(\boldsymbol{a}'). \tag{8.2}$$

The operator $P_s : L^1 \to L^1$ is known as the Frobenius–Perron operator corresponding to the measure-preserving map $s \in \mathcal{S}$ (see, for example, [16]). This operator tells us precisely how a given density transforms under the flow map

s, and so belongs to the Eulerian picture. The Frobenius–Perron operator P_s is a bounded linear Markov operator on L^1, i.e. $P_s \in \Gamma$ where

$$\Gamma = \{P : L^1 \to L^1 \ : \ \|Pf\| = \|f\| \text{ for all} f \geq 0, f \in L^1\}, \tag{8.3}$$

and $\|f\| = \int |f(\boldsymbol{a})| \, d\mu(\boldsymbol{a})$. One can show [16, Corollary 3.2.1] that $P_s f(\boldsymbol{a}) = f(s^{-1}(\boldsymbol{a}))J^{-1}(\boldsymbol{a})$, where $J^{-1}(\boldsymbol{a}) = d\mu(s^{-1}(\boldsymbol{a}))/d\mu(\boldsymbol{a})$. Since here s is a μ-measure preserving map of D onto itself, $J^{-1} \equiv 1$ and so $P_s\varphi(\boldsymbol{a}) = \varphi(s^{-1}(\boldsymbol{a}))$. Dual to the Frobenius–Perron operator w.r.t. the standard scalar product $\langle \cdot, \cdot \rangle : L^1 \times L^\infty \to \mathbb{R}$ defined by $\langle f, g \rangle = \int f(\boldsymbol{a}) \, g(\boldsymbol{a}) \, d\mu(\boldsymbol{a})$, is the Koopman operator $U_s : L^\infty \to L^\infty$ defined as the adjoint of P_s:

$$\langle P_s f, g \rangle = \langle f, U_s g \rangle, \forall f \in L^1, g \in L^\infty. \tag{8.4}$$

The Koopman operator satisfies $U_s g = g \circ s$ for all $g \in L^\infty$[2]. When s is measure-preserving, $U_s g$ is a rearrangement of $g \in L^\infty$.

We will define the configuration space Ω of the fluid to be the set $\mathcal{P}$ consisting of all Frobenius–Perron operators P_s defined by (8.2) as s ranges over the set $\mathcal{S}$ of all measure-preserving maps of the fluid domain D into itself. Let $\overline{\text{conv}}(A)$ denote the closed convex hull of a set A, and $\text{Ex}(A)$ denote the extreme points of A. The set $\mathcal{K} = \overline{\text{conv}}(\mathcal{P})$ is a closed convex subset of the space Γ of bounded linear operators of L^1 into itself. The work of Ryff [26] regarding the extremal structure of the weak closure of rearrangements suggests that each extreme point of $\mathcal{K}$ must be a Frobenius–Perron operator. It is known [21] that the extreme points of the convex space Γ satisfy the *multiplicative property*, i.e. $T \in \Gamma$ is extreme in Γ if and only if $T(fg) = (Tf)(Tg)$ for all $f, g \in L^1$. Now for $f \in L^1$, and P_s the Frobenius–Perron operator corresponding to some $s \in \mathcal{S}$, $P_s f = f \circ s^{-1}$. Hence if we also have $g \in L^1$, P_s satisfies $P_s(fg) = (fg) \circ s^{-1} = (f \circ s^{-1})(g \circ s^{-1}) = (P_s f)(P_s g)$. Thus every Frobenius–Perron operator is extreme in $\mathcal{K}$. We are thus lead to conclude that the Frobenius–Perron operators are the extreme points of $\mathcal{K}$. This construction allows us to identify the extreme points of $\mathcal{K} = \overline{\text{conv}}(\mathcal{P})$ with fluid configurations, and any point in K formally defines a probability distribution on these fluid configurations for the ensemble.

We now seek an integral representation for a given point in $\mathcal{K}$ as the convex combination of extreme points. Now $\mathcal{K}$ is a compact and convex metric space (since it is a closed subset of Γ, which is compact in the weak operator topology), so that by Choquet's theorem [22] (see also A.5 in the Appendix), for

[2]Those familiar with differential geometric terminology will have recognised that the Frobenius–Perron operator is identical to the push forward operator on smooth functions, whereas the Koopman operator is the corresponding pull back operator. Note that our definition of the push forward operation on functions here differs from some texts by the factor J^{-1}.

each $P \in \mathcal{K}$ there is a probability measure ν_P supported on $\mathcal{P} = \mathrm{Ex}(\mathcal{K})$ such that for any continuous linear functional L on Γ we have the representation

$$L(P) = \int_{Q \in \mathcal{P}} L(Q) d\nu_P(Q). \tag{8.5}$$

In (8.5), ν_P is a probability measure over the set of configurations $\mathcal{P}$. For example, when L is the identity, (8.5) gives the decomposition of P as a convex combination of Frobenius–Perron operators. We let $\mathcal{M}_0$ be the set of all such measures ν_P as P ranges over $\mathcal{K}$.

We now define the statistical entropy of each $\nu \in \mathcal{M}_0$ relative to the uniform distribution $\nu_0 \in \mathcal{M}_0$ as

$$\eta(\nu) = -\int_{\mathcal{P}} \log\left(\frac{d\nu}{d\nu_0}\right) d\nu(P), \tag{8.6}$$

if the Radon-Nikodym derivative $\frac{d\nu}{d\nu_0}$ exists, and $\eta = -\infty$ otherwise. Suppose that for $\nu \in \mathcal{M}_0$ we are given the ensemble mean of the continuous energy functional $h : \mathcal{P} \to \mathbb{R}$:

$$\bar{h} = \int_{\mathcal{P}} h(P)\, d\nu(P). \tag{8.7}$$

Here $h(P) = \int_D f^2 |\boldsymbol{X}(\boldsymbol{a}) - s(\boldsymbol{a})|^2 \, d\mu(\boldsymbol{a})$, where s is the measure-preserving map of D into itself corresponding to the Frobenius–Perron operator P.

The MEP then says that the equilibrium distribution of fluid configurations is any measure $\nu^* \in \mathcal{M}_0$ that maximises $\eta(\nu)$ subject to the constraint $h(\nu) = \bar{h}$.

We now find the conditions for which this maximiser ν^* exists and is unique. Rather than appeal to Lagrange multipliers to deal with the constraint (8.7), we follow the proof for the case where D is finite [14] and apply Jensen's inequality (see A.6 in the Appendix). Much of the proof in [14] carries across from the finite dimensional case. Let $Z(\beta) = \int_{\mathcal{P}} \exp(-\beta h(P))\, d\nu_0(P)$. Then we have, for $\nu \in \mathcal{M}_0$, and $\beta \in \mathbb{R}_+$,

$$\begin{aligned}
\eta(\nu) - \beta h(\nu) &= -\int_{\mathcal{P}} \left\{ \log\left(\frac{d\nu(P)}{d\nu_0(P)}\right) + \beta h(P) \right\} d\nu_\beta(P) \\
&= \int_{\mathcal{P}} \left\{ \log\left(\frac{\exp(-\beta h(P))}{d\nu(P)/d\nu_0(P)}\right) \right\} d\nu(P) \\
&\leq \log\left(\int_{\mathcal{P}} \left\{ \frac{\exp(-\beta h(P))}{d\nu(P)/d\nu_0(P)} \right\} d\nu(P) \right) \qquad (8.8) \\
&= \log Z(\beta).
\end{aligned}$$

Equality holds only if the integrand of (8.8) is a constant function, and so any maximiser must take the form $d\nu^* = \exp(-\beta h)/Z(\beta)\, d\nu_0$ for some $\beta \in \mathbb{R}$ and then $\eta(\nu^*) - \beta h(\nu^*) = \log Z(\beta)$.

We turn to the existence of such maximisers, so that we have to show that there exists a $\beta \in \mathbb{R}$ for which $\bar{h} = \int_{\mathcal{P}} h(P)\, d\nu_\beta(P)$, and $\eta(\nu) \leq \eta(\nu_\beta)$ for all $\nu \in \mathcal{M}_0$. First note that the function $\beta \mapsto h(\nu_\beta)$ is a continuous strictly decreasing function of $\beta > 0$. Let $\Theta(\beta) = h(\nu_\beta) = \int h(P) \exp(-\beta h(P))/Z(\beta) d\nu_0(P)$. Then since the integrand of Θ is a continuous function of β, so too is Θ. Furthermore, a straight computation shows that $\Theta'(\beta)$ is minus the variance of h w.r.t. the probability measure ν_β, which shows that Θ is a continuous decreasing function. We also note that $h(P)$ is not constant, and $\exp(-\beta h(P))/Z(\beta) > 0$, which shows that the variance is non-zero and thus Θ is strictly decreasing in β.

We consider two limiting cases where $\beta \to +\infty$ and $\beta \to -\infty$. We have

$$h(\nu_\beta) = \int_{\mathcal{P}} h(P) \frac{\exp(-\beta h(P))}{Z(\beta)}\, d\nu_0(P), \tag{8.9}$$

and we wish to find the limit of this integral as $\beta \to +\infty$. We note that

$$0 \leq (h(P) - h(Q)) \exp[-\beta(h(P) - h(Q))] \leq \frac{1}{\beta}$$

for all $P, Q \in \mathcal{P}$ such that $h(P) \geq h(Q)$. Hence, in particular,

$$0 \leq \left(h(P) - \inf_{Q \in \mathcal{P}} h(Q) \right) \exp(-\beta h(P)) \leq \frac{1}{\beta} \exp\left(-\beta \inf_{Q \in \mathcal{P}} h(Q) \right),$$

for all $P \in \mathcal{P}$. Dividing by $Z(\beta)$ and integrating over $P \in \mathcal{P}$, we obtain

$$0 \leq \int_{\mathcal{P}} \left(h(P) - \inf_{Q \in \mathcal{P}} h(Q) \right) d\nu_\beta \leq \frac{1}{\beta} \frac{\exp\left(-\beta \inf_{Q \in \mathcal{P}} h(Q)\right)}{Z(\beta)} < \frac{1}{\beta}.$$

Now let $\beta \to +\infty$ to conclude

$$\lim_{\beta \to \infty} \int_{\mathcal{P}} h(P) d\nu_\beta = \inf_{Q \in \mathcal{P}} h(Q). \tag{8.10}$$

In a similar way we can show that

$$\lim_{\beta \to -\infty} \int_{\mathcal{P}} h(P) d\nu_\beta = \sup_{Q \in \mathcal{P}} h(Q). \tag{8.11}$$

Notice that the extrema are taken over the set $\mathcal{P}$ of Frobenius–Perron operators, and not their closed convex hull $\mathcal{K}$. This happens because the measure ν_β is supported by $\mathcal{P}$. Finally Theorem 1 (or the CNP) tells us that the infimum in (8.10) is attained, and a suitable modification of the theorem's proof with inf and sup interchanged, and inequalities reversed, gives that the supremum in (8.11) is also attained.

Hence we have shown the existence of maximum entropy solutions in both limits as $|\beta| \to \infty$, and that the function $\beta \mapsto h(\nu_\beta) = \int_{\mathcal{P}} h(P)\, d\nu_\beta(P)$ is a strictly decreasing continuous function of $\beta \in \mathbb{R}$. Hence, by the intermediate value theorem, there is a unique $\bar{\beta}$ such that $\int_{\mathcal{P}} h(P)\, d\nu_{\bar{\beta}}(P) = \bar{h}$ provided that $\min_{P\in\mathcal{P}} h(P) \leq \bar{h} \leq \max_{P\in\mathcal{P}} h(P)$. The maximum entropy distribution for this $\bar{\beta}$ is given by

$$d\nu_{\bar{\beta}}(P) = \frac{\exp(-\bar{\beta}h(P))}{Z(\bar{\beta})}\, d\nu_0, \tag{8.12}$$

and the most likely configuration of a system chosen at random from the ensemble is that configuration of least energy.

9 The MEP lattice model for balance

We will now apply the above ideas to a Lattice model for the flow described by the Boussinesq equations for 3 dimensional flow summarised by equations (2.10) to (2.14).

Recall that we are treating the momentum $\boldsymbol{X}$ as a slow variable and the position $\boldsymbol{x}$ as a fast variable. On the fast timescale, we may assume that the momenta are fixed, and that the particles rapidly approach, or oscillate about, an 'equilibrium' state. The aim is to find this equilibrium state using the maximum entropy principle.

The fluid domain D is partitioned into n open domains D_i with centroids $\boldsymbol{a}_i \in \mathbb{R}^3$, $i = 1, \ldots, n$. Let $\Gamma = \{\boldsymbol{a}_1, \ldots, \boldsymbol{a}_n\}$. We approximate the fluid motion by allowing permutations of fluid 'parcels' amongst these centroids. Let S_n denote the group of permutations σ which act on the fluid parcels according to $\sigma \circ (\boldsymbol{a}_1, \ldots, \boldsymbol{a}_n) = (\boldsymbol{a}_{\sigma(1)}, \ldots, \boldsymbol{a}_{\sigma(n)})$. The set of configurations of the fluid parcels is thus $\Omega = S_n$.

We will consider an ensemble of these n parcel systems, and seek the probability distribution of the configurations at equilibrium defined as that distribution that maximises entropy subject to a fixed ensemble average of the energy. We define a probability vector p with $n!$ components $p(\sigma)$ each equal to the fraction of systems in the ensemble in the state $\sigma \in S_n$, so that $0 \leq p(\sigma) \leq 1$ and $\sum_\sigma p(\sigma) = 1$.

The energy of the fluid in the configuration $\sigma \in S_n$ is $h(\sigma) = \frac{f^2}{2}\sum_{i=1}^n |\boldsymbol{X}_i - \boldsymbol{a}_{\sigma(i)}|^2$. The canonical equilibrium distribution p^* for the ensemble with mean energy $\bar{h}$ is obtained by finding the probability distribution p^* that maximises $\eta(p) = -\sum_{\sigma\in S_n} p(\sigma)\log p(\sigma)$ subject to the mean energy $h(p) = \sum_{\sigma\in S_n} p(\sigma)$ $[\sum_{i=1}^n \frac{f^2}{2}|\boldsymbol{X}_i - \boldsymbol{a}_{\sigma(i)}|^2] = \bar{h}$. By Theorem 2, the unique optimal p^* is given by $p^*_{\bar{\beta}}$ where

$$p^*_{\beta}(\sigma) = \frac{1}{Z(\beta)} \exp\left(-\frac{\beta f^2}{2}\sum_{i=1}^n |\boldsymbol{X}_i - \boldsymbol{a}_{\sigma(i)}|^2\right), \tag{9.1}$$

with $Z(\beta) = \sum_{\sigma'\in S_n} \exp(-\frac{\beta f^2}{2}\sum_{i=1}^n |\boldsymbol{X}_i - \boldsymbol{a}_{\sigma'(i)}|^2)$ the partition function, and $\bar{\beta}$ the unique real number for which the energy constraint is satisfied.

We are now in a position to relate the CNP minimisation principle with the maximum entropy approach. Note that the ensemble mean energy for a (not necessarily optimal) distribution p is

$$\begin{aligned}
\sum_{\sigma\in S_n} p(\sigma)h(\sigma) &= \frac{1}{2}\sum_{\sigma\in S_n} p(\sigma)\sum_{i=1}^{n}|\boldsymbol{X}_i-\boldsymbol{a}_{\sigma(i)}|^2 \\
&= \frac{1}{2}\sum_{\sigma\in S_n} p(\sigma)\sum_{i,j=1}^{n}|\boldsymbol{X}_i-\boldsymbol{a}_j|^2\delta_{j\,\sigma(i)} \\
&= \frac{1}{2}\sum_{i,j=1}^{n}|\boldsymbol{X}_i-\boldsymbol{a}_j|^2\sum_{\sigma\in S_n} p(\sigma)\delta_{j\,\sigma(i)} \\
&= \frac{1}{2}\sum_{i,j=1}^{n}|\boldsymbol{X}_i-\boldsymbol{a}_j|^2 k_{ij} \qquad (9.2)
\end{aligned}$$

where $k_{ij} = \sum_{\sigma\in S_n} p(\sigma)\delta_{j\,\sigma(i)}$. It is easy to show by summation that $\sum_{i=1}^{n} k_{ij} = 1 = \sum_{j=1}^{n} k_{ij}$, i.e. $k = ((k_{ij}))$ is a doubly stochastic matrix, and that for this finite problem every doubly stochastic matrix can be expanded in this form.

Now suppose that we take the equilibrium ensemble with distribution $p^*_{\bar{\beta}}$ given by (9.1) and choose a system at random from the ensemble. Then the most likely state selected, say σ^*, is that state which maximises the equilibrium distribution $p^*_{\bar{\beta}}(\sigma)$ over all $\sigma \in S_n$. Since Z is independent of σ, this is just the same as finding the configurations σ^* of minimum energy $h(\sigma^*)$. In other words, the most likely state is a solution σ^* to

$$\sum_{i=1}^{n}\frac{1}{2}|\boldsymbol{X}_i-\boldsymbol{a}_{\sigma^*(i)}|^2 = \min_{\sigma\in S_n}\sum_{i=1}^{n}\frac{1}{2}|\boldsymbol{X}_i-\boldsymbol{a}_{\sigma(i)}|^2. \qquad (9.3)$$

Furthermore, we are now in a position to interpret the relaxation process (4.4) and (6.1). In the light of (9.2), we can rewrite (4.4) as

$$\begin{aligned}
\sum_{i=1}^{n}\frac{1}{2}|\boldsymbol{X}_i-\boldsymbol{a}_{\sigma^*(i)}|^2 &= \min_{k\in\mathcal{D}}\sum_{i,j=1}^{N}\frac{1}{2}|\boldsymbol{X}_i-\boldsymbol{a}_{\sigma(i)}|^2 k_{ij} \\
&= \min_{p\in\mathcal{M}}\sum_{\sigma\in S_n} p_\sigma\left(\sum_{i=1}^{n}\frac{1}{2}|\boldsymbol{X}_i-\boldsymbol{a}_{\sigma(i)}|^2\right).
\end{aligned}$$

Thus the relaxation process (4.4) and (6.1) simply says that the minimum energy is equal to the mean ensemble energy as the statistical temperature tends to zero, because with probability one, every replica in the ensemble is in the minimum energy state. This suggests that an appropriate interpretation of the doubly stochastic measure is as a probability distribution for an ensemble of identical systems.

Recall that each energy $h(\sigma)$ is actually also a function of the momentum $\boldsymbol{X}$. Let us now make this explicit by writing $h(\boldsymbol{X},\sigma)$ for the energy of configuration σ when the momenta are $\boldsymbol{X} = (\boldsymbol{X}_1, \dots, \boldsymbol{X}_n)$. The partition function Z will also depend on $\boldsymbol{X}$ so we will denote it by $Z(\boldsymbol{X},\beta)$.

Instead of fixing $\bar{h}$ and finding the corresponding unique $\bar{\beta}$, suppose we fix β. Then we can calculate (where the differentiation is with β held fixed)

$$\begin{aligned}
\frac{\partial \log Z(\boldsymbol{X},\beta)}{\partial \boldsymbol{X}_i} &= \frac{1}{Z(\boldsymbol{X},\beta)} \frac{\partial Z(\boldsymbol{X},\beta)}{\partial \boldsymbol{X}_i} \\
&= \frac{1}{Z(\boldsymbol{X},\beta)} \sum_{\sigma} (-\beta f^2) \exp(-\beta h(\boldsymbol{X},\beta)) \frac{\partial h(\boldsymbol{X},\sigma)}{\partial \boldsymbol{X}_i} \\
&= \frac{1}{Z(\boldsymbol{X},\beta)} \sum_{\sigma} (-\beta f^2) \exp(-\beta h(\boldsymbol{X},\beta)) (\boldsymbol{X}_i - \boldsymbol{a}_{\sigma(i)}) \\
&= -\beta f^2 (\boldsymbol{X}_i - \boldsymbol{x}_i^{\beta}),
\end{aligned} \tag{9.4}$$

where we have defined

$$\boldsymbol{x}_i^{\beta} = \sum_{\sigma \in S_n} p_{\beta}^{*}(\sigma)\, \boldsymbol{a}_i, \tag{9.5}$$

which is the expected position of parcel i for the distribution p_{β}^{*}. Hence we have the gradient relation:

$$\boldsymbol{x}^{\beta} = \nabla \phi(\boldsymbol{X}), \text{ where } \phi(\boldsymbol{X}) = \frac{1}{2} \|\boldsymbol{X}\|^2 + \frac{1}{f^2 \bar{\beta}} \log Z(\boldsymbol{X}, \bar{\beta}), \tag{9.6}$$

which should be compared with the first equation in (4.9). Note that here the function ϕ is smooth (for finite β), whereas its counterpart in the limit $\beta \to \infty$ (4.9) is only differentiable almost everywhere.

10 Conclusions

This aim of this chapter was to look at balance from a statistical viewpoint. Our first angle came from a probabilistic interpretation of the CNP minimisation principle for defining geostrophic balance, in which CNP was directly related to the construction of a minimal L^2 probability metric. Our second angle stemmed from an application of the maximum entropy principle to the 3D Boussinesq equations for the dry atmosphere written in terms of fast and slow components. Here the principle was used to derive an ensemble probability distribution for the fast variables, given data on the slow variables. The balance manifold was then defined as a mapping from the slow variables (the geostrophic momenta) to the expected values of the fast variables (the particle positions). In this second view, the CNP is regained as the statistical temperature of the system tends to zero, where minimising the energy over function rearrangements is equivalent to finding the system ground state. Our main conclusion, therefore, is that geostrophic balance and CNP can be understood from a statistical viewpoint, and we suggest that there may be advantages in treating applied problems of rearrangement theory as limiting cases of statistical mechanical problems, particularly when the system has other conserved integrals in addition to the energy.

Acknowledgments

SB would like to thank Rua Murray and Kalvis Jansons for helpful discussions. The greater part of this research was carried out during the 'Atmosphere and Ocean Dynamics' program held at the Isaac Newton Institute for Mathematical Sciences at Cambridge University.

A Appendix

First we recall some definitions (see, for example, [1]). Let V be a Banach space with norm $|\cdot|$ and dual V^*, related by the inner product $\langle , \rangle$.

A.1 Semicontinuity

A function $f : V \to \mathbb{R} \cup \{+\infty\}$ is lower semicontinuous (l.s.c.) if its *epigraph*

$$\text{epi } (f) = \{(x, r) \in V \times \mathbb{R} \ : \ r \geq f(x)\} \tag{A.1}$$

is a closed set. A function ψ is upper semicontinuous (u.s.c.) if $-\psi$ is l.s.c..

A.2 Convexity

A function $f : V \to \mathbb{R} \cup \{+\infty\}$ is convex iff whenever $\lambda \in [0, 1]$ and $x, y \in V$ we have

$$f(\lambda x + (1 - \lambda)y) \leq \lambda f(x) + (1 - \lambda) f(y).$$

A convex function is proper if $f(x) < +\infty$ for at least one $x \in V$. Similarly, a function $\psi : V \to \mathbb{R} \cup \{+\infty\}$ is concave if $-\psi$ is convex.

The set epi (f) is a non-empty convex set if and only if f is a proper convex function, and epi (f) is a closed convex set if and only if f is a proper convex l.s.c. function.

A.3 Subdifferentiation

The subdifferential of a proper convex function $f : V \to \mathbb{R} \cup \{+\infty\}$ at the point $x \in V$ is the set

$$\partial f(x) = \big\{x^* \in V^* \ : \ f(x') \geq f(x) + \langle x' - x, x^* \rangle, \ \forall x' \in V\big\}. \tag{A.2}$$

The subdifferential may be empty, but is guaranteed to be non-empty when f is continuous at x. Note that proper convex functions are continuous where bounded.

A.4 Fenchel conjugation

The Fenchel conjugate $f^* : V^* \to \mathbb{R} \cup \{+\infty\}$ of a function $f : V \to \mathbb{R} \cup \{+\infty\}$ is defined by

$$f^*(x^*) = \sup_{x \in V} \{\langle x, x^* \rangle - f(x)\} \quad \text{for each } x^* \in V^*. \tag{A.3}$$

When f is a proper convex function $f^{**} = (f^*)^* = f$.

A.5 Choquet's Theorem

Theorem 3 (Choquet) *Let C be a compact convex metrizable subset of a locally convex Hausdorff topological vector space Y, and suppose that $x \in C$. Then there is a probability measure P on (C, Borel subsets of C) such that P is supported by the set Ex(C) of extreme points of C, and $\int_C f dP = f(x)$ for each $f \in Y^*$.*

See for example [22].

A.6 Jensen's Inequality

Theorem 4 (Jensen's inequality) *Let $I \subset \mathbb{R}$ be an open interval and assume that $\varphi : I \to \mathbb{R}$ is convex. If $f \in L^1$ takes values in I, then the integral of $\varphi \circ f$ is well defined and $\varphi \left(\int f d\mu \right) \leq \int \varphi \circ f d\mu$ with equality if and only if f is constant μ-a.e.*

See for example [14].

References

[1] J.-P. Aubin and I. Ekeland. *Applied Nonlinear Analysis.* Wiley-Interscience, 1984.

[2] F. Baer and J. Tribbia. On complete filtering of gravity modes through nonlinear initialisation. *Mon. Weather Rev.*, 105:1536–1539, 1977.

[3] S. Baigent. Applications of vector-valued rearrangements to modelling the weather. Master's thesis, Oxford University, 1987.

[4] S. Baigent. *On the Integration of the Semi-Geostrophic Equations.* PhD thesis, Oxford University, 1995.

[5] S. A. Baigent and J. Norbury. Two discrete models for semi-geostrophic dynamics. *Physica D*, 109:333–342, 1997.

[6] J.-D. Benamou. *Transformations conservant la mesure, mécanique des fluides incompessibles et modèle semi-géostrophique en météorologie.* PhD thesis, INRIA, Grenoble, 1992.

[7] Y. Brenier. Polar factorisation and monotone rearrangements of vector-valued functions. *Comm. Pure Appl. Mathem.*, 44:375–417, 1991.

[8] M. J. P. Cullen. Solutions to a model of a front forced by deformation. *Quart. J. Roy. Meteor. Soc.*, 109:565–573, 1983.

[9] M. J. P. Cullen, J. Norbury, and R. J. Purser. Generalised Lagrangian solutions for atmospheric and oceanic flows. *SIAM J. Appl. Math.*, 51:20–31, 1991.

[10] M. J. P. Cullen, J. Norbury, R. J. Purser, and G. J. Shutts. Modelling the quasi-equilibrium dynamics of the atmosphere. *Quart. J. Roy. Meteor. Soc.*, 113:735–758, 1987.

[11] M. J. P. Cullen and R. J. Purser. An extended Lagrangian theory of semi-geostrophic frontogenesis. *J. Atmos. Sci.*, 41:1477–1497, 1984.

[12] D. D. Holm and V. Zeitlin. Hamilton's principle for quasigeostrophic motion. *Physics of Fluids*, 10:800–806, 1998.

[13] B. J. Hoskins. The geostrophic momentum approximation and the semi-geostrophic equations. *J. Atmos. Sci.*, 32:233–242, 1975.

[14] G. Keller. *Equilibrium States in Ergodic Theory.* London Mathematical Society Student Texts **42**, 1998.

[15] M. Knott and C. S. Smith. On the optimal mapping of distributions. *J. Opt. Theor. Appl.*, 43:39–49, 1984.

[16] A. Lasota and M. C. Mackey. *Chaos, Fractals and Noise. Stochastic Aspects of Dynamics.* Springer-Verlag, 2nd edition, 1994.

[17] C. E. Leith. Nonlinear normal mode initialisation and quasi-geostrophic theory. *J. Atmos. Sci.*, 37:958–969, 1980.

[18] B. Machenbaer. On the dynamics of gravity oscillations in a shallow water model, with applications to normal mode initialisation. *Beitr. Phys. Atmos.*, 50:253–271, 1977.

[19] R. E. O'Malley. *Singular Perturbation Methods for Ordinary Differential Equations.* Springer-Verlag, 1991.

[20] J. Pedlosky. *Geophysical Fluid Dynamics.* Springer, New York, 2nd edition, 1987.

[21] R. R. Phelps. Extreme positive operators and homomorphisms. *Trans. Amer. Math. Soc.*, 108:265–274, 1963.

[22] R. R. Phelps. *Lectures on Choquet's Theorem.* Van Nostrand, Princeton, N.J., 1966.

[23] S. T. Rachev. The Monge–Kantorovich mass transference problem and its stochastic applications. *Theor. Prob. Appl.*, 29:647–676, 1985.

[24] I. Roulstone and J. Norbury. A Hamiltonian structure with contact geometry for the semi-geostrophic equations. *J. Fluid Mech.*, 272:211–233, 1994.

[25] L. Ruschendorf and S. T. Rachev. A characterisation of random variables with minimum l^2-distance. *J. Mult. Anal.*, 32:48–54, 1990.

[26] J. V. Ryff. Orbits of L^1-functions under doubly stochastic transformations. *Trans. Amer. Math. Soc.*, 117:92–100, 1965.

[27] J. A. Sanders and F. Verhulst. *Averaging Methods in Nonlinear Dynamical Systems.* Applied Mathematical Sciences 59, Springer-Verlag, 1985.

[28] R. Temam. *Infinite Dimensional Dynamical Systems in Mechanics and Physics, Applied Mathematics Series.* Springer-Verlag, New York, 1988.

[29] K. Trustrum. *Linear Programming.* Library of Mathematics, Routledge and Kegan Paul, 1971.